ENVIRONMENTAL
SOIL CHEMISTRY

ENVIRONMENTAL SOIL CHEMISTRY

THIRD EDITION

DONALD L. SPARKS
University of Delaware

BALWANT SINGH
The University of Sydney

MATTHEW G. SIEBECKER
Texas Tech University

ACADEMIC PRESS
An imprint of Elsevier

ELSEVIER

Academic Press is an imprint of Elsevier
125 London Wall, London EC2Y 5AS, United Kingdom
525 B Street, Suite 1650, San Diego, CA 92101, United States
50 Hampshire Street, 5th Floor, Cambridge, MA 02139, United States
The Boulevard, Langford Lane, Kidlington, Oxford OX5 1GB, United Kingdom

Notices

Knowledge and best practice in this field are constantly changing. As new research and experience broaden our understanding, changes in research methods, professional practices, or medical treatment may become necessary.

Practitioners and researchers must always rely on their own experience and knowledge in evaluating and using any information, methods, compounds, or experiments described herein. In using such information or methods they should be mindful of their own safety and the safety of others, including parties for whom they have a professional responsibility.

To the fullest extent of the law, neither nor the Publisher, nor the authors, contributors, or editors, assume any liability for any injury and/or damage to persons or property as a matter of products liability, negligence or otherwise, or from any use or operation of any methods, products, instructions, or ideas contained in the material herein.

ISBN: 978-0-44-314034-1

For information on all Academic Press publications visit our website at https://www.elsevier.com/books-and-journals

Publisher: Katey Birtcher
Editorial Project Manager: Sara Valentino
Publishing Services Manager: Shereen Jameel
Senior Project Manager: Manikandan Chandrasekaran
Cover design: Miles Hitchen

Printed in India
Last digit is the print number: 9 8 7 6 5 4 3 2

Dedication

DLS

To the over 100 graduate students and postdoctoral researchers whom I have had the great privilege to advise and mentor.

BS

To my wife, Sunita, for her unconditional support and love, and to our children, Neha, Divya and Neil, for their encouragement and support. And to our grandchildren, Isha, Aalia and Oscar, who were no help in the writing but a great distraction and joy.

MGS

To my wife, Maria Eugenia, and our daughter, Gabriela.

Contents

Preface

Since the publication of the second edition of *Environmental Soil Chemistry* in 2003, global challenges, many of which are tied to climate change, have catapulted the central role that soils play in our lives. Some of these challenges include, carbon cycling and sequestration, sea level rise and flooding, soil contamination, water quality and quantity, droughts, fires, food production and security, and even national and international security. These global challenges are inextricably tied to soil chemical processes and reactivity. Accordingly, in the third edition, we discuss the significant role that soil mineralogical and organic matter chemistry, ion exchange and sorption processes, redox dynamics, and kinetics play in impacting the above challenges. Rapid advances in technology have also provided valuable analytical, computational, and modeling tools, which have provided important new insights into organic matter chemistry, real-time chemical reaction kinetics, sorption mechanisms, and direct speciation of soil nutrients, carbon, and metal(loid) contaminants. Chapter 1 contains a new section dedicated to the general chemistry principles that are critical to grasp for soil chemical reactions, including periodic table trends. New content and figures illustrate the mineralogical structure and mineral–organic matter interactions (Chapters 2 and 3). Updated content on surface complexation modeling and sorption reactions, as well as examples to teach data analysis for X-ray absorption spectroscopy, are also given (Chapter 5). New examples for environmentally relevant kinetics and redox reactions are presented (Chapters 7–10).

In addition to placing environmental soil chemistry in the context of contemporary issues, the Third Edition provides rigorous yet clear coverage of classic soil chemistry principles and processes. These include the inorganic and organic components of soil; soil porewater chemistry; interfacial chemical reactions between soils and inorganic, organic, and microbial species; kinetics and thermodynamics of soil chemical processes; and redox, acidity, and salinity chemistry of soils. Other useful features in the Third Edition include problem sets in each chapter for enhanced learning and comprehension, color figures, sidebars that provide key points, suggested readings, case studies that show the application of state-of-the-art techniques, and a new online course supplement for instructors.

Environmental Soil Chemistry, Third Edition, provides upper-level undergraduate and graduate students in soil science, geochemistry, and environmental chemistry and engineering with sound training in the basics of soil chemistry and applications to real-world environmental concerns. It offers a competitive advantage for those students looking to incorporate novel, advanced tools into their research.

DLS is grateful to students in his classes, and to former graduate students and postdoctoral researchers who taught the environmental soil chemistry course, for their helpful discussions and suggestions for improvement of the third edition. DLS also thanks Balwant Singh and Matt Siebecker, coauthors of the third edition, for their major contributions. It has been a great pleasure to work with them.

BS is grateful to Don Sparks for involving in the revision of the book and for his patience and guidance in getting the revised edition completed. He is indebted to Jon-Petter Gustafsson and Rai Kookana for their contribution, reviews, and suggestions. BS would like to thank Bob Gilkes, his former Ph.D. supervisor at the University of Western Australia, for continued support and encouragement throughout his career. He acknowledges encouragement from Cliff Johnston in getting the revised edition completed. Finally, BS is grateful to all his former and current students and colleagues at the University of Sydney for many useful discussions and interactions that are reflected in the revised edition of the book.

MGS is grateful to Don Sparks for his constant support and guidance for the past 15 years as both mentor and colleague. Don is an exemplary scientist and person, and he is a role model for numerous colleagues. MGS is also indebted to Balwant Singh, whose patience and insights are invaluable. MGS is grateful to the administrators, his colleagues, and students at Texas Tech University, who have provided a wonderful environment to teach and conduct research.

Teacher and Student Resources

Helpful ancillaries have been prepared to aid learning and teaching.

Please visit the student companion site for more details: https://www.elsevier.com/books-and-journals/book-companion/9780128158807

For qualified professors, additional, instructor-only teaching materials can be requested here: https://educate.elsevier.com/9780128158807

About the Cover

The cover art is designed to capture the modern challenges that threaten soil. The effects of climate on soil are a major focus of our discipline and the selected images represent the challenges highlighted in Sparks, 2020 (https://doi.org/10.1186/s12932-020-00068-6). The image of wheat represents food production and food security, which are directly impacted by changes in climate and soil conditions. Those changes are illustrated in each image. Soil contamination (top right) with trace elements (often called heavy metals) represents a major challenge to food production and environmental health. This image is from a soil contaminated with arsenic. Groundwater enriched in arsenic and ferrous iron come to the sediment surface, where the ferrous iron oxidizes to ferric iron. The oxidized iron (orange rills) contains up to 5% arsenic by weight. Soil salinization (lower right panel) has several major drivers, such as sea level rise, saltwater intrusion, and a continual decline in irrigation water quality and quantity from aquifers. The impacts of drought and fire on soils (bottom panel) often results in major physical and chemical changes to soil properties and is a critical problem. Organic carbon loss (bottom left panel) is directly linked to the chemistry of soil organo-mineral complexes; soils serve as a major yet extremely sensitive sink (and source) of carbon. An important goal is to increase soil organic carbon which will decrease atmospheric carbon dioxide concentration and improve soil health. Lastly, soil and water quality impact both the environment and human health (top left panel). Runoff from agricultural fields and many forms of soil erosion continue to create problems of eutrophication and loss of top soil.

About the Authors

DONALD L. SPARKS is the Unidel S. Hallock du Pont Chair and Francis Alison Professor at the University of Delaware. He is internationally recognized for his research in the areas of the kinetics of biogeochemical processes and surface chemistry of natural materials. His research has focused on the fate and transport of trace metals in soil and water, soil remediation, water quality, and carbon sequestration in soils. Dr. Sparks is the author of the two previous editions of *Environmental Soil Chemistry* and more than 350 refereed papers and book chapters. He is fellow of five scientific societies, and he has been the recipient of major awards and lectureships, including the Geochemistry Medal from the American Chemical Society, the Liebig Medal from the International Union of Soil Sciences, the Einstein Professorship from the Chinese Academy of Sciences, and the Philippe Duchaufour Medal from the European Geosciences Union. Dr. Sparks served as president of the Soil Science Society of America and the International Union of Soil Sciences, has served on advisory committees for several national laboratories and national and international centers and institutes, and served as chair of the U.S. National Academy of Sciences (NAS) Committee for Soil Sciences and other NAS Committees.

BALWANT SINGH is Professor of Soil Science at the University of Sydney. He has over 35 years of research and teaching experience in Australia, India, the U.K., and the U.S. His research interests include understanding the structural and chemical properties of soil clay minerals; adsorption reactions of trace elements at the mineral–water interface; phytoremediation of toxic elements: biochar and organic carbon interaction with soil minerals; and micronutrients availability in soils. Dr. Singh has coedited two books (*Biochar: A Guide to Analytical Methods* and *Synchrotron-Based Techniques in Soils and Sediments*) and authored over 200 research papers and book chapters. He is fellow of four scientific societies and the recipient of various awards, including The Marion L. and Chrystie M. Jackson Soil Science Award by the Soil Science Society of America and The Marion L. and Chrystie M. Jackson Mid-Career Clay Scientist Award by Clay Minerals Society. Dr. Singh has served as a Chair of the Soil Mineralogy Commission of the International Union of Soil Sciences, President of the Australian Clay Minerals Society, and President of the Australian Society of Soil Science, NSW Branch.

MATTHEW G. SIEBECKER is an Assistant Professor of Applied Environmental Soil Chemistry at Texas Tech University. He was a postdoctoral researcher at the University of Delaware. Broadly, his research interests focus on the chemical reactions that take place between dissolved ions and molecules and the solid components of soil, including clay mineral, metal oxides, and soil organic matter. These reactions include adsorption, desorption, dissolution, surface precipitation, and redox. Dr. Siebecker is especially interested in the rapid formation of surface precipitates that are enriched with trace metals. The main themes of his research include the cycling of heavy metal contaminants and plant nutrients in soil and mineral systems, as well as the impacts of climate change and the application of animal waste products on soil. He has expertise in both the traditional wet-chemical laboratory techniques as well as advanced synchrotron-based techniques used in soil chemical research, and he has developed research collaborations internationally with scientists in Australia, Brazil, China, Costa Rica, France, and Italy. At Texas Tech University, Dr. Siebecker teaches an introductory undergraduate soils course entitled Urban Soils as well as an advanced upper-level course entitled Environmental Soil Chemistry. He is an active member of the American Chemical Society and the Soil Science Society of America.

1

An Introduction to Environmental Soil Chemistry

1.1 Environmental Soil Chemistry: An Interdisciplinary Science

Key points: Environmental soil chemistry is the study of how elements and compounds, including plant nutrients, potentially toxic metals, metalloids, radionuclides, and organic compounds, react with the solid, aqueous, and gaseous components of soil. Many scientific disciplines utilize principles central to environmental soil chemistry.

Soil refers to the layers of generally loose mineral and/or organic material that are affected by physical, chemical, and/or biological processes at or near the planetary surface and usually holds liquids, gases, and biota and support plants (van Es, 2017). Environmental soil chemistry focuses on the chemical composition, chemical properties, and reactions in soils and other terrestrial ecosystems. The chemical reactions involve minerals and organic materials in soils, the soil solution, and gases in the soil pore space. These reactions impact water quality, contaminant transport, plant growth, and soil biota. The early history of environmental soil chemistry is rooted in soil science, where the focus was and still remains on enhancing food production. However, increasingly, the attention has shifted toward understanding and predicting the fate of contaminants in soils and sediments.

Because soils impact so many facets of human life, there are many topics in environmental soil chemistry that are also important to other disciplines. This interdisciplinary nature is what gives environmental soil chemistry its broad appeal to scientists from many different areas. Sometimes in nonobvious ways, aspects of environmental soil chemistry are relevant to other disciplines, such as environmental engineering, geochemistry, and materials science. Professionals and academics in the fields of environmental, marine, water, and climate sciences can relate various aspects of their work to the processes that occur in soils.

Environmental scientists and engineers are often concerned about the transport and leaching of toxic compounds from contaminated or remediation sites. The processes that control contaminant leaching are often related to the soil properties, for example, clay mineralogy, organic carbon concentration, soil structure, and porosity, at the site. The reactive soil surfaces are constituted by clay mineral, organic matter, or mixed organomineral systems, where geochemical reactions take place. Geochemistry, a diverse field in itself, has branches that focus on reactions at mineral surfaces, that is, the

Environmental Soil Chemistry, Third Edition
https://doi.org/10.1016/B978-0-12-815880-7.00001-8

mineral—water interface or solid—solution interface. Often the purpose of research in this area is to understand how elements react and bind to minerals.

Sidebar 1.1 — Iron and Manganese Oxides

Iron and manganese oxides are soil minerals that play critical roles in the cycling of environmental contaminants, often involving redox and adsorption reactions.

Geochemical reactions involving common soil minerals can even be important in materials science and energy research. For example, metal oxides (Sidebar 1.1), which are common in both natural and contaminated soils, have electrochemical properties that make them useful for use in batteries. Thus there is a mineralogical link between the disciplines of soil chemistry, mineralogy, and energy research. These mineralogical properties are important controllers of contaminant mobility. In water and environmental sciences contamination, groundwater flow, and contaminant mobility are critical and largely dependent upon soil chemical and physical properties.

1.2 Major Developments in Soil Chemistry

Key points: The focus of soil chemistry has transitioned over time from agriculture (aimed at food production) to environmental contaminants (to evaluate their bioavailability and impact) and, more recently, to carbon cycling in relation to climate change.

This section highlights some key milestones in the progression of soil chemistry; more detailed descriptions of historical developments in soil chemistry are given by Thomas (1977) and Sparks, (2006).

For over 2000 years, it was believed that plants obtained nutrients from organic carbon from the soil. This belief was termed *humus theory*. According to this theory, humus-derived extracts contained water-soluble compounds of C, H, O, and N for plant nourishment. Plants built other nutrients from these four elements using internal vital sources. Humus theory persisted for a long time but was finally disproven by Phillip Carl Sprengel (1787–1859) and later by Justus von Liebig (1803–1873) (Browne, 1944; van der Ploeg et al., 1999; Sparks, 2006).

Sprengel was an instructor in agriculture and agricultural chemistry at the University of Gottingen, Germany (Brown, 1944). By studying the water-soluble components of soil humus extracts, he found that soluble salts (e.g., nitrate and phosphate) were plant nutrients. Sprengel developed the "Law of Minimum," which in many ways stands true today (Sprengel, 1838). It states that the plant will suffer if one of the essential plant elements is not present in sufficient quantity. Today, common effects of the "Law of Minimum" are termed nutrient deficiencies and commonly result in, for example, chlorosis in the plant. Interestingly, early on, Sprengel discovered that some elements, such as As and Pb, were harmful to plants (Sparks, 2006). Since then, As and Pb have been implicated worldwide as extremely hazardous and toxic to both humans and plants. As and Pb are among the most highly researched topics in environmental soil chemistry.

The work of Sprengel laid an important foundation for Liebig, whose publications and books, particularly "Organic Chemistry in Its Applications to Agriculture and Physiology," led Liebig to be considered one of the founders of modern agricultural chemistry (Sparks, 2006).

1.2.1 Ion Exchange

Key point: Ion exchange is perhaps the earliest and most important discovery in environmental soil chemistry.

One of the first major developments in soil chemistry was the recognition of the cation exchange process in soil by a British farmer by the name of H.S. Thompson in 1850. However, an agricultural chemist, J. Thomas Way (1850), conducted the first comprehensive studies of cation exchange and some of his main conclusions that are valid even today include: (i) cation exchange is very rapid, almost instantaneous; (ii) cations are exchanged on an equivalent basis (*stoichiometrically*); (iii) the adsorption of cations increases with an increasing concentration of added salts; and (iv) clays are responsible for cation exchange reactions.

The basic process of cation exchange is the replacement of cations on soil solids with cations in the soil solution (Figure 1.1). The amount of cations or anions that a soil can exchange is a specific quantity referred to as the cation exchange capacity (CEC) or anion exchange capacity (AEC), respectively. The CEC and AEC are two of the most important soil properties for retaining and exchanging important plant nutrients and environmental contaminants. Sidebar 1.2 further elaborates on the terms cations/anions and adsorption/absorption.

Sidebar 1.2 — Ions in the Soil

The term "ion" includes both cations and anions. Cations are positively charged (+) and anions are negatively charged (−). One might think that "absorption" and "adsorption" are the same thing; but in fact, they are quite different processes. In absorption molecules are drawn into the solid, while in adsorption (including cation exchange and other processes, see Chapters 5 and 6) ions stick to the surface of the solid.

1.2.2 Soil Acidity

Key point: Aluminum, and not H, is the major cation that controls soil acidity. Aluminum hydrolyzes and produces H^+, which in turn lowers acidity.

Edmund Ruffin, a farmer from Virginia, measured $CaCO_3$ in his soils and found none; and in this process he discovered the remediation measure for acid soils by adding partly decomposed oyster shells (Ruffin, 1832). Over 70 years after Ruffin's work, Vietch (1902) proposed a method, using $Ca(OH)_2$ equilibration, to determine the lime requirement for crops grown in acid soils. Many researchers extracted Al using neutral salt solutions from acid soils during 1910–1920 and related it to plant growth. However, the existence of exchangeable Al remained a controversial topic for over two decades before being settled by the publications by Coleman and Harward (Coleman and Harward, 1953; Harward and Coleman, 1954). Aluminum toxicity to crop plant roots due to the presence of Al^{3+} ions in the soil solution is

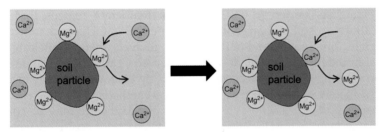

FIGURE 1.1   A schematic drawing of ion exchange. Cations in solution (e.g., Ca^{2+}) exchange with cations on the soil particle (e.g., Mg^{2+}). The exchange is stoichiometric and charge equivalent.

an important growth-limiting factor in acid soils having pH below 5.5.

1.2.3 Alkali Soils

Key point: Excess amounts of Na^+ ions on soil colloids cause soil dispersion.

Eugene Woldemar Hilgard (1906) identified alkali soils in California and concluded that their properties were determined by the presence of Na_2CO_3. He suggested leaching for the reclamation of sodic soils. Subsequently, a number of researchers independently established that alkali soils were formed by the exchange of cations by Na, so exchangeable Na^+ became the cation of importance. It was also suggested that excess Na was responsible for the poor physical properties, such as poor drainage and decreased porosity, of the alkali soils. Kelley and Brown (1925) presented a detailed overview of alkali soils and their remediation (involving the addition of Ca, leaching, and drainage) after 10 years of research.

1.2.4 Selectivity of Ions in the Cation Exchange Process

Key points: Ion exchange is a physicochemical process and there is a selectivity of cations on soil exchange sites, which depends on the cation's size and hydration.

Soon after the discovery of the cation exchange process, researchers started to observe the selectivity of cations on soil exchange sites. However, G. Wiegner from Switzerland and G. Gedroitz from Russia independently made the conclusion that ion exchange is a physicochemical process and not merely a chemical reaction (Thomas, 1977; Sparks, 2006).

In the early 1900s German chemist G. Wiegner and his student Hans Jenny performed experiments on the cation exchange behavior with respect to particles in solution. They emphasized the importance of cation size and hydration in ion exchange selectivity, with smaller ions having greater selectivity and replacing power. Gedroitz suggested the selectivity or replacing power of cations to be in the following order: $Li^+ < Na^+ < K^+ < Mg^{2+} < Rb^+ < NH_4^+ < Co^{2+} < Al^{3+}$.

1.2.5 Isoelectric Point and Variable Charge

Key points: Some soil components or soils carry variable charge: at low pH values, the charge of the component is positive, and at high pH values, the charge is negative.

Sante Mattson (1886–1980), a soil chemist from Sweden, proposed the transformational idea that oxides of Al and Fe were amphoteric, that is, positively charged at low pH and negatively charged at high pH. He also proposed the concept of isoelectric points (IEPs: the pH at which a particular soil or mineral carries no net charge, i.e., net charge is zero) for soil components. Mattson pioneered the variable charge concept in soils and demonstrated that soils from the southeastern part of the US contain variable charge minerals that are positively charged at low pH (Mattson, 1927, 1931, 1932).

Unfortunately, Mattson's work went unrecognized until the 1950s, and until that time, most soil chemistry research was carried out on soils from temperate regions that were dominated by permanent charged minerals. The concept was revived after Schofield and Samson (1953) observed Cl adsorption on the edges of kaolinite. Also, by then, with the advancement in mineralogical methods, it became clear that the crystalline phyllosilicates may be coated with Fe and Al oxides and positively charged surfaces of these oxides could react with anions (Thomas, 1977).

Sidebar 1.3 — **Variable Charge Soils**

Variable charge is used to describe a soil or a mineral whose charge varies with pH, ionic concentration, and composition of the soil solution.

During the 1960s and 1970s, increasing attention was given to highly weathered soils with variable charge minerals (Sidebar 1.3). Such research significantly contributed to the development of analytical methods of variable charge soils and our understanding of the charge characteristics and adsorption–desorption behavior of cations and anions (Sumner, 1963a,b; Hingston et al., 1967; van Raij and Peech, 1972; Gillman, 1974, 1979; Uehara and Gillam, 1981; Fey and Roux, 1976).

1.2.6 Clay Mineralogy

Key points: Molecular structure of minerals affects the behavior of solutes in the soil solution and the ion exchange process.

Until the discovery of X-rays and their applications to determine the structure of crystals or molecules, soil colloids were considered chemical mixtures of SiO_2, Al_2O_3, and Fe_2O_3. Hendricks and Fry (1930) and Kelley et al. (1931) were pioneers in this area, and they independently established that soil clays were largely crystalline. Within a decade after the discovery of the crystalline nature of soil clays, structural data for common soil clay minerals, including kaolinite, montmorillonite, and vermiculite, were obtained. Isomorphous substitution in the structure of montmorillonite was also established in some independent studies (Marshall, 1935; Kelley, 1945; Ross and Hendricks, 1945).

Although not appreciated at the time, Rich and Obenshain (1955) discovered the presence of Al-hydroxy-interlayered vermiculite in the clay fraction of soils formed from mica schist in Virginia. This mineral has been subsequently found in many highly weathered acid soils, contributing to their reactivity toward various ions.

Standard procedures for soil analysis, including particle size fractionation and separation, along with the X-ray diffraction identification of clay minerals, have been developed and the contributions of Marion L. Jackson and his coworkers are noteworthy (Jackson, 1956, 1985).

1.2.7 Equilibrium Constants and Adsorption Isotherms

Key points: Adsorption isotherms are one of the most common experiments used in the laboratory to study the reactivity of contaminants and nutrient ions with soils and soil components, that is, minerals and organic materials.

For several decades after 1928, a number of researchers developed equations for binary exchange reactions in soils. The aim of all these equations was to derive an equilibrium constant (Sidebar 1.4) to predict the activity (or concentration) of ions on exchange sites from their solution activities (or concentrations) or vice versa. Also, the equilibrium constant helps determine the selectivity of an ion on a soil surface. These equations were named after the researchers who developed them and some of the notable equations include Kerr, Vanselow, Gapon, Schofield, Krishnamoorthy–Overstreet, and Gaines–Thomas equations (Sparks, 2006). Among these equations, the Vanselow equation (Vanselow, 1932) is consistent with the thermodynamics of chemical reactions because it employs the activity of ions in the solution phase and the exchanger phase (i.e., the solid); this equation has been most commonly used in the literature. However, the US Salinity Laboratory successfully used the Gapon equation (Gapon, 1933) for predicting the exchange or solution concentration of Na. None of the abovementioned equations gives an equilibrium constant that is true for a range of concentrations and soil systems, and therefore they have found limited applications and are presently not used extensively.

Sidebar 1.4 — Equilibrium Constant and Adsorption Isotherms

Equilibrium constants are values for a given soil or mineral system that describe the ion concentrations in solution and on the solid phase as they reach equilibrium in laboratory experiments.

Adsorption isotherms are a type of experiment conducted at a set temperature (i.e., no temperature changes during the experiment) to determine the amounts of a cation or anion adsorbed with its increasing concentration in the equilibrium solution. The term "iso" means equal and the term "therm" means heat. Three common variables that are tested in adsorption isotherm experiments are pH, ionic strength, and ion concentration.

Adsorption isotherms, or a reaction carried out at a constant temperature, are common to determine the amounts of a cation or anion adsorbed with its increasing concentration in the equilibrium solution. Since the late 1950s, the Langmuir and Freundlich equations have been used to fit adsorption data for nutrients and contaminant ions. Although these equations do not provide any information about the adsorption mechanisms, they provide parameters or coefficients, which make it easy to compare different soils or minerals. Some authors have related the parameters to the maximum adsorption capacity, adsorption strength or energy, and even derived adsorption mechanisms. Such interpretations are not largely valid and have been questioned in many studies (e.g., Veith and Sposito, (1977); Harter and Smith, 1981). In fact, Veith and Sposito, (1977) illustrated that the Langmuir equation could equally well describe both adsorption and precipitation.

1.2.8 Surface Complexation Models

Key points: These models were developed to describe the binding of ions at mineral–water interfaces at the molecular scale. The models provide insight into the binding mechanisms involved in the adsorption reactions.

To provide a molecular description of the adsorption process at the solid–solution interface, a number of surface complexation models (SCMs) have been developed. SCMs describe the formation of complexes (e.g., covalent, ionic, or electrostatic bonds between ions and the mineral surface) at the solid–solution interface using an equilibrium approach and adsorption data. These models define surface species, chemical reactions, equilibrium constant expressions, and surface activity coefficient expressions. Charge balance equations are used in the model to balance the charge for both the ion bound to the solid (adsorbate) and the solid (adsorbent) itself. Starting in the 1970s, many SCMs were developed, including the constant capacitance model, triple-layer model, Stern variable surface charge-variable surface potential model, general two-layer model, and the one-pK and CD-MUSIC models. SCMs are discussed in more detail in Chapter 5 and thorough reviews of these SCMs can be found elsewhere (Davis and Kent, 1990; Goldberg, 1992; Lutzenkirchen, 2006).

Although the SCMs have proved useful in describing the chemical reactions of many key contaminants and nutrients in soils, these models have some drawbacks. One of the major drawbacks is that all models use multiple adjustable parameters to fit the experimental adsorption data. Consequently, it is possible to fit a number of the models equally well to the same data set (Westall and Hohl, 1980). Additionally, in some experiments both adsorption and surface precipitation of metals can occur on soil minerals and most of these models do not consider surface precipitation as a possible sorption mechanism. It is now known that several trace metals, such as Ni, Zn, and Co, as well as other metals such as Fe^{2+}, can react to form surface precipitates, whose solubility data are often not incorporated into the database used in aqueous equilibrium modeling.

Over the last three decades, SCMs have provided useful insight into surface adsorption complexes; however, the conclusions about the mechanisms are speculative and should be supported with direct empirical data. The advent of in situ molecular-scale spectroscopic techniques, such as Fourier transform infrared (FTIR), nuclear magnetic resonance (NMR), and X-ray absorption spectroscopies (XASs), has made them the preferred tools in modern environmental chemistry, and these are discussed below and throughout this textbook.

1.2.9 Kinetics of Sorption Reactions

Key points: The rates of diffusion-controlled soil chemical processes and of reactions involving nanoparticles heavily influence the mobility of contaminants and nutrients in soils.

Until around 1950, cation exchange reactions in soils were considered to be almost instantaneous, probably because of the initial exchange experiments conducted by J. Thomas Way and other researchers. However, Kelley (1948) suggested that ion exchange and sorption reactions may not be rapid in certain soil mineral systems. In diffusion-limited systems involving minerals, such as with mica and vermiculite, there would be slower reaction rates. Ion diffusion into minerals, such as intraparticle diffusion into the micropores of soil clay minerals and diffusion into soil organic matter (SOM), decreases reaction rates.

Sidebar 1.5 — Reaction Kinetics

The rate at which a chemical reaction occurs is called reaction "kinetics." Reaction kinetics is different from the "kinetic energy" studied in an introductory physics course.

The importance of the kinetics of diffusion-limited reactions (Sidebar 1.5) of environmental contaminants and plant nutrients in soils has been realized since the mid-1970s (e.g., Sparks, 1999). Chemical relaxation methods, such as pressure-jump and concentration-jump, can be used to identify reaction rates on millisecond timescales. More recently, quick-scanning X-ray absorption spectroscopy (Q-XAS) (e.g., Ginder-Vogel et al., 2009; Siebecker et al., 2014) and FTIR (e.g., Parikh et al., 2008) have been important tools to study the kinetics of metalloid oxidation, oxyanion sorption, and trace metal precipitation kinetics on mineral surfaces. Details on these methodologies are presented in Chapter 7.

Many common soil minerals, such as ferrihydrite, occur as nanophases. Because of their small size, the physical and chemical properties of nanoparticles are different from those of their larger "bulk" phases. The reaction kinetics involving nanoparticles can increase, and nanophase solubility is predicted to depend on size (Hochella et al., 2008). As particles get smaller, their solubilities increase exponentially. These reactions are important because solubility and sorption are crucial properties in predicting the fate of minerals and dissolved species in the environment. Nanophase minerals influence the movement of harmful heavy metals.

1.2.10 Soil Organic Matter

Key points: SOM is a continuum of organic materials, ranging from plant and animal tissues to biomolecules and organic acids. SOM components are at different stages of decomposition and are constantly disintegrating (mineralizing). Environmental and biological factors affect organic matter turnover more than its molecular structure.

Perhaps one of the most controversial and discussed themes in soil chemistry is the topic of

SOM. Throughout the decades, no topic has received the quantity or type of polemic discussion and debate more than SOM. Organic matter is important in soils for many processes, including cation exchange, soil structure, and carbon retention. It is a key part of the global carbon cycle. Organic C stored in soil is the balance between plant inputs and microbially mediated metabolic losses as CO_2 (Amundson et al., 2015).

Debates continue today on refining the definition of what actually constitutes SOM. There is increased acceptance of the view that the traditional, operationally defined categories of SOM, such as humic and fulvic acids, may not be naturally occurring in the environment and are a product of their chemical extraction procedures. SOM is now viewed to comprise a multitude of organic materials without defined separations between molecular weight and a continuum of organic material decomposing from fresh plant and animal tissue or introduced into the soil as biomolecules exuded from biological processes (Lehmann and Kleber, 2015).

In the historical view above-ground plant carbon inputs and organic matter in the top 30 cm were considered to be the most important location of soil organic carbon. It was also thought that molecular structure determined the recalcitrance of SOM. Now it is understood that the molecular structure of organic material does not necessarily determine its stability in soil (Dungait et al., 2012). SOM cycling is controlled by many processes including accessibility and enzymes from decomposers. SOM can be physically disconnected in the soil because spatially and temporally diverse habitats cause fragmentation, which restricts carbon turnover. Plant roots and the rhizosphere continually make large contributions to SOM, which is composed of partially degraded products more than humic substances. Lastly, it is now recognized that deep soil carbon is a critical carbon pool in soils (Schmidt et al., 2011). More than half of the global soil carbon stocks can be found in subsoil horizons. However, the deep soil carbon pool is particularly vulnerable to microbial degradation with a warming climate.

1.2.11 Advanced In Situ and Molecular-Scale Techniques

Key points: The arrival of advanced in situ and molecular-scale techniques fundamentally changed the direction of environmental soil chemistry. Infrared, NMR, and synchrotron-based spectroscopies, along with computational approaches such as density functional theory, provided new experimental and theoretical methods to analyze the surface chemistry and reactions of ions with soil mineral surfaces.

The advent of synchrotron-based and other spectroscopic techniques, such as XAS, FTIR, and NMR spectroscopies, have made it possible to distinguish between inner- and outer-sphere surface complexes and adsorption and surface precipitation reactions. The use of molecular-scale spectroscopies has been an important step in environmental soil chemistry because of its implications for the availability and transport of nutrients and contaminants.

Early in situ studies used FTIR spectroscopy focused on oxyanions adsorption onto metal oxide surfaces (e.g., Tejedor-Tejedor and Anderson, 1986; Parfitt et al., 1976). However, one of the most significant tools in the last 30 years has been synchrotron X-rays. With the availability of high brilliance, wide tunability, and high polarization and collimation from synchrotron sources, huge progress has been made in the determination of the reaction mechanisms at the soil—water interface and the structure of soil minerals.

Beginning in the late 1970s and early 1980s, synchrotron radiation made its debut in the scientific literature to assess the structural configuration of trace metals in geological materials (see Brown and Calas [2013] for a review of early literature). XAS was used to describe, for example, Fe, Mn, and Ni coordination inside the structure of solids, such as in natural and silicate glasses, phyllosilicates, and metal oxides. Since the early 1990s, the number of peer-reviewed scientific publications that utilize some type of synchrotron radiation has increased at an exponential rate. XAS and other high-energy techniques have essentially become common tools in environmental soil chemistry research. Most current research on ion adsorption onto soil minerals normally includes spectroscopic data along with quantitative aqueous data.

The principal reason for the ubiquitous nature and rise in the usage of in situ spectroscopic techniques is that chemical reactions can be measured in the presence of water. Samples from the natural environment or laboratory experiments can be examined directly to determine whether a trace element is adsorbed or precipitated. For redox-sensitive elements, oxidation states can be readily ascertained using these techniques. In addition, microfocused techniques have allowed the identification of minerals and spatial distribution of elements at the micron scale (Figure 1.2). These micron-scale studies help the identification of potentially reactive soil components that are not readily identifiable via bulk techniques.

1.3 Soil: A Three-Phase System

Key point: In soils minerals and organic matter are inextricably linked together to form a porous medium that hosts numerous organisms and is a hub of biogeochemical processes.

Soils are complex assemblies of inorganic and organic solids that vary in their chemical structures, chemical compositions, size, and shape. The arrangement and association of solids of different sizes and shapes create reactive surfaces and porosity that allow for the storage and exchange of ions, molecules, water, and gases. The soil pore space also serves as a habitat for wide varieties of soil organisms. Typically, the surface horizon of an agricultural soil consists of between 50% and 65% (on a volume basis) of soil solids and the rest is the pore space (Figure 1.3). Porosity can be calculated from soil bulk density as illustrated in Sidebar 1.6. Soil porosity generally increases with increasing clay and organic matter content in the soil.

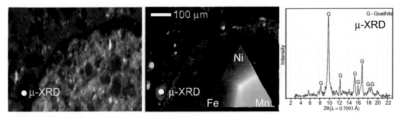

FIGURE 1.2   Synchrotron-based micro-X-ray fluorescence mapping with complementary micro-X-ray diffraction. A petrographic thin section (thickness = 30 μm) of serpentine soil was made (left panel); the particles are examined in the tricolored fluorescence map (middle panel), where Ni is red, Fe is blue, and Mn is green. The purple particle indicates the presence of both Fe and Ni, and the diffraction pattern (right panel) obtained, as highlighted by the white dot, clearly indicates the presence of goethite, indicating that in this particle Ni is incorporated into the goethite structure. Adapted from Siebecker et al. (2018), with permission.

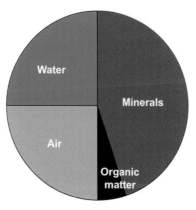

FIGURE 1.3   Soil: a three-phase system − approximate proportions of solid (minerals and organic matter), liquid, and gas phases in an examplary surface soil.

Sidebar 1.6 − Soil Bulk Density and Pore Space

Bulk density is the mass of a unit volume of oven dry soil (105°C), as given below:

$$\text{Bulk density (g cm}^{-3} \text{ or Mg m}^{-3}) = \frac{\text{Mass (g or Mg)}}{\text{Volume (cm}^{-3} \text{ or m}^{-3})}$$

The particle density of the solid minerals in soils is assumed to be 2.65 g cm^{-3}, which is the density of a common soil mineral quartz and close to other common primary and secondary minerals.

The pore space or soil porosity in soils can be calculated from bulk density and particle density, as follows:

$$\text{Pore space(\%)} = \left(1 - \frac{\text{Bulk density}}{\text{Particle density}}\right) \times 100$$

Solid (Rock) mass = 2.65 g
Volume = 1 cm^{-3}
Particle density = 2.65 g cm^{-3}

Solid (Soil) mass = 1.32 g
Volume = 1 cm^{-3}
Bulk density = 1.32 g cm^{-3}
Pore space (%) = (1 -(1.32/2.65))×100 = 50%

1.3.1 Solids

Key points: The amount and type of minerals present in a soil depend on its parent material and the extent of soil weathering. Being the predominant solid fraction, soil minerals, particularly those in the clay fraction, have a major bearing on soil properties and soil ecosystem services. Organic matter, though a small component, is crucial for many key soil properties and processes.

The solid phase of the soil is largely (95% −99%) composed of primary and secondary minerals with approximately 1%−5% of organic matter. The proportion and type of primary and secondary minerals that exist in a soil are dependent on the soil's parent material, the duration and extent of soil weathering, and soil environmental conditions. The common soil parent materials include rocks, glacial till (materials moved by glacial advance and retreat), alluvium (riverborne sediment), aeolian (wind-borne sediment), colluvium (hillslope sediment), lacustrine (lakebed sediment), and volcanic ash. In soils enriched in organic matter (e.g., Histosols) plant material is also considered parent material. The Earth's crust is made up of approximately 95% igneous rocks or their metamorphic equivalents, with sedimentary rocks forming the rest (4% shale, 0.75% sandstone, and 0.25% limestone). Sedimentary rocks, however, cover 75%−80% of the Earth's surface and form parent materials for a large majority of soils.

The composition of the parent material is very important to the properties and attributes of the soil, in particular, the suite of minerals present and the size of the soil particles. The mineralogical composition of soils derived from common parent rocks is given in Table 1.1. In general, younger or less weathered soils tend to have a relatively larger proportion of primary minerals, and conversely, more weathered and mature soils contain abundant amounts of secondary minerals. Therefore the abundance of primary

TABLE 1.1 A generalized mineralogical composition of soils derived from different parent rocks.

Rock	Minerals in fresh rocks	Common primary and secondary minerals in soils
Granite	*Major:* quartz, orthoclase, and plagioclase *Minor:* muscovite, biotite, amphiboles, zircon, rutile	Quartz, illite, goethite, kaolinite, gibbsite, hematite, anatase, rutile, zircon
Diorite	*Major:* plagioclase, amphiboles *Minor:* biotite, pyroxenes, alkali feldspar, olivines, magnetite, quartz, apatite, zircon	Kaolinite, smectite, goethite, halloysite, quartz, apatite, zircon
Basalt	*Major:* pyroxenes, plagioclase, olivines *Minor:* biotite, magnetite, calcite, amphiboles, chlorite, ilmenite, quartz, apatite	Smectite, kaolinite, halloysite, goethite, hematite, quartz, anatase, apatite
Shale	*Major:* illite, smectite, kaolinite, quartz *Minor:* feldspars, calcite, dolomite, chlorite, pyrite, goethite	Illite, smectite, kaolinite, quartz, goethite, hematite
Limestone	*Major:* calcite, aragonite, dolomite *Minor:* kaolinite, illite, illite—smectite mixed layer, quartz, siderite, chert, feldspars	Quartz, kaolinite, goethite, hematite, chlorite, smectite
Sandstone	*Major:* quartz *Minor:* orthoclase, plagioclase, biotite, hornblende, chlorite, glauconite, muscovite, calcite, hematite	Quartz, kaolinite, goethite, illite

and secondary minerals correlates well with the particle size distribution in soils, with the proportion of primary minerals increasing with the sand or sand and silt contents and secondary minerals' abundance increasing with clay content.

Organic matter concentration in most soils usually varies from 1% to 5% except for in Histosols (peat, muck, or bog soils), which contain at least 20% organic matter, and it can be up to 100%. Despite its relatively low concentration in most soils, SOM has a tremendous influence on many soil properties and processes and the ecosystem functioning of the soil. SOM serves as a metabolic energy resource for decomposer communities (fungi, microbes, and fauna) in soils and the mineralization of organic matter provides plant-available forms of nitrogen, phosphorus, sulfur, and many other nutrients. Additionally, SOM contributes to soil aggregation, water retention, CEC, buffering capacity, and the sorption of organic compounds.

Primary minerals are formed at very high temperatures and are inherited (unaltered) from igneous rocks and metamorphic rocks. These minerals usually occur in the sand (0.05—2.0 mm or 0.02—2.0 mm) and silt (0.05—0.02 mm or 0.002—0.02 mm) size fractions of the soil. Secondary minerals are formed from the weathering of primary minerals either through crystallization from soil solution or via the alteration of primary minerals. These minerals often are dominant in the clay fraction (<0.002 mm or 2 µm) of a soil and are often referred to as clay minerals.

1.3.2 Solution

Key points: The soil solution is the medium for plant nutrients. Many important chemical reactions involve the soil solution. Nutrients are released from the soil solids into the solution for absorption by plants.

Soil water or soil solution is the liquid phase that contains dissolved ions, gases, and other water-soluble compounds. Soil solution is the hub of biogeochemical activities in the soil because most chemical reactions involve the soil solution. Most common inorganic cations and anions in the soil solution are the same ones that occur on the exchange sites of soil colloids. The most prevalent cations are Ca^{2+}, Mg^{2+}, Na^+, and K^+ (additionally Al^{3+} in highly acidic soils), and the dominant anions are HCO_3^-, Cl^-, NO_3^-, and SO_4^{2-}. Ions of other nutrients usually exist in trace concentrations in the soil solution. In addition to the inorganic ions, soluble organic ions (e.g., aliphatic organic acids with carboxylic groups and aromatic amino acids) exuded by plant roots and soil microbes are also present in the soil solution. The total amounts of dissolved cations and anions in the soil solution are usually between 10^{-2} and 10^{-4} mol L^{-1}, the larger values being more common for less weathered soils and smaller values for more leached and highly weathered soils.

1.3.3 Gases

Key points: The composition of soil air reflects key processes and environmental conditions in the soil. The composition of soil air is also significantly different than atmospheric air, with the soil air having a greater concentration of CO_2 due to plant and microbial respiration.

Soil pore space is occupied by gases (soil air) and water (soil solution), with varying proportions depending upon the soil water content. After irrigation or a rainfall event, soil may be fully saturated with nearly all of the soil pore space filled up with water. On the other end, when soil is nearly completely dry, the pore space is mostly occupied by soil air. Atmospheric air at sea level contains 780.84 mL N_2, 209.48 mL O_2, 9.34 mL Ar, 0.417 mL CO_2, and 0.002 mL CH_4 in 1 L (dm^3) dry air. The value of 0.314 mL CO_2 has been used in many texts, but the updated value is used here.

Soil air composition can vary considerably from that of the atmospheric air. The concentration of O_2 in the soil air is usually less than that in the atmosphere because O_2 is continuously consumed by plant roots and in microbial respiration. Typically, well-aerated soil contains over 150 mL L^{-1} O_2; however, O_2 concentrations can vary between 140 mL L^{-1} and 210 mL L^{-1}. When soil water content exceeds 35%, O_2 concentrations can fall below 50 mL L^{-1} in the soil air, and under prolonged anaerobic or anoxic conditions, O_2 concentration can decrease to zero. Prolonged exposure of plant roots to anoxic soil solution can be detrimental to plant health.

The CO_2 concentration in soil air is much greater ($\sim$2–30 mL L^{-1}) than in atmospheric air (0.417 mL L^{-1}) due to microbial and root respiration and restrictive diffusion in the soil. In extreme cases, such as at depths, in the vicinity of plant roots and under low or trace oxygen conditions, CO_2 concentration can be as high as 115 mL L^{-1}. The extremely high CO_2 concentration in soil air versus the atmospheric air has implications for soil mineral weathering and carbonate chemistry.

The concentration of N_2 in the soil air is usually similar to that in the atmospheric air, varying only slightly depending upon the proportion of other gases in the soil air, particularly O_2. In wetlands and other oxygen-deficient soil environments, other gases, including nitrous oxide (N_2O, up to 0.09 mL L^{-1}), nitric oxide (NO), methane (CH_4), and hydrogen sulfide (H_2S), make important contributions to soil air.

(a) Henry's Constant

Key points: The tendency of soil gases to distribute themselves between soil solution and soil air can be expressed by their respective values of Henry's constant.

The equilibrium state of the partitioning of gases between soil air and soil solution (or water) can be characterized by Henry's law. According to Henry's law, the amount of dissolved gas in water is proportional to the partial pressure of the gas in the atmosphere, as described below:

$$K_H = C_{aq}/P_g \qquad (1.1)$$

where K_H is Henry's law constant (mol m^{-3} atm^{-1}), C_{aq} is the concentration (mol dm^{-3} or mol L^{-1} or M) of a gas species in the aqueous phase (soil solution), and P is the partial pressure (atm) of that gas species in the atmosphere (soil air) under equilibrium conditions. The SI unit for K_H is mol m^{-3} Pa^{-1}; however, the unit mol L^{-1} atm^{-1} is more commonly used, where 1 mol m^{-3} = 1000 mol L^{-1} and 1 atm = 101,325 Pa. Sanders (2015) provides a detailed compilation of many variants of Henry's law that exist in the literature.

Henry's law constant values for selected gases are given in Table 1.2. The values are characteristic of the given gas and solvent, as well as the temperature. Henry's law predicts the solubility of simple gases (e.g., N_2 or O_2), which are unreactive (i.e., do not undergo any chemical reaction with water). However, some gases, including CO_2, sulfur dioxide (SO_2), and nitrogen dioxide (NO_2), react with water, forming acids and this can increase their solubility well exceeding the value that would be predicted by Henry's law. Thus it is necessary to consider the influence of such reactions in addition to Henry's law calculations in determining the total solubility of reactive gases. An example of using

TABLE 1.2   Henry's law constants for selected gases in water at 25°C.

Gas	K_H (mol m^{-3} atm^{-1})
O_2	1.32
N_2	0.65
CH_4	1.42
CO_2	33.44
SO_2	1.32×10^3
NH_3	5.98×10^4
N_2O	24.32
NO	1.93
H_2S	1.01×10^2
NO_2	12.16

Henry's Law in calculating O_2 concentration in natural water is given in Box 1.1.

(b) Gas Transport Mechanisms in Soils

Key point: Diffusion of gases such as CO_2 and O_2 is directly impacted by the amount of water in the soil pores.

Sidebar 1.7 — Advection and Dispersion

Advection refers to the transport mechanism of a substance (or conserved property i.e., mass) by a fluid due to the fluid's bulk motion (i.e., to the flow).

Diffusion refers to the net movement of molecules from a high concentration to a low concentration, which occurs without any motion of the fluid's bulk. It is a time-dependent process that finds its physical origin in the random motion of individual particles.

There are two major mechanisms by which gas is transported in soils, namely, diffusion and advection (Sidebar 1.7); however, diffusion is the principal mechanism for the exchange of gases between the soil and atmosphere (Rolston and Moldrup, 2012). When there is a buildup of CO_2 in the soil due to respiration, CO_2 diffuses out of the soil and into the atmosphere. This process is important, for example, in the warming of Histosols and Gelisols (or Cryosols). Gelisols, which contain permafrost, are beginning to thaw. The warming of previously frozen soils increases the rates of microbial metabolism and allows carbon stored as SOM to be oxidized and released into the atmosphere. Previously inaccessible soil organic carbon is now available for oxidation into CO_2, which diffuses into the atmosphere. Gases can be produced in soil via multiple biological processes, such as fermentation, nitrification, and denitrification. Pressure gradients, such as those caused by wind blowing across the surface of the soil and landscape, can drive advection (Rolston and Moldrup, 2012).

BOX 1.1

The concentration of oxygen in natural waters in equilibrium with air at 1 atm at 25°

According to Henry's law: $K_H = C_{aq}/P_g$

or $C_{aq} = P_g \cdot K_H$

First, the total pressure of dry air is corrected for moist air by subtracting the partial pressure of water at 25°, which is 0.0313 (the values are available in standard handbooks). Dry air contains 20.95% oxygen (fraction of oxygen = 0.2095) on a volume basis, so the partial pressure of oxygen in moist air is:

$$P_{O_2} = (1.0 - 0.0313) \times 0.2095 = 0.2029$$

$$K_H = 1.32 \times 10^{-3} \text{mol L}^{-1}\text{atm}^{-1}$$

$$C_{O_2} = \left[1.32 \times 10^{-3}\right] \times 0.2029$$
$$= 2.67 \times 10^{-4} \text{ mol L}^{-1}$$

$$C_{O_2} = 2.67 \times 10^{-4} \text{ mol L}^{-1} \times 32 \text{ g mol}^{-1}$$
$$\times 10^3 \text{ mg g}^{-1}$$
$$= 8.57 \text{ mg L}^{-1}$$

The solubility of oxygen and other gases decreases with an increasing temperature. In soil environments dissolved oxygen is essential for plants and soil microbes, and prolonged water-logged conditions may deplete the dissolved oxygen completely, causing the reduction of various ions. Further discussion on the oxidation–reduction reactions in soil environments is in Chapter 8.

There are several ways to characterize gas diffusion in soils, and thorough discussions can be found in books by Glinski and Stepniewski (1985), Scott (2000), Jury and Horton (2004), and Rolston (2005). Fick's law is one way to model soil gas transport, and there are several variations of this law that can be considered:

$$\frac{\partial C}{\partial t} = \frac{\partial}{\partial z}\left[D_s \frac{\partial C}{\partial z}\right] + \alpha \qquad (1.2)$$

where the variable C is the concentration of a gas in the soil air (e.g., CO_2 or O_2), t is time, D_s is the diffusion coefficient of the gas in the soil air, z is the depth in the soil, and α is the rate of gas production or consumption. Gas diffusion in soils is driven by a difference in the concentration gradients of gases. Another variation of Fick's law used for a two-component system of gases is given as:

$$N_i^F = -D_{ij}C\nabla y_i \qquad 1.3$$

where N_i^F is the molar diffusive flux of gas i in mol m^{-2} s^{-1}, C is the total molar density in mol m^{-3}, is the binary diffusion coefficient in m^{-2} s^{-1} for the gas pair, and ∇y_i is the gradient in the mole fraction of gas i.

1.4 Soil: The Role of Soils in Environmental and Ecosystem Services

Key points: An ecosystem service is a benefit provided to humans by a particular ecosystem. Soil is one of the ecosystems that provide numerous life-sustaining benefits.

Soil contains biologically active components and dissipates mass fluxes and energy. In its natural state soil is a self-regulating dynamic body that operates and controls many key biogeochemical processes, including the regulation and supply of water and nutrients to

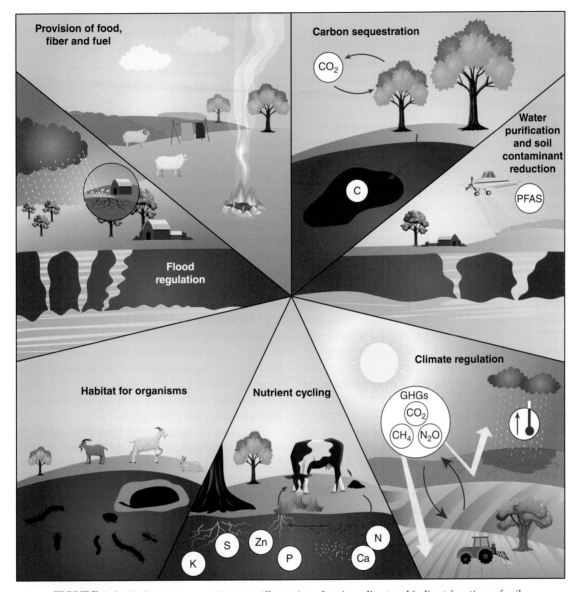

FIGURE 1.4   Soil ecosystem services — an illustration of various direct and indirect functions of soils.

terrestrial fauna and flora. Soil is the foundation of terrestrial life, and humans have used soils for producing food, fiber, and shelter for thousands of years. However, in the last few decades the importance and role of soil in many other aspects have been recognized, as depicted in Figure 1.4. Soil is now recognized to play a central role in many global issues, including climate change.

Sidebar 1.8 — **Ecosystem Services**

Ecosystems provide a range of services that are of fundamental importance to human well-being, health, livelihood, and survival (Costanza et al., 2014). An ecosystem is a community of living (humans, animals, or any other organism) and nonliving things (such as soil, water, and air) that work together as a system.

Soils have been defined and categorized as important to multiple aspects of human life, including food and nutritional security, water quality and renewability, climate change mitigation and adaptation, human health and biological diversity, and in fact, national security (NASEM, 2017). Soils of natural and managed ecosystems perform a multitude of functions, which have been placed into four main ecosystem services (see Sidebar 1.8): (i) provisioning food, fiber, fuel, and water; (ii) regulating climate, flood, disease, and water quality; (iii) supporting soil formation, nutrient cycling, and primary production; and (iv) providing a cultural medium aesthetically, spiritually, educationally, and recreationally (MEA, 2005).

There is an increasing trend to consider natural ecosystems as capital assets, which requires proper monetary evaluation of the ecosystem services from an asset and any investment needed for its conservation or protection. In relation to soils, so far, most of the focus has been on cataloging and classifying the numerous ecosystem services provided by soils. The progress on the monetary values of soil ecosystem services has been limited. Current approaches largely focus on evaluating the processes and functions in relation to services and do not require monetization. Several international organizations and national regulatory agencies are increasingly adopting the "ecosystem services framework" for the conservation and sustainable use of soils (Turner and Daily, 2008).

Soil ecosystem services depend on inherent (e.g., mineralogical, organic matter) and dynamic (e.g., water content, temperature) soil properties, their interactions, and management practices. The scale of soil ecosystem services varies from a micrometer (e.g., habit for soil microorganisms) to a landscape (flood mitigation) to a global (air quality) scale.

1.4.1 The Water—Energy—Food Nexus

Key points: Soil is a key component of the water—energy—food nexus because of its central role in water quality and use, energy production, and food production.

The "Water—Energy—Food" nexus concept was first introduced at the World Economic Forum in 2011. It is a focused approach to evaluating natural resources and their interactions with respect to the growing challenges of securing food, water, and energy resources for a growing world population. Most existing nexus documents ignore the role of soil; however, soil is important not only to produce food but also for energy (e.g., biofuels or shale gas) and water supply (reservoirs, underground water storage). The overexploitation of soil resources leads to soil degradation and diminishing soil resources, which will limit our ability to produce the quantity and quality of food required for an increasing future population.

The area of agricultural land per capita has declined by about 50% over the last five decades, from 0.415 ha in 1961 to 0.214 ha in 2007 (Smith et al., 2010), and is continuing to decline. Increasing population, the use of agricultural land for industries, urban and suburban housing developments and other infrastructures (particularly in developing countries), and land degradation (e.g., soil erosion, soil salinization, soil acidification) are some of the main causes of the rapidly declining per-capita land area of agricultural land. The average cereal yields will need to be increased by 25%, from the average of 3.23 t ha^{-1} in 2005—2007 to 4.34 t ha^{-1} in 2030, to feed a growing population (Smith et al., 2010). The increased food production will have to come largely from increased crop yields and greater cropping intensity. Thus the key role of soil must be recognized. Some of the important

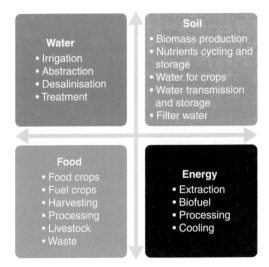

FIGURE 1.5 Water—Energy—Food—Soil nexus and key processes and functions involved in the linkages.

processes and functions of the soil in the Water—Energy—Food—Soil nexus are shown in Figure 1.5.

The Water—Energy—Food—Soil nexus plays out differently in different parts of the world. Countries and regions with the highest rural poverty, malnutrition, and food insecurity often experience the greatest water loss and soil degradation. Thus increasing productivity will require better management to preserve soil ecosystem services, particularly in more vulnerable parts of the world. Soil structure needs to be preserved and maintained to minimize the degradation of agricultural lands and nearby water bodies. There is an urgent need to develop a deeper understanding of the complex dynamics of water, energy, food and soil systems, and their feedback mechanisms for sustainable use of the terrestrial ecosystems.

1.4.2 Plant Nutrients

Key point: Storage, regulation, and provision of plant nutrients are important ecosystem services provided by soils.

The provision, regulation, and storage of plant nutrients, which support primary production, are some of the most important functions of the soil. The capacity of soils to provide plant essential nutrients can be highly varied and depends on several factors, including soil parent material, the extent of soil weathering, SOM concentration, environmental conditions, and agronomic practices. Soil parent material has a major influence on the nutrient element concentrations in the soil; this influence is more pronounced in young soils and diminishes with increasing soil age and soil weathering.

Major nutrients, for example, nitrogen, phosphorus, and potassium, are often applied as chemical fertilizers in cropping systems to meet crop requirements and optimize crop yields. Soil colloids retain these and other nutrients through sorption processes and control their supply to crop plants.

Nitrogen is one of the principal elements required by most crop plants in large quantities. Soil parent materials contain very low concentrations of this nutrient. The continuous supply of plant-available forms of nitrogen (NH_4^+ and NO_3^-) is met through the microbial mineralization of SOM and nitrogen fixation. The mineralization of SOM also releases plant-available forms of phosphorus, sulfur, and other nutrients. Mineral weathering, adsorption—desorption, and precipitation—dissolution reactions are important in the regulation and provision of plant essential nutrients in the soil.

1.4.3 Water

Key point: Water storage, provision, regulation, and improvement of water quality are significant ecosystem services provided by soil.

Soil, being a porous medium, has a tremendous influence on the water cycle. Water from precipitation or irrigation infiltrates into the soil, and water in the soil profile undergoes evapotranspiration and is used for the growth of crop plants. The retention of water in the soil prevents erosion. Soil pore water regulates plant-essential nutrients and contaminant flow in the rhizosphere and beyond. The excess water from precipitation can be distributed as overland flow (surface water runoff). Subsurface lateral flow and percolation through the soil profile into groundwater also occurs. Wetlands exhibit a unique case of soil's role in regulating services such as water quality improvement, flood control, carbon management, and habitat provision to a range of biota, including fungi, fauna, and other microorganisms.

If rainfall, irrigation water, or wastewater contains organic and/or inorganic contaminants, pathogens, or excess nutrients (e.g., N, P), the soil acts as a filter. Many of these substances are adsorbed by soil particles and/or entrapped in the soil matrix. The purification of contaminated water occurs through physical, chemical, and biological processes. There are enormous environmental, economic, and recreational benefits to humans from this process. With increasing population and enhanced pressure on land, this particular role of soil will become increasingly more important.

Sidebar 1.9 — Adhesion and Cohesion

Adhesion is the attraction between water molecules and the solid particles. The intermolecular attractive force between water molecules is called cohesion.

The capacity of soil to store water depends on several factors, including soil porosity, depth or thickness of the soil layer, and the interaction of the soil matrix with water. The interaction of water and the soil matrix involves a combination of adhesive and cohesive forces (Sidebar 1.9) and the osmotic binding of water in the diffuse electric double layers (Verhoef and Egea, 2013). Plant roots have to overcome these forces to take up water from soil, and by a similar token, water is protected against evaporation and drainage. Soil water content and the forces holding it in place are related quantitatively and graphically through a water retention curve plot or a water release curve.

The water retention curve is primarily influenced by the particle size distribution in the soil (i.e., sand, silt, and clay contents). Medium-textured soils (loamy or silty loam) often have the largest available water capacities (Verhoef and Egea, 2013) Although clayey soils have more pore space than the medium-textured soils that have higher sand content, some of the water is retained at greater potential (i.e., held more tightly), so it is not available to plants. In sandy soils pores are mostly large (>50 μm), and they drain water readily with a small amount of plant-available water retained. Soil texture (particularly clay content), a well-aggregated soil structure, and organic matter concentration are important for water infiltration in the soil and the water retention capacity of the soil.

1.4.4 Carbon Storage and Cycling

Key points: Soils serve as both a source and a sink of carbon to and from the atmosphere and play a major role in global carbon cycling.

The solid and gas phases of C in soil are major components in the C cycle. Solid-phase C consists of both organic matter and inorganic minerals. Of the gas-phase C species, two important gases are CO_2 and CH_4. Carbon storage in soil, particularly SOM, is of critical importance because of the detrimental effects of increasing CO_2 levels in the atmosphere, which contribute to Earth's changing climate. Not only is the current quantity of soil carbon present on the Earth's surface important, but so is the flux of C to and from this critical pool of solid-phase carbon. During intense agricultural operations, however, SOC can become depleted, and C oxidizes to CO_2 and transfers to the atmosphere. This process, in addition to adding CO_2 to the atmosphere, is detrimental to several soil properties that are critical to plant growth.

1.4.5 Contaminants

Key point: Soils are important for the storage and regulation of contaminants, including organic and inorganic hazardous wastes.

Soil has been used as a repository for all kinds of waste materials since time immemorial. All dead terrestrial organisms and plants are disposed to soils where biogeochemical processes mineralize these dead materials into smaller components, which can be utilized again over time. This is nature's way of recycling. Biota in soils, including microorganisms, fungi, and fauna, through a complex set of steps, carry out the decomposition of natural and anthropogenic organic wastes. The decomposition processes are also regulated by different biophysical conditions in the soil, such as temperature, water availability, aeration, and soil pH.

Because soil is a heterogeneous mix of inorganic and organic matter, it has the capability to bind with many different types of contaminants. SOM is particularly important for the adsorption of organic pollutants, such as polyaromatic hydrocarbons (PAHs), as the hydrophobic portions of PAHs are attracted to SOM (Chapter 3).

Hazardous industrial wastes have been disposed into soils without any long-term consideration. However, national regulatory agencies have put much stricter regulations in place for the disposal of waste materials into soil. When hazardous wastes are deposited in properly designed landfills, which use impermeable liners and leachate and gas collection systems, environmental contamination is minimized. The role of soils in the management and storage of waste is still very crucial. Clayey soils with very low permeability are preferred for the storage, management, and disposal of hazardous wastes.

1.4.6 Earth's Critical Zone

Key points: The Earth's critical zone is where rock, soil, water, air, and living organisms interact. It includes all of the areas where humans inhabit.

The critical zone supports life for all organisms and includes the vital process of food production. The interactions between rock, soil, water, air, and organisms are interdependent, one affecting and depending upon the other. These interactions provide humans with their fundamental needs for life, including food, water, energy, and shelter.

Because of the interdependent nature of the many processes occurring in the critical zone and the indispensable ecosystem services provided by it, a global network of critical zone observatories (CZOs) has been created to investigate these integrated processes. Soil is considered the central reactive layer that receives inputs of energy, mass, and biodiversity. It transmits and transforms these components across the reacting layer. Soil produces output flows both above- and belowground (Banwart and Sparks, 2017).

The stations at various CZOs have been used to simulate the biophysical flows and transformations to quantify multiple soil functions. The CZOs provide data for a range of soils, landscapes, and environmental conditions to help us better understand both sustainable and disruptive actions that influence our wellbeing. Based on the understanding of key processes in the critical zone, new practices and policy options can be designed to enhance soil functions and reduce soil threats.

1.4.7 Biodiversity

Key point: Soils provide a three-dimensional, highly dense, and heterogeneous environment that hosts a multitude of plants, animals, fungi, and other microorganisms.

Soils are one of the most biologically diverse ecosystems on Earth. In addition to nematodes and microorganisms that live in soils, such as bacteria, fungi, and protozoa, soil fauna, including earthworms, termites, springtails, ants, spiders, and beetles, are important to soil biodiversity. The roots of plants are also part of the living ecosystem in soils and contribute to maintaining biodiversity (Sidebar 1.10) in soils. Near the roots of plants is a unique zone called the rhizosphere; this is the volume of soil adjacent to a root that is influenced directly by root exudates and is where the plant obtains nutrients from the soil. Plant roots are commonly associated with a symbiotic relationship with mycorrhizal fungi.

Sidebar 1.10 − Biodiversity

Biodiversity is defined as "the variability among living organisms from all sources, including interalia, terrestrial, marine and other aquatic ecosystems and the ecological complexes of which they are part; this includes diversity within species, between species and of ecosystems" (Convention on Biological Diversity, Article 2).

Increasing land pressure, soil contamination, and other human-induced degradation tend to reduce soil biodiversity in tandem with aboveground biodiversity. Some soil processes, such as organic matter decomposition, are carried out by a range of soil organisms that operate over a diverse range of environmental conditions. Despite some extreme soil environmental conditions, such as extreme dry or cold conditions, or metal contamination, the mineralization of organic matter is maintained, although the microbial biomass is severely decreased (Hopkins and Gregorich, 2013) Contrary to this, other processes (e.g., symbiotic nitrogen fixation, nitrification, the decomposition of selected xenobiotic compounds) are carried out by a limited group

of microorganisms that exist and operate under a highly specific set of soil environmental conditions. These processes are much less resilient and can be easily reduced or lost under adverse soil environmental conditions (Bardgett et al., 2005). The conservation of soil biodiversity is paramount to maintaining soil ecosystem services. Additionally, biodiversity provides stability and resilience to perturbations in soil environments.

Soil biodiversity may be extremely important for pharmaceuticals, as most of the existing antibiotics have come from cultured soil bacteria. With the vast majority of soil bacteria not culturable using current methods, there is a huge potential for finding valuable new drugs, and thus it is vital to maintain greater biodiversity in soils.

1.4.8 Culture

Key points: As a public and social service, soils preserve a wide range of archaeological and cultural heritage objects and buried infrastructures.

The existence and distribution of archaeological artifacts in different soils and landscapes provide information about the past, including cultural practices, land management, and human activities. Artifacts, such as stone, ceramics, bone, metals, wood, skins, glass, burial mounds, and cultivation terraces, have been discovered in many parts of the world. The knowledge about which and how soils preserve various artifacts is valuable for the management of heritage and buried infrastructures and also useful to assess the long-term impact on soil properties of certain artifacts that contain contaminants.

Sidebar 1.11 − Cultural Heritage

UNESCO defines cultural heritage as "the legacy of physical artifact and intangible attributes of a group or society that are inherited from past generations, maintained in the present and bestowed for the benefit of future generations."

Certain natural or seminatural features may be associated with the identity of an individual, a community, or a society and thus provide cultural information associated with or shared across generations. Visible material representations of cultural activities on the landscape (such as rice paddies, viticulture terraces) and ecosystem features (e.g., certain types of landscapes, forests, prairies, or deserts) may be used as indicators for cultural heritage services. Cultural heritage (Sidebar 1.11) is inextricably linked with historical relationships between human societies and ecosystems. There are many examples, such as the Champagne region in France, Tuscany in Italy, the Napa Valley in the US, and the Darjeeling region in India, where the cultural landscapes of a region (or products derived from it) may serve as an important trademark for tourism and marketing. The preservation of cultural heritage therefore can have considerable synergy with preserving other ecosystem services.

1.4.9 Human Health

Key points: Soil quality is of utmost concern to the public in both urban and rural areas. Exposure to contaminated soil and remediation of contaminated soil are both risks and priorities that weigh heavily on human health and the economic development of contaminated land.

Above environmental, ecosystem, and animal wellbeing, human health is of the utmost public concern when discussing and acting on environmental issues. Major environmental policy changes tend to occur when there is imminent danger to humans from anthropogenic activities. Often, these policy changes are a great challenge to pass through legislative procedures. However, the real-world issues that affect many

different populations of people are some of the signature topics that make environmental soil chemistry relevant both scientifically and socially.

The applicability of soil chemistry to human health, via either a particular contaminated site or basic science research, is a critical part of maintaining an engaged public. Soils play an important role in both urban and rural environments for the health of nearby residents. This role can range from the soil quality or contamination of an urban community garden or urban soils to soil erosion and nutrient management in rural areas. Particularly, with these diverse ecosystem settings, the idea of environmental justice with respect to soil contamination and quality is an important catalyst to motivate research. Part of the focus of environmental justice is that there should be a fair distribution of both the negative and positive environmental consequences and benefits, respectively, for people.

1.4.10 Infrastructure

Key points: Soils provide both a foundation and a source for construction and construction materials, respectively. Local roads, interstate highways, bridges, houses, and skyscrapers are several examples of constructions that are built upon soil.

The physicochemical properties of soil, as discussed above, are important fundamental properties affecting soil chemistry. Several soil properties, however, are directly related to its usefulness for the built environment (anthropogenic ecosystems). These properties affect the physical structure of soil and are carefully considered for the physical and industrial uses of soils. Whether it is, for example, constructing new houses or buildings, creating new roads or artificial wetlands, or engineering landfills, there are several soil properties that determine which type of soil will be most appropriate.

Particle size distribution, soil water content, porosity, and clay mineralogy all affect a soil's ability to compact and hydraulic conductivity. Soil compaction and hydraulic conductivity are two critical characteristics, for example, in the construction of buildings and the engineering of landfills. Clay mineralogy affects the shrinking and swelling of the clay fraction. If a soil shrinks and swells upon hydration, the foundation upon which a structure is built will be unstable. High clay content reduces the hydraulic conductivity of the soil, thus reducing water flow. The physical properties of soil make it an indispensable ingredient for geotechnical engineering and construction projects in urban ecosystems. Soils are also used as a source of primary substances; for example, gravel and clay are used for the construction of buildings and roads, and clays are used for making bricks.

1.5 The Periodic Table

Key points: Elements in the periodic table are organized based on atomic numbers and electron configurations. As the foundation and organizational scheme for our chemical knowledge, the periodic table of elements is an essential tool and reference for soil chemistry.

1.5.1 The Periodic Table of Elements

Devised by Russian chemist Dmitri Mendeleev in 1869, the periodic table (Figure 1.6) places elements in order of increasing atomic number into horizontal rows called periods (1–7) and vertical columns called groups (1–18). Elements in a group have the same number of valence electrons, that is, the number of electrons in the outer shell, and elements in a period have the same principal quantum number of valence shells or the number of electron shells.

Electrons within an atom are assigned to shells, subshells, and orbitals. Each shell has a different energy level, which is represented by a number called the principal quantum number (n). The principal quantum number serves to describe the size of the orbital or the distance between the electron and the nucleus. The closest shell has a value of $n = 1$, the next shell has a value of $n = 2$, and so on. Shells increase in size and energy as they extend further away from the nucleus.

Shells are further subdivided into subshells, with the maximum number of subshells equivalent to the shell number. For example, if $n = 1$ (first shell), then only one subshell is possible and when $n = 2$ (second shell), two subshells are possible. There are four subshells, designated by the letters $s, p, d,$ and f. Each subshell can hold a certain maximum number of electrons, that is, $s - 2$ electrons, $p - 6$ electrons, $d - 10$ electrons, and $f - 14$ electrons. Each $s, p, d,$ and f subshell has one, three, five, and seven orbitals, respectively. The s subshell has the lowest energy subshell and the f subshell has the highest energy subshell. Electrons in an atom occupy orbitals of the lowest available energy levels before additional electrons begin to occupy higher-energy-level orbitals. Elements in the periodic table are grouped into four blocks, namely, $s, d, p,$ and f blocks (Figure 1.7A). The elements in the s and p blocks are together referred to as the main-group elements. The order of subshell filling is presented in Figure 1.7B.

1.5.2 General Trends in the Periodic Table

Key point: The periodic table of the elements is organized such that elemental trends can be readily understood.

Some general trends in the properties of elements in the periodic table are presented in Figure 1.8. With elements within a given period, with increasing atomic number, that is, from left to right, the atomic radius and metallic properties of the elements decreases, because atoms

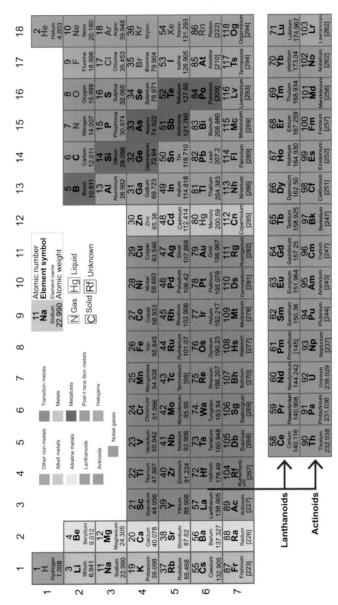

FIGURE 1.6 The periodic table of elements showing the periods (1–7 shown horizontally) and groups (1–18 shown vertically). Various categories of the metals and nonmetals and forms (solid, liquid, gas) of the elements are also indicated in the table.

(A)

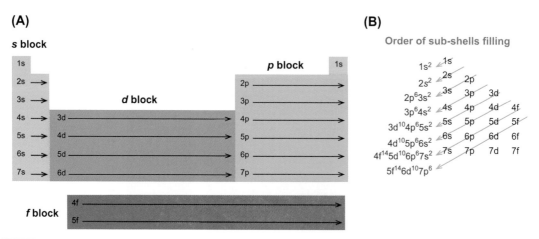

(B)

FIGURE 1.7 (A) Elements grouped into four blocks, that is, s, d, p, and f, in the periodic table; (B) the general order of the subshell filling of elements.

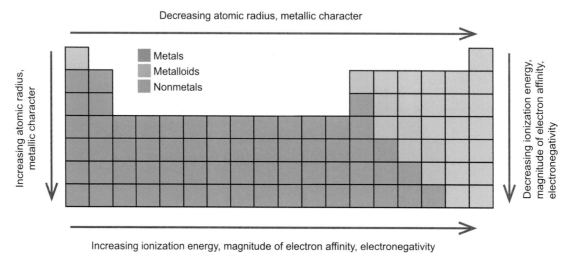

FIGURE 1.8 General trends in some key properties of elements in the periodic table; the lanthanoids and actinoids are excluded from this part due to the lack of data, particularly for the actinoids. There is an increase in the ionization energy, magnitude of electron affinity, and electronegativity of elements and a decrease in the atomic radius and metallic character from left to right, that is, Group 1→18. There is a decreasing trend in the ionization energy, magnitude of electron affinity, and electronegativity of elements and an increasing trend in the atomic radius and metallic character from top to bottom, that is, Period 1→7.

with more electrons and protons have a greater attraction that brings atoms together, hence a decreasing trend for the atomic size; and atoms do not lose their valence electron as readily as with larger atoms. Conversely, moving down (i.e., increasing period) within a group, the atomic size and metallic properties of elements have an increasing trend because of the addition of extra outer energy levels and valence electrons are lost more readily.

Sidebar 1.12 — Some Useful Definitions

Atomic radius is half the distance between the nuclei of identical neighbouring atoms in the solid form of an element, that is, the distance between an atom's nucleus and its valence electrons.

Electronegativity is a measure of an atom's ability, when in a molecule, to attract the shared pair of electrons in a bond (covalent) to itself.

Metallic character is the set of chemical properties associated with elements that are metals. These chemical properties result from how readily metals lose their electrons to form cations (positively charged ions).

Electron affinity is defined as the change (gain or loss) in energy (kJ mol^{-1}) of a neutral atom (in the gaseous phase) when an electron is added to a neutral atom or molecule to form a negative ion. In other words, the neutral atom's likelihood of gaining an electron.

Ionization energy or ionization potential is the energy (kJ mol^{-1}) required to remove the outermost (highest energy) electron from an atom in the gaseous state.

The trends for the ionization energy, magnitude of electron affinity, and electronegativity are opposite to the atomic radius and metallic properties (Sidebar 1.12) in both directions (Figure 1.8). Group 1 elements have low ionization energies because the loss of an electron forms a stable octet. Moving down within a group, the ionization energy decreases because electrons are farther away from the nucleus, so the attraction is weaker, and thus it is easier to remove the outermost electron. Metals have a tendency to lose electrons and have relatively low electronegativities; conversely, nonmetals tend to gain electrons and have high electronegativities. The metalloids along the diagonal line dividing metals and nonmetals show properties and reactivities intermediate between those of

metals and nonmetals. Fluorine in Group 17 and Period 2 has the highest electronegativity. Within a group and moving downward, electronegativity decreases because atoms have a bigger radius (more shells) and the positive charge of the nucleus is further away from the bonding electrons and is shielded by the extra electron shells, whereas there is an increasing trend in electronegativity moving across a period. Electron affinity becomes less negative moving down within a group and decreases or increases across a period depending on the electronic configuration.

1.5.3 Properties of Elements in the Blocks

Key points: In soils some elements are more abundant than others. Some elements are essential for plant growth, and others are highly toxic. Elements can have multiple oxidation states.

The s block elements are those in which the outermost or valence electron(s) is (are) in the s orbital with the exception of He. Most s block alkali metals (i.e., Li, Na, K, Rb, Cs, and Fr) in Group 1 and alkaline earth metals (i.e., Be, Mg, Ca, Sr, Ba, and Ra) in Group 2 are highly reactive due to the ease with which they lose their outer s orbital electrons and interact to form mostly ionic compounds (e.g., NaCl, KCl, $CaCl_2$). The first period elements in this block, however, are nonmetals. Hydrogen is highly chemically reactive similar to the other s block elements, but He is a virtually unreactive noble gas.

Alkali metals in Group 1 occur as monovalent cations (+1 oxidation state), whereas alkaline earth metals exist as divalent cations (+2 oxidation state). The alkali and alkaline metals are commonly present in many primary minerals and released during the weathering process. In soils Ca^{2+}, Mg^{2+}, Na^+, and K^+ occur as free cations and do not generally form soluble complexes with ligands (Sidebar 1.13). These cations usually dominate on the soil exchange sites and in the soil solution. The alkali and alkaline metals form soluble minerals or salts in soils, including halite (NaCl), gypsum ($CaSO_4.2H_2O$), and calcite ($CaCO_3$), under certain environmental conditions, such as restricted drainage, low-lying poorly drained areas, and regions with little or no rainfall.

Sidebar 1.13 — What Is a Ligand?

A ligand is an ion or molecule that binds to a central metal cation by donating or sharing electrons (i.e., a Lewis base) to form a coordination complex.

The p block elements are those in which the last electron (valence) occupies the p orbitals. The elements from Groups 13 to 18 belong to the p block except for He, which is part of the s block. The p block has the most diverse range of elements with three types of elements, namely, metals, nonmetals, and metalloids, present in this block. The p orbital can hold up to a maximum of six electrons and the p subshell is progressively filled in the p block elements, moving from Group 13 to Group 18. The properties of p block elements can be best described in terms of their group or element type. The oxidation state of elements in the p block is maximum when it is equal to a total number of valence electrons, that is, the sum of s and p electrons.

Group 13 elements (B, Al, Ga, In, and Tl) have diverse physical and chemical properties. Boron is the only nonmetal in the group, and then there is an increased metallic character from Al to Tl in the group. Due to their electronic configuration ($ns2 \, np1$), Group 13 elements form electron-deficient compounds that act as Lewis acids (Sidebar 1.14). Except for B, Group 13 elements are all relatively electropositive; that is, they tend to lose electrons in chemical reactions rather than gain them.

Sidebar 1.14 — Lewis Acid and Base

A Lewis acid is a species (molecule or ion) that contains an empty orbital that is capable of accepting an electron pair (i.e., an electrophile). A Lewis base is an atom, molecule, or functional group that donates an electron pair (i.e., a nucleophile) and has a filled orbital containing an electron pair that is not involved in bonding.

Aluminum is the most abundant metallic element in the Earth's crust (and soils) and is a structural component of most common clay minerals in the soils. Aluminum occurs in the +3 oxidation state in soils, and in highly acidic soils $\left(pH_{H_2O} < 5.5 \right)$ free Al^{3+} exists in the soil solution, which inhibits root growth and can be highly toxic to crop plants. The concentration of Al in the soil solution is highly dependent on pH and is controlled by precipitation–dissolution of Al hydroxides as well as the concentration of other inorganic and organic ions (e.g., F^-, OH^-, SO_4^{2-}, oxalate, citrate) that form aqueous complexes with Al.

Boron is a micronutrient and it exists in the +3 oxidation state in soils. Boric acid ($B(OH_3)^0$) is the main species in the soil solution, and the availability of B in soils is mostly controlled by adsorption–desorption reactions.

Group 14 elements (C, Si, Ge, Sn, and Pb) vary considerably in their properties. Carbon and Si are nonmetals, Ge is a metalloid, and Sn and Pb are metals. Carbon is the building block of life and exists in a number of oxidation states in soil and terrestrial environments (Figure 1.9). Both organic and inorganic forms of C are common in soils. Silicon is the second most abundant element in the Earth's crust, and it exists in the +4 oxidation state. It is a structural component of most primary and secondary soil minerals. Silicic acid (H_4SiO_4) is the dominant species in the soil solution, and its concentration is controlled by poorly crystalline forms of SiO_2.

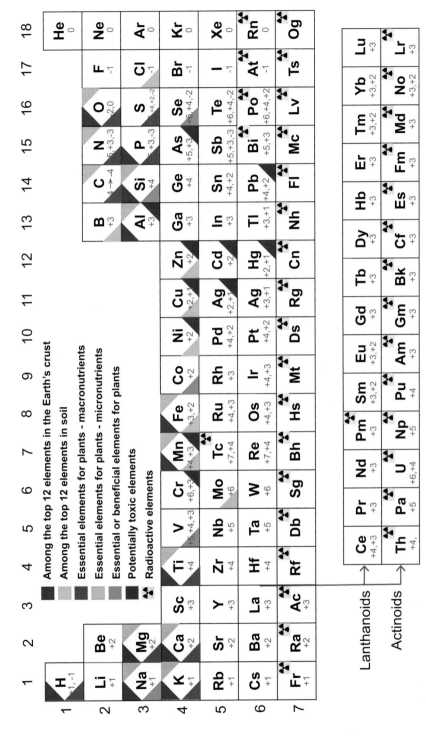

FIGURE 1.9 The environmental periodic table indicating the 12 most abundant elements in the Earth's crust (in purple) and in the soil (in yellow). Plant essential nutrients are indicated in the table in dark green (macronutrients), light green (micronutrients), and blue (elements that are necessary or beneficial for certain crop plants). Potentially toxic elements (in red) and radioactive elements (radioactive symbol) are also indicated in the table. Common oxidation states are also listed in red below the elemental symbols.

Lead is among the most toxic metals to humans, and it occurs in the +2 oxidation state in soils. Natural concentrations of Pb in soils are usually small; however, there has been significant soil contamination from past and current uses of Pb in paints, pigments, gasoline, used engine oil, pesticides, batteries, and in industries, such as foundries or smelters. Organic compounds form strong complexes with Pb in soil solution and Pb is also strongly adsorbed by SOM. Most inorganic ligands also make complexes with Pb in solution.

Similar to Groups 13 and 14, the properties of Group 15 elements (N, P, As, Sb, and Bi) are also diverse, particularly their wide range of oxidation states and complex compounds they form with O. Nitrogen occurs in the N_2 gaseous state, which is the most abundant gas in the atmosphere. Nitrogen is a building block of life and is essential for growth and reproduction in both plants and animals. Nitrogen forms gases and oxyanions in its oxidation states ranging from +5 to +1 and −3.

Similar to N, P is also an essential element for plants and animals. Phosphorus occurs in a range of organic and inorganic compounds, and the main oxidation state in soils is +5. In acidic soils the solubility of P in soils is controlled by adsorption reactions onto Fe and Al oxides and phosphate minerals formed with Fe and Al. Calcium phosphates and adsorption reactions on calcite mostly control the P solubility in neutral and alkaline soils.

In contrast to N and P, As is a highly toxic element, and Bi is a radioactive element. Arsenic and Bi also form oxides and oxyanions in a range of oxidation states; however, the +5 and +3 oxidation states, respectively, are common in soils. Both of these elements occur in the structure of primary minerals as trace impurities. The solution concentrations of As and Bi are controlled by adsorption reaction onto Fe oxides.

Group 16 elements (O, S, Se, Te, and Po) consist of a combination of nonmetals, metalloids, and metals. The elements in the group are called the chalcogens or oxygen family. The properties of O, the head of the group, vary significantly from those of the other members. Oxygen is the only group member that exists as a diatomic molecule and generally forms compounds with small coordination numbers. Oxygen is the most important element for all forms of terrestrial life, and O_2 makes up 21% of the atmosphere by volume. Oxygen is a reactive element; it generally occurs in the −2 oxidation state and forms oxides with most other elements. The element and its compounds or minerals make up nearly half of the Earth's crust on a weight basis and over 90% on a volume basis. Like O, S is also essential to all living things; however, some of the S compounds (e.g., carbon disulfide, hydrogen sulfide, and sulfur dioxide) are toxic. Sulfur exists in several oxidation states, and it easily combines with most other elements forming a range of oxyanions. Selenium resembles S in terms of chemical activity and physical properties and occurs in several oxidation states (Figure 1.9). It is a trace nutrient element for some living species but, in high concentrations, can be toxic. Tellurium and Po are among the rarest elements in the Earth's crust and have properties similar to those of Se. Polonium is a radioactive element and thus is highly toxic.

Group 17 elements (F, Cl, Br, I, At) are called halogens, and it is the only group that contains elements in all three states of matter at room temperature and pressure. Fluorine and Cl are gases; Br is a liquid, and I is a solid. Astatine is a radioactive element with properties similar to I. Halogens have seven valence electrons, and they only require one additional electron to form a full octet. This characteristic makes them more reactive than other nonmetal groups. Elemental halogens are diatomic molecules; however, they are never found in nature in native forms due to their high reactivity.

Group 18 elements (He, Ne, Ar, Kr, Xe, Ra) in the *p* block are noble gases. These are the least reactive elements; they have been called rare gases and inert gases and are currently designated as noble gases. Noble gases have the most stable electron configuration since their p orbitals are completely filled with electrons. Noble gases occur as monatomic gases at room temperatures. Among the noble gases, only Xe forms a range of compounds with F and O. Radon, which is a by-product of the radioactive decay of Ra, is a radioactive and highly mobile gas.

The d block consists of 40 elements from Groups 3, 4, 5, 6, 7, 8, 9, 10, 11, and 12 of the periodic table (Figure 1.9). All d block elements are metals; hence they are often referred to as the d block metals. The metals in the groups from 3 to 11 have partially filled d subshells in the free elements or their cations, and these are called transition metals. The metals from Group 12 (Zn, Cd, and Hg) do not have partially filled d shells, so they are part of the d block but are not transition metals.

Unlike the s and p block elements, the d block elements exhibit significant similarities in chemistry in their respective groups and periods. Moving from left to right in the d block (e.g., $Sc \rightarrow Zn$, $Y \rightarrow Cd$, $La \rightarrow Hg$), the electronegativity and ionization energy of the elements slightly increase, and the hydration energy of the metal cations slightly decreases. Similarly, moving from top to bottom in the d block (e.g., $Sc \rightarrow La$, $Zn \rightarrow Hg$), the electronegativity of the elements increases, and the hydration energy of the metal cations decreases. The increased electronegativity and decreased hydration energy make the metals (e.g., Pt, Au) in the lower right corner of the d block unreactive: so-called *noble metals*.

The transition metal cations exist in several oxidation states and are formed by the initial loss of ns electrons. The common oxidation states of these metals in soils are given in Figure 1.9.

The d block metals are widely distributed in the Earth's crust, and many of them are important for industrial applications, for example, making materials from their alloys and compounds. Iron is the most abundant of these metals in soils, and it occurs in many primary and secondary

minerals. In highly weathered soils goethite (α-FeOOH) and hematite (α-Fe_2O_3) are often the dominant secondary minerals. The ionic form of Fe in soil solution is highly varied depending on the solution pH and Eh. Titanium, Zr, and V are present in significant amounts in soils, mostly bound in the structure of primary and secondary minerals.

Manganese occurs in the $+2$, $+3$, and $+4$ oxidation states in soils. Secondary minerals of Mn are often poorly crystalline and rarely present in significant amounts in soils. The presence of Mn minerals in mottles, nodules, or gravels is more common. Under highly acidic conditions, the concentration of Mn in soil solution could be very large and toxic to plants. Trace amounts of several first-row transition metals are essential for plants and humans, whereas many other metals, including Cr, Ag, Cd, and Hg, are hazardous. Even the biologically essential trace elements (e.g., Mn, Ni, Cu, Zn) are hazardous if they are present in larger concentrations in soils from industrial and anthropogenic activities. The availability of most transition metals in soil solution is largely governed by adsorption–desorption reactions on clay and oxide minerals, notably Fe oxides. The elements in the fourth row (Period 7) are synthetic and radioactive, with very limited chemical characterization of these elements.

The f block contains two series, lanthanoids (also called lanthanides) (14 elements following La) and actinoids (also called actinides) (14 elements following Ac). In the lanthanoids $4f$ orbitals are progressively filled from Ce to Lu in Period 6, and in actinoids $5f$ orbitals are filled from Th to Lr in Period 7. The chemical properties of the f block elements are largely determined by outer s orbital electrons since the f orbital electrons are less active in determining the chemistry of these elements. The lanthanoid series elements have predominantly one oxidation ($+3$) state and very uniform chemical properties. Lanthanoid elements, along with Y and Sc, have been referred to as "rare earth elements

(REEs)," although many of these elements exist in soils at concentrations (e.g., median concentrations of 40 and 50 mg kg^{-1} for Ce and La, respectively) similar to or greater than concentrations of some of the transition metals and micronutrients (median concentrations of Cu, Mo, Cd, and Pb are 30, 1.2, 0.35, and 35 mg kg^{-1}, respectively).

Actinoids have a greater diversity in their chemical properties due to their several oxidation states; however, limited data are available for most of these elements due to their associated radioactivity. Because of their natural abundance and low levels of radioactivity of the common nuclides of Th and U, the chemical properties of these elements have been extensively studied. In oxidized soil environments U exists as UO_2^{2+} (uranyl ion, oxidation state +6), which is the most stable and mobile form. Under anoxic conditions, UO_2^{2+} may be reduced to an oxidation state forming $U(OH)_4$ and $U(OH)_3$ or react with a sulfide. Under aerobic conditions, UO_2^{2+} forms soluble complex compounds with SOM, carbonates, phosphate, and sulfates. The mobility of U in soils is mostly determined by adsorption–desorption reactions with Fe oxides.

1.6 Units, Conversions, and Constants

Key points: Because soils are diverse in their physical states, including solid, liquid, and gaseous phases, environmental soil chemistry involves the use of multiple units and constants. Often it is necessary to convert units, and many acronyms are also used. This section serves as a source of information to help the reader navigate the multitude of terms used in this textbook.

There is an internationally agreed system of units built from seven base units, one each for seven dimensionally independent base quantities. There is only one SI unit for each physical quantity, which is either the appropriate SI base unit itself or an appropriate SI derived

unit. *Le Système International d'Unités* (abbreviated SI), or the international system of units, was adopted in 1960. The system is built from seven independent base quantities that are listed in Table 1.3.

To express decimal multiples and submultiples of the SI units, the following prefixes should be used (Table 1.4). The prefix symbols should be printed in roman (upright) type with no space between the prefix and the unit symbol (e.g., km, mmol). A prefix should never be used on its own (e.g., μ, G) and prefixes are not to be combined into compound prefixes (e.g., pm and not μμm).

Some non-SI units are frequently used in environmental soil chemistry. For example, the land

TABLE 1.3   Basic SI units.

Quantity	Unit	Symbol
Length	meter	M
Mass	kilogram	Kg
Time	second	S
Electric current	ampere	A
Thermodynamic temperature	kelvin	K
Quantity of substance	mole	Mol
Luminous intensity	candela	Cd

TABLE 1.4   Prefixes for SI units.

Submultiple or fraction	Prefix	Symbol	Multiple	Prefix	Symbol
10^{-1}	deci	d	10^1	deca	da
10^{-2}	centi	c	10^2	hecto	h
10^{-3}	milli	m	10^3	kilo	k
10^{-6}	micro	μ	10^6	mega	M
10^{-9}	nano	n	10^9	giga	G
10^{-12}	pico	p	10^{12}	tera	T
10^{-15}	femto	f	10^{15}	peta	P

TABLE 1.5 Commonly used constants or relations in environmental soil chemistry.

Constants	Definition or symbol	Formula or value
Atmospheric pressure	P	101.325 kPa
0°C (zero on the Celsius temperature scale)	T	273.15 K
Atomic mass unit	u	1.66054×10^{-27} kg
Gas constant (molar, universal, or ideal gas constant)	R	8.31446 J mol^{-1} K^{-1}
Avogadro's number	N_A or L	6.02214×10^{23} mol^{-1}; number of atoms or molecules in one mole of a substance
Boltzmann's constant	k_B or k	$k = R/N_A$; 1.38065×10^{-23} J K^{-1}

area is often expressed in hectares (1 ha = 10,000 m^2). For the expression of concentration on a volume basis, liter (1 L = 1 dm^3 = 10^{-3} m^3) is often used. Soil bulk density is generally expressed in g cm^{-3} (1 g cm^{-3} = 1 Mg m^{-3}). The Celsius scale is very commonly used to express temperature in the literature; zero on the Celsius scale (°C) is equivalent to 273.15 K. Another temperature unit, Fahrenheit, is used for temperature in the US, with C = (F − 32)/1.8. The use of Angstroms is common for atomic spacings and in X-ray diffraction analysis, and for FTIR spectroscopy analysis, wave number is reported as a reciprocal of centimeter (cm^{-1}).

Some important physical constants, units, and conversions that are used in environmental soil chemistry are listed in Tables 1.5−1.7.

1.6.1 Molarity

Key points: The use of molarity to describe aqueous concentrations of ions is commonly used in soil chemistry. Molarity is based on the volume of the solvent. Molality is based on the mass of the solvent.

The mole (or mol) is defined as the amount of a chemical substance that contains as many number of atoms (or molecules or ions or charges) as there are number of atoms in exactly 12 g of carbon-12 (an isotope of carbon); the number is 6.02214×10^{23} (Avogadro's number) of atoms of carbon-12. The number of atoms (or other particles) in a mole is the same for all substances.

Sidebar 1.15 − **Molality and Molarity**

Molality is the number of moles per kilogram of solvent (e.g., water). This is different from molarity, where concentration is measured with respect to the volume of the solvent, that is, 1 mole (gram-molecular weight) of solute in 1 L of the solution. For example, the density of pure water is close to 1 g cm^{-3}, or 1 g mL^{-1}, at normal temperatures. However, a solution becomes denser when solutes are added (e.g., saltwater), that is, density increases, and 1 mL of the solution becomes greater than 1 g.

The mole is related to the mass of an element, so one mole of carbon-12 atoms has 6.02214×10^{23} atoms and a mass of 12 g (called gram atomic weight); similarly one mole of nitrogen consists of the same number of atoms as carbon-12, but it has a mass of 14 g. The same logic can be applied to molecules or formula weights to get gram molecular weights or gram formula weights.

To convert the mass (g) of an element or a molecule to moles (mol), simply divide by the

TABLE 1.6 Units that are commonly used in environmental soil chemistry.

Quantity	Unit name	Symbol	Units	Examples and/or Alternate Units
Concentration in the aqueous phase	• molarity (moles per liter) • millimoles per liter • micromoles per liter	• M • mM • µM	• $mol\ L^{-1}$ • $mmol\ L^{-1}$ • $µmol\ L^{-1}$	• The soil solution had a concentration of 0.002 M Ca (0.002 $mol\ L^{-1}$). • 0.002 M Ca = 2 mM Ca = 2000 µM Ca • Often, 2 mM Ca is written as 2 $mmol\ L^{-1}$ Ca (or 2 $µmol\ mL^{-1}$)
Concentration in the solid phase	• milligram per kilogram • parts per million		• $mg\ kg^{-1}$ • ppm	• Pb concentration in a soil was 60 $mg\ kg^{-1}$ or $µg\ g^{-1}$. • Therefore there was 60 ppm Pb in the soil.
Bulk density	• grams per cubic centimeter • megagrams per cubic meter		• $g\ cm^{-3}$ • $Mg\ m^{-3}$	• A surface soil sample had a bulk density of 1.35 $g\ cm^{-3}$ or 1.35 $Mg\ m^{-3}$.
Specific surface area	• square meters per gram • square meters per kilogram		• $m^2\ g^{-1}$ • $m^2\ kg^{-1}$	• Ferrihydrite has a specific surface area of 200 $m^2\ g^{-1}$ or 200,000 $m^2\ kg^{-1}$.
Volume	• liter	• L	• $0.001\ m^3$	• $1\ L = 1\ dm^3 = 1000\ mL = 10^{-3}\ m^3$
Cation exchange capacity	• millimoles of charge per kilogram soil • centimoles of charge per kilogram soil	• CEC	• $mmol_c\ kg^{-1}$ • $cmol_c\ kg^{-1}$	• The CEC of a soil sample was 250 $mmol_c\ kg^{-1}$. • Also written as $mmol(+)\ kg^{-1}$ • 250 $mmol_c\ kg^{-1}$ = 25 $cmol_c\ kg^{-1}$ = 25 meq 100 g^{-1}
Electrical conductivity	• DeciSiemens per meter	• EC	• $dS\ m^{-1}$	• The EC of the soil solution extract was measured to be 1.7 $dS\ m^{-1}$ = 1.7 $mS\ cm^{-1}$.
Interatomic distances	• nanometers • Angströms	• nm • Å	• $1 \times 10^{-9}\ m$ • $1 \times 10^{-10}\ m$	• An interatomic distance of arsenate bound to ferrihydrite (As-Fe) was found to be 2.850 Å or 0.2850 nm.
d-spacing	• Angströms	• nm • Å	• $1 \times 10^{-9}\ m$ • $1 \times 10^{-10}\ m$	• The distance between parallel planes of atoms in a mineral (crystal) is 3.256 Å (0.3256 nm).

TABLE 1.7 Conversions commonly used or required in environmental soil chemistry.

Context	Example
• Dilutions: In the laboratory it is common to prepare stock solutions for experiments. The relationship between the volume and concentration of the stock solution and the volume and concentration of the desired diluted solution can be expressed as: $C_1V_1 = C_2V_2$, where C_1 and V_1 indicate the concentration and volume of the stock solution (initial concentration and volume) and C_2 and V_2 indicate the concentration and volume of the dilute solution (final concentration and volume).	• What volume of a stock solution of 10 mM Mn^{2+} is needed to prepare 50 mL of 50 μM Mn^{2+}. First make sure to convert the two concentrations into the same unit, so either change the unit of the stock solution to μM ($10 \times 1000 = 10,000$ Mm) or change the unit of the dilute solution to Mm ($50/1000 = 0.05$ Mm). • Then using the formula calculate V_1: $(10,000 \text{ Mm})(V_1) = (50 \text{ μM})(50 \text{ mL})$ $V_1 = (50 \text{ μM})(50 \text{ mL})/(10,000 \text{ μM}) = 0.25$ mL. • If the volume is too small to measure accurately with a pipette or analytical balance, intermediate dilute solutions should be prepared.
• Conversion in a solid phase: Concentrations of some elements in the solid phase (e.g., soils or plants) are often expressed in non-SI units, such as percent (%) or ppm. These can be converted to appropriate SI units. The term "ppm" should be avoided in favor of concentration expressed in a mass fraction (such as mg kg^{-1} or μg g^{-1}). Thus the preferred statement would be: The Pb concentration in the soil was 3000 mg kg^{-1}, which is a weight per weight unit (w/w). Recall that 1% = 10,000 ppm.	• A contaminated soil sample contains 3000 ppm (mg kg^{-1}) Pb. • $3000 \text{ ppm} = 3000 \text{ mg kg}^{-1} = \dfrac{3000}{1,000,000} \times 100 = 0.3\%$ • $3000 \text{ mg kg}^{-1} = \dfrac{3000}{\text{Atomic weight of Pb}} = \dfrac{3000}{207.2}$ $= 14.48 \text{ mmol kg}^{-1}$
• Conversion in an aqueous phase: Often, aqueous concentrations can be given in the ppm unit, which can be converted to appropriate SI units as in the solid phase.	• The soil solution of a contaminated soil is 50 ppm As. • $50 \text{ ppm As} = 50 \text{ mg L}^{-1} = 50 \text{ μg mL}^{-1}$ $50 \text{ mg L}^{-1} = \dfrac{50}{\text{Atomic weight of As}} = \dfrac{50}{74.922}$ $= 0.667 \text{ mmol L}^{-1} = 0.667 \text{ μmol mL}^{-1} = 667 \text{ μmol L}^{-1}$
• Conversion from the aqueous phase to solid phase: Often, soil chemists are interested in the total concentration of an element, such as lead, in the soil. To determine this value, there are many methods, commonly including microwave-assisted acid digestion followed by inductively coupled plasma (ICP) techniques. Thus often it is important to convert between ICP values (liquid) to the soil values (solid). In this case the precise volumes and masses must be written, recorded, and used in the conversion. A common protocol to follow is to digest 0.5 g soil in 12 mL concentrated acid. The concentrated acid is then diluted 10× for analysis via ICP.	• The ICP instrument reports a value of 12.5 mg Pb L^{-1} in a liquid sample (from the soil that was digested). • $12.5 \text{ mg Pb L}^{-1} \times 10 = 125 \text{ mg Pb L}^{-1}$ • $125 \text{ mg Pb L}^{-1} \times 0.012 \text{ L} = 0.15 \text{ mg Pb}$ • $0.15 \text{ mg Pb}/0.0005 \text{ kg soil} = 3000 \text{ mg Pb kg soil}^{-1}$
• Hectare: The land area is expressed in a non-SI unit: hectare. In some countries, including the United States, the use of a non-SI unit, acres, is common.	• 1 ha = 10,000 m^2 • 1 ha = 2.471 acres • To convert acres to ha, multiply the value by 0.405.

(Continued)

TABLE 1.7   Conversions commonly used or required in environmental soil chemistry.—cont'd

Context	Example
• Cation exchange capacity units: Historically, units for CEC (and exchangeable cations) of soils were given in meq 100 g^{-1}. However, this unit is no longer used. The correct units are $mmol_c$ kg^{-1} or $cmol_c$ kg^{-1}. In the units for CEC ($mmol_c$ kg^{-1}) the subscript "c" represents "+" charge or charge on a proton. In some publications "+" or "p" have been used for the same purpose. Importantly, equivalent fractions or equivalents of charge are no longer an accepted unit for CEC (or exchangeable cations) of soils or clay minerals. It is a historical unit, where moles of charge were given as "equivalents" (eq) per unit mass of soil. Commonly, CEC values were expressed as milliequivalents (meq) per 100 g soil. The equivalent weight of an atom is the molecular weight divided by the valance state of the atom.	• For example, a montmorillonite sample has the CEC of 80 meq 100 g^{-1} = 80 $cmol_c$ kg^{-1} = 800 $mmol_c$ kg^{-1}. • 1 eq = 1000 meq = 1 mol_c = 1000 $mmol_c$ = 100 $cmol_c$
• Conductivity and resistivity: Electrical conductivity (EC) is a measure of the ability of the solution to conduct electricity. It is a common measurement to determine the concentration of dissolved salts in soils. Conductivity meters measure EC values in the units of siemens (S). However, laboratory water filtration systems often give the resistance of water in ohms (Ω). The ohm is the SI unit for electrical resistance, while Siemens (S) is the SI unit for electrical conductivity. In some older literature mho ($\mho$) has been used, which is equivalent to Siemens but should be avoided.	• Electrical conductivity (S) is the reciprocal of electrical resistance (Ω); thus $S = \Omega^{-1}$. A measurement of resistance of ultrapure water at 18.2 M$\Omega\cdot$cm, which is equal to 1/18.2 = 0.055 μS cm^{-1}. • 0.055 μS cm^{-1} • 1 dS m^{-1} = 100 mS m^{-1} = 1 mS cm^{-1} = 1000 μS cm^{-1}
• Logarithms: When converting from the base e natural logarithm (ln) to base 10 logarithm (log), a conversion factor of 2.303 is used. When measuring reaction kinetics, it is common to use either unit.	• ln x = 2.303 log x

molar mass of the element or molecule (g mol^{-1}). The unit generally used to express aqueous concentration is molarity. However, molality (Sidebar 1.15) is also employed by computational software (such as Visual Minteq). Water is generally the solvent when considering environmental soil chemical reactions. In dilute solutions of water molality and molarity are nearly the same value. This is because the mass of the solvent (water) is approximated to be the same as the mass of the solution (water with solute). This is only true for dilute solutions.

Suggested Reading

Brown, G.E., Calas, G., 2013. Mineral-aqueous solution interfaces and their impact on the environment. Geochem. Perspect. 1 (4–5) 483–742.

Henderson, G.S., Neuville, D.R., Downs, R.T. (Eds.), 2014. Spectroscopic Methods in Mineralogy and Materials Sciences. Vol. 78. Mineralogical Society of America, Chantilly, pp. i-xviii + 800 pages.

Shriver, D., Atkins, P., 2010. Inorganic Chemistry. Oxford University Press.

Sparks, D.L., 2006. Historical aspects of soil chemistry. In: Warkentin, B.P. (Ed.), Footprints in the Soil: People and Ideas in Soil History. Elsevier Science, pp. 307–337.

Problem Set

Q1. Soil samples were extracted from two depths, namely, 0 to 15 cm (S1) and 15 to 30 cm (S2), of an agricultural field using a cylindrical core sampler (height = 15 cm; inner radius = 5 cm). The samples were placed in two separate aluminum containers and brought back to the laboratory. After keeping the samples in an oven for 24 h at 105°C, the dry weight of the soil samples was recorded, which was S1 = 1476.2 g and S2 = 1671.6 g.

  (i) Calculate the bulk density of the two samples. Are the bulk density values for the two samples different? If so, why?

  (ii) Calculate the soil weight in the top 15 cm layer of soil in one hectare area.

Q2. The soil solution of a highly weathered soil was obtained by centrifugation, and the following chemical composition was obtained:

Ion	Concentration (mg L^{-1})
Ca^{2+}	25.6
Mg^{2+}	10.7
K^+	11.7
Na^+	1.6
HCO_3^-	129.3
Cl^-	12.1
SO_4^{2-}	10.6

  (i) Calculate the concentration in $mmol_c$ L^{-1} of all the cations and anions in the soil solution.

  (ii) If all the cations and anions in the solution have been accounted for the sums of cations ($mmol_c$ L^{-1}) should be equal to the sum of anions ($mmol_c$ L^{-1}); comment on the results obtained in this example.

  (iii) Calculate the values of N, C, and S in mg L^{-1} in the soil solution.

Q3. A soil sample was extracted from the top 30 cm of an agricultural soil using a cylindrical core sampler (height = 10 cm; inner radius = 4 cm). The samples were placed in an aluminum container (weight with lid = 154.6 g), sealed to prevent any water loss, and brought back to the laboratory. The container with field moist soil was weighed in the laboratory (field moist weight of soil with container + lid = 937.3 g). The sample was then dried in an oven for 24 h at 105°C, and the dry weight of the soil sample was recorded (dry weight of soil with container + lid = 829.9 g).

  (i) Calculate the gravimetric water content (the mass of water per unit mass of dry soil, w) in the field moist soil.

  (ii) Calculate the bulk density of the soil.

  (iii) Calculate the volumetric water content (the volume of water per unit volume of dry soil) in the field moist soil.

  (iv) Calculate the volumetric water content (the volume of water per unit volume of dry soil) and depth of water in the top (0–30 cm) soil. (Hint: The volumetric water content (m^3 m^{-3}) is equivalent to the depth of water [m m^{-1}].)

Q4. Soil analysis of an agricultural soil (0–30 cm) showed a total N concentration of 1.3 g kg^{-1} and exchangeable Ca, Mg, and K values of 118, 79, and 13 $mmol_c$ kg, respectively. Calculate the total amount of N and exchangeable Ca, Mg, and K in the 0 to 30 cm soil layer of one hectare area; the bulk density of the soil is 1.28 g cm^{-3} (Mg m^{-3}).

Q5. Demonstrate the trend in the atomic radii of Group 1 and 17 elements and explain the reasons for the observed trends in the atomic radii of the elements in these two groups.

Q6. Determine the amount of ferrous ammonium sulfate $(NH_4)_2Fe(SO_4)_2 \cdot 6H_2O$ (MW = 392.14 g mol^{-1}) necessary to make 20 mL of 0.010 M Fe(II) stock solution in anoxic ultrapure water.
Describe how to dilute this stock solution to 10 mL of 500 µM Fe(II) for use in experiments.

Q7. A highly contaminated soil sample (750 mg) was digested in a 10 mL mixture of aqua regia (1:3 mixture of concentrated HNO_3 and HCl). Following digestion, the volume was made to 25 mL. The sample was analyzed using inductively coupled plasma atomic mass spectrometry (ICP-MS) after diluting 10 times. The concentrations of Cd, Cr, Pb, and Zn in the diluted samples were 0.02, 0.14, 0.38, and 1.1 mg L^{-1}, respectively. Calculate the concentrations of these metals in the soil in mg kg^{-1}.

Q8. Why is it important to specify the difference between molarity and molality? What unit is more commonly used to describe concentrations in dilute solutions and why?

Q9. Why does SOM have a large impact on soil given that it normally constitutes 1%–5% of the soil? How does SOM provide ecosystem services?

Q10. Convert the atmospheric concentrations of N_2, O_2, Ar, CO_2, and CH_4 given in Section 1.4.3 to percentage. What is their total in percentage?

Q11. What are two gases that vary significantly in the soil solution compared to the atmosphere in terms of concentration? Why is there such a difference between these gas concentrations, and what are the implications for plants and minerals?

Q12. What is the water–energy–food nexus? Why does soil play a major role in the water–energy–food nexus? Describe some of the major challenges related to soils and population growth.

Q13. What is ion exchange, and why is it an important soil process from crop production or environmental points of view?

Q14. Give an example of environmental justice and how it is related to soil chemistry. How is gentrification related to environmental justice?

Q15. Describe how climate change is related to soils by giving two examples. How are "positive" and "negative" feedbacks related to the soil carbon cycle?

References

Amundson, R., Berhe, A.A., Hopmans, J.W., Olson, C., Sztein, A.E., Sparks, D.L., 2015. Soil and human security in the 21st century. Science 348, 6.

Banwart, S.A., Bernasconi, S.M., Blum, W.E.H., de Souza, D.M., Chabaux, F., Duffy, C., et al., 2017. Soil functions in Earth's Critical Zone: key results and conclusions. Adv. Agron. 142, 1–27.

Bardgett, R.D., Bowman, W.D., Kaufmann, R., Schmidt, S.K., 2005. A temporal approach to linking aboveground and belowground ecology. Trends Ecol. Evol. 20, 634–641.

Baveye, P.C., Baveye, J., Gowdy, J., 2016. Soil "ecosystem" services and natural capital: critical appraisal of research on uncertain ground. Front. Environ. Sci. 4, 41.

Brown, G.E., Calas, G., 2013. Mineral-aqueous solution interfaces and their impact on the environment. Geochem. Perspect. 1 (4–5), 483–742.

Browne, C.A., 1944. A Source Book of Agricultural Chemistry. Verdoorn, F. (Ed.), In: Chronica Botanica The Chronica Botanica Co., Waltham.

Coleman, N.T., Harward, M.E., 1953. The heats of neutralization of acid clays and cation exchange resins. J. Am. Chem. Soc. 75, 6045–6046.

Costanza, R., de Groot, R., Sutton, P., van der Ploeg, S., Anderson, S.J., Kubiszewski, I., Farber, S., Turner, R.K., 2014. Changes in the global value of ecosystem services. Glob. Environ. Change 26, 152–158.

Davis, J.A., Kent, D.B., 1990. Surface complexation modeling in aqueous geochemistry. In: Hochella, M.F., White, A.F. (Eds.), Mineral-Water Interface Geochemistry, Vol. 23. Mineralogical Society of America, Washington, DC, pp. 177–260.

Dungait, J.A.J., Hopkins, D.W., Gregory, A.S., Whitmore, A.P., 2012. Soil organic matter turnover is governed by accessibility not recalcitrance. Glob. Change Biol. 18, 1781–1796.

Fey, M.V., Roux, J.L., 1976. Electric charges on sesquioxidic soil clays. Soil Sci. Soc. Am. J. 40, 359–364.

Gapon, E.N., 1933. Theory of exchange adsorption in soils. J. Gen. Chem. (USSR) 3 (2), 144–152.

Gillman, G.P., 1974. The influence of net charge on water dispersible clay and sorbed sulphate. Aust. J. Soil Res. 12, 173–176.

Gillman, G.P., 1979. Proposed method for the measurement of exchange properties of highly weathered soils. Aust. J. Soil Res. 17, 129–139.

Ginder-Vogel, M., Landrot, G., Fischel, J.S., Sparks, D.L., 2009. Quantification of rapid environmental redox processes with quick-scanning x-ray absorption spectroscopy (Q-XAS). Proc. Natl. Acad. Sci. U. S. A. 106, 16124–16128.

Glinski, J., Stepniewski, W., 1985. Soil Aeration and Its Role for Plants. CRC Press, Boca Raton, FL.

Goldberg, S., 1992. Use of surface complexation models in soil chemical systems. Adv. Agron. 47, 233–329.

Harter, R.D., Smith, G., 1981. Langmuir equation and alternate methods of studying "adsorption" reactions in soils. In: Dowdy, R.H., et al. (Eds.), Chemistry in the Soil Environment. ASA Special Publications 40. ASA and SSSA, Madison, WI, pp. 167–182.

Harward, M.E., Coleman, N.T., 1954. Some properties of hand Al clays and exchange resins. Soil Sci. 78, 181–188.

Hendricks, S.B., Fry, W.H., 1930. The results of X-ray and mineralogical examination of soil colloids. Soil Sci. 29, 457–476.

Hilgard, E.W., 1906. Soils, Their Formation, Properties, Composition, and Relations to Climate and Plant Growth. MacMillan Co., New York, NY, p. 593.

Hingston, F.J., Atkinson, R.J., Posner, A.M., Quirk, J.P., 1967. Specific adsorption of anions. Nature (London) 215, 1459–1461.

Hochella Jr., M.F., Lower, S.K., Maurice, P.A., Penn, R.L., Sahai, N., Sparks, D.L., et al., 2008. Nanominerals, mineral nanoparticles, and earth systems. Science 319, 1631–1635.

Hopkins, D.W., Gregorich, E.G., 2013. Managing the soil-plant system for the delivery of ecosystem services. In: Gregory, P.J., Nortcliff, S. (Eds.), Soil Conditions and Plant Growth. John Wiley & Sons, New York, NY, pp. 390–416.

Jackson, M.L., 1956. Soil Chemical Analysis – Advanced Course. Author, Madison, WI.

Jackson, M.L., 1985. Soil Chemical Analysis – Advanced Course, Second edition. University of Wisconsin, Madison, WI.

Jury, W.H., Horton, R., 2004. Soil Physics, Sixth edition. John Wiley & Sons, Hoboken, NJ.

Kelley, W.P., 1945. Calculating formulas for fine grained minerals on the basis of chemical analyses. Am. Mineral. 30, 1–26.

Kelley, W.P., 1948. Cation Exchange in Soils. Reinhold Pub. Corp., New York, NY.

Kelley, W.P., Brown, S.M., 1925. Base exchange in relation to alkali soils. Soil Sci. 20, 477–495.

Kelley, W.P., Dore, W.H., Brown, S.M., 1931. The nature of the base exchange material of bentonite, soils and zeolites, as revealed by chemical investigation and X-ray analysis. Soil Sci. 31, 25–55.

Lehmann, J., Kleber, M., 2015. The contentious nature of soil organic matter. Nature 528, 60–68.

Lutzenkirchen, J., 2006. Surface Complexation Modelling. First editon, vol. 11. Academic Press, Amsterdam.

Marshall, C.E., 1935. Layer lattices and the base-exchange clays. Z. Kristallgr. 91, 433–449.

Mattson, S., 1927. Anionic and cationic adsorption by soil colloidal materials of varying $SiO_2/Al_2O_3 + Fe_2O_3$ ratio. In: Trans. 1st. Int. Cong. Soil Science. Vol. 2. USDA, Washington, DC, pp. 199–211.

Mattson, S., 1931. The laws of soil colloidal behavior: VI. Amphoteric behavior. Soil Sci. 32, 343–365.

Mattson, S., 1932. The laws of soil colloidal behavior: VII. Proteins and proteinated complexes. Soil Sci. 33, 41–72.

MEA (Millennium Ecosystem Assessment), 2005. Ecosystems and Human Well-Being: Synthesis. Island Press, Washington, DC.

NASEM, 2017. National Academies of Sciences, Engineering, and Medicine, 2017. Soils: The Foundation of Life: Proceedings of a Workshop – In Brief. The National Academies Press, Washington, DC.

Parfitt, R.L., Russell, J.D., Farmer, V.C., 1976. Confirmation of surface-structures of goethite (alpha-FeOOH) and phosphated goethite by infrared spectroscopy. J. Chem. Soc. Faraday Trans. I 72, 1082–1087.

Parikh, S.J., Lafferty, B.J., Sparks, D.L., 2008. An ATR-FTIR spectroscopic approach for measuring rapid kinetics at the mineral/water interface. J. Colloid Interface Sci. 320, 177–185.

Rich, C.I., Obenshain, S.S., 1955. Chemical and clay mineral properties of a red-yellow podzolic soil derived from muscovite schist. Soil Sci. Soc. Am. Proc. 19, 334–339.

Rolston, D.E., 2005. Aeration. In: Hillel, D. (Ed.), Encyclopedia of Soils in the Environment. Elsevier, Amsterdam, pp. 17–21.

Rolston, D.E., Moldrup, P., 2012. Gas transport in soils. In: Huang, P.M., Li, Y., Sumner, M.E. (Eds.), Handbook of Soil Sciences. CRC Press, Boca Raton, FL.

Ross, C.S., Hendricks, S.B., 1945. Minerals of the montmorillonite group, their origin and relation to soils and clays. U. S. Geol. Surv., Tech. Pap. 205B.

Ruffin, E., 1832. An Essay on Calcareous Manures. J.W. Campbell, Petersburg, VA.

Sander, R., 2015. Compilation of Henry's law constants (Version 4.0) for water as solvent. Atmos. Chem. Phys. 15, 4391–4981.

Schmidt, M.W.I., Torn, M.S., Abiven, S., Dittmar, T., Guggenberger, G., Janssens, I.A., et al., 2011. Persistence of soil organic matter as an ecosystem property. Nature 478, 49–56.

Schofield, R.K., Samson, H.R., 1953. The deflocculation of kaolinite suspensions and the accompanying changeover from positive to negative chloride adsorption. Clay Miner. Bull. 2, 45–51.

Scott, H.D., 2000. Soil Physics: Agricultural and Environmental Applications. Iowa State University Press, Ames, IA.

Siebecker, M.G., Chaney, R.L., Sparks, D.L., 2018. Natural speciation of nickel at the micrometer scale in serpentine (ultramafic) topsoils using microfocused X-ray fluorescence, diffraction, and absorption. Geochem. Trans. 19 (1), 14.

Siebecker, M., Li, W., Khalid, S., Sparks, D., 2014. Real-time QEXAFS spectroscopy measures rapid precipitate formation at the mineral − Water interface. Nat. Commun. 5, 6003.

Smith, P., Gregory, P.J., van Vuuren, D., Obersteiner, M., Havlik, P., Rounsevell, M., et al., 2010. Competition for land. Philos. Trans. R. Soc. B 365, 2941–2957.

Sparks, D.L., 1999. Kinetics and mechanisms of chemical reactions at the soil mineral/water interface. In: Sparks, D.L. (Ed.), Soil Physical Chemistry. CRC Press, Boca Raton, FL, pp. 135–191.

Sparks, D.L., 2006. Historical aspects of soil chemistry. In: Warkentin, B.P. (Ed.), Footprints in the Soil: People and Ideas in Soil History". Elsevier Science, Amsterdam, pp. 307–337.

Sparks, D.L., 2006. Historical Aspects of Soil Chemistry. In: Warkentin, B.P. (Ed.), Footprints in the soil: people and ideas in soil history. Elsevier, Amsterdam, pp. 307–337.

Sprengel, C., 1838. Die Lehre von den Urbarmachungen und Grundverbesserungen (The Science of Cultivation and Soil Amelioration). Immanuel Müller Publ. Co., Leipzig.

Sumner, M.E., 1963a. Effect of alcohol washing and pH value of leaching solution on positive and negative charges in ferruginous soils. Nature (London) 198, 1018–1019.

Sumner, M.E., 1963b. Effect of iron oxides on positive and negative charge in clays and soils. Clay Miner. Bull. 5, 218–226.

Sumner, M.E., 1998. Soil chemistry: past, present and future. In: Huang, P.M., Sparks, D.L., Boyd, S.A. (Eds.), Future Prospects for Soil Chemistry. SSSA Special Publication No. 55. Soil Science Society of America, Madison, WI, pp. 1–38.

Tejedor-Tejedor, M.I., Anderson, M.A., 1986. "In situ" ATR-Fourier transform infrared studies of the goethite (α-FeOOH)-aqueous solution interface. Langmuir 2, 203–210.

Thomas, G.W., 1977. Historical developments in soil chemistry: ion exchange. Soil Sci. Soc. Am. J. 41, 230–238.

Turner, R.K., Daily, G.C., 2008. The ecosystem services framework and natural capital conservation. Environ. Res. Econ. 39, 25–35.

Uehara, G., Gillman, G., 1981. The Mineralogy, Chemistry, and Physics of Tropical Soils With Variable Charge Clays. Westview Press, Boulder, CO.

van der Ploeg, R.R., Böhm, W., Kirkham, M.B., 1999. On the origin of the theory of mineral nutrition of plants and the law of the minimum. Soil Sci. Soc. Am. J. 63, 1055–1062.

van Es, H., 2017. A new definition of soil. CSA News 62, 20–21.

van Raij, B., Peech, M., 1972. Electrochemical properties of some Oxisols and Alfisols of the tropics. Soil Sci. Soc. Am. Proc. 36, 587–593.

Vanselow, A.P., 1932. Equilibria of the base exchange reactions of bentonites, permutites, soil colloids, and zeolites. Soil Sci. 33, 95–113.

Veith, J.A., Sposito, G., 1977. On the use of the Langmuir equation in the interpretation of "adsorption" phenomena. Soil Sci. Soc. Am. J. 41, 697–702.

Verhoef, A., Egea, G., 2013. Soil water and its management. In: Gregory, P., Nortcliff, S. (Eds.), Soil Conditions and Plant Growth. John Wiley & Sons, Chichester, pp. 269–322.

Vietch, F.P., 1902. The estimation of soil acidity and the lime requirements of soils. J. Am. Chem. Soc. 24, 1120–1128.

Way, J.T., 1850. On the power of soils to absorb manure. J. R. Agric. Soc. Engl. 11, 313–379.

Westall, J.C., Hohl, H., 1980. A comparison of electrostatic models for the oxide/solution interface. Adv. Colloid Interface Sci. 12, 265–294.

2

Soil Minerals

2.1 Inorganic Soil Fraction

Key point: The inorganic solid fraction of soils is largely composed of minerals in which major elements are bonded together into different structures.

The inorganic fraction of soils exists in two forms: (i) minerals and (ii) amorphous solids. A mineral is a naturally occurring solid with a highly ordered atomic arrangement and a definite (but not necessarily fixed) homogenous chemical composition and is usually formed via inorganic processes. Contrary to minerals, amorphous solids lack an ordered atomic arrangement (Klein and Dutrow, 2008).

The crystalline or mineral fraction, which is the dominant fraction ($>95\%$) in most soils, consists of a mixture of minerals inherited from the soil parent material and secondary minerals formed by the weathering processes within the soil or elsewhere. The amorphous fraction in soils is usually small; however, it can be significant in certain soil types (e.g., Andisols or Andosols). The amorphous soil fraction is usually composed of a mixture of variously hydrated oxides of Fe, Al, Mn, and Si.

Minerals, being the predominant fraction of soil solids, exert much influence on soil chemical, physical, and biological properties. The weathering of primary minerals releases plant essential and nonessential elements into the soil solution. The potential capacity of a soil to release elements through the weathering of primary minerals diminishes with the increasing extent of soil weathering. Secondary minerals, particularly clay minerals (Sidebar 2.1), due to their small size, large specific surface area, and charge characteristics greatly influence many important equilibrium and kinetic reactions and processes in soils.

Sidebar 2.1 — Clay Minerals

The term "clay minerals" refers to phyllosilicates and to minerals (such as iron oxyhydroxides/oxides) that impart plasticity (ability of the material to be molded into any shape) to clay and that harden upon drying or firing (Guggenheim and Martin, 1995). Clay minerals should not be equated solely to phyllosilicates.

The concentration of different elements in soils largely follows the trends of their concentrations in the Earth's crust and sediments (Table 2.1). Additionally, soils are enriched with respect to C and N due to the increased content of organic matter in the surface soil. Oxygen is the most prevalent element in the Earth's crust and in soils. It comprises about 47% of the Earth's crust by weight and greater than 90% by volume (Mason and Moore, 1982). Silicon, Al, Fe, C, Ca, Mg, Na, and K are generally present at concentrations greater than approximately 0.5%, and these

Environmental Soil Chemistry, Third Edition
https://doi.org/10.1016/B978-0-12-815880-7.00002-X

TABLE 2.1 Concentration ($mg\ kg^{-1}$) of 13 major elements in soils, the Earth's crust and sediments.

Element	Soils					Earth's crust[c]	Sediments[c]
	median[a]	Median[b]	Range[b]	Median[c]	Range[c]	mean	mean
O	450,906	–	–	490,000	–	474,000	486,000
Si	337,510	310,000	16,000-450,000	330,000	250,000–410,000	277,000	245,000
Al	49,218	72,000	700–>10,000	71,000	10,000–300,000	82,000	72,000
Fe	23,780	26,000	100–>100,000	40,000	2000–550,000	41,000	41,000
C	–	25,000	600–370,000	20,000	7000–500,000	480	29,400
K	13,282	15,000	50–63,000	14,000	80–37,000	21,000	20,000
Ca	5718	24,000	100–320,000	15,000	700–500,000	41,000	66,000
Mg	4222	9000	50–100,000	5000	400–9000	23,000	14,000
Ti	3596	2900	70–20,000	5000	150–25,000	5600	3800
Na	4080	12,000	<500–100,000	5000	150–25,000	23,000	5700
N	–	–	–	2000	200–5000	25	470
P	524	430	<20–6800	800	35–5300	1000	670
Mn	465	550	<2–7000	1000	20–10,000	950	770

[a] de Caritat et al. (2012);
[b] Shacklette and Boerngen (1984); .
[c] From Bowen (1979) and references therein.

elements, along with O, constitute up to 99% of the soils and sediments.

2.2 Chemical Bonds in Minerals

Key point: Minerals are formed by chemical bonds between cations, such as Si^{4+}, Al^{3+}, Mg^{2+}, and O^{2-}.

Minerals are stable structures resulting from the interatomic forces holding atoms in the structure together. When two atoms (or ions) come close to each other, both attractive and repulsive forces are exerted by the atoms. At larger distances, the interactions are negligible. The origin of an attractive force depends on the particular type of bonding that exists between the two atoms. The attractive forces may be electrostatic (such as in ionic, metallic, van der Waals, and hydrogen bonding) or result from sharing valence electrons to fill their outer shells, as in covalent bonding. The repulsive forces result from the interaction of outer-shell electrons and from the opposition of the positively charged nuclei of the atoms. The distance at which the attractive and repulsive forces are balanced is the characteristic equilibrium bond length or interionic distance for a pair of atoms or ions (Figure 2.1; Sidebar 2.2).

In the bonding of two atoms or ions the outmost electrons (*valence electrons*) become perturbed by the close proximity of the atoms (or ions) and are involved in the chemical bonding that forms a mineral structure. The process of bonding and redistribution of electrons leads to a stable configuration of atoms in a structure and energy configuration is lowered. The energy of atoms in a mineral structure is lower than the energy of free or unordered atoms.

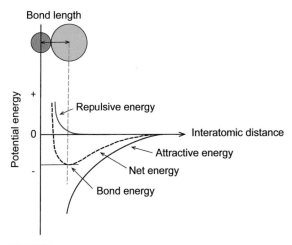

Bond length

Potential energy

Repulsive energy

Interatomic distance

Attractive energy

Net energy

Bond energy

FIGURE 2.1 A typical curve representing the potential energy and interaction between two atoms. The distance corresponding to the minimum potential energy (or bond energy) is the equilibrium bond length or interionic distance.

Sidebar 2.2 — Bond Length

The equilibrium bond length is the distance between the centers of the two bonded atoms.

Ionic, covalent, van der Waals, and hydrogen bonds are important in relation to environmental soil chemistry. The atoms of all elements have a strong tendency to achieve a noble gas configuration, that is, to have eight electrons in their valence shell: *the octet rule*. This electronic configuration is the most stable and has the lowest energy; the atoms of an element achieve this configuration by gaining, losing, or sharing their valence electrons through bonding with other or the same atoms.

2.2.1 Ionic Bond

Key points: Bond resulting from the loss or gain of electrons between two atoms. It is a type of electrostatic bond where oppositely charged ions are attracted to each other.

An ionic bond is formed when atoms interact with one another and, in this process, either lose or gain electrons. The atoms acquire a net positive charge (cations) after they have lost electrons, and those that have gained electrons acquire a net negative charge (anions). The number of electrons gained or lost by an atom commonly conforms with Lewis's valence octets, and thus both atoms (or elements) achieve an inert gas configuration. Halite (sodium chloride) is a classic example of ionic bonding. In a crystal of halite each ion will have six neighboring ions of opposite charge, which makes ionic bonding nondirectional and strong. The energy of the ionic bonds depends on the interionic distance of the charges of the interacting ions.

2.2.2 Covalent Bond

Key point: Bond resulting from sharing of electrons between two atoms.

In covalent bonding two atoms share two valence electrons, one from each of the atoms. For example, Si ($1s^2 2s^2 2p^6 3s^2 3p^2$) has four electrons in the outer sp^3 *hybrid orbital* (Sidebar 2.3), which is half-filled. By forming covalent bonds, an Si atom shares one valence (outer-shell) electron with each of its four nearest neighbor atoms (e.g., oxygen) in a tetrahedron. Covalent bonds are highly directional and the strongest of the chemical bonds.

Sidebar 2.3 — Hybrid Orbital

Hybrid orbitals are formed by the interference or mixing of atomic orbitals of the same atom. For example, in the case of Si mixing between one s and three p orbitals forms a hybrid orbital (sp^3).

2.2.3 van der Waals and Hydrogen Bonds

Key point: Weaker bonds formed by electrostatic attractions between two atoms.

The *van der Waals and hydrogen bonds* result from relatively weak electrostatic forces resulting from an asymmetric charge distribution. These bonds do not involve the transfer of valence electrons between atoms. In neutral and uncharged molecules with no permanent dipole a distortion in the distribution of electric charge produces instantaneous dipoles. This results in one side of a molecule becoming somewhat positive and the opposite side somewhat negative. The van der Waals bond is an extremely short-range ($\sim$0.5 nm) and weak chemical bond (0.4–4 kJ mol^{-1}). Graphite is an example that consists of covalently bonded sheets of carbon atoms linked only by van der Waals bonds; the weak bonds between carbon sheets allow the use of graphite as the "lead" in pencil. Similar bonds exist between unit layers of talc that give a soft feel when the talcum powder is pressed between a finger and thumb.

The hydrogen bond results from an electrostatic attraction between two polar groups of molecules that display dipole behavior. In this bonding a hydrogen ion (H^+) that is covalently bound to a highly electronegative atom, such as oxygen (O), experiences the electrostatic field of another highly electronegative atom nearby.

The energy of a hydrogen bond varies between 4 and 167 kJ mol^{-1}, which makes it somewhat stronger than a van der Waals bond and weaker than a covalent or ionic bond. Kaolinite unit layers are bonded together via hydrogen bonding.

2.2.4 Bond Enthalpy and Bond Characteristics

Key point: Electronegativity values of elements can be used to estimate the proportion of ionic bonding between elements.

The strength of a chemical bond is measured by its dissociation enthalpy, which is the standard reaction enthalpy (Sidebar 2.4). In simpler terms, bond dissociation enthalpy is the energy required to break one mole of the bond to give separated atoms in a gaseous state and at a temperature of 298 K. Bond order is the number of chemical bonds between a pair of atoms; the greater the bond order between a given pair of atoms, the bond strength increases and bond length decreases.

Sidebar 2.4 — Enthalpy

Enthalpy (H) is a thermodynamic property of a system. It is the sum of the internal energy added to the product of the pressure and volume of the system. H (joules) = E + PV. It is not possible to directly measure the H of a system, but the change in enthalpy, ΔH, of a system is measured. The ΔH is a positive change in endothermic reactions and negative in heat-releasing exothermic processes.

Linus Pauling used the difference in electronegativity values of elements to estimate the proportion of ionic bonds between elements (Table 2.2) using the following relationship: fraction of ionic character = $1 - e^{-0.25 \times (\Delta\chi)2}$, where $\Delta\chi$ is the difference in Pauling's electronegativity of constituent elements.

According to this relationship, binary compounds with differences in the electronegativity of constituent elements greater than 1.7 are regarded as predominantly ionic. This concept was further refined by Anton van Arkel and Jan Ketelaar in the 1940s by using an equilateral triangle and placing ionic, metallic, and covalent bonds on the three ends. Gordon Sproul (2001) produced a modern version of the van Arkel–Ketelaar triangle that classifies binary compounds into their predominant bonding type, that is, ionic, covalent, or metallic, using $\Delta\chi$ (the differences in Pauling's electronegativity of elements) and χ_{mean} (mean electronegativity of the constituent elements) in binary compounds (Figure 2.2). Structures of soil minerals generally consist of a mixture of different bonds,

TABLE 2.2 Pauling's electronegativity (χ_P), bond enthalpy, and proportion of ionic character for various elements and oxygen bonds that are commonly present in the soil environment.

Element	χ_P	$-$O bond enthalpy (kJ mol^{-1})	$\Delta\chi$ between O and element	Proportion of ionic bond (%)
H	2.20	463	1.34	36
C	2.55	360	0.89	18
N	3.04	157	0.40	4
O	3.44	146	$-$	
Na	0.93	270	2.51	79
Mg	1.31	358	2.13	68
Al	1.61	502	1.83	57
Si	1.90	466	1.54	45
P	2.19	335	1.25	32
S	2.58	523	0.86	17
K	0.82	271	2.62	82
Ca	1.00	383	2.44	77

predominantly a mixture of ionic and covalent bonds. The stability of minerals in soils is dependent on the structural bonds, and the Si$-$O bond (mostly covalent) is stronger than the ionic bond between metal and oxygen atoms.

2.3 Pauling's Rules

Key point: Ionic radius and charge are important characteristics of an atom in mineral structural coordination.

The chemical bond is the main determinant of mineral structure, with both ionic and covalent bonds keeping cations and anions together in the structure. Most chemical bonds have a combination of ionic and covalent character. For example, the Si$-$O bond is equally ionic and covalent. The Al$-$O bond is approximately 40% covalent and 60% ionic (Sposito, 1984). These chemical bonds are a conceptual idealization of the real bonds in a structure. Most mineral structures in soils can be predicted based on Pauling's rules (Pauling, 1929).

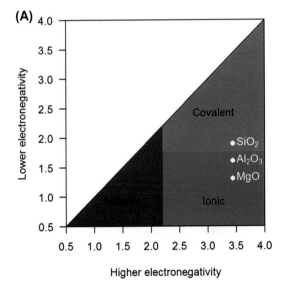

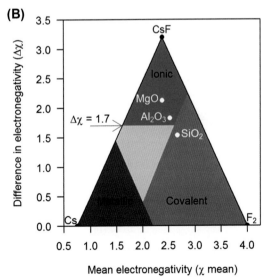

FIGURE 2.2 (A) The van Arkel$-$Ketelaar triangle showing a plot of mean electronegativity (χ_{mean}) against electronegativity difference ($\Delta\chi$) to classify bonds in binary compounds. The three endmembers are CsF, Cs, and F$_2$, which represent ionic, metallic, and covalent bonds, respectively. The $\Delta\chi$ demarcation value of 1.7, suggested by Pauling, misclassifies a significant number of compounds, and the gray region in the figure indicates the uncertainty zone. (B) Sproul (2001) proposed a plot of lower and higher electronegativities for binary compounds that separates the three bonding types better than the plot in (a).

2.3.1 Rule 1: The Coordination Principle

A coordinated polyhedron of anions is formed about each cation. The cation–anion distance is determined by the radius sum (sum of cation and anion radii), and the coordination number (CN) of the cation is determined by the radius ratio (Sidebar 2.5).

Sidebar 2.5 — Coordination Number

The coordination number (CN) is the number of nearest anions surrounding the cation in a mineral. The radius ratio is the ratio of the cation radius to the anion radius in a given geometry or CN.

The ionic radii (IR) of common soil ions, along with their CN and radius ratio of the cations in coordination with O, are given in Table 2.3. In soils cations in mineral structures have CNs of 4, 6, 8, or 12. Cations having IR less than a critical minimum radius ratio can fit between closely packed anions having different configurations (Table 2.4).

For elements in the same group of the periodic table, the IR increases as the atomic number increases; for example, when CN is 8, the IR of Li^+, Na^+, K^+, Rb^+, and Cs^+ are 0.092, 0.118, 0.151, 0.161, and 0.174 nm, respectively. For cations of the same electronic configuration or

TABLE 2.3 Ionic radii (IR), radius ratio (RR), and coordination numbers (CN) of ions of common elements in soil minerals. The numbers in the parentheses in the ionic radius column are the CN for ions in different configurations.

Ion	Ionic radius (nm)[a]	Radius ratio[b]	CN
O^{2-}	0.142 (8), 0.140 (6), 0.138 (4)	—	—
F^-	0.133 (6), 0.131 (4)	—	—
Cl^-	0.181 (6)	—	—
Si^{4+}	0.026 (4)	0.188	4
Al^{3+}	0.0535 (6), 0.039 (4)	0.382 (6), 0.283 (4)	6, 4
Fe^{2+}	0.078 (6)	0.557	6
Fe^{3+}	0.0645 (6)	0.461	6
Mg^{2+}	0.072 (6)	0.514	6
Ca^{2+}	0.112 (8), 0.100 (6)	0.789 (8), 0.714 (6)	8, 6
Na^+	0.118 (8), 0.102 (6)	0.831 (8), 0.729 (6)	8, 6
K^+	0.164 (12), 0.151 (8)	1.155 (12), 1.063 (8)	12, 8
Ti^{4+}	0.0605 (6), 0.042 (4)	0.432 (6), 0.304 (4)	6, 4
Mn^{2+}	0.083 (6)	0.593	6
Mn^{3+}	0.0645 (6)	0.461	6
Mn^{4+}	0.053 (6), 0.039 (4)	0.379 (6), 0.283 (4)	6, 4
Ba^{2+}	0.142 (8)	1.0	8
Rb^+	0.161 (8)	1.134	8

[a]From Shannon (1976);
[b]Ratio of cation radius to O^{2-} radius.

TABLE 2.4 Ideal radius ratio (RR) and coordination number (CN) of different packing geometries and some example minerals with different geometries and their RR.

Packing geometry	CN	Ideal radius ratio (R_C/R_A)	Mineral	Bond (C−A)	R_C (nm)	R_A (nm)	R_C/R_A
Linear	2	<0.155	Does not occur in soil minerals	−	−	−	−
Triangle	3	0.155−0.225	Calcite, $CaCO_3$	C−O	<0.01	0.136	−
			Borax, $Na_2(B_4O_5)(OH)_4.8H_2O$	B−O	0.01	0.136	0.073
Tetrahedron	4	0.225−0.414	Barite, $BaSO_4$	S−O	0.012	0.138	0.087
			Quartz, SiO_2	Si−O	0.026	0.138	0.188
			Orthoclase, $KAlSi_3O_8$	Al−O	0.039	0.138	0.283
			Sphalerite, ZnS	Zn−S	0.060	0.184	0.326
Octahedron	6	0.414−0.732	Rutile, TiO_2	Ti−O	0.0605	0.140	0.432
			Hematite, Fe_2O_3	Fe−O	0.0645	0.140	0.461
			Diopside, $CaMgSi_2O_6$	Mg−O	0.072	0.140	0.514
			Halite, NaCl	Na−Cl	0.102	0.181	0.564
			Galena, PbS	Pb−S	0.119	0.184	0.647
Cube	8	0.732−1.000	Fluorite, CaF	Ca−F	0.112	0.133	0.842
Cuboctahedron	12	>1.000	Muscovite, $KAl_2(Si_3Al)O_{10}(OH)_2$	K−O	0.164	0.142	1.155

(After Wenk and Bulakh, 2004.)

structure, the IR decreases with increasing valence; for example, in 8 coordination the IR of K^+ and Ca^{2+} are 0.151 and 0.112 nm, respectively. For elements (e.g., Mn) that exist in multiple valence states, the IR decreases with increasing positive valence, with IR of 0.083, 0.0645, and 0.053 nm for Mn^{2+}, Mn^{3+}, and Mn^{4+} nm, respectively.

Table 2.4 shows the relationship of RR, CN, and the geometrical arrangements of nearest anions around a central cation. The radius ratio rule can only be used as a guideline, and it is not strictly followed in predicting the CN and structure types. For example, the Si^{4+} cation occurs in a fourfold or tetrahedral coordination with O^{2-} in quartz and many other minerals; however, the radius ratio (0.188) is smaller than the ideal range (0.225−0.414) for a tetrahedron

geometry. In situations where a cation may "rattle" inside the polyhedron of anions the structure is considered to be unstable.

The radius ratio approach ignores the fact that the Si−O bond is highly covalent and Si^{4+} atoms cannot be considered spherically symmetrical. In covalent bonding orbitals from the cation and anion overlap and oxygens are drawn closer to the cation than predicted by IR. The Mg^{2+} ion has a radius ratio of 0.514 (Table 2.3), which is within the limits of an octahedral coordination (0.414−0.732). The Al^{3+} ion, with a radius ratio in the fourfold coordination of 0.283, fits well in the tetrahedron geometry, and the ratio (0.382) in the sixfold coordination is close to the lower limit of the octahedron geometry. In fact, Al^{3+} occurs in both four- and sixfold coordinations, depending on the temperature during the

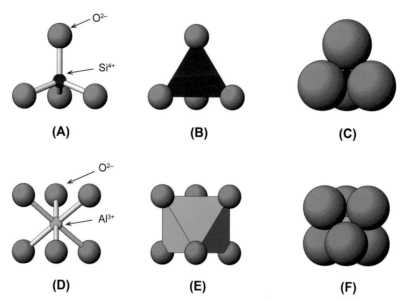

FIGURE 2.3 Structural representations of a tetrahedron (made up of Si and O) and an octahedron (Al and O) using ball-and-stick (A, D), polyhedral (B, C), and sphere-packing (C, F) models.

crystallization of minerals. High temperatures cause small CNs, that is, fourfold coordination, while at low temperatures, a sixfold coordination is favored.

Based on the information given in Tables 2.1 and 2.2, Fe^{2+} (0.078 nm), Fe^{3+} (0.0645 nm), Mn^{2+} (0.083 nm), Mn^{3+} (0.0645 nm), and Ti^{4+} (0.0605 nm) would be expected to occur in an octahedral coordination. As noted above, Al^{3+} and Mg^{2+} also occur in octahedral coordination. The larger cations with a small charge, such as Ca^{2+}, Na^+, and K^+, are found at high CN sites.

Soil minerals mostly consist of tetrahedral and octahedral coordinated structures, as depicted in Figure 2.3.

This can be expressed as $s = Z/CN$, where s is the electrostatic bond strength to each coordinated anion, Z is the valence of the cation, and CN is the CN (Pauling, 1929). The resulting number from this calculation is called electrostatic valency (e.v.). Thus for Si^{4+} in a tetrahedral coordination with O^{2-} atoms, the electrostatic bond strength is equal to 1, and for Al^{3+} in octahedral coordination with O^{2-} atoms, the electrostatic bond strength is equal to 0.5 (Figure 2.4). If Al^{3+} substitutes for Si^{4+} in the tetrahedral sheet, the electrostatic bond strength would be Z(3+) divided by CN(4) or 0.75, not 1. On the other hand, if Mg^{2+} substitutes for Al^{3+} in the octahedral layer, then the electrostatic bond strength is 2/6 or 0.33, not 0.5.

2.3.2 Rule 2: The Electrostatic Valency Principle

In a stable coordination structure the total strength of the valency bonds that reaches an anion or a cation from all neighboring cations or anions is equal to the charge of the anion or cation.

2.3.3 Rule 3: Sharing of Polyhedral Elements-I

The existence of edges, and particularly of faces, common to the anion polyhedra in a coordinated structure, decreases its stability; this effect is large for cations with a high valency and

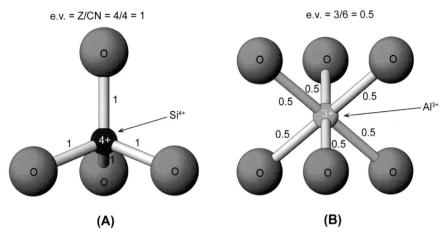

FIGURE 2.4   An illustration of the charge neutralization of (A) Si^{4+} in a tetrahedral coordination and (B) Al^{3+} in an octahedral coordination with O^{2-} atoms.

small CN and is especially large when the radius ratio approaches the lower limit of stability of the polyhedron.

Rule 3 is a statement of Coulomb's law for cations and indicates that there are three ways for tetrahedral (Figure 2.5) and octahedral polyhedra to bond: (i) corner sharing, the most stable configuration; (ii) edge sharing; and (iii) face sharing, which is the least stable configuration. Tetrahedra, with high valence actions, such as Si^{4+}, are bonded by sharing corners, and octahedra, with lower valence actions, such as Al^{3+}, are bonded through edge sharing. Polyhedra are not bonded via face sharing. The bonding by corner sharing in Si tetrahedra places the Si^{4+} cations at the greatest distance from each other, as shown in Figure 2.5. In this example for tetrahedral coordination if the distance between the cations in the polyhedra that share corners is taken as 1.0, an arbitrary unit, then sharing edges reduces the distance to 0.58, and the sharing of faces reduces the distance to 0.38.

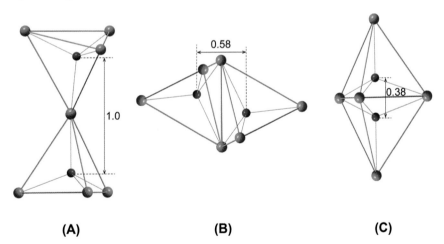

FIGURE 2.5   An illustration of (A) corner- (B) edge-, and (C) face-sharing between tetrahedra. The corner-sharing form occurs commonly in minerals. The cation–cation distance is reduced substantially (1 to 0.38 arbitrary units) in the sequence from corner, edge, and face sharing.

2.3.4 Rule 4: Sharing of Polyhedral Elements-II

In a crystal containing different cations, those of high valency and small CN tend not to share polyhedron elements with each other. This rule is saying that highly charged cations stay as far from each other as possible to lessen their contribution to the crystal's Coulomb energy. Supporting this rule, there are no silicate minerals that have edge-sharing or face-sharing Si tetrahedra. However, edge-shared octahedra are common (TiO_2), and even face-shared octahedra are found (Fe_2O_3).

2.3.5 Rule 5: The Principle of Parsimony

The number of essentially different kinds of constituents in a crystal tends to be small. This is because all substances tend to form the lowest possible potential energy. Many kinds of constituents would result in a complex structure characterized by strains, which would cause a high potential energy and instability. Different kinds of constituents in this principle refer to crystallographic configurations, such as tetrahedra and octahedra. In a mineral with a complex chemical composition many ions may occupy the same structural positions; for example, amphiboles have up to 14 different ions (Si^{4+}, Al^{3+}, Mg^{2+}, Fe^{2+}, Mn^{2+}, Fe^{3+}, Ti^{4+}, Ca^{2+}, Na^+, K^+, O^{2-}, OH^-, Cl^-, and F^-) but only five different crystallographic configurations (i.e., tetrahedral, octahedral, eightfold coordination, a large irregularly coordinated site, and OH locations). In soil minerals two to three configurations are most common.

2.4 Elements of Crystal Structures

Key points: A unit cell is the smallest repeating unit in a mineral structure, which is defined by repeat distances along the three axes and the angles between different axes. The atoms in a mineral structure are represented by an array of points in crystallography and the representation of lattice in the three dimensions is called space lattice. All mineral structures are defined in 14 unique ways, in which lattice points can be arranged in three dimensions.

2.4.1 Crystal Periodicity, the Unit Cell, and Crystal Systems

Mineral structures consist of a long-range three-dimensional order; that is, atoms are arranged in an order that is repeated over and over along the three axes, namely, X, Y, and Z. We observe repeat patterns in two dimensions in many familiar objects, such as patterned curtains, wallpapers, tile floors, tile walls, and brick walls, and each of these objects consists of a basic repeating unit or a building block. In Figure 2.6A a group of atoms are represented in a two-dimensional pattern; the periodicity of the pattern in this figure can be defined by three parameters: the repeat distance (b) along the Y-axis, the repeat distance (c) along the Z-axis, and the angle (α) between the Y- and Z-axes.

In crystallography the array of atoms or groups is represented by an array of points in space, with each point having an identical surrounding, which is called a lattice (Figure 2.6B). The representation of lattice in two dimensions is called a *plane lattice*, and when the planar lattices are stacked along a third direction (X), the periodic array in the three dimensions results, which is called a *space lattice* (Figure 2.7). Crystals (including minerals) are described using the space lattice concept; essentially, it is a mathematic abstraction that describes the translational symmetry of a periodic structure.

Linking the lattice points produces a series of parallel-sided unit cells. A unit cell is the smallest unit of a structure, which is repeated to generate the entire structure. There may be several ways to define a unit cell of a given structure; three examples of the various possible ways are

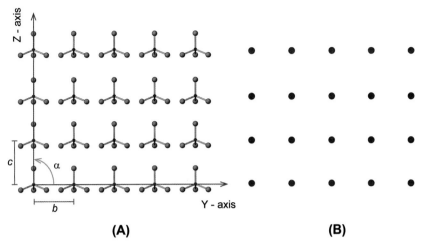

FIGURE 2.6   (A) A two-dimensional representation of atoms and (B) a plane lattice of points. The arrangement of atoms (and lattice) in two dimensions can be defined as the repeat distance *b* along the Y-axis, the repeat distance *c* along the Z-axis, and the angle α between the two axes.

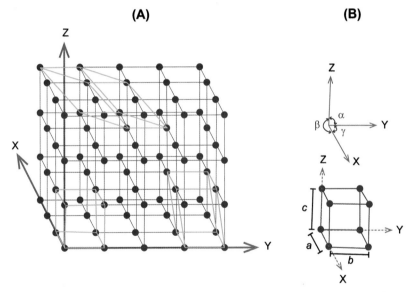

FIGURE 2.7   (A) A space lattice depicting three different unit cells; (B) the unit cell parameters, i.e., the repeat distance along the three axes and the angles between the axes, for one of the unit cells are presented.

highlighted in Figure 2.7A. In restricting the choice of the number of unit cells it is considered that unit cell edges, where possible, coincide with symmetry axes; and the edges should be related to each other by the symmetry of the lattice. Based on these guiding principles, the smallest possible unit cell is chosen to define a structure. Occasionally, the same mineral structure has been defined by using two different unit cells. The unit cell is defined by parameters

a, b, and c, which are repeat distances along crystallographic axes X, Y, and Z, respectively, and the angles between different axes, that is, α between Y and Z, β between X and Z, and γ between X and Y (Figure 2.7B).

Sidebar 2.6 — Hexagonal System

A hexagonal system has four crystallographic axes, three equal horizontal axes ($a_1 = a_2 = a_3$) that lie on a plane intersecting each other at 120°, and the fourth vertical axis (c) is of a different length that is perpendicular to the other three axes.

All structures can be defined by using 14 Bravais lattices, which are 14 unique ways in which lattice points can be arranged periodically in three dimensions. The 14 Bravais lattices can be classified among the seven crystal systems; there is a primitive (P) lattice for each of the seven crystal systems and a few lattices have additional points on the center of all faces (F, face-centered) or centered within the body (B, body-centered) or centered along a particular axis (C) (Table 2.5).

Sidebar 2.7 — Parallelepiped

Parallelepiped is a solid with six faces, each a parallelogram and each being parallel to the opposite face. A parallelogram is a quadrilateral with two pairs of parallel sides; it relates to a parallelogram just as a cube relates to a square.

2.4.2 Miller Indices

William Hallowes Miller, a British mineralogist and crystallographer, in 1839, devised a scheme to define the orientation of a plane or set of parallel planes of atoms in a mineral structure, which is now known as Miller indices.

By convention, the number of intercepts of planes per unit length along the X, Y, and Z axes are termed h, k, and l, respectively. To determine Miller indices for planes: (i) the plane or face intercepts with the crystallographic axes are identified first; (ii) intercepts in fractional coordinates are determined; (iii) reciprocals of the fractional intercepts of each unit length for each axis are taken; and (iv) finally the fractions are cleared by making 1 as the common denominator.

TABLE 2.5   Definitions of seven crystal systems based on the relative length (unit cell dimensions) of the crystallographic axes and the angles between axes.

Crystal system	Unit dimension relationship	Angles between crystal axes	Bravais lattice[b]
Isometric (cubic)	$a = b = c$	$\alpha = \beta = \gamma = 90°$	P, F, B
Tetragonal	$a = b \neq c$	$\alpha = \beta = \gamma = 90°$	P, B
Orthorhombic	$a \neq b \neq c$	$\alpha = \beta = \gamma = 90°$	P, F, B, C
Trigonal or rhombohedral	$a = b = c$	$\alpha, \beta, \gamma \neq 90°$	P
Hexagonal[a]	$a = b \neq c$	$\alpha = \beta = 90°, \gamma = 120°$	P
Monoclinic	$a \neq b \neq c$	$\alpha = \gamma = 90°, \beta > 120°$	P, C
Triclinic	$a \neq b \neq c$	$\alpha \neq \beta \neq \gamma = 90°$	P

[a] Includes the trigonal or rhombohedral system, which is considered a seventh system by some authors (Sidebar 2.6).
[b] P is a primitive space lattice that is a parallelepiped (Sidebar 2.7) with lattice points only at corners. Additional points may occur on the center of all faces (F, face-centered) or centered within the body (B, body-centered) or centered along a particular axis (C).

For example, if a plane cuts the X-axis at one unit length and runs parallel to the two axes, the intercepts are $x = 1, y = \infty, z = \infty$; taking reciprocals: $1/1 = 1, 1/\infty = 0, 1/\infty = 0$; and then cleared fractions: 1, 0, and 0; so the Miller indices $= (100)$. Some other examples are also demonstrated in Figure 2.8. In hexagonal and trigonal systems, with four axes, the convention is $(hkil)$ where $h = a_1, k = a_2, i = a_3$, and $l = c$.

2.5 Primary Minerals in Soils

Silicate minerals constitute over 90% of the Earth's crust, and a similar mineral composition can be expected in soils considering the similarity in the chemical composition of soils with the Earth's crust. A tetrahedron (Figure 2.3) is the fundamental structural unit of all silicates, where a strong bond between Si and O keeps together Si^{4+} and O^{2-} ions. The sharing of oxygens in the tetrahedron with another tetrahedron results in six different structural configurations (Figure 2.9), which are briefly described below.

2.5.1 Types of Silicate Minerals

Key points: Silicates constitute over 90% of the Earth's crust. A tetrahedron is the fundamental structural unit of silicate minerals and different mineral structures are created by sharing oxygens among tetrahedra.

(a) Nesosilicates

Silicates contain independent or isolated tetrahedra (SiO_4^{4-}), which are held together by ionic bonds with interstitial cations such as Fe^{2+} and Mg^{2+} (Figure 2.9A). Olivines (e.g., forsterite, Mg_2SiO_4; fayalite, Fe_2SiO_4) and garnets (e.g., zircon, $ZrSiO_4$) are a common group of nesosilicates.

(b) Sorosilicates

In this group of silicates two SiO_4 tetrahedra ($Si_2O_7^{6-}$) are linked together (Figure 2.9B). The disilicate units are held together by metal cations in an octahedral coordination. The epidote group of minerals (e.g., epidote $Ca_2(Al,Fe^{3+}(Al_2O)(SiO_4)$ $(Si_2O_7)(OH))$ exists in this group.

(c) Cyclosilicates

Silicates are formed by a ring of linked SiO_4 tetrahedra ($Si_6O_{18}^{12-}$); mostly rings are formed by 6 tetrahedra but minerals containing 3 and 4 linked tetrahedra also exist in this group (Figure 2.9C). The beryl ($Be_3Al_2Si_6O_{18}$) and tourmaline groups belong to this group of minerals.

(d) Inosilicates

Silicates formed by the linking of tetrahedra to form infinite single or double chains are known as inosilicates (Figure 2.9D). Inosilicates include two important groups of primary minerals: pyroxenes and amphiboles. Pyroxenes are formed

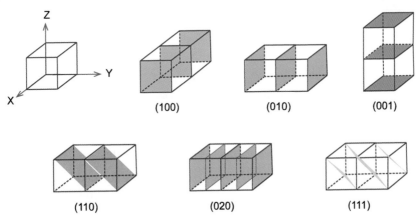

FIGURE 2.8 An illustration of different planes in the cubic system along with their Miller indices.

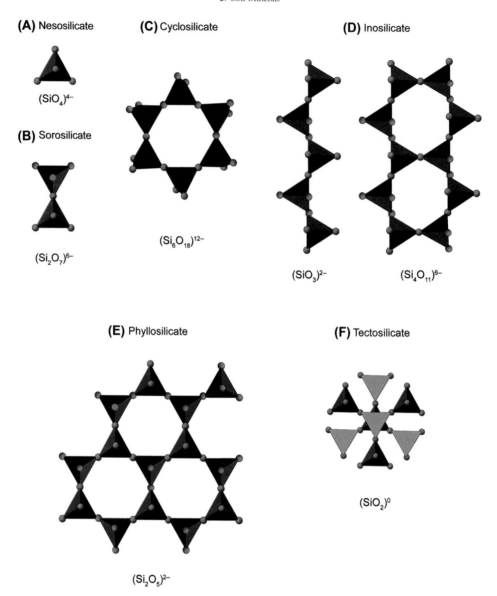

(A) Nesosilicate

$(SiO_4)^{4-}$

(B) Sorosilicate

$(Si_2O_7)^{6-}$

(C) Cyclosilicate

$(Si_6O_{18})^{12-}$

(D) Inosilicate

$(SiO_3)^{2-}$ $(Si_4O_{11})^{6-}$

(E) Phyllosilicate

$(Si_2O_5)^{2-}$

(F) Tectosilicate

$(SiO_2)^0$

FIGURE 2.9   Structural representation of six classes of silicate minerals: (A) nesosilicate, (B) sorosilicate, (C) cyclosilicate, (D) inosilicate, (E) phyllosilicate, and (F) tectosilicate. The structural formula and the net charge for the unsubstituted repeat units are also given for each group of the silicates.

by a single chain of tetrahedra (SiO_3^{2-}) formed by sharing two O atoms, whereas the structure of amphiboles consists of a double chain of tetrahedra ($Si_4O_{11}^{6-}$) linked by the alternate sharing of two and three O atoms.

(e) Phyllosilicates

Silicates are formed by linking all three basal oxygens of each tetrahedron to form an infinite tetrahedral sheet ($Si_2O_5^{2-}$), as shown in Figure 2.9E. Muscovite ($KAl_2(AlSi_3)O_{10}(OH)_2$)

and biotite ($K(Mg, Fe^{2+})_3(AlSi_3)O_{10}(OH)_2$) micas are important examples of this group of silicates.

(f) Tectosilicates

Tectosilicates are formed by linking all four oxygen ions of each tetrahedron with neighboring tetrahedra, thus forming a three-dimensional framework of tetrahedra (SiO_2) (Figure 2.9F). This type of bridging leads to a strongly bonded structure, and the minerals of this group are very stable. Approximately two-thirds of the Earth's crust is made up of tectosilicate minerals, which include quartz (SiO_2), feldspars, zeolites, and feldspathoids.

2.5.2 Weathering of Primary Minerals

The primary mineral composition of soils is highly variable, depending on the parent material and the extent and duration of weathering. The average mineralogical composition of Earth's crust is approximately 39% plagioclase feldspars, 12% K-feldspars, 12% quartz, 11% amphiboles, 5% pyroxenes, 5% micas, 3% olivines, 4.6% clay minerals, and 8.4% nonsilicate minerals. Among the nonsilicates, calcite (including dolomite) and magnetite comprise about 24% and 18%, respectively. A general description of primary minerals that frequently occur in soils is provided in Table 2.6. The most common and abundant primary minerals and their weathering reactions are described in the following sections.

Primary minerals are unstable or weather in terrestrial environments as they form under conditions vastly different from soil environmental conditions. Mineral weathering is a complex process of disintegration and decomposition of minerals and rocks through various physical, chemical, and biological processes.

The weathering of primary minerals results in the formation of secondary minerals and releases many essential plant nutrients (e.g., K, Ca, Mg, Na, Si, Fe, and Mn) and nonessential and potentially toxic elements (e.g., Ni, Cr, Pb, Co, As) in the soil.

While physical weathering processes cause the disintegration of rocks via mechanical processes, such as freezing and thawing, plant roots, living organisms, and chemical weathering produce the decomposition of minerals that involves chemical and mineralogical changes in the structure of primary minerals. The rate of chemical weathering is determined by many factors, including the structure and composition of the mineral, specific surface area of the mineral, availability of water, temperature, and hydrology. Water is the primary driver of the chemical weathering reactions of minerals. Water alone and in combination with oxygen and acids drives chemical weathering reactions that include dissolution, hydration, hydrolysis, carbonation, and oxidation. Some minerals, such as halite and gypsum, dissolve readily when they come in contact with rain and percolating waters, while other minerals (e.g., quartz, muscovite) dissolve at much slower rates, and it may take hundreds of thousands or millions of years.

(a) Hydration

The absorption of water into a mineral structure is called hydration; for example, gypsum formation from anhydrite ($CaSO_4 + 2H_2O \leftrightarrow CaSO_4.2H_2O$) and goethite formation from hematite ($Fe_2O_3 + H_2O \leftrightarrow 2FeOOH$) occur via the process of hydration. These reactions are reversible, and thus dehydration will result in the formation of the original mineral phases.

(b) Carbonation

Carbon dioxide (CO_2) dissolved in rainwater produces carbonic acid ($CO_2 + H_2O \leftrightarrow H_2CO_3 \leftrightarrow H^+ + HCO_3^-$), which is a weak acid that can cause chemical weathering; the reaction is known as carbonation. The dissolution of calcite or limestone in acidic water ($CaCO_3 + H_2CO_3 \rightarrow Ca^{2+} + 2HCO_3^-$) on land surfaces and percolating waters that create caves and sinkholes are classic examples of the carbonation reaction.

The concentration of CO_2 in the soil environment is typically two to three orders of magnitude

TABLE 2.6 Common primary minerals in soils presented in order of their decreasing stabilities in the soil environment.

Mineral	Ideal chemical formula	Importance and possible weathering products
Zircon	$ZrSiO_4$	Most stable mineral in soils and common in the heavy mineral fraction of soils. Often used as a weathering index mineral in pedological studies.
Rutile	TiO_2	Highly resistant to weathering, common residual mineral in highly weathered soils.
Tourmaline	$(Na,Ca)(Mg,Fe^{2+},Mn^{2+},Li,Al)_3(Al,Mg,Fe^{3+})_6$ $(Si_6O_{18})(BO_3)_3(O,OH)_3(O,F)$	Highly resistant to weathering, main primary source of boron.
Ilmenite	$FeTiO_3$	Resistant to weathering, common residual mineral in highly weathered soils, weathers to anatase or pseudorutile.
Garnet	$(Ca,Mg,Fe^{2+},Mn^{2+})_3(Al,Fe^{3+},Cr^{3+})_2(SiO_4)_3$	Gibbsite, goethite.
Quartz	SiO_2	Almost ubiquitous in the coarser fraction of soils, highly resistant to weathering.
Epidote	$Ca_2(Al,Fe)(Al_2O(SiO_4)(Si_2O_7)(OH)$	Common heavy mineral in soils developed on metamorphic rocks, highly resistant to weathering.
Sphene	$CaTiO(SiO_4)$	Widespread accessory mineral in soils formed on igneous rocks, may weather to anatase.
Magnetite	$(Fe^{2+}Fe_2^{3+}O_4)$	Common accessory mineral in soils and is resistant to weathering, may weather to hematite and maghemite.
Muscovite	$KAl_2(AlSi_3)O_{10}(OH)_2$	Widespread in the silt and size fractions of a variety of soils, weathers to illite, vermiculite, other 2:1 phyllosilicates, and kaolinite.
Microcline	$KAlSi_3O_8$	Very widespread in the coarser fraction of a variety of soils, quite resistant to weathering. Commonly weathers to kaolinite, halloysite, and smectite.
Orthoclase	$KAlSi_3O_8$	
Albite	$NaAlSi_3O_8$	
Anorthite	$CaAlSi_2O_8$	
Hornblende	$Ca_2Na(Mg,Fe)_4(Si,Al)_8O_{22}(OH,F)_2$	Common in the heavy mineral fraction of silt and sand-size fractions of moderately weathered soils, weathers to vermiculite and other 2:1 phyllosilicates, and goethite.
Chlorite	$(Mg,Fe^{2+})_3(Si,Al)_4O_{10}(OH)_2(Mg,Fe^{2+})_3(OH)_6$	Present in relatively younger soils formed on basic or ultrabasic parent materials, weathers to vermiculite and other 2:1 phyllosilicates.
Augite	$(Ca,Na)(Mg,Fe,Al)(Si,Al)_2O_6$	Common accessory mineral in the heavy mineral fraction of younger soils, weathers to smectite and vermiculite.
Biotite	$K(Mg,Fe^{2+})_3(AlSi_3)O_{10}(OH)_2$	Present in significant amounts in relatively younger soils on acidic and intermediate parent materials, weathers to vermiculite, other 2:1 phyllosilicates, kaolinite, and goethite.
Serpentine	$Mg_3Si_2O_5(OH)_4$	Present only in recent soils developed on serpentinite rock, commonly weathers to smectites.
Olivines	$(Mg,Fe)_2SiO_4$	Least stable of the major rock-forming silicates, found in young soils on mafic rocks, weathers to serpentine, Fe-rich smectite, and Fe oxides.

greater than that in the atmosphere due to microbial activity and respiring plant roots, so water in the near-surface environments of soils and sediments can become significantly more acidic. The acidity from dissolved CO_2 also drives the hydrolysis process of primary minerals, which is perhaps the most important chemical weathering reaction of the primary silicate minerals.

(c) Oxidation

Iron in primary minerals is largely in the Fe^{2+} (reduced) form and oxidation reactions occur when Fe-bearing minerals are exposed to air and water. The oxidation of the Fe^{2+}-iron in ferromagnesian silicates often initiates the chemical weathering of these minerals, and examples of the oxidation reactions are discussed in the following sections.

(d) Hydrolysis

In hydrolysis a water molecule or acidic water breaks chemical bonds in the structure of a primary mineral. Often such reactions cause incongruent dissolution of minerals. Feldspar weathering often occurs through the hydrolysis process and the chemical reactions involved in the process are given in the following sections.

2.5.3 Common Primary Minerals in Soils and Their Weathering Products

(a) Feldspars

Feldspars are the most abundant rock-forming minerals. These minerals have a three-dimensional tetrahedral framework, that is, tectosilicates, with 25% to 50% Al^{3+} substitution for Si^{4+} in their structures, and consequently, large monovalent (K^+ and Na^+) and divalent (Ca^{2+}) cations are present in the structure to balance the negative charge. The chemical composition of feldspars is highly varied and expressed using the three endmembers, namely, orthoclase

($KAlSi_3O_8$), albite ($NaAlSi_3O_8$), and anorthite ($CaAl_2Si_2O_8$). The minerals in the series between $KAlSi_3O_8$ and $NaAlSi_3O_8$ are referred to as alkali feldspars, whereas the minerals between $NaAlSi_3O_8$ and $CaAl_2Si_2O_8$ are called plagioclase feldspars. Feldspars are very common primary minerals in soils, and they weather more rapidly in comparison to quartz (also a tectosilicate) because of the structural substitutions and/or imperfections. Feldspar weathering rates and products are variable depending on the environmental conditions and the removal of weathering products from the system.

The overall chemical reaction of feldspar weathering in an open environment, that is, in the presence of CO_2 and other acidifying agents, is generally written as follows; for example, for albite:

$$2NaAlSi_3O_8 \ (albite) \ + \ 9H_2O$$
$$+ \ 2H^+ \rightarrow 2Na^+$$
$$+ \ Al_2Si_2O_5(OH)_4 \ (kaolinite)$$
$$+ \ 4H_4SiO_{4(aq)} \ or \ 4Si(OH)_4$$
$$(both \ called \ "silicic \ acid") \qquad (2.1)$$

The above reaction occurs in two steps: (i) the initial incongruent dissolution (Sidebar 2.8) of feldspar into solution and (ii) the subsequent precipitation of kaolinite and other secondary phases from solution.

Sidebar 2.8 − Congruent vs Incongruent Dissolution of Minerals

In congruent dissolution elements are released into solution in proportion to their mole fraction in the mineral structure (i.e., stoichiometrical), whereas in incongruent dissolution some elements are released faster, i.e., out of proportion to their mole fractions in the mineral structure (i.e., nonstoichiometrical). For example, in the dissolution of K-feldspar K is released at a much higher rate than Si and Al.

Similarly, anorthite weathers to form kaolinite:

$$CaAl_2Si_2O_8 \ (anorthite) + H_2O$$
$$+ 2H^+ \rightarrow Al_2Si_2O_5(OH)_4 \ (kaolinite) + Ca^{2+}$$
$$(2.2)$$

These reactions are typically referred to as silicate hydrolysis reactions. The weathering of feldspars releases Na, K, and Ca ions and soluble Si in the case of the alkali feldspars but not via the weathering of anorthite. Kaolinite is the most common weathering product of feldspars; however, other secondary minerals, such as halloysite, smectite, mica, and gibbsite, have also been observed.

(b) Olivines

Olivines have independent Si tetrahedra, that is, nesosilicates, and divalent cations, most frequently Mg^{2+} and Fe^{2+}, are surrounded by six oxygens of tetrahedra to balance the excess negative charge of tetrahedra. Forsterite-fayalite series $((Mg,Fe)SiO_4)$ olivines are olive green in color, with their chemical composition varying between the two endmembers of the solid solution series (Sidebar 2.9): Forsterite-fayalite series $((Mg,Fe)SiO_4)$ olivines are olive green in color and their chemical composition vary between the two endmembers of the solid solution series (Sidebar 2.9), i.e., forsterite (Mg_2SiO_4) and fayalite $(Fe_2^{2+}SiO_4)$.

Sidebar 2.9 — Solid Solution

Solid solutions occurring complex crystalline solids, i.e., minerals, when different ions occupy the same structural position. For example, in the forsterite-fayalite series of the olivine group, members have an identical crystal structure and a range of Mg and Fe contents varying from 100% Mg in forsterite (Mg_2SiO_4) to 100% Fe in fayalite (Fe_2SiO_4).

Olivines weather rapidly in soil and occur only in recently developed soils formed on mafic and ultramafic rocks. The weathering of olivine involves hydrolysis (and carbonation) and oxidation reactions, as shown below:

$$Mg_2SiO_4 \ (forstertite) + 4CO_2$$
$$+ 4H_2O \rightarrow 2Mg^{2+} + 4HCO_3^- + H_4SiO_4(aq)$$
$$(2.3)$$

$$Fe_2SiO_4 \ (fayalite) + 4H_2CO_3 \rightarrow 2Fe^{2+}$$
$$+ 4HCO_3^- + H_4SiO_4(aq)$$
$$(2.4)$$

$$2Fe^{2+} + 1/2O_2 + 2H_2O$$
$$+ 4HCO_3^- \rightarrow Fe_2O_3 \ (hematite) + 4H_2CO_3$$
$$(2.5)$$

Under similar conditions, fayalite weathers faster (up to six times) than forsterite, which is consistent with bond-length/bond-strength principles with Fe^{2+} having a larger ionic radius than Mg, and thus it forms longer and weaker bonds with oxygen (Wogelius and Walther, 1991). Additionally, the structural Fe^{2+} is prone to an oxidation reaction.

Olivine weathers to various secondary minerals depending on the local environmental conditions; the usual weathering products include serpentine, saponite, nontronite, brucite, hematite, magnetite, and magnesite. The weathering of olivine to serpentine can be generalized by using the following weathering reaction:

$$2Mg_2SiO_4 \ (forstertite)$$
$$+ 3H_2O \rightarrow Mg_3Si_2O_5(OH)_4 \ (serpentine)$$
$$+ Mg(OH)_2 \ (brucite)$$
$$(2.6)$$

(c) Quartz

The purest form of a SiO_2 framework exists in quartz and other polymorphs (Sidebar 2.10), that is, tridymite, cristobalite, of SiO_2. Quartz is ubiquitous in soils, and it is one of the last minerals to crystallize from magma, formed under conditions closest to the present terrestrial conditions; therefore it is highly resistant to weathering. Being highly resistant to weathering, quartz is present in significant amounts in most soils including highly weathered soils. The other polymorphs of quartz are rare in soils. Under specific conditions, quartz dissolves (hydration) by the following reaction:

Sidebar 2.10 − Polymorphs

Minerals with the same chemical formula or chemical composition but different structures are called polymorphs, e.g., calcite ($CaCO_3$) and aragonite ($CaCO_3$); in calcite Ca exists in a sixfold coordination with O, whereas in aragonite Ca is present in a ninefold coordination.

$$SiO_2 \ (quartz) + 2H_2O \rightarrow H_4SiO_4(aq) \quad (2.7)$$

(d) Pyroxenes and Amphiboles

Pyroxenes and amphiboles have single- and double-chains of SiO_4 tetrahedra, respectively, and both of these minerals are classified as inosilicates (Figure 2.8). Despite their easy weatherability, these minerals are common accessory minerals in many soil types since they exist in a variety of parent rocks.

Augite [$Ca(Mg,Fe,Al)(Si,Al)_2O_6$], a clinopyroxene, that is, monoclinic system pyroxene, and hypersthene , an orthopyroxene, that is, orthorhombic system pyroxene, are the two most common pyroxenes. Amphibole [$Ca_2Na(Mg,Fe)_4(Si,Al)_8O_{22}(OH,F)_2$] is the most common hornblende mineral in soils. The weathering of pyroxenes and amphiboles forms many secondary clay minerals, including chlorite, vermiculite, smectites, talc, calcite, and Fe oxides. Based on different minerals and weathering products, various idealized weathering reactions have been proposed, for example, the following reaction, largely a solid-state transformation,

where an orthopyroxene weathers to talc via amphibole, with acid input and Mg^{2+} ions released in the process:

$$Mg_8Si_8O_{24} \ (orthopyroxene)$$
$$+ \ 2H^+ \rightarrow Mg_7Si_8O_{22}(OH)_2 \ (amphibole)$$
$$+ \ Mg^{2+}$$
$$(2.8)$$

$$Mg_7Si_8O_{22}(OH)_2 \ (amphibole)$$
$$+ \ 2H^+ \rightarrow Mg_6Si_8O_{20}(OH)_4 \ (talc) + Mg^{2+}$$
$$(2.9)$$

Pyroxene minerals contain some structural Fe, and therefore the weathering pathways and reactions are more complex than described above and involve the oxidation of Fe^{2+}. Eggleton and Boland (1982) summarized the overall reaction (not balanced) of an orthopyroxene as:

$$(Mg_{3.18}Fe^{2+}_{0.40}Fe^{3+}_{0.25}Cr_{0.02}Ca_{0.07})$$
$$(Si_{3.89}Al_{0.11})O_{12} \ (orthopyroxene) \ + \ 0.2O_2$$
$$+ \ 0.86 \ H_2O \rightarrow 0.86 \ (Ca_{0.05}(Fe^{3+}_{0.65}Cr_{0.02}Mg_2)$$
$$\times (Si_{3.89}Al_{0.11})O_{10}(OH)_2 \ (talc)$$
$$+ \ 0.26 \ MgAl_2O_4 \ (spinel) \ + \ 0.68MgO$$
$$+ \ 0.07CaO \ + \ 0.45SiO_2$$
$$(2.10)$$

Augite weathers to kaolinite, as described below:

$$(Mg_{0.7}CaAl_{0.3})(Al_{0.3}Si_{1.7})O_6 \ (augite) \ + \ 3.4H^+$$
$$+ \ 1.1 \ H_2O \rightarrow 0.3Al_2Si_2O_5(OH)_4 \ (kaolinite)$$
$$+ \ Ca^{2+} \ + \ 0.7Mg^{2+} + 1.1 \ H_4SiO_4(aq)$$
$$(2.11)$$

Hornblende weathering occurs via dissolution−reprecipitation reactions, where hornblende dissolves stoichiometrically, and goethite, gibbsite, and kaolinite have been identified as among the weathering products. The following weathering reaction has been proposed for hornblende weathering:

$$Na_{0.5}Ca_2(Fe^{2+}_{1.3}Mg_{2.6}Al_{1.1})$$

$$(Al_{1.6}Si_{6.4})O_{22}(OH)_2 \ (hornblende) + 9.7H^+$$

$$+ 11.65H_2O + 0.325O_2 \rightarrow 0.5Na^+$$

$$+ 2Ca^{2+} + 2.6Mg^{2+} + 1.3FeOOH(s)$$

$$+ xAl_2Si_2O_5(OH)_4 \ (kaolinite) + (2.7$$

$$- 2x)Al(OH)_3(s) + (6.4 - 2x)H_4SiO_4(aq)$$

(e) Micas   (2.12)

Micas belong to the phyllosilicate group, where all three basal oxygens are shared with neighboring tetrahedra, forming a sheet structure. The apical oxygens are shared with an octahedral sheet in the phyllosilicates. In micas two tetrahedral sheets are sandwiched with an octahedral sheet in between, with an apical oxygen of the one tetrahedral sheet pointing upward and a second tetrahedral sheet with the apical oxygen pointing downward. Muscovite ($KAl_2[AlSi_3]O_{10}(OH)_2$) and biotite ($K(Mg,Fe^{2+})_3$ $[AlSi_3]O_{10}(OH)_2$) are the two most important minerals of this group. Muscovite is a dioctahedral mica, whereas biotite is a trioctahedral mica (Sidebar 2.11). In both these micas one out of every four Si^{4+} ions in the tetrahedral sheet is substituted by Al^{3+} ions, which generates an excess negative charge in the structure.

Sidebar 2.11 — Dioctahedral and Trioctahedral

In the octahedral sheet of phyllosilicates, when two out of the total three positions are occupied, the minerals are classified as dioctahedral minerals (e.g., muscovite); the cations in dioctahedral minerals are mostly trivalent, e.g., Al^{3+}, Fe^{3+}. However, when divalent cations, such as Mg^{2+} or Fe^{2+} occupy the octahedral positions, all three positions in the octahedral sheet are occupied, and the minerals are classified as trioctahedral minerals, e.g., biotite.

The layer charge in mica ranges from ≈ 0.85 to 1.0 per formula unit and $\geq 50\%$ of the interlayer cations that balance the layer charge must be monovalent cations (Table 2.8). With the exception of paragonite, an Na-bearing mica, the other micas have K^+ in the interlayer positions, and thus micas are major K-bearing minerals in soils. As they weather, the nonexchangeable K is released into the soil solution for plant uptake.

In contrast to other primary mineral weathering, where minerals dissolve and released ions crystallize to form new minerals and some ions may be removed from the soil profile by leaching, the weathering of micas occurs mostly via solid-state transformations (Sidebar 2.12). It is well established that biotite and other trioctahedral micas weather much more rapidly than the dioctahedral mica muscovite. The transformation of micas involves two main steps: (i) the exchange of interlayer K^+ by hydrated cations, such as Ca^{2+} and Mg^{2+}, and (ii) a decrease in the structural layer charge. The removal of interlayer K is a diffusion-controlled process and depends on several factors, such as mineral structure and particle size, solution K concentration, and the presence of other ions.

Sidebar 2.12 — Neoformation and Solid-State Transformation

Neoformation is the crystallization of a new mineral structure from ions in solutions, and there is no inheritance of a preexisting mineral structure; e.g., kaolinite is formed from the weathering of a feldspar. Solid-state transformation is the formation of a new mineral in which part or all of a preexisting structure is inherited, e.g., biotite weathering to vermiculite.

The main processes for the decrease of layer charge and other structural changes in the weathering of micas are oxidation of structural

Fe^{2+} with the subsequent ejection of Fe from the octahedral sheet, loss of Al from the tetrahedral sheet and increase in Al in the octahedral sheet, and the incorporation of protons in the mineral structure.

The weathering of micas leads to the formation of many secondary phyllosilicates, such as vermiculite, smectite, kaolinite, and halloysite, which have the same structural coordination as Si tetrahedra. Goethite and gibbsite have also been formed from the weathering of micas.

Some general weathering reactions of micas are given below:

(i) Weathering of biotite to vermiculite

$K(Mg_2Fe^{2+})[Si_3Al]O_{10}(OH)_2$ (*biotite*)

$+ \ 1.3Mg^{2+} + 0.1H_4SiO_4 + H_2O + 0.3O_{2(g)} \rightarrow$

$Mg_{0.4}(Mg_{2.9}Fe^{3+}_{0.1})[Si_{3.1}Al_{0.9}]O_{10}(OH)_2$ (*vermiculite*)

$+ \ Fe^{3+}_{0.9}Al_{0.1}OOH$ (*goethite*)

$+ \ K^+(aq) + 1.4H^+(aq)$

$$(2.13)$$

(ii) Weathering of muscovite to vermiculite

$2.7KAl_2(Si_3Al)O_{10}(OH)_2$ (*muscovite*)

$+ \ 0.6H^+ + 1.05Ca^{2+} + 1.8H_4SiO_4 \rightarrow$

$3Ca_{0.35}(Al_2)[Si_{3.3}Al_{0.7}]O_{10}(OH)_2$ (*vermiculite*)

$+ \ 2.7K^+ + 3.6H_2O$

$$(2.14)$$

(iii) Weathering of muscovite to kaolinite

$2KAl_2(Si_3Al)O_{10}(OH)_2$ (*muscovite*) $+ \ 2H^+$

$+ \ 3H_2O \rightarrow 1.5Al_4Si_4O_{10}(OH)_8$ (*kaolinite*)

$+ \ 2K^+$

$$(2.15)$$

2.5.4 Weathering Sequence of Primary Minerals

A general weathering sequence of the most common primary minerals is presented in Figure 2.10. The mineral stability series proposed by Goldich (1938) coincides with Bowen's solid solution series that was formulated to explain the order of crystallization of silicate minerals from a cooling magma (Bowen, 1922). Based on the solid solution series, the minerals formed at higher temperatures, such as olivines, pyroxenes, and Ca plagioclase, weather more rapidly when exposed to low-temperature terrestrial conditions that are farthest from their crystallization temperatures. Conversely, the minerals formed at relatively lower temperatures from cooling magma, such as quartz, are stable or weather very slowly under terrestrial conditions that are closer to their crystallization temperatures.

In addition to the minerals listed in Figure 2.10, Goldich (1938) classified zircon to be the most stable mineral and ilmenite and magnetite as moderately stable minerals. Ions released from mineral dissolution are released in the water, which are then washed away or precipitate to form a new mineral or minerals.

Values for the lifetime of 1-mm crystals of common silicates in Table 2.7 correspond well with the mineral stability series in Figure 2.10.

Under aerobic, near-surface conditions, secondary phyllosilicates as end products of weathering are about equally as stable as the structurally related muscovite.

Polynov (1937) used the average composition of igneous rocks (lithosphere) and river waters (hydrosphere) for the geochemical classification of landscapes and mobility of chemical elements. Hudson (1995) has modified Polynov's mobility of ions and categorized the main elements into five mobility phases. Phase I: Cl, SO_4; Phase II: Na; Phase III: Ca, Mg; Phase IV: Si; and Phase V: Fe and Al. The mobility sequence of the element is proposed as Cl > SO_4 > Na > Ca > Mg > K > Si > Fe > Al.

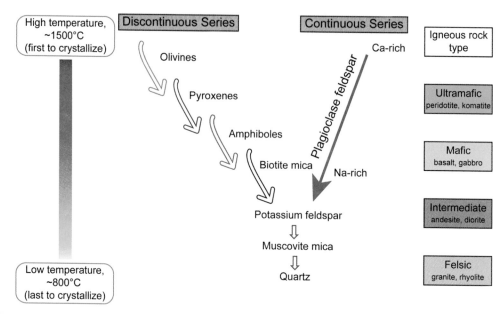

FIGURE 2.10   The stability of primary minerals on the Earth's surface based on Bowen's reaction series (Bowen, 1922) and Goldich's mineral stability series (Goldich, 1938).

TABLE 2.7   Average time to dissolve 1-mm-diameter primary mineral crystals in dilute solution at pH 5.0 and 25°C.

Mineral	Mineral class/group	Log rate (mol m^{-2} s^{-1})	Lifetime (years)
Quartz	Tectosilicate	−13.39	34×10^6
Muscovite	Phyllosilicate	−13.07	2.6×10^6
Epidote	Sorosilicate	−12.61	923,000
Microcline	Tectosilicate, K-feldspar	−12.50	921,000
Albite	Tectosilicate, Na-feldspar	−12.26	575,000
Enstatite	Inosilicate, pyroxene	−10.00	10,100
Diopside	Inosilicate, pyroxene	−10.15	6800
Forsterite	Nesosilicate, olivine	−9.50	2300
Anorthite	Tectosilicate, Ca-feldspar	−8.55	112

(From Lasaga, 2014)

The mobility of ions from weathering rocks to river systems is a simple and logical concept to understand soil genesis from the combination of different parent rocks and weathering intensities. Ion mobility is thus a key indicator of mineral weathering. In highly weathered and free leaching soil environments elements in Phase V will be expected to be present, for example, Fe and Al oxides, and contrastingly in arid regions, with minimal or no leaching, elements from Phase I to III are increasingly present in soils.

2.6 Secondary Minerals in Soils

Secondary minerals are weathering by-products of primary minerals, and these are formed either via the crystallization of ions in solution or through the transformation of primary minerals. These minerals predominantly exist in the clay fraction (<2 μm) of soils, as demonstrated by a generalized relationship between the abundance of primary and secondary minerals with particle size in Figure 2.11. The proportion of both primary and secondary minerals in different size fractions of soils also depends on the extent of weathering. In highly weathered soils quartz, being highly resistant to weathering, is the most abundant primary mineral and kaolinite and Fe/Al oxides are the most abundant secondary minerals.

From a soil environmental chemistry perspective, the most important secondary minerals are (i) phyllosilicate minerals and (ii) Fe, Al, and Mn oxides (used for brevity, including oxides, oxyhydroxides, and hydroxides). Without question, the secondary minerals play a profound role in affecting numerous soil chemical reactions and processes, as we shall see in this chapter and in the following chapters.

2.6.1 Phyllosilicates

Phyllosilicates (Sidebar 2.13) are assemblages of tetrahedral and octahedral sheets (Figure 2.12). The Si tetrahedral sheets (Figure 2.12A) are arranged so that all tips (apical oxygen atoms) are pointing in the same direction and the bases of all tetrahedra are in the same plane, and tetrahedra are bonded together by sharing all three basal oxygens with the other three tetrahedra.

Sidebar 2.13 — Phyllosilicates

The word "phyllo" originates from the Greek word "phyllon," meaning leaf. Such minerals contain a continuous sheet-like arrangement of SiO_4 tetrahedra, with all three basal oxygens shared among tetrahedra and the apical oxygen pointing in the same direction. These minerals are also called sheet silicates or layer silicates.

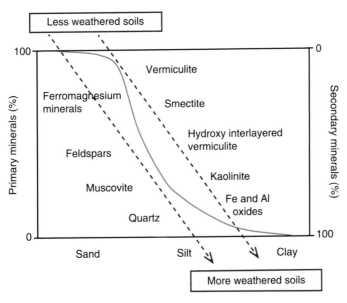

FIGURE 2.11 A generalized relationship between the abundance of primary and secondary minerals with the soil particle size fractions (based on McBride, 1994). With the increasing intensity of weathering, the amount of primary and secondary minerals increases in the direction of the arrow.

(A) The tetrahedral sheet: top and edge views

(B) The dioctahedral sheet: top and edge views

(C) The trioctahedral sheet: top and edge views

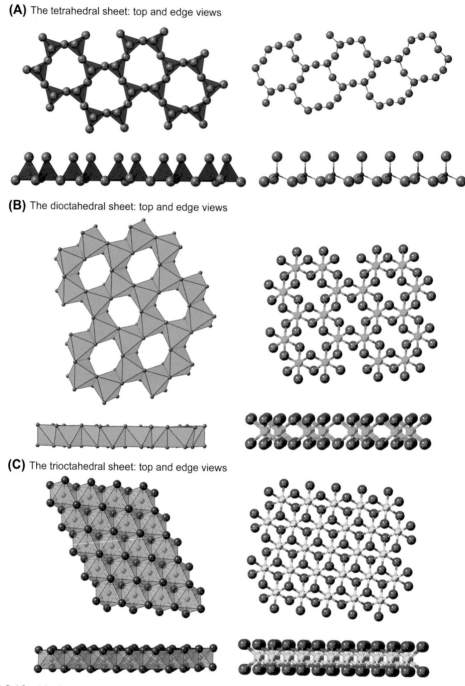

FIGURE 2.12 Idealized representations of the top and side views of tetrahedral (A) and octahedral sheets that together constitute phyllosilicates minerals. The octahedral sheet can be either a dioctahedral sheet (B) when a trivalent cation (e.g., Al^{3+}) occupies the octahedral sites or a trioctahedral sheet (C) when a divalent cation (e.g., Mg^{2+}) occupies the octahedral sites.

The octahedral sheet consists of cations in a sixfold coordination, and the sheet is constituted by edge-sharing octahedra. There are two ways to fill the octahedral sites: (i) when only two-thirds of the sites are occupied by cations (e.g., Al^{3+}), with each octahedral cation sharing two oxygens with only three neighboring octahedral cations (*dioctahedral sheet*, Figure 2.12B), and (ii) if all three octahedral sites are occupied by cations (e.g., Mg^{2+}), with each octahedral cation sharing two oxygens with each of the six neighboring octahedral cations (*trioctahedral sheet*, Figure 2.12C).

There are six possible chemical compositions of the octahedral sheet in different minerals: (i) octahedral sheets with $Al_2^{3+}O_2(OH)_4$ or $Mg_3^{2+}/Fe_3^{2+}O_2(OH)_4$ composition, for example, in the dioctahedral or trioctahedral sheet of 1:1 layer minerals (e.g., kaolinite and antigorite, respectively); (ii) octahedral sheets with $Al_2^{3+}O_4(OH)_2$ or $Mg_3^{2+}/Fe_3^{2+}O_4(OH)_2$ composition, for example, in the dioctahedral or trioctahedral sheet of 2:1 layer minerals (muscovite and biotite, respectively); and (iii) composition with $Al_2^{3+}(OH)_6$ or $Mg_3^{2+}/Fe_3^{2+}(OH)_6$, for example, in the interlayer of dioctahedral or trioctahedral chlorites, respectively.

Different phyllosilicate minerals are formed from the linkage through the apical (free) oxygens of the tetrahedral sheet with the octahedral sheet. The two common structures are a 1:1 layer formed by linking a tetrahedral sheet with an octahedral sheet and a 2:1 layer that links two tetrahedral sheets with an octahedral sheet. In order to accomplish the 2:1 layer linkage of sheets, the upper tetrahedral sheet is inverted so that apical oxygens point down to form a shared plane with the octahedral sheet.

In ideal structures of 1:1 and 2:1 layer minerals the charge in the structural assemblage is fully balanced and the interlayer is unoccupied and weakly bonded together, for example, 1:1 layer minerals – kaolinite (dioctahedral) and chrysotile (trioctahedral) – and 2:1 layer minerals – pyrophyllite (dioctahedral) and talc (trioctahedral). However, frequently many structures have isomorphous substitution(s), which is the substitution of a cation by another of similar size with a different or the same charge in the crystal lattice without disrupting the crystal structure of the mineral. The substitution of a cation in the tetrahedral and octahedral sheets is determined by the cation size (Table 2.3). In the tetrahedral sheet Al^{3+} usually substitutes for Si^{4+}, and in the dioctahedral sheet Fe^{2+}, Fe^{3+}, and Mg^{2+} frequently substitute for Al^{3+}. The isomorphous substitution can result in no change or excess negative or positive charge in individual octahedral and tetrahedral sheets. However, overall the substitution always leads to no net charge or the development of excess negative charge in the structural assemblage of phyllosilicate minerals. The structural negative charge is balanced by the presence of unhydrated or hydrated cations (e.g., micas, vermiculites, and smectites) or hydroxides sheets (e.g., chlorites) in the interlayer space of the mineral.

For the linking of tetrahedral and octahedral sheets, the shared plane must have the same lateral dimension (b) in a mineral structure. Free octahedral sheets in dioctahedral (e.g., in gibbsite $b = 8.64$ Å or 0.864 nm) and trioctahedral (in brucite $b = 9.36$ Å or 0.936 nm) coordinations have a mismatch in the lateral parameter size to the ideal tetrahedral sheet with hexagonal symmetry ($b = 9.15\pm0.06$ Å or 0.915 ± 0.006 nm), as shown in Figure 2.13A. Despite the uncertainties in these values, the tetrahedral sheets generally have a larger b-dimension than the octahedral sheet, and the hexagonal symmetry is distorted by the alternate clockwise and anticlockwise rotation of the tetrahedra about axes perpendicular to the basal plane (Figure 2.13B).

Phyllosilicate minerals have been classified based on the layer type, layer charge, and the interlayer material characteristics (Table 2.8). There are two main layer types, namely, 1:1

(A)

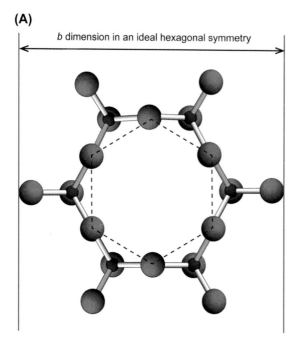

b dimension in an ideal hexagonal symmetry

(B)

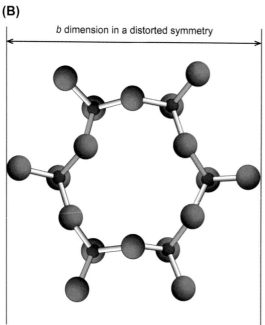

b dimension in a distorted symmetry

FIGURE 2.13   Schematic representations of (A) an ideal tetrahedral unit with hexagonal symmetry of oxygen atoms and (B) a distorted symmetry that occurs when the slightly larger tetrahedral sheet combines with an octahedral sheet along the lateral (*b*) dimension.

and 2:1 (Figure 2.14), and some mineral structures consist of regular or irregular interstratification of these layers. *Layer charge* is the amount of net negative charge per formula unit, which is expressed as a positive number. The negative charge in the layer is balanced by positively charged interlayer material, which can be hydrated or unhydrated cations or a positively charged metal hydroxide sheet that is not bound in the layer structure of a mineral.

(a) The 1:1 Phyllosilicates

Serpentine-kaolin is the only group in this class of minerals, which comprises both dioctahedral and trioctahedral species (Table 2.8). The most common dioctahedral kaolin is the mineral kaolinite. The structure of kaolinite consists of an Si tetrahedral sheet bonded to an Al octahedral sheet (Figure 2.15), with some structural strain in the tetrahedral sheet to combine with the octahedral sheet as described earlier (Figure 2.13). The 1:1 layers of kaolinite are held together via hydrogen bonding between oxygens of the tetrahedral sheet and hydroxyls of the octahedral sheets, and thus there is no interlayer space present in kaolinite.

The ideal formula unit or chemical formula (Sidebar 2.14) of kaolinite is $Al_2Si_2O_5(OH)_4$, and there is little, if any, isomorphic substitution in kaolinite. The unit structure of kaolinite is about 0.72 nm (7.2 Å) thick along the Z-axis (perpendicular direction) and thus identified by using X-ray diffraction (XRD) by the presence of a (*001*) peak at the d-spacing of 0.72 nm. Kaolinite is the most ubiquitous clay mineral in soils, and it is a common weathering product of feldspars and other primary minerals. In intense weathering environments, such as in humid tropics and humid temperate zones, kaolinite formation is most pronounced, and it is the most abundant phyllosilicate in highly weathered soils. The substitution of Fe^{3+} for Al in the octahedral sheet of kaolinite is common in highly weathered soils, with reported Fe_2O_3 values of about 2.5% (Singh and Gilkes, 1992b).

TABLE 2.8 Classification and generalized structural formulae of phyllosilicate minerals.

Layer type	Interlayer material and layer charge[a]	Group	Octahedral character[b]	Species and formula of soil relevant minerals[c]
1:1	None or H_2O only	Serpentine-kaolin	Dioctahedral	Kaolinite, $Al_2Si_2O_5(OH)_4$ Halloysite, $Al_2Si_2O_5(OH)_4 \cdot 2H_2O$
			Trioctahedral	Chrysotile, $Mg_3Si_2O_5(OH)_4$ Lizardite, $[Mg_{2.8}Al_{0.2}](Si_{1.8}Al_{0.2})O_5(OH)_4$ Antigorite, $[Mg_{2.72}Al_{0.09}Fe^{3+}_{0.04}Fe^{2+}_{0.15}](Si_2)O_5(OH)_4$
2:1	None ($x \approx 0$)	Talc-pyrophyllite	Dioctahedral	Pyrophyllite, $Al_2Si_4O_{10}(OH)_2$
			Trioctahedral	Talc, $Mg_3Si_4O_{10}(OH)_2$
	Hydrated exchangeable cations ($x \approx 0.2-0.6$)	Smectite	Dioctahedral	Montmorillonite, $M^+_{0.5}[Al_{1.5}Mg_{0.5}](Si_4)O_{10}(OH)_2$ Beidellite, $M_{0.5}^+[Al_2](Si_{3.5}Al_{0.5})O_{10}(OH)_2$ Nontronite, $M^+_{0.5}[Fe^{3+}_2](Si_{3.5}Al_{0.5})O_{10}(OH)_2$
			Trioctahedral	Hectorite, $M^+_{0.4}[Mg_{2.6}Li_{0.4}](Si_4)O_{10}(OH)_2$ Saponite, $M^+_{0.4}[Mg_3](Si_{3.6}Al_{0.4})O_{10}(OH)_2$ Sauconite, $M^+_{0.4}[Zn_3](Si_{3.6}Al_{0.4})O_{10}(OH)_2$
	Hydrated exchangeable cations ($x \approx 0.6-0.9$)	Vermiculite	Dioctahedral vermiculite	$M^+_{0.75}[Al_2](Si_{3.25}Al_{0.75})O_{10}(OH)_2$
			Trioctahedral vermiculite	$M^+_{0.65}[Mg_3](Si_{3.35}Al_{0.65})O_{10}(OH)_2$
	Nonhydrated mono- or divalent cations ($x \approx 0.6-0.85$ for dioctahedral)	Interlayer-cation-deficient mica	Dioctahedral	Illite, $K_{0.65}[Al_2](Si_{3.35}Al_{0.65})O_{10}(OH)_2$ Glauconite, $K_{0.8}[Al_{0.50}Fe^{3+}_{0.99}Fe^{2+}_{0.21}Mg_{0.30}](Si_{3.71}Al_{0.29})O_{10}(OH)_2$ Brammallite, $Na_{0.65}[Al_2](Si_{3.35}Al_{0.65})O_{10}(OH)_2$
			Trioctahedral	Wonesite, $Na_{0.5}[Mg_{2.5}Al_{0.5}](Si_3Al)O_{10}(OH)_2$
	Nonhydrated monovalent cations ($\geq 50\%$ monovalent cations; $x \approx 0.85-1.0$)	Mica	Dioctahedral	Muscovite, $K[Al_2](Si_3Al)O_{10}(OH)_2$ Celadonite, $K[(Al,Fe^{3+})_1(Mg,Fe^{2+})_1]Si_4O_{10}(OH)_2$ Paragonite, $Na[Al_2](Si_3Al)O_{10}(OH)_2$
			Trioctahedral	Phlogopite, $K[Mg_3](Si_3Al)O_{10}(OH)_2$ Biotite, $K[(Fe,Mg)_3](Si_3Al)O_{10}(OH)_2$ Annite, $K[Fe^{2+}_3](Si_3Al)O_{10}(OH)_2$
	Nonhydrated divalent cations ($\geq 50\%$ divalent cations; $x \approx 1.8-2.0$)	Brittle mica	Dioctahedral	Margarite, $Ca[Al_2](Si_2Al_2)O_{10}(OH)_2$

(Continued)

TABLE 2.8   Classification and generalized structural formulae of phyllosilicate minerals.—cont'd

Layer type	Interlayer material and layer charge[a]	Group	Octahedral character[b]	Species and formula of soil relevant minerals[c]
			Trioctahedral	Clintonite, $Ca[Mg_2Al](SiAl_3)O_{10}(OH)_2$
	Hydroxide sheet ($x \approx$ variable)	Chlorite	Dioctahedral (dioctahedral in both the 2:1 layer and the interlayer hydroxide sheet)	Donbassite, $(Al_{2.27})(OH)_6[Al_2](Si_{3.2}Al_{0.8})O_{10}(OH)_2$
			Trioctahedral (trioctahedral in both the 2:1 layer and the interlayer hydroxide sheet)	Clinochlore, $(Mg_2Al)(OH)_6[Mg_3](Si_3Al)O_{10}(OH)_2$
			Di-, trioctahedral (dioctahedral in the 2:1 layer but trioctahedral in the interlayer hydroxide sheet)	Cookeite, $(LiAl_2)(OH)_6[Al_2](Si_3Al)O_{10}(OH)_2$ Sudoite, $(Mg_2Al)(OH)_6[Al_{1.7}Mg_{0.3}](Si_{3.3}Al_{0.7})O_{10}(OH)_2$
			Tri-, dioctahedral (trioctahedral in the 2:1 layer but dioctahedral in the interlayer hydroxide sheet)	Fraklinfurnacite, $Ca_2Fe^{3+}Mn_3^{2+}Mn^{3+}Si_2Zn_2O_{10}(OH)_8$
2:1	Regularly interstratified ($x \approx$ variable)[d]	Variable	Dioctahedral	Corrensite, aliettite, hydrobiotite, kulkeite
			Trioctahedral	Rectorite, tosudite, brinrobertsite
1:1, 2:1	Regularly interstratified ($x \approx$ variable)	Variable	Trioctahedral	Dozyite

[a] x = net layer charge
[b] 2.5 cations per formula unit is the boundary for dioctahedral (<2.5 cations per formula unit) and trioctahedral (>2.5 cations per formula unit) minerals
[c] M^+ represents one or more exchangeable cations
[d] Interstratifications of alternating layers in 50/50 proportions.

Sidebar 2.14 — Unit Cell Formula and Formula Unit

The chemical formula of the repeating units of atoms, i.e., the unit cell, of a mineral is called the *unit cell formula*. *Formula unit* is the empirical formula of a mineral where elements are expressed as the lowest whole number ratio of ions present in the mineral structure. In 1:1 phyllosilicates the unit cell formula and formula unit consist of $O_{10}(OH)_8$ and $O_5(OH)_4$, respectively, and in 2:1 phyllosilicates these are $O_{20}(OH)_4$ and $O_{10}(OH)_2$, respectively.

Kaolinite has very little or no structural charge and thus has the lowest cation exchange capacity (CEC; $10-100$ mmol$_c$ kg^{-1}) among the phyllosilicates. Kaolinite has a platy morphology with plates varying from hexagonal and pseudohexagonal to subrounded and rounded particles. The specific surface area of soil kaolinite ranges

(A)

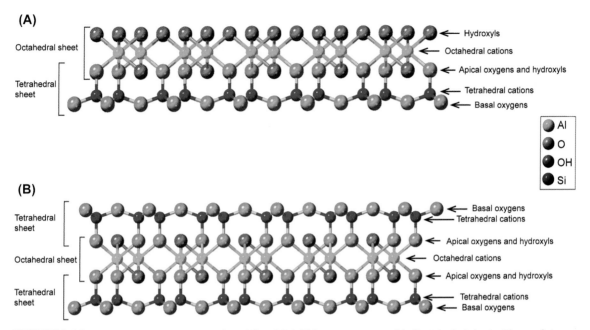

(B)

FIGURE 2.14   Schematic representations of 1:1 (A) and 2:1 (B) layer structures with dioctahedral sheets. Planes of atoms, sheets, and layers are indicated in the structures.

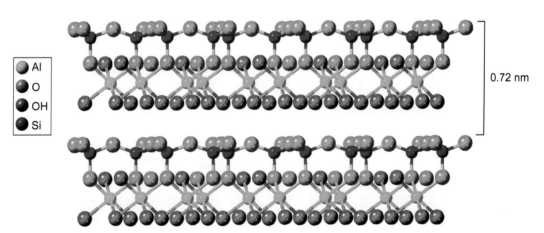

FIGURE 2.15   Kaolinite structural model showing the elemental composition of tetrahedral and octahedral sheets; the repeat unit distance (i.e., d-spacing) along the Z-axis is 0.72 nm.

TABLE 2.9   Cation exchange capacity and specific surface area of selected secondary minerals.

Mineral or mineral group	CEC (mmol$_c$ kg^{-1})	Total SSA (m^2 kg^{-1})
Kaolinite	10–100	5–55 × 10^3
Halloysite	20–600	20–60 × 10^3
Mica (and illite)	100–400	30–120 × 10^3
Vermiculite	1200–2070	300–500 × 10^3
Smectite	800–1500	600–800 × 10^3
Chlorite	100–400	10–50 × 10^3
Palygorskite-sepiolite	50–450	800–900 × 10^3
Allophane-imogolite	100–400	400–1100 × 10^3
Zeolite	200–4600	Variable, >1000 × 10^3

from 5 to 55 × 10^3 m^2 kg^{-1} depending on the mineral particle size (Table 2.9). In highly weathered soils kaolinite particles have subrounded to rounded crystal morphologies and particles are much smaller with larger specific surface area.

Halloysite is another dioctahedral kaolin that exists mostly in recent soils derived from volcanic ash and lower parts of the soil profiles of deeply weathered granite. The structure and chemical composition of halloysite are similar to those of kaolinite, except that halloysite (Figure 2.16) has a layer of water molecules between each 1:1 layer. The presence of a water monolayer in the interlayer increases the thickness of the unit structure to about 1.0 nm (10 Å). Drying will cause the water molecules to be removed and the layers to collapse together with a thickness (of about 0.72 nm) similar to kaolinite. In contrast to kaolinite, halloysite occurs in contrasting morphologies, including tubular, spheroidal, and rolled-up plates. The structural substitutions of Al in the tetrahedral sheet and Fe^{3+} in the octahedral sheet have been suggested reasons for the observed morphologies, which help correct the misfit between the octahedral and tetrahedral sheets.

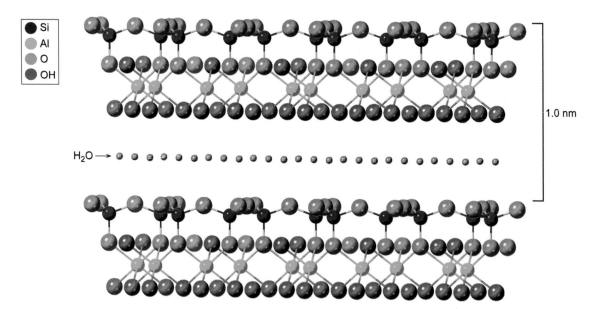

FIGURE 2.16   Halloysite structural model showing the elemental composition of tetrahedral and octahedral sheets and the interlayer water. The repeat unit distance along the Z-axis is 1.0 nm.

Due to some Al substitution in the tetrahedral sheet, the CEC of halloysite is greater than kaolinite, with values up to 600 mmol$_c$ kg^{-1}. The specific surface area of halloysite (50–60 × 10^3 m^2 kg^{-1}) is also greater than that of kaolinite (Table 2.9).

The trioctahedral 1:1 layer minerals are the serpentines (Table 2.8). Serpentine minerals have all three octahedral positions filled with Mg or other divalent cations (e.g., Fe^{2+}, Ni^{2+}, Mn^{2+}). The ideal trioctahedral sheet is larger than the tetrahedral sheet, which is corrected by the substitution of Al for Si in the tetrahedral sheet and Al for Mg in the octahedral sheet while maintaining the neutral layer charge. Other mechanisms to correct the misfits include rolling of the individual sheets that produce fibrous morphology (e.g., chrysotile), a very small particle size (e.g., lizardite), and reversing the tetrahedra direction along a crystallographic direction (e.g., antigorite). Serpentine minerals weather easily and are rarely found in soils except in younger soils developed on serpentine or ultramafic soils. Due to their fibrous morphology, chrysotile has been used in asbestos in the past and it is extremely important to identify the mineral if present in building wastes because of it's cancer causing effects to humans.

(b) The 2:1 Phyllosilicates

The 2:1 layer minerals have considerable diversity in their chemical compositions compared to 1:1 layer minerals. These minerals are discussed under different groups, which are based on the layer charge and interlayer materials (Table 2.8).

(b-1) Pyrophyllite-Talc Group

The two representative minerals of the group are pyrophyllite ($Al_2Si_4O_{10}(OH)_2$), a dioctahedral mineral, and talc ($Mg_3Si_4O_{10}(OH)_2$), a trioctahedral mineral (Table 2.8). The minerals of this group have negligible or no layer charge and adjacent 2:1 layers are held together by weak van der Waals forces. Because of their ideal structure and compositions, pyrophyllite and talc are considered the prototypes of the 2:1 layer minerals. These minerals rarely occur in soils and, if present, only occur in trace to small amounts. Since there is little or no permanent negative charge on pyrophyllite and talc minerals, they are useful models for the edge charge reactivity associated with 2:1 phyllosilicates. The edge sites of pyrophyllite can have a significant influence on the retention of metals and on various physical properties of the clay (Keren and Sparks, 1994).

(b-2) Smectite Group

Minerals in the smectite group are characterized by a layer charge of 0.2–0.6 mol per formula unit Table 2.8. The group includes both dioctahedral (e.g., montmorillonite, beidellite, and nontronite) and trioctahedral (e.g., hectorite and saponite) subgroups. The trioctahedral members of the smectite group (Table 2.8) are not common in soils and will not be discussed further.

The major difference between montmorillonite and the other two common dioctahedral smectites, beidellite and nontronite, is that isomorphous substitution in these minerals occurs in the tetrahedral sheet (i.e., Al^{3+} substitutes for Si^{4+}) whereas in montmorillonite the substitution occurs in the octahedral sheet (i.e., Mg^{2+} substitutes for Al^{3+}). Nontronite is an Fe-bearing mineral, with Fe^{3+} being the predominant cation in the octahedral sheet. The net negative charge in all these minerals is balanced by hydrated cations (e.g., Ca^{2+}, Mg^{2+}, Na$^+$, K$^+$). Since the negative charge in smectites is relatively small (0.2–0.6 mol per formula unit), the interlayer cations are not able to hold the 2:1 layers together tightly. The separation of the 2:1 layer depends on the interlayer cation (and associated water) and the location of the negative charge, that is, the octahedral sheet or tetrahedral sheet (Figure 2.17). The presence of water molecules in the interlayer

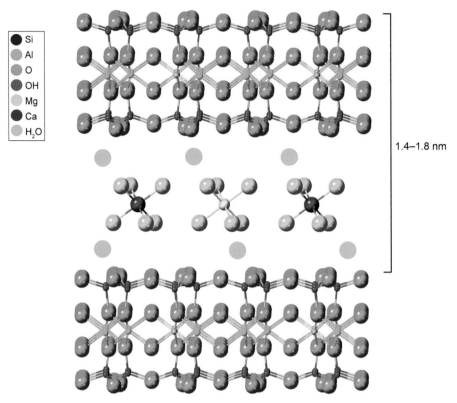

Si
Al
O
OH
Mg
Ca
H₂O

1.4–1.8 nm

FIGURE 2.17   A general structural model for dioctahedral smectites showing the elemental composition of tetrahedral and octahedral sheets, the exchangeable cations, and interlayer water. The repeat unit distance along the Z-axis is variable but generally between 1.4 and 1.8 nm.

space gives the shrink—swell characteristics to smectite minerals.

The basal spacing (001) of smectite minerals varies depending on the interlayer cation or other material and relative humidity. In Mg-saturated samples the basal XRD peak of smectites is present at ~1.5 nm at 54% relative humidity, which shrinks to 1.25 nm in K-saturated samples at 54% relative humidity and further decreases to 1.0 nm in a K-saturated and 110°C-heated sample. The d-spacing of Mg-saturated smectites expands to 1.7–1.8 nm after ethylene glycol or glycerol solvation.

Smectite minerals in soils and sediments can originate from three pathways: inheritance (e.g., in sedimentary rocks), transformation (e.g., from the weathering of micas), and neoformation (e.g., in soils formed in young glacial and alluvial materials). Smectites in soils occur frequently under environmental conditions that include low-lying topography, poor drainage, and ferromagnesian mineral-rich parent rocks. These systems provide favorable chemical conditions, namely, high Si and basic cation activity and high pH, for the formation and preservation of smectites in soils.

Smectites are common minerals in many soils from the temperate regions and in alluvial plain soils, particularly vertisols. Smectite-rich soils are often highly fertile due to their large CEC and water holding capacity. Soil smectites are often different from the three ideal dioctahedral

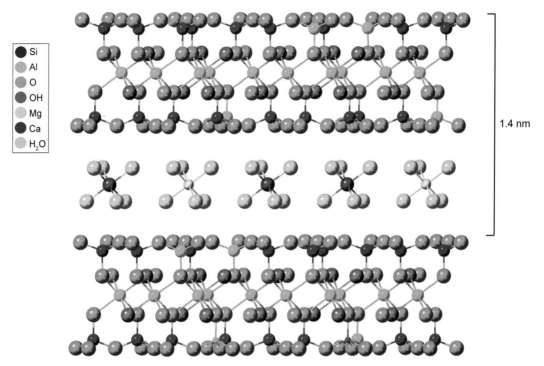

FIGURE 2.18 A general structural model for a dioctahedral vermiculite showing the elemental composition of tetrahedral and octahedral sheets and the exchangeable cations and interlayer water. The repeat unit along the Z-axis is approximately 1.4 nm.

smectites and other geological deposit smectites, with the layer charge originating in both octahedral and tetrahedral sheets in the soil smectites. Furthermore, soil smectites have greater amounts of structural Fe than montmorillonite and beidellite, *sensu stricto*. The CEC value of smectites in soils ranges from 800 to 1500 mmol$_c$ kg^{-1}, with an average value of about 1100 mmol$_c$ kg^{-1}. Smectites have a large specific surface area ($600-800 \times 10^3$ m^2 kg^{-1}) due to their small particle size and large internal surface area (Table 2.9).

(b-3) Vermiculite Group

Vermiculite group minerals have a layer charge of 0.6–0.9 mol per formula unit, which is higher than smectites. Dioctahedral vermiculite (Figure 2.18) is characterized by substitution in both the tetrahedral and octahedral sheets, while trioctahedral vermiculite has substitution only in the tetrahedral sheet, and all three of the octahedral cation positions are filled with Mg (Table 2.8).

Vermiculites in soils mainly originate from the weathering of micas, particularly trioctahedral vermiculites that derive from the weathering of biotite and phlogopite micas. While vermiculites are characterized by a platy morphology, similar to that of micas, they have a layer charge lower than that of micas, Fe^{2+} in the octahedral sheet is oxidized to Fe^{3+}, and the interlayer K$^+$ is replaced by hydrated cations, such as Ca^{2+} and Mg^{2+}.

Trioctahedral vermiculite in soils often occurs in all size fractions, that is, sand, silt, and clay, whereas dioctahedral vermiculite is concentrated in the clay fraction. Dioctahedral vermiculites are frequently found in highly weathered soils. When Al-hydroxy species or Al-hydroxy polymers are present in the interlayer space of the mineral, it is referred to as hydroxy-interlayered vermiculite (HIV) or inhibited vermiculite or pedogenic chlorite. The interlayering with Al-hydroxy species limits the expansion and collapse of the interlayer space and decreases the CEC of HIV versus other vermiculites.

Vermiculites have the highest CEC among phyllosilicate minerals, with values between 1200 and 2070 $mmol_c\ kg^{-1}$, with an average value of about 1600 $mmol_c\ kg^{-1}$. The SSA of vermiculites is slightly lower than that of smectites, with values ranging from 300×10^3 to 500×10^3 $m^2\ kg^{-1}$ (Table 2.9).

XRD identification of vermiculites is confirmed by the presence of basal d-spacing at ~1.45 nm in Mg-saturated samples that collapses to 1.0 nm in K-saturated samples and there is no or very little expansion in the spacing after ethylene glycol or glycerol solvation of Mg-saturated dioctahedral vermiculites. However, trioctahedral vermiculites expand to ~1.6 nm in Mg-saturated and ethylene glycolated samples. Trioctahedral vermiculites can be distinguished from smectites by the glycerol solvation of Mg saturated samples where only smectites expand (~1.8 nm) and vermiculites' basal peak remains at ~1.4 nm. Al-HIV does not collapse to 1.0 nm on K saturation and even after heating to 550°C, with the peak becoming broad with maximum intensity at ~1.2 nm.

(b-4) Interlayer-Cation-Deficient Mica Group

The interlayer-cation-deficient mica or interlayer-deficient mica group contains minerals in which more than 50% of the interlayer cations are monovalent and the layer charge ranges from ≥0.6 to <0.85 mol per formula unit (Table 2.8). Illite and glauconite are important minerals of this group; both are dioctahedral

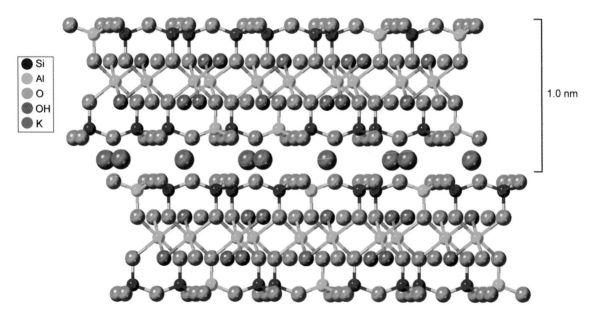

FIGURE 2.19 A general structural model of a dioctahedral mica (muscovite or illite) showing the elemental composition of tetrahedral and octahedral sheets and the exchangeable cations and interlayer water. The repeat unit along the Z-axis is approximately 1.0 nm.

minerals, with illite being Al rich while glauconite is a Fe-rich mineral. These minerals often occur in the clay fraction of soils and sedimentary rocks, and both contain more water and less K (80% of true micas) than the true micas. Illite may possibly be derived from the weathering of muscovite. Several other terms have been used in lieu of illite in the past literature, including hydromica, hydromuscovite, hydrous illite, hydrous mica, K-mica, micaceous clay, and sericite. Illite has more Si^{4+}, Mg^{2+}, and H_2O but less tetrahedral Al^{3+} and K^+ than muscovite. While K^+ is the predominant interlayer cation along with divalent ions such as Ca^{2+} and Mg^{2+}, NH_4^+ can also occur in the interlayers of illite. The structural model of an ideal dioctahedral mica is presented in Figure 2.19; illite has a similar structure except for a few small differences, as described above.

Minerals in the interlayer-cation-deficient mica group, similar to mica group minerals, show the basal d-spacing at ∼1.0 nm that does not change with pretreatments (e.g., interlayer cation, relative humidity, solvation with glycerol or ethylene glycol, and heat treatment); however, the peak is much broader and weaker (lower intensity) than micas (e.g., muscovite and biotite).

The CEC of interlayer-deficient mica (and micas) is limited because the interlayer cations are largely nonexchangeable, and it ranges from 100 to 400 $mmol_c$ kg^{-1} (Table 2.9). The CEC values are generally greater for the interlayer-deficient mica minerals (e.g., illite) than micas. Illite and other interlayer-deficient micas exist in the soil clay fraction compared to mica that may be present in all particle size fractions of soils; hence the specific surface area is larger for the interlayer-deficient micas than micas. Similar to K-bearing micas, illite serves as a long-term source of K for field crops in agricultural soils.

The release of K^+ from layer and edge weathering of illite results in frayed edges and "wedge zones" that play profound roles in the selectivity

and fixation of K^+ and NH_4^+ ions in the interlayers of illite (Sparks, 1987).

(b-5) Chlorite Group

Chlorite group minerals have a layer structure similar to other 2:1 layer minerals, such as micas, smectites, and vermiculites; however, chlorites differ from other 2:1 minerals in that the interlayer region is occupied by a positively charged hydroxide sheet of Mg, Fe, and Al instead of cations (Figure 2.20a). Minerals in this group are further classified into four subgroups depending on the dioctahedral or trioctahedral characteristics of the octahedral sheet in the 2:1 layer structure and the interlayer region (Table 2.8).

Chlorites are both primary and secondary; however, the primary minerals, inherited mostly from metamorphic and igneous rocks, are more common in soils. These minerals are easily weatherable and thus less abundant in soils than the other 2:1 layer minerals. The weathering of chlorites releases significant amounts of Mg and other elements and produces other 2:1 layer minerals, such as smectites and vermiculites.

Pedogenic chlorites, also referred to as Al-hydroxy vermiculites (HIV), exist in highly weathered acidic soils, where Al-hydroxy polymers occupy the interlayer space of 2:1 layer minerals (Figure 2.20B). The structure and chemical composition of pedogenic chlorites vary depending upon the composition of the 2:1 layer and the type and extent of filling of the interlayer space filling by hydroxy polymers.

Chlorites' basal d-spacing ranges from 1.40 to 1.44 nm depending on the mineral species, and the spacing does not change with various pretreatments. Similar to micas, the CEC of chlorites is limited, ranging between 100 and 400 $mmol_c$ kg^{-1}, and the specific surface area varies from 10×10^3 to 50×10^3 m^2 kg^{-1} (Table 2.9).

(b-6) Interstratified Layer Silicates

Since the 2:1 and 1:1 layers of clays are strongly bonded internally but are weakly bonded to each other, layers can stack together

(A)

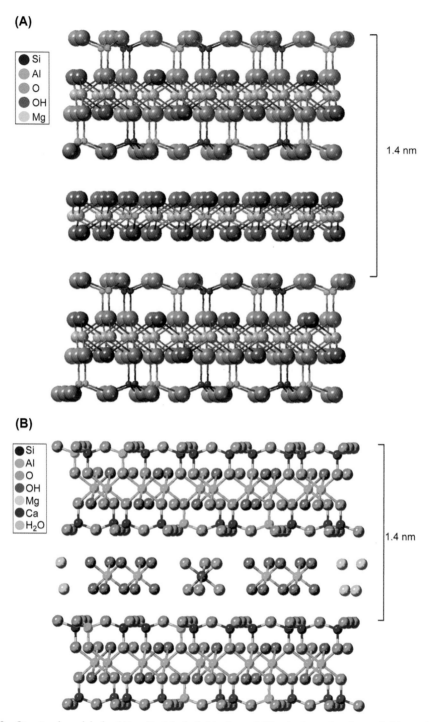

(B)

FIGURE 2.20   Structural models for (A) a dioctahedral chlorite and (B) a hydroxy-interlayered chlorite (or vermiculite) showing the elemental composition of tetrahedral and octahedral sheets. The interlayer part is filled with a continuous Al, Mg hydroxide sheet in chlorite and Al-hydroxy polymers in hydroxy-interlayered vermiculite (or chlorite). The repeat unit along the Z-axis is approximately 1.4 nm in both minerals.

to form interstratified layer silicates. These minerals have also been called mixed-layered or intergrade minerals. The interstratification can be regular, that is, alternating layers in equal proportions, or nonregular, which can have layers of different minerals with some order or randomly stacked or layers being segregated or zoned. Examples of regularly interstratified minerals containing 2:1 layers include (i) trioctahedral − corrensite (chlorite-smectite or chlorite-vermiculite), aliettite (talc-smectite), hydrobiotite (biotite-vermiculite), and kulkeite (talc-chlorite) − and (ii) dioctahedral species − rectorite (mica-smectite), tosudite (chlorite-smectite), and brinrobertsite (pyrophyllite-smectite). Dozyite is a regularly interstratified mineral that has 1:1 and 2:1 layers (serpentine-chlorite).

Regularly interstratified minerals are identified by the basal spacing, that is, the sum of the two mineral components. For example, rectorite, a mica-smectite regularly interstratified mineral, shows the basal peak at 2.4 nm in an Mg-saturated sample that expands to 2.7−2.8 nm after ethylene glycol or glycerol solvation.

The nonregularly interstratified minerals show the basal peak at an intermediate d-spacing between the two mineral components, and the exact position depends on the proportion of the layers of the two minerals that constitute the interstratified mineral; for example, the basal peak for an Mg-saturated sample is at 1.18 nm for an interstratified mineral consisting of 75% biotite and 25% vermiculite layers.

2.6.2 Nonphyllosilicates

(a) Allophane and Imogolite

These minerals have short-range order in their structures and therefore often are referred to as short-range ordered (SRO) minerals or poorly crystalline minerals. Allophanes form from volcanic ash materials and thus are major components in the clay fraction of volcanic-derived soils. They may also occur in the clay fraction of many non−volcanically derived soils. Volcanic soils containing allophane usually contain significant organic matter and have low bulk densities. The SiO_2/Al_2O_3 ratio of allophanes varies from 0.8 to nearly 2, the characteristic ratio for kaolinite. Aluminum is in both tetrahedral and octahedral coordination. Imogolite has an almost constant SiO_2/Al_2O_3 ratio of 1, and Al is present in octahedral coordination only. While it has a little charge resulting from the isomorphous substitution, imogolite can adsorb substantial quantities of monovalent cations, similar to halloysite (Newman and Brown, 1987).

Microscopic analyses of imogolite reveal thread-like particles that are bundles of parallel tubes about 20 nm in diameter. Allophane exhibits spherical particles 30−50 nm in diameter (Brown, 1980). Due to their large specific surface area and variable surface charge, the soils containing these minerals have some unique properties, such as a very high adsorption capacity for P and trace metals and strong interaction with organic matter.

(b) Palygorskite and Sepiolite

Palygorskite (previously referred to as attapulgite) and sepiolite are fibrous minerals that do not have continuous octahedral sheets. The structure of these minerals can be considered ribbons of 2:1 layers, where one ribbon is linked to the next via the inversion of SiO_4 tetrahedra along a set of Si−O−Si bonds (Sidebar 2.15). The width of the ribbon is different for the two minerals, with the number of octahedral-cation position per formula unit being five and eight in the ideal structure of palygorskite $(Si_8Mg_5O_{20}(OH)_2(OH_2)_4.4H_2O)$ and sepiolite $(Si_{12}Mg_8O_{30}(OH)_4(OH_2)_4.8H_2O)$, respectively. Since the octahedral sheets are discontinuous at each inversion of tetrahedra, the coordination and charge balance along the channel side are completed by the presence of additional H_2O, H+, and exchangeable cations, such as Ca^{2+} and Mg^{2+}. Due to their structures and morphologies, palygorskite and sepiolite have unique sorption properties for both organic and inorganic compounds (Mo et al., 2021).

Sidebar 2.15 — Structure of Palygorskite

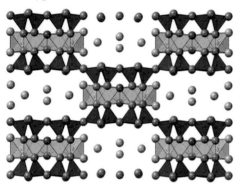

The inversion of SiO_4 tetrahedra can be seen clearly here. The yellow octahedra are Mg, the green octahedra are Al, the blue tetrahedra are Si, the light and dark brown spheres represent oxygens and hydroxyls, respectively, and the blue spheres represent water molecules. In nonideal palygorskite Al occupies 28%–59% of the occupied octahedral sites.

Palygorskite and sepiolite commonly exist in soils of arid and semiarid regions. These minerals are largely inherited from the parent materials. However, neoformation from smectite and illite dissolution has also been observed.

(c) Zeolites

Zeolites have three-dimensional framework structures (tectosilicates) with interconnected channels. Negative charge in the framework or cage is balanced by large alkali and alkaline earth cations. Saline and alkaline soils provide environmental conditions for the neoformation and occurrence of zeolites; these minerals rarely occur in significant amounts in soils. Zeolites may also exist via inheritance from the soil parent materials. Analcime ($Na_{16}(Al_{16}Si_{32}O_{96}).16H_2O$) and clinoptilolite ($Na_3K_3(Al_6Si_{30}O_{72}).24H_2O$) are the two most common zeolites found in soils.

Zeolites have a unique set of chemical and physical properties, including a very high CEC. Their framework and structure with different size channels and cages have many environmental and industrial applications.

2.6.3 Oxides, Hydroxides, and Oxyhydroxides

The general term "oxides" is used for brevity to describe Al, Fe, and Mn oxides; hydroxides; oxyhydroxides; and hydrous oxides (where nonstoichiometric water is in the structure). In most oxides the metallic cations are octahedrally coordinated and the differences in the structures of different oxide minerals result from differences in the stacking arrangements of oxygen and hydroxy planes, the continuity of the structure in the direction of stacking, and the extent and the way the octahedral sites are filled (Taylor, 1987).

Oxides are ubiquitous in soils, and the minerals found in soils are listed in Table 2.10. They may exist as discrete crystals; coatings on minerals (e.g., quartz), ped faces, and organic materials; or in discrete features including ferricretes, mottles, and nodules (Schwertmann and Taylor, 1989).

Aluminum, Fe, and Mn oxides play extremely important roles in the chemistry of many soils. While they may not be found in large quantities in many soils, oxides have significant effects on many soil chemical processes, such as sorption and redox, because of their high specific surface area and chemical reactivity. The point of zero charge (PZC) of Fe and Al oxides ranges between 7 and 9. These minerals particularly contribute to the sorption of trace metal cations (e.g., Cd, Pb, Zn) and oxyanions (e.g., phosphate, molybdate,

TABLE 2.10   Metal oxides in soils.

Oxide group	Mineral	Ideal chemical formula
Fe	Goethite	α-FeOOH
	Hematite	α-Fe$_2$O$_3$
	Ferrihydrite	Fe$_5$HO$_8$.4H$_2$O (approx.)
	Lepidocrocite	γ-FeOOH
	Magnetite	Fe$_3$O$_4$
	Maghemite	γ-Fe$_2$O$_3$
	Schwertmannite	Fe$_8$O$_8$(OH)$_6$SO$_4$
	Akaganéite	β-FeOOH
	Green rust	Fe(OH)$_2$
Al	Gibbsite	Al(OH)$_3$
	Boehmite	γ-AlOOH
	Nordstrandite	Al(OH)$_3$
	Bayerite	α-Al(OH)$_3$
	Diaspore	α-AlOOH
	Corundum	α-Al$_2$O$_3$
Mn	Birnessite	(Na$_{0.7}$Ca$_{0.3}$)(Mn^{3+}, Mn^{4+})$_7$O$_{14}$.2.8H$_2$O
	Vernadite	δ-MnO$_2$
	Lithiophorite	(Al$_2$Li)(Mn$_2^{4+}$Mn^{3+})$_7$O$_6$(OH)$_6$

arsenate, and sulfate) in tropical and subtropical soils.

(a) Iron Oxides

Iron oxides are widespread in most soils and often dominate the clay fraction of highly weathered soils in tropical regions. Iron present in oxides in soils has been released from the weathering of Fe-containing silicate minerals and other primary minerals (e.g., biotite, amphiboles, pyroxenes, olivine), in which Fe is mostly in the divalent (Fe^{2+}) form. Under oxidized conditions, Fe^{3+} ions are hydrolyzed and polymerized to produce different crystalline forms of Fe

oxides in soils. Fe oxides in soils have very low solubility, with the solubility product (K_{sp} = (Fe)(OH)$_3$) of common Fe oxides ranging from 10^{-37} to 10^{-44} (Schwertmann and Taylor, 1989). The acid dissolution of Fe oxides may not occur under normal soil conditions. However, dissolution through reduction reactions occurs easily, and reduction−oxidation reactions often cause the redistribution of Fe oxides in soil horizons or within a soil horizon.

(a-1) Goethite

Goethite (α-FeOOH) is the most common and thermodynamically stable Fe oxide in soils. The structure of goethite consists of double chains of edge-shared octahedra that are joined together by sharing corners (Fe−O−Fe) and weak hydrogen bonds (Figure 2.21). The stacking sequence of O/OH in the goethite structure forms hexagonal close packing (hcp), hence termed the α phase. Goethite gives a yellow to yellow-brown color in soils. The substitution of Al for Fe^{3+} in the goethite structure is very common in soils, and in highly weathered soils the substitution can be as high as 35 mole% Al. Many other cations, such as Ti, V, Cr, Mn, Co, Ni, Cu, Zn, Cd, and Pb, also substitute for Fe^{3+} in the goethite structure; however, the extent of such substitutions is often minor in most cases (Singh and Gilkes, 1992a; Singh et al., 2002; Cornell and Schwertmann, 2003; Kaur et al., 2009a, b, c).

Goethite occurs in wide soil environments ranging from cool, wet climates to subtropical and tropical regions. The morphology of goethite particles varies from needles, lath, rods, and subrounded plates to rounded plates, with the particle size ranging from 5 to 150 nm and with a specific surface area of 10 × 10^3−200 × 10^3 m^2 kg^{-1}.

(a-2) Hematite

Hematite (α-Fe$_2$O$_3$) is the second most common Fe oxide in soils, and it gives a red color

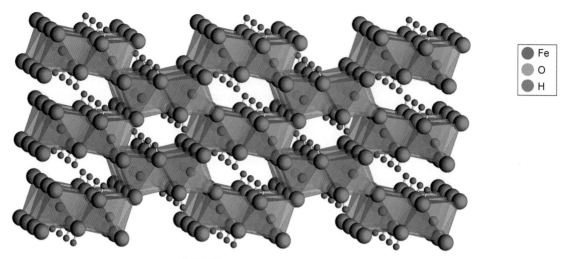

	Fe
	O
	H

FIGURE 2.21　Structural model of goethite.

to soils. The structure of hematite consists of FeO_6 octahedra connected by edge sharing with three neighboring octahedra in the same plane and face sharing with an octahedron in an adjacent plane (Figure 2.22; Bigham et al., 2002). Aluminum substitution also occurs in hematite in soils; however, the maximum substitution is in the range of 15–18 mole% Al, which is about half of that observed in goethite. Similar to goethite, several other cations, including Mn, Ni, Ti, and Cr, may substitute for Fe^{3+} in the hematite structure (Singh et al., 2000).

Hematite is common in highly weathered soils, Mediterranean soils, and soils of warm-dry regions. Hematite commonly occurs in association with goethite in soils. The morphology of hematite particles is varied, ranging from hexagonal plates, cubes, spindles, and rods to spheres, and the crystal size and specific surface area are similar to that of goethite.

(a-3) Ferrihydrite

Ferrihydrite (approx. $Fe_5HO_8.4H_2O$) is a SRO mineral that has an ill-defined composition and exits in a nanocrystalline form. Because of its very small crystal size and the lack of long-range structural order, the structural model of ferrihydrite continues to remain a contentious issue. However, there is mounting evidence for the occurrence of up to 20% Fe in tetrahedral coordination in ferrihydrite and the rest in two different types of octahedral coordination.

Ferrihydrite has a reddish-brown color, and it exists in many soil environments, particularly in young Fe accumulations, such as in ochreous precipitates from Fe-rich waters (e.g., acid mine drainage), bog Fe ores, podzol B horizons, and lichen weathering crusts. It is unstable in soils and transforms (via dissolution or solid-state) to more stable phases, such as goethite or hematite. The extremely high specific surface area (100×10^3–700×10^3 m^2 kg^{-1}) of spherical particles (c. 3–7 nm) and reactivity make ferrihydrite an important mineral for the sequestration of contaminants and other compounds in soils and sediments.

(a-4) Lepidocrocite

Lepidocrocite (γ-FeOOH) is a polymorph of goethite, and its structure consists of double chains of octahedra ($Fe(O,OH)_6$) along the c-axis, where octahedra share edges by forming H bonds with adjacent double-chains of

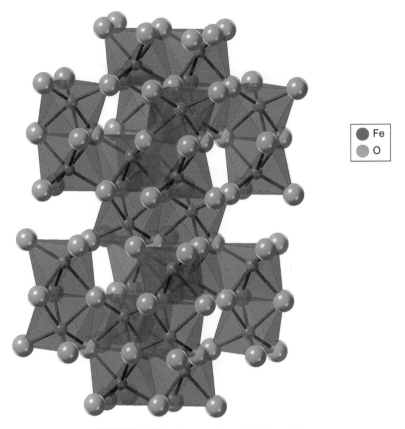

Fe
O

FIGURE 2.22 Structural model of hematite.

octahedral, making corrugated layers of octahedra (Figure 2.23).

It is widespread in noncalcareous, seasonally waterlogged soils that experience alternate oxidizing and reducing conditions. Lepidocrocite is also common in mottles, bands, pipestems, and concretions and is often recognized from its typical orange color. Lath or raft-shaped particles of lepidocrocite vary in size from ~ 10 to 500 nm and the specific surface area ranges from 15 to 260×10^3 m^2 kg^{-1}.

(a-5) Other Fe Oxides

Magnetite (Fe$_3$O$_4$) and maghemite (γ-Fe$_2$O$_3$) have similar structures, having Fe in both octahedral (Fe(VI)) and tetrahedral (Fe(IV))

coordination. Magnetite contains a significant amount of structural Fe^{2+} that is confined to octahedral sites, and the ideal formula of magnetite is (Fe^{3+}(IV)(Fe^{2+}(VI)Fe^{3+}(VI)O$_4$). Magnetite is a black-colored primary mineral that is inherited from the parent material and is usually found in the heavy-mineral fraction of the soils.

Maghemite forms from the oxidation of magnetite or via microbial processes. Maghemite may also form during bushfires due to the dehydroxylation of goethite in the presence of organic matter. The oxidation of Fe in soil maghemites is not complete and they may contain up to 30 g kg^{-1} Fe^{2+}, compared to 241 g kg^{-1} Fe^{2+} in magnetite. Maghemite particles are laths or cube shaped with a brownish-red color and

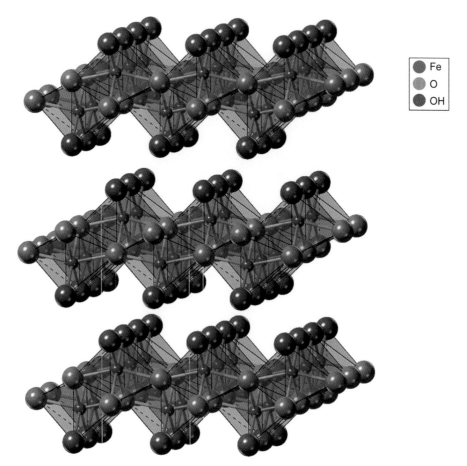

FIGURE 2.23   Structural model of lepidocrocite.

have specific surface areas ranging from 8 to 130×10^3 m^2 kg^{-1}.

Schwertmannite, akaganéite, and green rust are found in restricted soil environments and are not stable under common environmental conditions. Schwertmannite ($Fe_8O_8(OH)_6SO_4$) has yellowish-brown color and it forms under acidic conditions (pH 3–4) and high SO_4 concentrations, which are typically found in acid mine drainage environments. Akaganéite (β-FeOOH) has been identified in acid sulfate soils formed under highly acidic (pH 3–4) and saline environments with high Cl concentrations (Bigham et al., 2002; Bibi et al., 2011). Green rusts have a distinctive blue-green color, and their occurrence is restricted to reductomorphic soils. The mineral is highly unstable under oxidizing conditions yet has significant reactivity with trace metals such as Ni (Elzinga, 2021).

(b) Aluminum Oxides

Unlike Fe oxides, the occurrences of Al oxides are mainly limited to warm and humid climates in highly weathered soils. The surface chemical

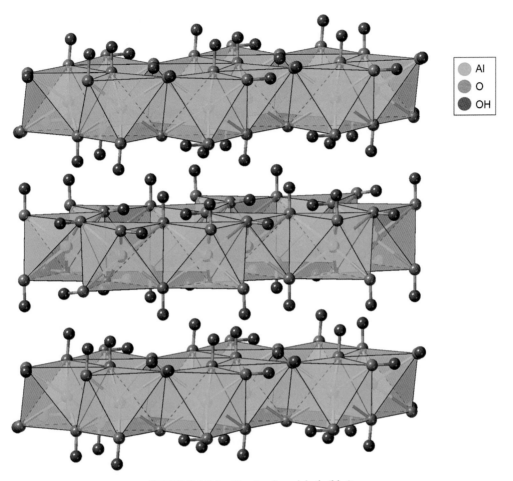

FIGURE 2.24   Structural model of gibbsite.

properties of Al oxides are quite similar to those of Fe oxides.

Gibbsite (γ-Al(OH)$_3$) is the most common Al oxide in soils. The structure of gibbsite (Figure 2.24) is analogous to the dioctahedral sheet of phyllosilicates, with two planes of closely packed OH$^-$ ions with Al^{3+} ions between the planes. The Al^{3+} atoms are found in two of the three octahedral positions and are in hexagonal rings. In the interior of gibbsite each Al^{3+} ion shares six OH$^-$ with three other Al^{3+} ions and each OH$^-$ ion is bridged between two Al^{3+}. However, at the terminal edge sites, each Al^{3+} ion shares only four OH$^-$ with two other Al^{3+} ions, and the other two coordination sites are filled by one OH$^-$ and one H$_2$O; neither is bridged between Al^{3+} ions. In the gibbsite structure the OH$^-$ ions in a particular unit are stacked directly above the OH$^-$ ions in another unit and form H bonds (Hsu, 1989).

Boehmite (γ-AlOOH) is isostructural with lepidocrocite, where the structure consists of double-chain octahedra that are joined together by sharing edges (Figure 2.23). Both gibbsite and boehmite are common in highly weathered soils of tropical regions. Gibbsite occasionally occurs in soils formed on volcanic ash.

Nordstrandite $(Al(OH)_3)$ and bayerite $(\alpha\text{-}Al(OH)_3)$ have been found occasionally in soils, mostly in association with abundant gibbsite. Both these minerals are natural polymorphs of gibbsite. Diaspore $(\alpha\text{-}AlOOH)$ and corundum $(\alpha\text{-}Al_2O_3$, an anhydrous Al oxide) rarely exist in soils (Table 2.10). The anhydrous Al oxides may be found in some igneous and metamorphic rocks and can also form from the dehydroxylation of aluminous goethite from bushfires.

(c) Manganese Oxides

Manganese oxides (Table 2.10) are usually a common and minor constituent in many soils. Often, they occur in association with Fe oxides and can have a significant influence on certain soil properties. Mn oxides serve as a source of Mn, an essential element for plants. They are effective sorbents for heavy metals and are a natural oxidant of some toxic elements, such as As^{3+} and Cr^{3+} in soils. Perhaps due to their precipitation mechanisms, Mn oxides often occur in soils as coatings on ped faces, in cracks and veins, and as nodules and concretions as large as 2 cm in diameter. Pedogenic Mn oxides have a PZC between 1.5 and 4.6; thus they carry a negative charge in most soils, which is quite different from Fe and Al oxides, where the PZC is in the range 7–9.

The mineralogy of Mn oxides is chemically complex. Mn in oxides largely exists in the Mn^{4+} state; however, there is some substitution of Mn^{3+} and Mn^{2+} in the structure that results in variable amounts of O^{2-} and OH^- (McKenzie, 1989). Manganese oxides in soils are often present as small crystals, typically in the 5–100 nm size range. Being present in small concentrations and poorly crystalline forms, Mn oxides are not identified in soils unless they are present in concentrated forms, such as nodules or mottles or coatings on peds with a distinct black to brown color.

Birnessite and vernadite are the most prevalent Mn oxides in soils, and lithiophorite has been identified in highly weathered acidic soils. Pyrolusite is the most stable form of Mn oxides, but it is uncommon in soils. The birnessite structure is comprised of layers of edge-sharing octahedra (MnO_6) that are about 0.7 nm apart and exchangeable cations and water molecules are present in the interlayer space. The structure of vernadite is poorly understood due to the lack of long-range structural order and complex structure; however, it has been suggested to consist of both edge-shared and corner-shared octahedra (Manceau et al. 1992). Lithiophorite has a layer structure similar to birnessite; however, in lithiophorite an MnO_6 octahedral sheet alternates with an octahedral sheet that contains Al and Li.

(d) Titanium Oxides

Rutile is the most common residual Ti primary mineral in soils. Anatase (TiO_2) is a secondary Ti oxide that frequently occurs in small amounts (<5%) in the clay fraction of highly weathered soils. It may have crystallized from Ti released from the weathering of primary silicates or an alteration product of sphene and ilmenite.

(e) Carbonate, Sulfate, and Evaporites

Calcite $(CaCO_3)$ is the most common and abundant carbonate mineral in soils of arid and semiarid regions. Mg calcite $(Ca_{1-x}Mg_xCO_3)$ and dolomite $(CaMg(CO_3)_2)$ are also found frequently in soils, particularly in soils of arid and semiarid regions associated with calcite. Aragonite $(CaCO_3)$ occurs occasionally in soils and siderite $(FeCO_3)$ is found under highly reduced soil conditions. Carbonates in soils can form from a solution-precipitation pathway or may be inherited from parent materials, such as limestone and dolomite. In arid and semiarid environments calcites are important scavengers of trace metals (e.g., Pb, Cd, Zn) and P.

The major sulfate mineral in soils is gypsum $(CaSO_4.2H_2O)$, and like carbonates, gypsum in soils may be derived from parent materials or precipitates from soil solution. Gypsum and calcite often exist together in soils, with gypsum being more soluble than calcite as it moves farther with water. High pH and low SO_4 concentration favor the formation of calcite over gypsum. Carbonate and sulfate minerals are highly soluble (e.g., the K_{sp} of calcite and gypsum is $10^{-8.48}$ and $10^{-4.64}$, respectively) compared to oxides (Fe oxides ranging from 10^{-37} to 10^{-44}) and phyllosilicate minerals (Doner and Lynn, 1989).

Many other soluble minerals or evaporites often occur in salt-affected soils or poorly drained soil environments. Common evaporites are a combination of Na^+, Mg^{2+}, K^+, and Ca^{2+} cations with Cl^-, SO_4^{2-}, HCO_3^-, and CO_3^{2-} anions. Sodium salts are most common among evaporites; for example, halite (NaCl), thenardite ($NaSO_4$), mirabilite ($NaSO_4.10H_2O$), and nahcolite ($NaHCO_3$) have been identified in efflorescent crusts in drainage systems and saline seeps in arid and semiarid regions.

2.6.4 The Origin and Pathways of Clay Minerals in Soils

Clay minerals in soils originate from three sources: (i) inheritance, (ii) neoformation by crystallization from ions in solutions, and (iii) transformation of preexisting minerals. Major pathways for the formation of soil clay minerals are described in Figure 2.25. The dissolution of primary minerals, such as feldspars, pyroxenes, amphiboles, and olivines, releases individual ionic components of the minerals. The neoformation of secondary minerals is governed by factors such as the ionic composition of the solution, solution pH, the solubility products of a mineral, and temperature. The transformation of 2:1 layer minerals is common in soil environments and common pathways are indicated in Figure 2.25. A more detailed discussion about the processes and pathways is presented in some review papers (Wilson,

2004; Churchman and Lowe, 2012) and mineralogy texts (Dixon and Weed, 1989; Dixon and Schulze, 2002).

2.7 X-Ray Diffraction Procedures for Mineral Identification in Soils

2.7.1 Principles of X-Ray Diffraction

X-ray diffraction (XRD) is the most widely used technique for the identification and characterization of clay minerals. It is a nondestructive method and requires only small amounts of material. The basic principle underlying this technique is that crystalline substances, including soil minerals, are made up of ordered arrays of atoms arranged in a periodic or repetitive way. Because crystals are composed of regularly spaced atoms, each crystal contains planes of atoms that are separated by a constant distance. The crystal planes in a mineral act as gratings and reflect X-rays in a characteristic pattern that depends on the atomic structure of the minerals. Another important feature of XRD is that the distance between scattering centers is about the same (i.e., 10^{-8} cm or 1 Å) as the wavelength of X-rays; therefore crystals can diffract the radiation when the diffracted beams are in-phase. It is not applicable to amorphous or glassy substances. The details of X-ray production, instrumentation, and principles of XRD are presented in several other texts (Moore and Reynolds, 1997; Cullity and Stock, 2001; Hammond, 2001; Pecharsky and Zavalij, 2009). Here, we present a brief background and aspects that are most relevant for the XRD analysis of soil minerals.

X-rays are a part of electromagnetic radiation with wavelengths of the order of 0.01 to 10 nm (0.1 to 100 Å). X-rays are generated within an X-ray tube by the bombardment of a target metal (anode) with high-velocity electrons. Monochromatic X-rays used for XRD have finite energy and wavelength characteristic for the particular

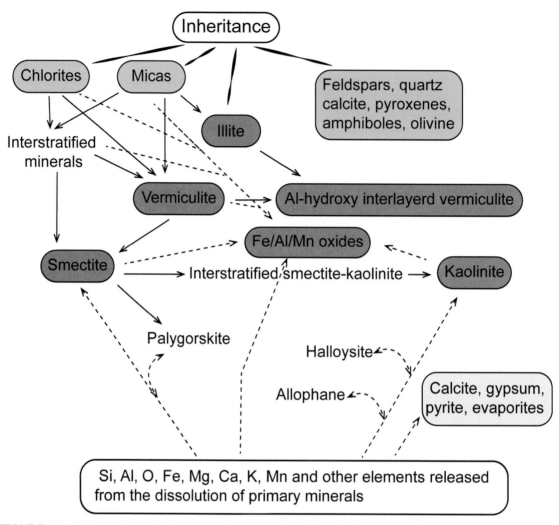

FIGURE 2.25   Main pathways for the origin of common minerals in soils. The neoformation and transformation mechanisms are indicated by dotted and solid lines, respectively, while the inheritance mechanism is indicated by uneven lines in the figure. The most common clay minerals are enclosed in brown boxes, inherited minerals are in green boxes, and soluble and unstable minerals are given in the yellow box. *Adapted from Wilson (2004); Churchman and Lowe (2012).*

target metal. When an incident X-ray beam impinges upon crystals, general scattering occurs (i.e., the electrons around the nuclei of the atoms scatter the X-rays). Most of the scattered waves are out of phase (destructive interference) and destroy one another (Figure 2.26a). There is constructive interference when the waves are in phase with each other (they have zero and maximum amplitude at the same point along the abscissa) and the addition of individual amplitudes of such waves produces a similar wave having a much larger amplitude (Figure 2.26b). The new wave front is called a diffracted beam that we observe and is measured and composed of an enormous number of constructively interfering rays.

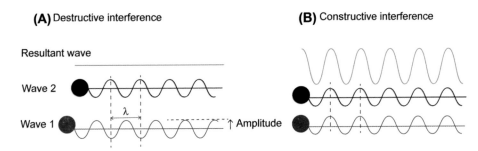

FIGURE 2.26   Destructive (wave out of phase) (A) and Constructive (wave in phase) (B) interferences of X-rays.

(a) Bragg's Law

In 1913 W.H. Bragg and his son W.L. Bragg applied XRD to determine the crystal structures of several crystalline compounds. They devised an approach based on the reflection of X-rays by a plane of atoms of the crystal lattice.

Consider a crystal made up of parallel planes of atoms periodically repeated by lattice translations and spaced by a distance "d" from each other (Figure 2.27). The figure shows a monochromatic beam of X-rays with wavelength (λ) impinging on a crystal surface at an angle θ and the atomic planes are represented by P, P_1, and P_2. The AB incident beam is reflected by plane P in the BC direction, and similarly, P_1

reflects beam DE in the EF direction. Although the X-ray beam penetrates several planes of atoms in a crystal, for simplicity, we are considering only the top two planes of atoms.

Since plane P_1 is lower than plane P, the beam path DEF is longer than ABC by the distance XE + EY. For the two reflected rays to arrive in phase, the extra distance or path difference must be equal to the wavelength or a whole number multiple of it as given below:

$$\text{path difference} = XE + EY = n\lambda,$$
$$\text{where } n = 1, 2, 3....$$

(2.16)

The two triangles XEB and YEB share a side EB, which is the spacing (d) between the plane

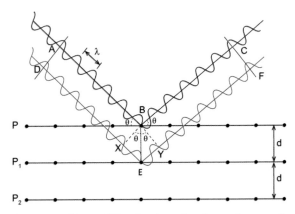

FIGURE 2.27   A schematic illustration of X-ray diffraction from the planes of atoms of a crystalline structure following Bragg's law.

of atoms. Also, if BX on one side of the XEB triangle is perpendicular to beam 1, that is, $\angle ABX = 90°$ and $\angle ABP = \theta$, then $\angle PBX = 90 - \theta$.

Since $\angle PBE = 90°$, $\angle XBE = \theta$, and similarly, we can show that $\angle YBE = \theta$. Additionally, from Δ XBE, $\sin\theta = XE/d$ (BE). Therefore $XE = d \sin\theta$. Similarly, from Δ YBE, $\sin\theta = EY/d$ (BE). Therefore $EY = d \sin\theta$.

Now, substituting the values of XE and EY in the path difference equation:

$$XE + EY = n\lambda$$

$$d \sin\theta + d \sin\theta = n\lambda,$$

or

$$n\lambda = 2d \sin\theta \qquad (2.17)$$

This equation is called Bragg's law or equation, where n is an integer called "the order of the reflection," λ is the wavelength of the X-rays, d is the distance between the diffracting planes of the crystal, and θ is the angle between the incident X-ray beam and the diffracting plane.

Diffraction will occur from a plane of atoms in a mineral structure only if the conditions of Bragg's equation are fulfilled. Since no two minerals have the same interatomic distances in three dimensions, the angles at which diffraction occurs will be distinctive for a mineral. The interatomic distances within a mineral crystal thus result in a unique array of diffraction maxima, which is used for the identification of that mineral.

The main technique for obtaining a diffraction pattern from clay and other size fractions of soils uses a diffractometer and a goniometer. The sample, mounted on a planar surface, is placed at the axis of the diffractometer and slowly rotated at a desired rate so as to increase the angle of incidence of the X-ray beam on the sample surface. The X-ray beam is collimated through a series of slits and directed onto the specimen at the center of diffraction. The detector, usually rotating at twice the rate of the sample, °2θ, continuously measures the intensity of the diffracted monochromatic beam. When the sample is being scanned, a plot of intensity (on the Y-axis) against °2θ (on the X-axis) is obtained. Once the scan is completed, the peaks on the diffraction pattern can be identified and d-spacing assigned, and minerals can be identified with the help of standard tables.

2.7.2 Sample Preparation for Clay Mineral Identification

For the positive identification of phyllosilicates, several diagnostic treatments are employed, and a sample needs to be analyzed more than one time by using XRD. The number of analyses and pretreatments (e.g., to remove organic carbon and calcium carbonate) depends on the information required and the mineral assemblages in a sample (Whittig and Allardice, 1986; Moore and Reynolds, 1997). Although it is impossible to physically separate mineral species from polycomponent systems in soils, the separation of the clay fraction ($<2 \mu m$) via sedimentation and/or centrifugation is usually sufficient for the identification of clay minerals present in a soil sample. Two common methods are employed for sample mounting for XRD analysis: (i) a basally oriented aggregate and (ii) a random powder.

(a) Basally Oriented Samples

Since most phyllosilicates are structurally very similar along the X- and Y-axes, hkl peaks are not very diagnostic. The diffraction pattern along the Z-axis that gives basal (00l or 001) reflections is the most diagnostic for the identification of phyllosilicate minerals. Basal reflections derive from the fundamental repeat layers of phyllosilicates. The basal orientation of samples is essential to enhance the intensity of basal (00l or 001) reflections. Most phyllosilicates are platy in shape and the intensity of the basal reflection is enhanced if the sample lies parallel to the plane of the diffractometer.

The d-spacing and intensity of the *001* XRD peaks of phyllosilicates can be influenced by various factors, such as the saturating cation, relative humidity, type of solvating liquid, and thermal treatment. The degree of hydration is most important to obtain the correct d-spacing of the *001* XRD peak of expanding minerals (MacEwan and Wilson, 1980); the basal spacing of smectites with the atmosphere humidity varies from about 1.0 nm under dry conditions to 1.8 nm under fully wet conditions. For obtaining reproducible results, oriented samples should be equilibrated at specific relative humidity over a saturated salt solution in desiccators. For example, a saturated solution of $Mg(NO_3)_2$ is used to obtain 54% relative humidity and is commonly used for obtaining the XRD analysis of air-dried samples.

The oriented clays can be prepared by depositing onto a ceramic tile under suction or by filtering a clay suspension onto a membrane filter and then transferring the clay film onto a glass slide or simply adding the clay suspension onto a glass slide followed by air-drying.

The basally oriented clay fraction is examined for the identification of phyllosilicates with the following treatments:

(i) Mg- or Ca-saturated and air-dried basally oriented aggregates;

(ii) Mg- or Ca-saturated and ethylene glycol−solvated, where the Mg or Ca air-dried sample is placed in a desiccator containing ethylene glycol and the desiccator is then transferred into an oven set at 60°C for 8−12 hours;

(iii) K-saturated, air-dried basally oriented aggregates;

(iv) K-saturated, heated at 335°C for 4 hours in a muffle furnace; and

(v) K-saturated, heated at 550°C for 4 hours.

Mineral identification is based on the presence of a peak at a d-spacing, the extent of swelling along the Z-axis (perpendicular to the (*001*) plane or the thickness of platy crystals), and contraction under the treatment used (Table 2.11; Figure 2.28).

(b) Randomly Oriented Samples

TABLE 2.11   The d-spacings (in nm; 1 nm = 10 Å) for common phyllosilicates in soils.

Mineral	Mg/Ca saturated and air-dried	K saturated and air-dried	Mg/Ca saturated and ethylene glycol treated	K saturated and 300−350°C heated	K saturated and 550°C heated
Kaolinite	0.72	0.72	0.72	0.72	No peak
Halloysite − dehydrated	0.72	0.72	0.72	0.72	No peak
Halloysite − hydrated	1.1	1.0	1.01−1.07	0.72	No peak
Mica (illite)	1.0	1.0	1.0	1.0	1.0
Smectite	1.4−1.5	1.2−1.3	1.7	1.0	1.0
Vermiculite[a]	1.45	1.05	1.45	1.0	1.0
Chlorite	1.4	1.4	1.4	1.4	1.4
Hydroxy-interlayered vermiculite	1.4	1.4	1.4	1.1−1.3	1.0−1.1

[a] *Trioctahedral vermiculites expand to ~1.6 nm spacing on the ethylene glycol solvation of Mg/Ca-saturated samples.*

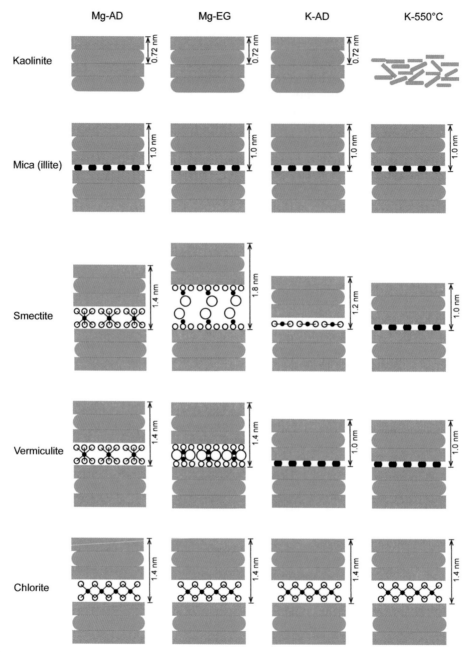

FIGURE 2.28   Schematic representation of the responses of phyllosilicates to diagnostic treatments that are used in the X-ray diffraction analysis of oriented clay samples.

In a randomly oriented sample crystals lie in all possible positions so that X-rays are diffracted off equally from all atomic planes of minerals in the sample. Random samples enable one to obtain XRD peaks from all minerals, including phyllosilicates and nonphyllosilicates, which are present in a soil or sediment sample. This procedure is also used to determine the bulk mineralogy of a soil/sediment.

Random powder XRD is also useful in determining the position of *060* reflections for phyllosilicates. The *060* reflection allows the distinction between dioctahedral and trioctahedral minerals; for example, *060* spacing for kaolinite, a dioctahedral 1:1 mineral, is 0.149 nm, and for serpentines and trioctahedral 1:1 minerals, it is between 0.1531 and 0.1538 nm. The *b* cell dimension is more sensitive to the size of the cations and to the site occupancy in the octahedral layer than are the *a* or *c* dimensions.

It is very difficult or nearly impossible to perfectly order a sample in random orientation and this is particularly true for platy phyllosilicates. Several methods have been suggested in the literature and the key is to minimize excessive force in packing the powdered sample into the sample holder. Sieving the finely ground sample directly into the sample holder and packing the powder using the razor blade edge is a good way to prepare a randomly oriented clay for XRD analysis.

Phyllosilicate minerals in soils are seldom found as monomineralic materials. Commonly, small amounts of nonphyllosilicates minerals like quartz, feldspars, and metal oxides are also present. Some of these minerals will produce much sharper peaks than phyllosilicates in the random powder diffraction patterns. However, the peaks of many of the nonphyllosilicate minerals may not be present in the XRD patterns of oriented aggregates.

Three main diagnostic peaks of nonphyllosilicate minerals that commonly occur in soils are given in Table 2.12. For complete information about the diffraction file of all minerals, refer to

TABLE 2.12 Strongest three reflections (in decreasing order of intensities) of commonly occurring nonphyllosilicate minerals in soils.

Mineral	Diagnostic reflection or peak (Å)
Quartz	3.34, 4.26, 1.817
Microcline	3.25, 3.29, 4.21
Albite	3.19, 4.03, 3.21
Anorthite	3.20, 3.18, 4.04
Hornblende	8.52, 3.16, 2.73
Calcite	3.03, 2.095, 2.285
Dolomite	2.89, 2.192, 1.804
Gypsum	7.56, 3.06, 4.27
Goethite	4.18, 2.45, 2.69
Al-goethite	4.12, 2.406, 2.645
Hematite	2.69, 2.514, 1.692
Lepidocrocite	6.27, 3.29, 2.473
Magnetite	2.532, 1.485, 2.97
Maghemite	2.514, 1.474, 2.95
Schwertmannite	2.55, 3.39, 4.86
Akaganéite	3.311, 7.40, 2.543,
Gibbsite	4.85, 4.37, 2.382
Rutile	3.25, 1.687, 2.487
Anatase	3.52, 1.892, 2.378

other sources such as specialized books (Brindley and Brown, 1980) and the Powder Diffraction File (PDF), which is a database of X-ray powder diffraction patterns maintained by the International Center for Diffraction Data (ICDD).

In routine XRD analysis the quantification of clay minerals, and particularly of the phyllosilicates, remains a major challenge due to several factors, including structural disorders, varied chemical compositions, and preferred orientation. However, significant progress has been made in various methods based on XRD analysis. The most common procedures either use the intensity of a single diffraction peak of the mineral with or without an internal or external standard or the whole XRD pattern of the sample. The Rietveld method (Rietveld, 1969) is the most sophisticated quantification method out of the whole pattern methods. It uses a least-squares approach to match the experimental XRD pattern with a theoretical XRD pattern, which is simulated from the crystallographic and other characteristics of minerals and instrumental factors. Common software tools that use the Rietveld method for soil mineral quantification include SIROQUANT (Taylor, 1991), BGMN (Bergmann et al., 1998), GSAS (General Structure Analysis System) (Larson and Von Dreele, 1994), Highscore (PANalytical™ Company), and TOPAS (TOtal PAttern Solution, Bruker™ Company). Unlike the Rietveld method, other whole pattern methods, such as RockJock and FULLPAT, do not require crystallographic data and match the observed XRD pattern with the combined XRD pattern made up of individual XRD patterns of phases that are present in a given sample (Eberl, 2003; Chipera and Bish, 2013).

The chemical formula for a clay mineral can be derived from the chemical composition (Box 2.3) as long as the sample does not contain any other impurities.

2.8 Specific Surface Area (SSA) of Soil Minerals

Specific surface area (Sidebar 2.16) is an important property of secondary soil minerals in relation to many physical and chemical properties of soils. In particular, SSA is very important in relation to the chemical reactions that occur at the mineral–water interface. Clay minerals in soils contribute most of the specific surface area. The total specific surface area of a mineral depends on both internal and external specific surface areas.

Sidebar 2.16 — Specific Surface Area

Specific surface area (SSA) is a physical property of minerals (or any solids) that is defined as the total surface area of a mineral per unit of mass ($m^2 \, kg^{-1}$ or $m^2 \, g^{-1}$) or volume ($m^2 \, m^{-3}$).

The SSA of individual soil minerals on a mass basis can be calculated from their particle surface area, particle volume, and particle density as follows:

SSA ($m^2 \, g^{-1}$) = particle surface area/(particle volume × particle density).

1 g quartz, (0.2-mm cubic crystals) =

$$\frac{(0.2 \times 10^{-3})^2 \times 6}{(0.2 \times 10^{-3})^3 \times (2.65 \times 10^6)} = 0.01 \, m \, g^{-1}$$

In quartz the numerator is multiplied by 6 because of the cubic structure.

1 g illite (1 μm × 1 μm × 0.1 μm) =

$$\frac{(1 \times 10^{-6})^2 \times 2}{(1 \times 10^{-6})^2 \times (1 \times 10^{-7}) \times (2.75 \times 10^6)} = 7.3 \, m \, g^{-1}$$

In illite the numerator is multiplied by 2 because illite is a sheet structure (phyllosilicate).

1 g goethite (0.05 μm × 0.01 μm × 0.01 μm) =

$$\frac{(1 \times 10^{-8}) \times (5 \times 10^{-8}) \times 4}{(1 \times 10^{-8})^2 \times (5 \times 10^{-8}) \times (3.8 \times 10^6)} = \sim 105.3 \, m \, g^{-1}$$

In goethite the numerator is multiplied by 4 because crystals have a lath or rod shape morphology.

2.8.1 External Surface Area Measurement

External surface area is normally determined by measuring the adsorption of N_2 gas (the adsorbate) at temperatures near the boiling point of liquid N_2 (77 K). At this temperature, the N_2 gas condenses on the surface mineral particles, that is, adsorbent. Based on the assumption that condensed N_2 molecules form

BOX 2.1

Balancing Chemical Weathering Reactions

Weathering reactions signify chemical and physical changes in soil minerals. Like any other chemical reaction, the mass (or mole or atoms) and charge of the reactants and products must be balanced in chemical weathering reactions. Some examples illustrating various steps needed to balance the chemical weathering reactions are given below.

(a) Weathering via the incongruent dissolution of potassium feldspar (a common primary mineral) to form kaolinite (a common secondary mineral)

   (i) Write the chemical weathering reaction using the chemical formula of the reactant and product:

$$KAlSi_3O_8(s) \rightarrow Al_2Si_2O_5(OH)_4(s) \qquad (B2.1a)$$

   (ii) Add expected soluble species in the reaction; here K and Si species are added on the products side:

$$KAlSi_3O_8(s) \rightarrow Al_2Si_2O_5(OH)_4(s)K^+(aq) + H_4SiO_4^0(aq) \qquad (B2.1b)$$

   (iii) Balance (mass) all elements except O and H:

$$2KAlSi_3O_8(s) \rightarrow Al_2Si_2O_5(OH)_4(s) + 2K^+(aq) + 4H_4SiO_4^0(aq) \qquad (B2.1c)$$

   (iv) Add water to balance O atoms. There are 16 mol of O in feldspar (reactant) and 25 mol O in products, including 13 mol O in kaolinite and 16 mol O in silicic

acid; thus to mass balance, we must add 9 mol of O on the reactant side:

$$2KAlSi_3O_8(s) + 9H_2O(l) \rightarrow Al_2Si_2O_5(OH)_4(s) + 2K^+(aq) + 4H_4SiO_4^0(aq) \qquad (B2.1d)$$

   (v) Lastly, add protons to balance H atoms:

$$2KAlSi_3O_8(s) + 9H_2O(l) + 2H^+(aq) \rightarrow Al_2Si_2O_5(OH)_4(s) + 2K^+(aq) + 4H_4SiO_4^0(aq) \qquad (B2.1e)$$

(b) Congruent dissolution of forsterite (Mg_2SiO_4), an olivine mineral

   (i) Write the chemical weathering reaction using the chemical formula of reactant and products:

$$Mg_2SiO_4(s) \rightarrow 2Mg^{2+}(aq) + H_4SiO_4^0(aq) \qquad (B2.1f)$$

The mineral dissolves congruently to produce elements in the same proportion as they are present in the mineral, that is, 2 mol of Mg^{2+} and 1 mol of Si.

In the above equation O is also balanced, so we need to add protons on the reactant side to balance H atoms:

$$Mg_2SiO_4(s) + 4K^+(aq) \rightarrow 2Mg^{2+}(aq) + H_4SiO_4^0(aq) \qquad (B2.1g)$$

(c) Weathering of beidellite ($Na_{0.5}[Al_2](Si_{3.5}Al_{0.5})O_{10}(OH)_2$), a smectite, to form kaolinite, $Al_2Si_2O_5(OH)_4$

   (i) Write the chemical weathering reaction using the chemical formula of the reactant and product:

$$Na_{0.5}[Al_2](Si_{3.5}Al_{0.5})O_{10}(OH)_2(s) \rightarrow Al_2Si_2O_5(OH)_4(s) \qquad (B2.1h)$$

Continued

BOX 2.1   *(cont'd)*

Balancing Chemical Weathering Reactions

(ii) Add the expected soluble species in the reaction; in this example $Na^+_{(aq)}$ and $H_4SiO_4^0$ species are added:

$$Na_{0.5}[Al_2](Si_{3.5}Al_{0.5})O_{10}(OH)_2(s) \rightarrow$$
$$Al_2Si_2O_5(OH)_4(s) + Na^+(aq) + H_4SiO_4^0(aq)$$

$$(B2.1i)$$

(iii) Mass balance all elements, except O and H; to mass balance for Al, we need 2.5 mol of Al on each side; thus the stoichiometric coefficient for kaolinite should be $2.5/2 = 1.25$; mass balance for Na requires 0.5 mol on both sides, and Si has 3.5 mol on both sides.

$$Na_{0.5}[Al_2](Si_{3.5}Al_{0.5})O_{10}(OH)_2(s) \rightarrow$$
$$1.25Al_2Si_2O_5(OH)_4(s) + 0.5Na^+(aq) + H_4SiO_4^0(aq)$$

$$(B2.1j)$$

(iv) Add water to balance O atoms. There are 12 mol of O on the reactant side, and 15.25 mol O in products, including 11.25 mol O in kaolinite and 4 mol O in silicic acid, thus we need to add 3.25 mol of O on the reactant side to achieve mass balance, this is accomplished by adding 3.25 mol of water on the reactant side:

$$Na_{0.5}[Al_2](Si_{3.5}Al_{0.5})O_{10}(OH)_2(s) + 3.25H_2O(l) \rightarrow$$
$$1.25Al_2Si_2O_5(OH)_4(s) + 0.5Na^+(aq) + H_4SiO_4^0(aq)$$

$$(B2.1k)$$

(v) Now balance the H atoms by adding protons:

$$Na_{0.5}[Al_2](Si_{3.5}Al_{0.5})O_{10}(OH)_2(s) + 3.25H_2O(l) \rightarrow$$
$$1.25Al_2Si_2O_5(OH)_4(s) + 0.5Na^+(aq) +$$
$$H_4SiO_4^0(aq) + 0.5H^+(aq)$$

$$(B2.1l)$$

---

a monolayer on the mineral surface and a knowledge of the size of the gas atom/molecule, the amount of condensed or adsorbed gas is correlated to the total surface area of the mineral particles.

Surface area analysis is based on this correlation calculation and uses the Brunauer, Emmett, and Teller (BET; 1938) theory. The common linearized BET equation is as follows:

$$\frac{P}{V(P_0 - P)} = \left(\frac{1}{V_m c}\right) + \left[\frac{(c-1)}{V_m c} \cdot \left(\frac{P}{P_0}\right)\right] \quad (2.18)$$

where P = pressure (Pa) of a gas at equilibrium with a solid, P_0 = gas pressure (Pa) required for saturation at the temperature of the experiment, V = volume (m^3) of gas adsorbed at pressure P, V_m = volume (m^3) of gas required for monolayer coverage over the complete adsorbent surface, and c is the BET constant.

$$c = \exp\left(\frac{E_1 - E_L}{RT}\right) \quad (2.19)$$

where E_1 is the heat of adsorption of the first layer of adsorbate, E_L is the heat of adsorption for the second and higher layers and is equal to the heat of liquefaction of the adsorbent, R is the gas constant, and T is the absolute temperature.

BOX 2.2

Calculation of Layer Charge from the Chemical Formula of a 2:1 Clay Mineral

Let us take the formula unit (or half unit-cell) of a dioctahedral vermiculite that has substitutions in both the tetrahedral and octahedral sheets:

$$M_{0.74}\left(Si^{4+}_{3.56} + Al^{3+}_{0.44}\right)\left(Al^{3+}_{1.4} + Mg^{2+}_{0.3} + Fe^{3+}_{0.3}\right)O_{10}(OH)_2$$

The total positive charge in the tetrahedral sheet = 3.56 × 4 (Si^{4+}) + 0.44 × 3 (Al^{3+}) = 14.24 + 1.32 = 15.56

The ideal tetrahedral sheet in the formula unit has a net cation charge of +16.0; thus the net negative charge in the tetrahedral sheet is 16.0 − 15.56 = 0.44

The total positive charge in the octahedral sheet = 1.4 × 3 (Al^{3+}) + 0.3 × 2 (Mg^{2+}) + 0.3 × 3 (Fe^{3+}) = 4.2 + 0.6 + 0.9 = 5.7

The ideal octahedral sheet in the formula unit has a net cation charge of +6.0; thus the net negative charge in the tetrahedral sheet is 6.0 − 5.7 = 0.3

Total negative charge or net charge in the formula unit = 0.44 + 0.3 = 0.74

One can also determine the net charge from the difference in the total cationic and anionic charge.

The total negative charge = 10 × 2 (O^{2-}) + 2 × 1(OH^-) = −22; the total positive charge = 15.56 + 5.7 = 21.26

Thus the net layer charge is +21.26 + (−22) = −0.74.

This negative charge on the mineral is balanced out by metal cations in the interlayer space represented as $M_{0.74}$, where M represents a cation or cations, such as Mg^{2+}, Na^+, or K^+, with charge expressed on a +1 charge basis.

To obtain the values of V_m and c, $P/V(P_0 - P)$ versus P/P_0 are plotted typically at a relative pressure (P/P_0) between 0.05 and 0.35. The relationship between $P/V(P_0 - P)$ and P/P_0 is expected to be linear for this range; the slope would equal $(c - 1)/V_m c$ and the intercept would equal $1/V_m c$. The values for V_m and c are calculated as follows:

$$V_m = \left(\frac{1}{slope + intercept}\right); \quad c = \left(1 + \frac{slope}{intercept}\right)$$
$$(2.20)$$

Once the value of V_m is determined, the specific surface area can be calculated according to Brame and Griggs (2016):

$$S = \left(\frac{NAV_m}{V_0 \times m}\right) \quad (2.21)$$

where S = specific surface area ($m^2 \, g^{-1}$) of the adsorbent, N = Avogadro's number (6.02×10^{23} molecules mol^{-1}), A = cross-sectional area (m^2; for liquid nitrogen 16.2×10^{-20} m is a generally accepted value) of the adsorbate, V_0 = molar volume (m^3; at the standard temperature and pressure [STP] volume of one mole of gas = 0.022414 m^3), and m is the mass of adsorbent used in the measurement.

Other adsorbates can also be used in external surface area analysis, such as ethylene for solids having surface areas <0.1 $m^2 \, g^{-1}$ and krypton for solids with intermediate surface areas. Nitrogen is the most frequently used adsorbate, particularly with solids that have surface areas >10 $m^2 \, g^{-1}$. However, it is not good to use N_2 with a clay mineral like montmorillonite since its surface area is primarily internal and

BOX 2.3

Calculation of Structural Formulas and CEC of Clay Minerals.

X-ray diffraction (XRD) can be routinely used to identify and semiquantitatively determine the mineral suites in soils. The Rietveld analysis of XRD data or XRD in conjunction with thermal and specific surface area analyses can be used to quantify soil minerals. However, these methods will not provide structural formulae. Such information can only be gleaned through chemical analyses. To obtain the necessary data, the mineral must first be broken down using acid dissolution. The procedure outlined below was first introduced by Ross and Hendricks (1945). In calculating the structural formulae it is assumed that elemental concentrations reported in a mineral sample, such as Si and Al, are a part of the clay mineral structure. While this may not be completely valid for impure samples, it is usually true for clay minerals from natural deposits. Additionally, the elemental contents of a mineral are generally reported as % oxide (e.g., the SiO_2, Al_2O_3, etc.).

The analysis shown in Table 2.13 is for a 2:1 clay mineral with an ideal formula of $Si_8Al_4O_{20}(OH)_4$. Thus there are a total of 44 positive charges and 44 negative charges, assuming a full unit-cell chemical formula. The steps for the calculations in the table are:

(i) Convert the element oxides weight (%) to % moles of oxides per g by dividing the oxide % by their molecular weights (e.g., for $SiO_2 = 59.6/60.083$); this value yields the moles of the element of interest (e.g., Si) per unit mass.

(ii) Convert each element mole (%) to atomic (%) by multiplying with the number of atoms in each oxide form (e.g., in SiO_2 one atom of Si, and in Al_2O_3 there are two atoms of Al).

TABLE 2.13   Structural analysis[a] of clay, Spur, Wyoming, montmorillonite.

Oxide	Weight (%)	Molecular weight	Moles per unit	Atoms per unit	Cationic charge	Cation charge per 44	No. of atoms
SiO_2	59.60	60.083	0.992	0.992	3.968	30.72	7.680
Al_2O_3	22.17	101.961	0.217	0.435	1.305	10.10	3.367
Fe_2O_3	4.32	159.687	0.027	0.054	0.162	1.26	0.419
TiO_2	0.09	79.865	0.001	0.001	0.005	0.03	0.017
MgO	2.73	40.304	0.068	0.068	0.135	1.05	0.524
CaO	0.14	56.077	0.002	0.002	0.005	0.04	0.019
Na_2O	3.18	61.979	0.051	0.103	0.103	0.79	0.794
K_2O	0.03	94.195	0.000	0.001	0.001	0.00	0.005
Total					5.683		

[a] Earley et al. (1953).

BOX 2.3 *(cont'd)*

Calculation of Structural Formulas and CEC of Clay Minerals.

(iii) Now convert atomic (%) to cationic charge ($mol_{(c)}$) by multiplying with the cation valency (e.g., $Si = 4$, $Al = 3$), and add the charge of all cations.

(iv) Calculate the charge contributed by each cation in the formula unit (i.e., to balance 44 negative charge in the 2:1 phyllosilicate); this is done by multiplying with a normalization factor $= 44/(sum of charge)$ obtained in Step (iii); in our example it is 5.683).

(v) Calculate the number of atoms of each cation by dividing the formula charge obtained in Step (iv) by the cation valence.

(vi) Allocate the cations to their respective positions, i.e., the tetrahedral sheet, octahedral sheet, and interlayer position for exchangeable cations.

In allocating the cations the following assumptions are made: (a) a total of eight atoms can fit in the tetrahedral sheet, which can contain Si^{4+} and Al^{3+}; (b) a total of four atoms can fit in the octahedral sheet with Al^{3+}, Fe^{3+}, Fe^{2+}, Mg^{2+}, and Li^{+} able to fit in this sheet; and (c) larger cations such as K^{+}, Na^{+}, and Ca^{2+} occur as exchangeable cations and satisfy negative charge resulting from isomorphic substitution.

The data given in Table 2.13 indicate that there are only 7.68 Si^{4+} atoms present in the clay mineral; a total of eight atoms can fit in the tetrahedral sheet. Therefore the remaining 0.32 atoms in the tetrahedral layer will be Al^{3+}. Therefore this sheet has a cationic composition of $Si_{7.68}Al_{0.32}$.

The remaining Al^{3+} atoms ($3.367 - 0.32 = 3.047$) will go in the octahedral sheet along with Fe^{3+}, Ti^{4+}, and Mg^{2+} atoms, in that order, to yield four atoms.

Octahedral cations: $Al^{3+} = 3.047$, $Fe^{3+} = 0.419$, $Ti^{4+} = 0.017$, $Mg^{2+} = 0.524$, Total $= 4.008$.

Thus the octahedral layer would have the cationic composition ($Al^{3+} = 3.047$, $Fe^{3+} = 0.419$, $Ti^{4+} = 0.017$, $Mg^{2+} = 0.524$, Total $= 4.008$).

$Ca^{2+} = 0.019$, $Na^{+} = 0.794$, and $K^{+} = 0.005$ will go in the interlayer of the mineral as exchangeable cations to balance the negative charge created in the tetrahedral and octahedral sheets. The complete structural formula for the mineral would be:

$$(Ca_{0.019}Na_{0.794}K_{0.005})\left[Si^{4+}_{7.68}Al^{3+}_{0.32}\right]$$
$$\times \left(Al^{3+}_{3.047}Fe^{3+}_{0.419}Ti^{4+}_{0.017}Mg^{2+}_{0.524}\right)O_{20}(OH)_4$$

Negative charge in the tetrahedral sheet $= 0.32$ mol_c

Negative charge in the octahedral sheet $= 0.524 - 0.017 = 0.507$ mol_c

Total structural negative charge $= 0.32 + 0.507 = 0.827$ mol_c

Positive charge contributed by the exchangeable cations $= 2 \times 0.019 + 0.794 + 0.005 = 0.837$ mol_c.

Thus the charge balance is pretty good (0.827 mol ≈ 0.837) because the mineral sample was quite pure.

The formula can be used to determine the cation exchange capacity of the mineral.

The cation exchange capacity ($mmol_c$ kg^{-1}) of the clay can be calculated using the following formula:

CEC $=$

$$\frac{\text{Negative charge (mol) in the unit cell}}{\text{Molecular weight of the mineral in the unit cell}}$$

$\times$ 1000 (convert mol_c to $mmol_c$) $\times$ 1000 (convert CEC from g to kg)

$$= \frac{0.827}{731.51} \times 1000 \times 1000 = 1130.5 \text{ } mmol_c \text{ } kg^{-1}$$

nonpolar N_2 molecules do not penetrate the interlayer regions of the mineral.

2.8.2 Total Surface Area Measurement

The total surface area measurement of minerals (such as smectites) whose internal surfaces are inaccessible to N_2 gas is done by the adsorption of polar or other liquids.

Ethylene glycol monoethyl ether (EGME) is a common material used to measure the total surface area of a mineral. One applies excess EGME to the adsorbent that has been dried over a P_2O_5 desiccant in a vacuum desiccator and then removes all but a monolayer of the EGME from the mineral surface using a vacuum desiccator containing $CaCl_2$. Then, by plotting the weight of EGME retained on the solid as a function of time, one can determine when the weight of EGME becomes constant. At this point, one assumes that monolayer coverage has occurred. To calculate the specific surface, one can use (Carter et al., 1986):

$$S = W_g/[(W_s)(0.000286)] \qquad (2.22)$$

where S = specific surface ($m^2\ g^{-1}$), W_g = weight (g) of EGME retained by the sample after monolayer equilibration, and W_s = weight (g) of P_2O_5 (desiccant) dried sample; 2.86×10^{-4} g EGME are required to form a monolayer on 1 m^2 of surface. To obtain internal surface area (SI),

$$ST = SE + SI \qquad (2.23)$$

or

$$SI = ST - SE \qquad (2.24)$$

where ST = total surface area ($m^2\ g^{-1}$), SE = external surface area ($m^2\ g^{-1}$), and SI = internal surface area ($m^2\ g^{-1}$).

The adsorption of other liquids has also been used for determining surface area, for example, methylene blue (Hang and Brindley, 1970), water (Newman, 1983), cetyl pyridinium bromide (Greenland and Quirk, 1964), p-nitrophenol (pNP) adsorption (Ristori et al., 1989). Some of these methods give similar surface area values, and the choice of the method may depend on the sorbent/mineral properties and the availability of specialized equipment that is required for a certain method.

2.9 Surface Charge Characteristics of Soil Minerals

2.9.1 Types of Charge

Soil minerals can exhibit two types of charge: permanent (or "constant") charge and variable (or "pH-dependent") charge. In most soils there is a combination of permanent and variable charge.

Permanent charge is invariant with soil pH and other properties, and this charge results from the isomorphic substitutions of lower valence cations for higher-valence cations (e.g., Al^{3+} for Si^{4+} substitution) in the tetrahedral sheet and/or octahedral sheet (e.g., Mg^{2+} for Al^{3+} substitution) of phyllosilicates. Isomorphic substitutions occur both in primary and secondary soil minerals; however, the significant charge results only in 2:1 layer minerals. Clay minerals that most commonly exhibit permanent charge in soils are smectite, vermiculite, mica, and chlorite.

The constant (or permanent) charge terminology is attributed to Schofield (1949). He found that the cation exchange capacity (CEC) of a clay subsoil at the Rothamsted Experimental Station in England was relatively constant between pH 2.5 and 5, but CEC increased between pH 5 and 7. He also showed that the CEC of montmorillonite, a smectite mineral, did not vary below pH 6 but increased as pH increased from 6 to 9. He ascribed the constant charge component to isomorphous substitution and the variable component to the ionization of H^+ from SiOH groups on the clay surfaces. His conclusion about constant charge is still valid if the clays are relatively pure. However, we now know that the contribution of SiOH groups to the creation of negative charge is not significant in most soils since the pK_a (see Box 3.2 for the derivation and description of K_a and pK_a) of silicic acid is so high (about 9.5) that the ionization of the SiOH group would occur only at very alkaline pH. Recently, it has been shown that some silanol sites can have pK_a values that range down to 7, whereas aluminol groups can have considerably lower pK_a values (e.g., c. 3.7) (Tournassat et al., 2016). These pK_a values depend on the metal coordination at each site.

The variable charge in soils is derived from the protonation and deprotonation of surface functional groups and the adsorption and desorption of certain ions on mineral surfaces. The variable charge on soil minerals depends on solution pH and ionic strength; however, soil pH is the main determinant, and hence the charge is often called pH-dependent charge. Kaolinite, metal oxides, including oxyhydroxides and hydroxides, and amorphous materials are the main inorganic components in soils that contribute to this charge. A significant portion of the variable charge in soils is contributed by soil organic matter. However, the mechanism is somewhat different from that for surface functional groups of minerals. Soil organic matter has several different kinds of functional groups that dissociate at different pH values resulting in a negative charge on the groups; these are discussed in Chapter 3.

2.9.2 Cation Exchange Capacities of Secondary Soil Minerals

The negative charge, permanent and variable charge, on soil minerals is balanced by positive charge in the form of exchangeable cations. While cation and anion exchange phenomena will be discussed in detail later (Chapters 5 and 6), it would be instructive at this point to briefly describe the CEC of important secondary minerals in soils. A major component of a soil's CEC is attributable to the secondary clay minerals. The other major component is the organic matter fraction, which will be discussed in Chapter 3. The CEC of secondary minerals is very

important in affecting the retention of inorganic and organic species.

Table 2.9 shows the CEC for important secondary minerals. The measured CEC of kaolinite is variable, depending on the soil type, degree of impurities in the clay, and the pH at which the CEC is measured. If few impurities are present, the CEC of kaolinite should be very low, <20 mmol kg^{-1}, since the degree of isomorphic substitution in kaolinite is small. Kaolinites from highly weathered soils generally have a small amount of permanent charge from Al^{3+} for Si^{4+} substitution, which results in the CEC of about 50 mmol$_c$ kg^{-1}. Simple calculations show that a substitution of a single Al atom for every 400 Si atoms in the kaolinite structure can contribute a CEC of ~ 25 mmol$_c$ kg^{-1}. Additionally, impurities of 0.1 to 10% smectitic, micaceous, and vermiculitic layers occur in kaolinite (Moore and Reynolds, 1997). Minor amounts of anatase, rutile, feldspars, quartz, and iron oxides are often present in kaolinite dominant clay fractions of soils. Selective dissolution analysis of kaolinite and separation into fine fractions to determine the degree of impurities and whether isomorphous substitution is present can be employed (Newman and Brown, 1987). Another explanation for the often unexpected high CEC that is measured for kaolinite is negative charge resulting from surface functional edge groups, such as AlOH, which are deprotonated when neutral or alkaline pH-extracting solutions (e.g., ammonium acetate) are employed in the CEC measurements.

Halloysite, which also has little, if any, isomorphic substitution, can have a higher measured CEC than expected. This is attributable to the association of halloysite with allophane (which often has a high CEC) in soils and to the large quantities of NH_4^+ and K^+ ions (which will be assumed to be exchangeable) often present between the clay layers. These ions result in an inflation of the CEC when it is measured. When they are accounted for, the CEC of halloysite is close to that of kaolinite.

Smectites have a high CEC due to substantial isomorphic substitution and the presence of fully expanded interlayers that promote the exchange of ions. Even though the layer charge of dioctahedral vermiculite is higher than that for smectites, measured CEC is often less. This may be caused by the fixation of certain ions, such as K^+, NH_4^+, and Cs^+, used in the determination of CEC. These ions have a small hydrated radius compared to other cations (e.g., Ca^{2+} and Mg^{2+}) and can be selectively adsorbed in the interlayer region of vermiculites. Furthermore, the smaller cations fit snugly into the pseudo-hexagonal spaces in the interlayers of vermiculites. The trapped ions are not always accessible for exchange, and hence the CEC values are underestimated, particularly in the common procedure where NH_4^+ is used as the index cation to measure CEC.

The measured CEC for dioctahedral vermiculite is also often affected by the presence of Al-hydroxy polymers $(Al-(OH)_x)$ in the interlayer. This material often blocks the exchange of cations in the interlayer, and consequently, the measured CEC is lower than what would be expected based on the layer charge. In trioctahedral vermiculites the interlayer cations are usually Ca^{2+} or Mg^{2+}, and there are no restrictions to complete cation exchange.

While micas have a higher layer charge than vermiculites, the measured CEC is lower because the exchangeable K^+ in the interlayer, which satisfies the negative lattice charge, is usually inaccessible to index cations used in CEC analyses. Potassium ions are "fixed" tightly in the siloxane cavities that exist between adjacent 2:1 layers, where each K^+ is complexed by 12 oxygen atoms.

The $Al-(OH)_x$ or $Mg-(OH)_x$ interlayer sheet of chlorite is positively charged, satisfying some of the negative charge. The interlayer sheet also blocks the exchange of cations. Both of these factors cause the measured CEC of chlorite to be lower than one might predict based on layer charge. Allophane has Al^{3+} substituting for Si^{4+} and a lot of edge sites that can create a very high CEC.

2.10 Points of Zero Charge

One of the most useful and meaningful chemical parameters that can be determined for a soil or soil component is a point of zero charge, which is abbreviated to PZC to denote a general point of zero charge. The PZC can be defined as the suspension pH at which a surface has a net charge of zero (Parks, 1967). If the measured pH of a solution is lower than the PZC of a colloid, the colloid's surface is net positively charged; if the pH is greater than the PZC, the surface is net negatively charged.

Soil components have a wide range of PZC values (Table 2.14). Oxides (Fe and Al) have high PZC values, while silica and SOM have

TABLE 2.14   Point of zero charge (PZC) values of soil minerals.

Mineral	PZC
Quartz	2.0
Feldspars	2.0–2.4
Birnessite	2.3
Vernadite (δ-MnO_2)	2.8
Cryptomelane ($K_2Mn_8O_{16}$)	6.4
Pyrolusite (β-MnO_2)	7.2
Kaolinite	4.5–4.6
Montmorillonite	2.5
Goethite	7.8 (7.3–8.7)
Hematite	8.5 (8.3–9.5)
Ferrihydrite	8.5 (7.9–8.7)
Gibbsite	10.0 (8.4–9.3)
Corundum	9.4 (8.5–9.6)
Rutile	5.4 (5.5–5.9)
Anatase	6.0 (5.2–6.8)

Sources: Oscarson et al. (1983); Stumm (1992); Sverjensky (2005); Kosmulski (2016). The range of PZC values for a mineral reflects different synthesis procedures and measurement procedures used by investigators.

low PZC values. The PZC values for a whole soil are strongly reflective of the individual PZC values of the soil's components. Generally, surface soil horizons have a PZC lower than that of subsoil horizons since SOM is higher on the surface, and it typically has a low PZC. A low PZC would indicate that the soil has a net negative charge over a wide pH range and thus would have the ability to adsorb cations. The PZC of a soil generally increases with depth as clay and oxide contents increase, both of which

have PZC values higher than that of SOM. Thus the soil would exhibit some positive charge and could electrostatically adsorb anions.

There are a number of points of zero charge that can be experimentally measured, as shown in Figure 2.29. The PZC is determined by measuring electrophoretic mobility, which is the movement of the colloidal particles in an applied electric field. The PZC is the pH value at which there is no net movement of the particles or the point at which flocculation occurs. It

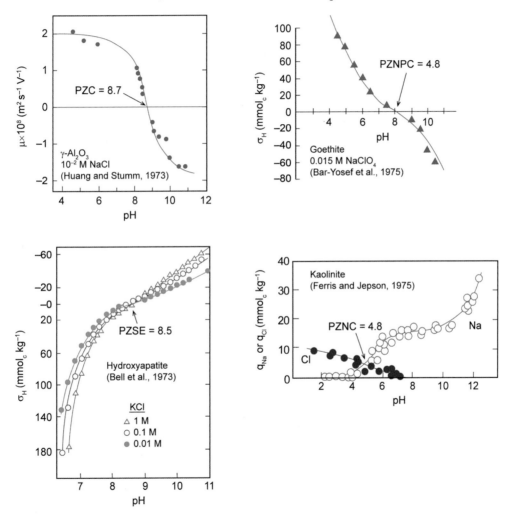

FIGURE 2.29 Experimental examples of the PZC of γ-Al$_2$O$_3$, the PZNPC of goethite, the PZSE of hydroxyapatite, and the PZNC of kaolinite (μ is the electrophoretic mobility). From Sposito (1984), with permission.

can be theoretically defined, based on the balance of surface charge, as the pH where the net total particle charge, σ_p, is 0 (Sposito, 1984).

Parks (1967) also defined the term "isoelectric point of solid" (IEPS) as the pH value at which the immersed solid oxide has a zero net charge; he considered the IEPS and the PZC as identical terms. Contemporary terminology restricts the use of the term IEP to the results obtained by using electrokinetic or electroacoustic measurement, while the term PZC is related to the surface charge density results obtained using acid–base titration (Kosmulski, 2016).

The point of zero net charge (PZNC) is the pH at which the difference in CEC and AEC (anion exchange capacity) is zero. One can experimentally measure PZNC by reacting the soil with an electrolyte solution such as NaCl and measuring the quantity of Na^+ and Cl^- adsorbed as a function of pH (Figure 2.29). The pH where the two curves intersect is the PZNC. If the electrolyte can displace only ions contributing to σ_{os} and σ_d, that is, nonspecifically adsorbed (outer-sphere complexed) ions, then the PZNC corresponds to the condition $\sigma_{os} + \sigma_d$ (Sposito, 1984).

The point of zero net proton charge (PZNPC) is another point of zero charge (Figure 2.29), which is the pH at which the net proton charge, σ_H, is 0 (Sposito, 1984). The point of zero salt effect (PZSE) is the pH at which two or more potentiometric titration curves intersect (Parker et al., 1979). This is referred to as the common intersection point (CIP), or the crossover $(\partial\sigma H/\partial I)T = 0$. This is the most commonly determined point of zero charge, and its experimental determination is shown in Figure 2.29. Details on methodologies for measuring the different PZCs can be found in Zelazny et al. (1996).

Suggested Reading

Brindley, G.W., Brown, G. (Eds.), 1980. Crystal Structure of Clay Minerals and Their X-Ray Identification. Monogr. 5. Mineralogical Society, London.

Cornell, R.M., Schwertmann, U., 2003. The Iron Oxides: Structure, Reaction, Occurrences and Uses. Wiley-VCH, Weinheim.

Dixon, J.B., Schulze, D.G. (Eds.), 2002. Soil Mineralogy with Environmental Applications. Soil Science Society of America, Madison, WI.

Dixon, J.B., Weed, S.B. (Eds.), 2002. Minerals in Soil Environments. SSSA Book Ser. 1. Soil Science Society of America, Madison, WI.

Moore, D.M., Schwertmann Jr., R.C., 1989. X-Ray Diffraction and the Identification and Analysis of Clay Minerals. Oxford University Press, New York, NY.

Newman, A.C.D. (Ed.), 1987. Chemistry of Clays and Clay Minerals, Mineral. Soc. Monogr. 6. Longman, Essex.

White, A.F., Brantley, S.L., 1995. Chemical weathering rates of silicate minerals. Rev. Miner, Vol. 31. Miner. Soc. Am., Washington, D.C.

Problem Set

Q1. Consider the data in Table 2.1 and calculate the soil enrichment factor (the ratio of the concentration of an element in the soil to its concentration in the Earth's crust) of all listed elements. Identify the outliers and explain the reasons for their different concentrations in the soil.

Q2. Based on Pauling's rules, why does Al often substitute for Si into the tetrahedral layer of phyllosilicates and Fe^{3+} does not?

Q3. Explain the following terms using schematic diagrams: (i) plane, (ii) sheet, (iii) layer, and (iv) unit structure.

Q4. Distinguish between a dioctahedral sheet and a trioctahedral sheet giving an appropriate example.

Q5. Define isomorphous substitution; what is its significance in relation to soil chemical properties?

Q6. What is the difference between clay and a clay mineral?

Q7. The classification of silicate minerals is based on the sharing of oxygens in the tetrahedra; using this principle, identify the silicate mineral class for the minerals listed below.
- **(i)** $Zn_4Si_2O_7(OH)_2.H_2O$ (hemimorphite)
- **(ii)** $MgSiO_3$ (enstatite)
- **(iii)** $Mg_7(Si_4O_{11})_2(OH)_2$ (anthophyllite)
- **(iv)** $NaAlSiO_3$ (nepheline)
- **(v)** $KMg_3(Si_3Al)O_{10}(OH)_2$ (phologopite)
- **(vi)** Zn_2SiO_4 (willemite)
- **(vii)** $Na(Mg,Fe)_3Al_6(BO_3)_3Si_6O_{18}(OH)_4$ (tourmaline)

Q8. Why don't 1:1 clay minerals expand when they are exposed to water?

Q9. Write balanced weathering reactions for the following:
- **(i)** Orthoclase $(KAlSi_3O_8)$ weathering to produce kaolinite $(Al_2Si_2O_5(OH)_4)$ and gibbsite $(Al(OH)_3)$
- **(ii)** A dioctahedral vermiculite (Ca) weathering to Al hydroxy-interlayered $\left(Al(OH)_{2.65}^{0.35+}\right)$ vermiculite
- **(iii)** Biotite weathering to produce kaolinite and goethite

Q10. How can one determine the internal surface area of clay minerals?

Q11. The substitution of a single Al atom for every 400 Si atoms in the kaolinite structure can contribute a CEC of $\sim 25\ mmol_c\ kg^{-1}$. Demonstrate this by calculating the CEC of kaolinite with this substitution, that is, one Al atom substituting for every 400 Si atoms.

Q12. Write the half unit-cell formula for montmorillonite (smectite) with a layer charge of 0.4 mol_c per half unit-cell, with 90% of the charge from the substitution in the octahedral sheet (Mg^{2+} for Al^{3+}) and 10% of the charge from the substitution in the tetrahedral sheet (Al^{3+} for Si^{4+}). The saturating or charge balancing cation is Na^+. Calculate the CEC of the mineral and express the results in millimoles(+) per kilogram. How does this value compare with values normally reported for montmorillonite?

Q13. X-ray diffraction analysis of basally oriented clay fractions of two soils showed the peaks given in the table below. Identify the phyllosilicates present in the two soils. Speculate on the presence of other minerals based on the unallocated peaks to phyllosilicates. The X-ray diffraction analysis was done using $CuK\alpha_{(1,2)}$ radiation ($\lambda = 1.5418\ Å$).

Pretreatment	Peaks (°2Θ) within the given range
Soil A	
Mg saturated and air-dried	3–30°2Θ: 6.10, 8.84, 12.36, 17.74, 20.85, 24.94, 26.67
Mg saturated and ethylene glycolated	3–15°2Θ: 5.20, 8.85, 10.40, 12.36
K saturated and air-dried	3–15°2Θ: 7.30, 8.85, 12.35
K saturated and 550°C heated	3–15°2Θ: 8.86
Soil B	
Mg saturated and air-dried	3–30°2Θ: 8.84, 12.36, 17.74, 21.26, 20.85, 24.94, 26.67
Mg saturated and ethylene glycolated	3–15°2Θ: 8.85, 12.36
K saturated and air-dried	3–15°2Θ: 8.85, 12.35
K saturated and 550°C heated	3–15°2Θ: 8.86

Q14. (i) What are Miller indices?
     (ii) What does the *hkl* (111) represent?
     (iii) The random powder diffraction peak of kaolinite shows (*hkl*) 060 at 1.49 Å; estimate the unit cell parameter "*b*" from this value.

Q15. Determine the external surface area of pyrophyllite using the BET isotherm with the following data and conditions:
- Saturated vapor pressure (P_0) = 775 atm
- Molar volume of adsorbate, N_2 = 22.4 L
- Cross-sectional surface area (σ) of N_2 = 1.62×10^{-19} m^2
- Temperature of the adsorbate = 77 K
- Sample weight = 0.854 g

P (atm)	V (L)
100.75	2.670×10^{-3}
201.50	6.275×10^{-3}
302.25	1.141×10^{-2}
403.00	1.932×10^{-2}
503.75	3.310×10^{-2}
604.50	6.316×10^{-2}
705.25	1.800×10^{-1}

References

Bergmann, J., Friedel, P., Kleeberg, R., 1998. BGMN – a new fundamental parameter-based Rietveld program for laboratory X-ray sources, its use in quantitative analysis and structure investigations. IUCr Commission on Powder Diffraction Newsletter No. 20, pp. 5–8.

Bibi, I., Singh, B., Silvester, E., 2011. Akaganéite (β-FeOOH) precipitation in inland acid sulfate soils of south-western New South Wales (NSW), Australia. Geochim. Cosmochim. Acta 75, 6429–6438.

Bigham, J.M., Fitzpatrick, R., Schulze, D.G., 2002. Iron oxides. In: Dixon, J.B., Schulze, D.G. (Eds.), Soil Mineralogy with Environmental Applications. Soil Science Society of America, Madison, pp. 323–366.

Bowen, H.J.M., 1979. Environmental Chemistry of the Elements. Academic Press, London, p. 333.

Bowen, N.L., 1922. The reaction principle in petrogenesis. J. Geol. 30, 177–198.

Brame, J., Griggs, C., 2016. Surface Area Analysis Using the Brunauer-Emmett-Teller (BET) Method. Scientific Operating Procedure Series. SOP-C. Environmental Laboratory U.S. Army Engineer Research and Development Center, Vicksburg, MS.

Brindley, G.W., Brown, G. (Eds.), 1980. Crystal Structures of Clay Minerals and Their X-Ray Identification. Mineralogical Society, London.

Brown, G., 1980. Associated minerals. In: Brindley, G.W., Brown, G. (Eds.). Crystal Structures of Clay Minerals and Their X-Ray Identification. Mineralogical Society, London, pp. 361–410.

Brunauer, S., Emmett, P.H., Teller, E., 1938. Adsorption of gases in multimolecular layers. J. Am. Chem. Soc. 60, 309–319.

Carter, D.L., Mortland, M.M., Kemper, W.D., 1986. Specific surface. In: Klute, A. (Ed.), Methods of Soil Analysis. Part 1. Physical and Mineralogical Methods. Agronomy No. 9. Second edition. ASA, SSSA Publ., Madison, WI, pp. 413–423.

Chipera, S.J., Bish, D.L., 2013. Fitting full X-ray diffraction patterns for quantitative analysis: a method for readily quantifying crystalline and disordered phases. Adv. Mater. Phys. Chem. 03, 47–53.

Churchman, G.J., Lowe, D.J., 2012. Alteration, formation, and occurrence of minerals in soils. In: Huang, P.M., Li, Y., Sumner, M.E. (Eds.), Handbook of Soil Sciences. Properties and Processes, Second edition, Vol. 1. CRC Press, Boca Raton, FL, pp. 20.1–20.72.

Cornell, R.M., Schwertmann, U., 2003. The Iron Oxides – Structure, Properties, Reactions, Occurrences and Uses, Second edition. Wiley-VCH, Weinheim.

Cullity, B.D., Stock, S.R., 2001. Elements of X-Ray Diffraction, Third edition. Prentice-Hall, Upper Saddle River, NJ.

de Caritat, P., Reimann, C., NGSA Project Team, and GEMAS Project Team, 2012. Comparing results from two continental geochemical surveys to world soil composition and deriving Predicted Empirical Global Soil (PEGS2) reference values. Earth Planet Sci. Lett. 319, 269–276.

Dixon, J.B., Schulze, D.G. (Eds.), 2002. Soil Mineralogy with Environmental Applications. SSSA Book Ser. 7. Soil Science Society of America, Madison, WI.

Dixon, J.B., Weed, S.B. (Eds.), 1989. Minerals in Soil Environments. SSSA Book Ser. 1. Soil Science Society of America, Madison, WI.

Doner, H.E., Lynn, W.C., 1989. In: Dixon, J.B., Schulze, D.G. (Eds.), Carbonate, halide, sulfate, and sulfide minerals. In Soil Mineralogy with Environmental Applications. Soil Science Society of America, Madison, WI, pp. 331–378.

Earley, J.W., Osthaus, B.B., Milne, I.H., 1953. Purification and properties of montmorillonite. Am. Mineral. 38, 707–724.

Eberl, D.D., 2003. User Guide to RockJock – Program for Determining Quantitative Mineralogy From X-Ray Diffraction Data, Report 2003-78. U.S. Geological Survey Open-File, p. 40.

Elzinga, E.J., 2021. Mechanistic study of Ni(II) sorption by green rust sulfate. Environ. Sci. Technol. 55, 10411–10421.

Fitzpatrick, R.W., Chittleborough, D., 2002. Titanium and zirconium minerals. In: Dixon, J.B., Schulze, D.G. (Eds.), Soil Mineralogy with Environmental Applications. Soil Science Society of America, Madison, WI, pp. 667–690.

Goldich, S.S., 1938. A study in rock-weathering. J. Geol. 46, 17–58.

Greenland, D.J., Quirk, J.P., 1964. Determination of the total specific surface area of soils by adsorption of cetyl pyridinium bromide. J. Soil Sci. 15, 178–191.

Guggenheim, S., Martin, R.T., 1995. Definition of clay and clay mineral: joint report of the AIPEA Nomenclature and CMS Nomenclature Committees. Clays Clay Miner. 43, 255–256.

Hammond, C., 2001. The Basics of Crystallography and Diffraction. Oxford University Press, New York.

Hang, P.T., Brindley, G.W., 1970. Methylene blue adsorption by clay minerals. Determination of surface areas and cation exchange capacities (Clay-organic studies XVIII). Clays Clay Miner. 18, 203–212.

Hsu, P.H., 1989. Aluminum hydroxides and oxyhydroxides. In: Dixon, J.B., Schulze, D.G. (Eds.), Soil Mineralogy With Environmental Applications. Soil Science Society of America, Madison, WI, pp. 331–378.

Huang, P.M., Wang, M.K., Kampf, N., Schulze, D.G., 2002. Aluminum hydroxides. In: Dixon, J.B., Schulze, D.G. (Eds.), Soil Mineralogy With Environmental Applications. Soil Science Society of America, Madison, WI, pp. 261–289.

Hudson, B.D., 1995. Reassessment of Polynov's ion mobility series. Soil Sci. Soc. Am. J. 59, 1101–1103.

Kaur, N., Gräfe, M., Singh, B., Kennedy, B.J., 2009a. Simultaneous incorporation of Cr, Zn, Cd, and Pb in the goethite structure. Clays Clay Miner. 57, 234–250.

Kaur, N., Singh, B., Kennedy, B.J., 2009b. Copper substitution alone and in the presence of chromium, zinc, cadmium and lead in goethite (α-FeOOH). Clay Miner. 44, 293–309.

Kaur, N., Singh, B., Kennedy, B.J., Gräfe, M., 2009c. The preparation and characterization of vanadium-substituted goethite: the importance of temperature. Geochim. Cosmochim. Acta 73, 582–593.

Keren, R., Sparks, D.L., 1994. Effect of pH and ionic strength on boron adsorption by pyrophyllite. Soil Sci. Soc. Am. J. 58, 1095–1100.

Klein, C., Dutrow, B., 2008. The Manual of Mineral Science, Twenty-Third edition. John Wiley and Sons, Hoboken, NJ.

Kosmulski, M., 2016. Isoelectric points and points of zero charge of metal (hydr)oxides: 50 years after Parks' review. Adv. Colloid Interfac. Sci. 238, 1–61.

Larson, A.C., Von Dreele, R.B., 1994. General Structure Analysis System (GSAS), Report LAUR 86-748. Los Alamos National Laboratory, Los Alamos, NM.

Lasaga, A.C., 2014. Kinetic Theory in the Earth Sciences. Princeton Series in Geochemistry. Princeton University Press, Princeton, NJ.

MacEwan, D.M.C., Wilson, M.J., 1980. Interlayer and intercalation complexes of clay minerals. In: Brindley, G.W., Brown, G. (Eds.), Crystal Structures of Clay Minerals and Their X-Ray Identification. Mineralogical Society, London, pp. 197–248.

Manceau, A., Gorshkov, A.I., Drits, V.A., 1992. Structural chemistry of Mn, Fe, Co, and Ni in Mn hydrous oxides. II. Information from EXAFS spectroscopy, electron and X-ray diffraction. Am. Mineral. 77, 1144–1157.

Mason, B., Moore, C.B., 1982. Principles of Geochemistry. John Wiley & Sons, New York, NY.

McBride, M.B., 1994. Environmental Chemistry of Soils. Oxford University Press Inc., New York, NY.

McKenzie, R.M., 1989. Manganese oxides and hydroxides. In: Dixon, J.B., Schulze, D.G. (Eds.), Soil Mineralogy with Environmental Applications. Soil Science Society of America, Madison, WI, pp. 437–465.

Mo, X., Siebecker, M.G., Gou, W., Li, W., 2021. EXAFS investigation of Ni(II) sorption at the palygorskite-solution interface: New insights into surface-induced precipitation phenomena. Geochim. Cosmochim. Acta 314, 85–107.

Moore, D.M., Reynolds, R.C., 1997. X-Ray Diffraction and the Identification and Analysis of Clay Minerals, Second edition. Oxford University Press, Oxford.

Newman, A.C.D., 1983. The specific surface of soils determined by water sorption. J. Soil. Sci. 34, 23–32.

Newman, A.C.D., Brown, G., 1987. The chemical composition of clays. In: Newman, A.C.D. (Ed.), Chemistry of Clays and Clay Minerals. Mineralogical Society, London, pp. 1–128.

Oscarson, D.W., Huang, P.M., Liaw, W.K., Hammer, U.T., 1983. Kinetics of oxidation of arsenite by various manganese dioxides. Soil Sci. Soc. Am. J. 47, 644–648.

Parker, J.C., Zelazny, L.W., Sampath, S., Harris, W.G., 1979. A critical evaluation of the extension of zero point of charge (ZPC) theory to soil systems. Soil Sci. Soc. Am. J. 43, 668–674.

Parks, G.A., 1967. Aqueous surface chemistry of oxides and complex oxide minerals. In: Stumm, W. (Ed.), Equilibrium Concepts in Natural Water Systems. American Chemical Society, Washington, DC, pp. 121–160.

Pauling, L., 1929. The principles determining the structure of complex ionic crystals. J. Am. Chem. Soc. 51, 1010–1026.

Pecharsky, V.K., Zavalij, P.Y., 2009. Fundamentals of Powder Diffraction and Structural Characterization of Materials, Second edition. Springer, New York, NY.

Polynov, B.B., 1937. The Cycling of Weathering. Translated From Russian by Alexander Muir. Thomas Murby and Co., London.

Rietveld, H., 1969. A profile refinement method for nuclear and magnetic structures. J. Appl. Crystallogr. 2, 65–71.

Ristori, G.G., Sparvoli, E., Landi, L., Martelloni, C., 1989. Measurement of specific surface areas of soils by p-nitrophenol adsorption. Appl. Clay. Sci. 4, 521–532.

Ross, C.S., Hendricks, S.B., 1945. Minerals of the montmorillonite group. U.S. Geological Survey Professional, pp. 23–47, Paper 205B.

Schofield, R.K., 1949. Effect of pH on electric charge carried by clay particles. J. Soil. Sci. 1, 1–8.

Schulze, D.G., 1989. An introduction to soil mineralogy. In: Dixon, J.B., Schulze, D.G. (Eds.), Soil Mineralogy With Environmental Applications. Soil Science Society of America, Madison, WI, pp. 1–34.

Schwertmann, U., Taylor, R.M., 1989. Iron oxides. In: Dixon, J.B., Schulze, D.G. (Eds.), Soil Mineralogy With Environmental Applications. Soil Science Society of America, Madison, WI, pp. 379–438.

Shacklette, H.T., Boerngen, J.G., 1984. Element concentrations in soils and other surficial materials of the conterminous United States: an account of the concentrations of 50 chemical elements in samples of soils and other regoliths. U.S. Geological Survey Professional, p. 105, Paper 1270.

Shannon, R.D., 1976. Revised effective ionic radii and systematic studies of interatomic distances in halides and chalcogenides. Acta Cryst. A32, 751–767.

Singh, B., Gilkes, R.J., 1992a. Properties and distribution of iron oxides and their association with minor elements in the soils of south-western Australia. J. Soil. Sci. 43, 77–98.

Singh, B., Gilkes, R.J., 1992b. Properties of soil kaolinites from south-western Australia. J. Soil. Sci. 43, 645–667.

Singh, B., Sherman, D.M., Gilkes, R.J., Wells, M., Mosselmans, J.F.W., 2000. Structural chemistry of Fe, Mn and Ni in synthetic hematites as determined by EXAFS spectroscopy. Clays Clay Miner. 48, 521–527.

Singh, B., Sherman, D.M., Gilkes, R.J., Wells, M., Mosselmans, J.F.W., 2002. Incorporation of Cr, Mn and Ni into goethite (α-FeOOH): mechanism from extended X-ray absorption fine structure spectroscopy. Clay Miner. 37, 639–649.

Sparks, D.L., 1987. Dynamics of soil potassium. Adv. Soil Sci. 6, 1–63.

Sposito, G., 1984. The Surface Chemistry of Soils. Oxford University Press, New York, NY.

Sproul, G., 2001. Electronegativity and bond type: predicting bond type. J. Chem. Educ. 78, 387–390.

Stumm, W., 1992. Chemistry of the Solid-Water Interface: Processes at the Mineral-Water and Particle-Water Interface. John Wiley & Sons Inc., New York, NY.

Sverjensky, D.A., 2005. Prediction of surface charge on oxides in salt solutions: revisions for 1:1 (M^+L^-) electrolytes. Geochim. Cosmochim. Acta 69, 225–257.

Taylor, R.M., 1987. Non-silicate oxides and hydroxides. In: Newman, A.C.D. (Ed.), Chemistry of Clays and Clay Minerals. Mineralogical Society, London, pp. 129–201.

Taylor, J.C., 1991. Computer programs for standardless quantitative analysis of minerals using the full powder diffraction profile. Powder Diffr. 6, 2–9.

Tournassat, C., Davis, J.A., Chiaberge, C., Grangeon, S., Bourg, I.C., 2016. Modeling the acid–base properties of montmorillonite edge surfaces. Environ. Sci. Technol. 50, 13436–13445.

Wenk, H.-R., Bulakh, A., 2004. Minerals: Their Constitution and Origin. Cambridge University Press, Cambridge.

Whittig, L.D., Allardice, W.R., 1986. X-ray diffraction techniques. In: Klute, A. (Ed.), Methods of Soil Analysis. Part 1. Physical and Mineralogical Methods. Agron. Monogr. 9, Second edition. ASA and SSSA, Madison, WI, pp. 331–362.

Wilson, M.J., 2004. Weathering of the primary rock-forming minerals: processes, products and rates. Clay Miner. 39, 233–266.

Wogelius, R.A., Walther, J.V., 1991. Olivine dissolution at 25°C: effects of pH, CO_2, and organic acids. Geochim. Cosmochim. Acta 55, 943–954.

Zelazny, L.W., Liming, H., Vanwormhoudt, A., 1996. Charge analysis of soils and anion exchange. In: Sparks, D.L. (Ed.), Methods of Soil Analysis. Part 3 – Chemical Methods. SSSA Book Series No. 5. SSSA and ASA, Madison, WI, pp. 1231–1253.

Chemistry of Soil Organic Matter

Key points: Organic matter exerts a strong control on the biological, chemical, and physical properties of the soil, which in turn affects soil fertility, soil productivity, and ecosystem services of the soil. Understanding the chemical and structural composition of SOM, its decomposition and stabilization processes, and the chemical reactions of SOM is paramount for sustainable soil management and stabilizing atmospheric CO_2 concentration.

Similar to the inorganic components of soil, soil organic matter (SOM) plays a significant role in affecting the chemistry of soils. Despite extensive and important studies, the molecular structure and chemistry of SOM is still not well understood. Advanced spectroscopic and spectrometric techniques utilized in the past decade have addressed some of these issues. However, because of the variability of SOM and its close relationship/association with clay minerals and metal oxides, the chemistry and reactions it undergoes with metals and organic chemicals are complex. In this chapter background discussions on SOM content and function in soils and its composition, fractionation, structure, and intimate association with inorganic soil components will be covered. Additionally, environmentally important reactions between SOM and metals and organic contaminants will be discussed. For further in-depth discussions on these topics, the reader is referred to the suggested readings at the end of this chapter.

3.1 Introduction

Key points: SOM is a complex mixture of organic compounds of plant and animal tissues at different decomposition stages, live faunal and microbial biomass, and black C derived from fires. Many soil properties, such as soil aggregation, nutrient cycling, cation exchange capacity, and the adsorption of organic compounds, are largely governed by SOM.

SOM has been defined as "the sum of all naturally derived organic materials present in soils" (Baldock and Broos, 2011). Other definitions of SOM include the total organic compounds in soils, excluding undecayed plant and animal tissues, their "partial decomposition" products, and the soil biomass (Stevenson, 1982). Schnitzer and Khan (1978) note that SOM is "a mixture of plant and animal residues in different stages of decomposition, substances synthesized microbiologically and/or chemically from the breakdown products, and the bodies of live and dead microorganisms and their decomposing remains." Humus includes humic substances (HS) plus resynthesized products of microorganisms that are stable and a part of the soil.

Common definitions and terminologies for these are given in Table 3.1. There is no simple method to measure SOM and it is estimated from the organic C concentration as explained in Sidebar 3.1.

Environmental Soil Chemistry, Third Edition
https://doi.org/10.1016/B978-0-12-815880-7.00003-1

TABLE 3.1   Definitions of soil organic matter (SOM), organic materials, and humic substances.

Term	Definition
Organic residues	Undecayed plant and animal tissues and their partial decomposition products.
Litter	Macroorganic matter (e.g., crop residues) that lies on the soil surface.
Light fraction	Plant residues and their partial decomposition products that reside within soil itself.
Microbial biomass	Organic matter present as live microbial tissue, notably bacteria, actinomycetes, and fungi.
Belowground plant constituents	Primarily live roots with lesser amounts of dead roots and exudates.
Humus	Total of the organic compounds in soil exclusive of undecayed plant and animal tissues, their "partial decomposition" products, and the soil biomass.
Humified products	Transformed organic materials, humus, which bear no morphological resemblances to the structures from which they are derived.
Soil organic matter	Same as humus.
Humic substances	A series of relatively high-molecular-weight, brown- to black-colored substances formed via secondary synthesis reactions. The term is used as a generic name to describe the colored material, or its fractions obtained on the basis of solubility characteristics. These materials are distinctive to the soil (or sediment) environment in that they are dissimilar to the biopolymers of microorganisms and higher plants (including lignin). Humic substances are differentiated based on their solubility characteristics into humic acid, fulvic acid, and humin.
Nonhumic substances	Compounds belonging to known classes of biochemistry, such as amino acids, carbohydrates, fats, waxes, resins, and organic acids. Humus probably contains most, if not all, of the biochemical compounds synthesized by living organisms.
Humin	The alkali-insoluble fraction of soil organic matter or humus.
Humic acid	The dark-colored organic material that can be extracted from soil by using various reagents and is insoluble in dilute acid.
Fulvic acid	The colored material that remains in solution after removal of humic acid by acidification.
Hymatomelanic acid	Alcohol-soluble portion of humic acid.
Water-soluble organics	Organic substances dissolved in the soil solution.

From Stevenson and Cole (1999), with permission. From Hayes and Swift (2018), with permission. From Stevenson (1982), with permission.

Sidebar 3.1 — SOM and SOC

Soil organic matter (SOM) is frequently used to describe all organic materials in soils in the literature. However, there is no simple procedure to measure SOM content. In practice, soil organic carbon (SOC) concentration is measured using a dry or wet oxidation method, which is then converted to SOM concentration using a conversion factor. A conversion factor of 1.724 is typically used to convert SOC concentration to SOM concentration. The origin of the factor is based on the analysis of "Humussäure" or humus acid (58% C) by Sprengel in 1826 and several other studies.

A comprehensive analysis of SOC data suggests that a conversion factor of 2.0 (i.e., 50% C in organic matter) is more accurate than the conventional factor of 1.724. However, it is best to report analyzed SOC values instead of SOM values.

Soil organic C (SOC) concentrations (and not content; see Sidebar 3.2) range from 1 to 10% on a weight basis in the surface horizon of mineral soils to 100% in organic soils (Histosols). In soils of the prairie regions, Mollisols, SOC concentration may be as high as 10%, while in sandy soils, for example, those of the Atlantic Coastal Plain of the United States, the concentration is often <2.0%. Even at these low levels, the reactivity of SOC is so high that it has a pronounced effect on soil chemical reactions.

Some of the general properties of SOM and its effects on soil chemical and physical properties are given in Table 3.2. It improves soil structure, water-holding capacity, aeration, and aggregation. It is an important source of macronutrients such as N, P, and S and of micronutrients such as B and Mo. It also contains large quantities of C, which provides an energy source for soil macroflora and microflora. The C/N ratio of soils is about 10–15:1.

SOM has a high specific surface area (as great as 800–900 $m^2 g^{-1}$) and a CEC that ranges from 1500 to 3000 $mmol_c kg^{-1}$. Thus the majority of a

TABLE 3.2 General properties of soil organic matter and associated effects in the soil.

Property	Remarks	Effect on soil
Color	The typical dark color of many soils is caused by organic matter.	May facilitate warming.
Water retention	Organic matter can hold up to 20 times its weight in water.	Helps prevent drying and shrinking. May significantly improve the moisture-retaining properties of sandy soils.
Combination with clay minerals	Cements soil particles into structural units called aggregates.	Permits exchange of gases. Stabilizes structure. Increases permeability.
Chelation	Forms stable complexes with Cu^{2+}, Mn^{2+}, Zn^{2+}, and other polyvalent cations.	May enhance the availability of micronutrients to higher plants.
Solubility in water	Insolubility of organic matter is because of its association with clay. Also, salts of divalent and trivalent cations with organic matter are insoluble. Isolated organic matter is partly soluble in water.	Little organic matter is lost by leaching.
Buffer action	Organic matter exhibits buffering in slightly acid, neutral, and alkaline ranges.	Helps to maintain uniform reactions and pH in the soil.
Cation exchange	Total acidities (charge) of isolated fractions of humus range from 3000 to 14,000 $mmol_c kg^{-1}$.	May increase the CEC of the soil. An increase from 20% to 70% in the CEC of many soils (e.g., Mollisols) is due to organic matter.
Mineralization	Decomposition of organic matter yields CO_2, NH_4^+, NO_3^-, PO_4^{3-}, and SO_4^{2-}.	A source of nutrient elements for plant growth, particularly the major nutrients N, P, and S.
Combines with organic chemicals	Affects bioactivity, persistence, and biodegradability of pesticides and other organic chemicals.	Modifies application rate of pesticides for effective control.

From Stevenson (1982) with permission.

surface soil's CEC is, in fact, attributable to SOM, particularly in lighter-textured soils. Due to the high specific surface area and CEC of SOM, it is an important sorbent of plant nutrients, trace element cations, and organic compounds such as pesticides. The uptake and availability of plant nutrients, particularly micronutrients such as Cu and Mn, and the effectiveness of herbicides are greatly affected by SOM. For example, manure additions can enhance micronutrient availability in alkaline soils where the precipitation of the micronutrients at high pH reduces their availability. The complexation of low-molecular-weight SOM components such as fulvic acids (FA) with metals such as Al^{3+} and Cd^{2+} can decrease the uptake of metals by plants and their mobility in the soil profile.

Sidebar 3.2 − SOC Content and Concentration

The results of SOC analysis are traditionally expressed as a percentage of the soil mass, i.e., g C 100 g^{-1} soil or mg C kg^{-1} soil. These are units of concentration and should not be referred to as C content. The content or stock expression is correct when the total quantity of C is determined in the soil mass for a given area and to a given depth, where SOC content is generally expressed Mg C ha^{-1} to the defined depth (such as 15 cm). Alternatively, SOC content may be expressed as a C stock per unit area.

3.2 Measurement of Soil Organic Carbon

Organic carbon in the soil is measured using the dry combustion method in well-equipped laboratories. In this method a small amount (<1 g) of dry and finely ground sample is combusted at temperatures exceeding 1000°C in the presence of purified oxygen and catalysts. All forms of carbon present in the sample are converted to CO_2, which is measured by using a

detector, such as a thermal conductivity detector, infrared, or sometimes mass spectrometer. For the samples containing inorganic carbon, that is, carbonates and bicarbonates, acid treatment is conducted to convert carbonate to CO_2 ($CaCO_3 + HCl = CaCl_2 + H_2O + CO_2$), and total carbon is measured again using the above procedure, or a separate procedure (e.g., calcimeter method) could also be used to determine inorganic carbon. Then the concentration of organic carbon is determined by subtracting inorganic carbon from the total carbon.

A wet chemical oxidation procedure is still used in many laboratories. In this procedure soil organic carbon is oxidized by mixing soil with potassium dichromate ($K_2Cr_2O_7$) and sulfuric acid, and in some procedures phosphoric acid is also used. Some variants in the wet oxidation method apply additional heating to obtain complete oxidation (Nelson and Sommers, 1996). The chemical reaction for the oxidation of organic carbon by chromate ($Cr_2O_7^{2-}$) can be written as follows:

$$2K_2Cr_2O_7 + 3C^0_{(organic)} + 8H_2SO_4 \rightarrow 4Cr_2(SO_4)_3$$
$$+ 3CO_2 + 2K_2SO_4 + 8H_2O$$

$$3.1$$

(The equation above shows that 2 moles $K_2Cr_2O_7$ oxidize 3 moles of carbon.)

Soil organic carbon is then calculated (i) by measuring the remaining (unreduced) dichromate by titrating with ferrous sulfate or ammonium ferrous sulfate in the presence of diphenylamine or o-phenanthroline-ferrous complex indicator, and (ii) the moles of $K_2Cr_2O_7$ consumed in the oxidation of organic carbon is calculated from the difference in titration volume of the blank and sample. The chemical reaction for the titration step is given below:

$$K_2Cr_2O_7 + 6FeSO_4 + 7H_2SO_4 \rightarrow Cr_2(SO_4)_3$$
$$+ K_2SO_4 + 3Fe_2(SO_4)_3 + 7H_2O \qquad 3.2$$

(The equation above shows that 1 mole $K_2Cr_2O_7$ oxidizes 6 moles of $FeSO_4$)

The main drawbacks of this method are the incomplete recovery of organic carbon and the use of hazardous, carcinogenic, and strong oxidant chemicals.

The loss of ignition (LOI) method has been used to estimate soil organic carbon concentrations in certain situations, such as in remote areas without laboratory access, comparative analysis of very large numbers of samples, or soil samples with relatively high OC concentrations. In this method OC is estimated based on the weight loss in the sample after heating at between 105°C and 550°C. The method relies on the assumption that the loss of weight is due entirely to the destruction of organic matter. This method gives erroneous results in soils that contain minerals, such as gypsum, gibbsite, ferrihydrite, and goethite, which lose structural water within this temperature range.

In the last few decades SOC (and other soil properties) estimation based on spectral reflectance in the visual (400–700 nm), near-infrared (NIR, 700–2500 nm), and mid-infrared (MIR, 2500–25,000 nm) regions have shown good promise. These methods are relatively cheap and fast and do not require expensive reagents or consumables and thus are useful in high-throughput scenarios. The interaction of electromagnetic radiation with molecular bonds in inorganic and organic soil components is reflected in the spectra. The MIR spectra provide information on the fundamental bands of molecular vibrations, and NIR spectra show weak overtones and combinations of fundamental bands and electronic excitations in the visible regions. Chemometric methods (e.g., partial least-squares regression) are used to predict SOC concentration from the spectral data. Spectral-based estimation of SOC requires building a prediction model consisting of many samples, where soil samples are from the same or similar region and spectra obtained on the same instrument. A major limitation in the application of spectroscopic techniques has been the transfer of calibration models between geographic areas, instruments, and the inconsistency of reference data (i.e., dry combustion, wet oxidation, or LOI) used in developing calibrations.

3.3 Effect of Soil Forming Factors on SOM Contents

Key points: Soil forming factors, namely, climate, organisms, relief, parent materials, and time, all have major influences on SOM content. Both the input and output of OM in the soil are influenced by these factors.

SOM stocks reflect the net difference between the total input and output of OM in a soil. The primary source of SOM input is plants that add above- (shoots) and belowground (roots) litter, root exudates, and C through symbiotic associations with soil microbes (e.g., nitrogen-fixing and mycorrhizal). The output or loss of SOM occurs through microbial mineralization, erosion, and surface runoff. The quantity of SOM in a soil depends on the five soil forming factors first espoused by Jenny (1941): time, climate, vegetation, parent material, and topography. These five factors determine the equilibrium level of SOM after a period of time. Of course, these factors vary for different soils, and thus SOM accumulates at different rates and therefore in varying quantities.

The accumulation rate of SOM is usually rapid initially, declines slowly, and reaches an equilibrium level varying from 110 years for fine-textured parent material to as high as 1500 years for sandy materials. The equilibrium level is attributed to organic colloids that are produced, which are resistant to microbial attack, and the stability of organic compounds via interactions with soil minerals, where SOM may be inaccessible to microorganisms and/or degrading enzymes (Stevenson and Cole, 1999). There is increasing evidence that attributes the long-term preservation of SOM to the sorption reactions of organic species with minerals, which protect its degradation through microbes and enzymes. Additionally, low concentrations of one or more essential nutrients such as N, P, and S may limit the quantity of stable humus that can be synthesized by soil organisms (Stevenson, 1982).

Climate is an extremely important factor in controlling SOM contents because it affects the type of plant species, the amount of plant material produced, and the degree of microbial activity. Climate and precipitation are two key drivers of SOM stocks at global and regional scales. Precipitation determines net primary productivity, that is, the input of OM into the soils, whereas temperature affects the microbial decomposition of SOM. Typically, a humid climate leads to a forest ecosystem, while a semi-arid climate creates a grassland ecosystem. Soils formed under grasses usually have the highest SOM content, while desert, semiarid, and tropical soils have the lowest quantities of SOM. However, some tropical soils often contain high quantities of OM, even though they are highly weathered. This is due to the formation of complexes between the OM and inorganic constituents, such as phyllosilicates, Fe and Al oxides, and amorphous materials (organo-mineral

complexes), which are quite stable. In a complexed form the SOM is less susceptible to microbial attack (Stevenson, 1982).

The SOM stocks are generally largest under cool, humid conditions and decrease under warmer and drier climates. Soils (0−100 cm) in north temperate and boreal regions store between 53% and 74% of the total SOM soil stocks (Table 3.3). However, with increasing warming in the Arctic region, there is increasing concern about the loss of SOM from permafrost soils, which could further increase levels of atmospheric carbon dioxide. While OM in permafrost soils is thought to be resistant to biodegradation, recent research has shown that fractions of OM are prone to decomposition and transport upon thawing (Schuur et al., 2008; Drake et al., 2015). Similar to soils of other regions, OM complexation with soil minerals is a primary mechanism for the preservation of C in soils of the Arctic region. Sowers et al. (2020) investigated C associations with minerals in a permafrost chronosequence in Alaska. Soil cores that ranged from 19,000 to 36,000 years in age were sampled and analyzed via STXM (scanning transmission X-ray microscopy) and C NEXAFS (near-edge extended X-ray absorption fine structure spectroscopy). A strong association was observed

TABLE 3.3   Total organic carbon stocks (Pg C) in the 0−100 cm soil layer at regional and global scales estimated using different global databases (Tifafi et al., 2018). The numbers in parentheses (percentage) are for the regional distribution of the global carbon stock.

Region	SoilGrids[a]	HWSD_SOTWIS[b]	HWSD_SAXTON[c]	NCSCD[d]
Global	3421	2439	2798	−
Boreal (60°N−90°N)	1161 (34%)	390 (16%)	807 (29%)	290
North Temperate (30°N−60°N)	1376 (40%)	890 (37%)	1227 (44%)	−
Tropical (30°S−30°N)	865 (25%)	1061 (44%)	696.6 (25%)	−

Vegetation also has a profound effect on SOM contents. The vegetation composition influences SOM input via above- and belowground productivity and allocation, rooting depth, and horizontal root expansion.
[a] SoilGrids is a global soil information system that contains spatial predictions for several soil properties, including organic carbon at seven standard depths (i.e., 0, 5, 15, 30, 60, 100, and 200 cm); the system uses approximately 110,000 world soil profiles to generate this data set.
[b] HWSD_SOTWIS — Harmonized World Soil Database where bulk density values were estimated by using soil type and depth.
[c] HWSD_SAXTON — Harmonized World Soil Database where bulk density values were calculated from equations developed by Saxton et al. (1986).
[d] NCSCD — Northern Circumpolar Soil Carbon Database.

between C and Fe-bearing minerals, with the Fe minerals appearing to provide a substantial framework for C association in the permafrost soils. Additionally, Ca seemed to be playing a role in bridging the C−Fe complexes (Sowers et al., 2020).

Grassland soils generally accumulate large amounts of SOM because large amounts of plants are produced in grassland settings, there is inhibition of nitrification that preserves N and C, and the high base saturation content of grassland soils promotes NH_3 fixation by lignin (Stevenson, 1982). Also, it has been suggested that a greater proportion of fine roots in graminoids compared to forbs or shrubs protects OM in soil aggregates (Garcia-Pausas et al., 2017). Vegetation change associated with long-term land use changes may result in chemical compositional shifts in SOM. The change in the vegetation composition, particularly the presence of legumes, can have a significant influence on the microbial activity and community structure in the soils. A global study comprising 818 plant species found an 18.4-fold range in the decomposition rate of plant residues, which suggests a predominant control of plant species traits on the rate of decomposition of SOM (Cornwell et al., 2008).

The main effect of parent material on SOM content is the manner in which it affects soil texture and mineral composition. Clayey soils have larger SOM stocks than sandy soils. Organic matter associated with clay and fine silt particles has longer residence times and is older than the organic matter in the coarser soil fractions. The type of clay mineral is also important. Different adsorption mechanisms are involved in the interaction of organic compounds with clay minerals. Permanent charge clay minerals, such as smectites, adsorb organic compounds by forming outer-sphere complexes involving polyvalent cation bridges. In the case of variable charge minerals (e.g., Fe and Al oxides) and variable charge surfaces of phyllosilicates inner-sphere complexation or ligand exchange is the main mechanism for the adsorption of organic matter compounds. Weak interactions involving H-bonding, van der Waals forces, and hydrophobic interactions are important in the interaction of nonpolar, uncharged, and high-molecular-weight organic compounds with soil clays.

Topography, or the lay of the landscape, affects the content of SOM via climate, runoff, evaporation, and transpiration. Moist and poorly drained soils are high in SOM since organic matter degradation is lessened due to the anaerobic conditions of wet soil. In the northern hemisphere soils on north-facing slopes, which are wetter and have lower temperatures, are higher in SOM than soils on south-facing slopes, which are hotter and drier, and the opposite effects are expected in the southern hemisphere (Stevenson, 1982; Bohn et al., 1985).

Cultivating soils also affects the content of SOM. When soils are first cultivated, SOM usually declines. In soils that were cultivated for corn production it was found that about 25% of the N was lost in the first 20 years, 10% in the second 20 years, and 7% during the third 20 years (Jenny et al., 1948). Long-term and more recent predictions show that 30%−70% of SOM is lost in the first 50 years after land use changes from prairie or forest to production agriculture in the United States. (Gollany et al., 2011). This decline is not only due to less input of plant residues but also due to improved aeration resulting from cultivation. The improved aeration results in increased microbial activity and greater mineralization of SOM. Wetting and drying of the soil also causes increased respiration, which decreases the amount of SOM (Stevenson, 1982).

3.4 Global Carbon Cycle and Carbon Sequestration

Key points: The concentration of CO_2 in the atmosphere has been increasing because of industrialization and other anthropogenic activities. Soil organic matter (SOM) is the largest terrestrial C pool and thus has a strong influence on atmospheric CO_2 concentration and climate change. Globally, there is a potential to increase SOM stocks in mineral soils between 0.4 and 8.64 Gt CO_2 year^{-1}.

Atmospheric C concentration was approximately 280 ppm in the preindustrial era (1750) and had increased to 417 ppm in 2021 based on the observation at Hawaii's Mauna Loa Atmospheric Baseline Observatory. There is a near-proportional relationship between global mean surface air temperature change and cumulative CO_2 emissions. Total anthropogenic greenhouse gas (GHG) emissions were estimated to be 52 ± 4.5 Gt CO_2-equivalents per year (IPCC, 2019). Of the total anthropogenic GHG emissions, CO_2 accounts for 75% (39.1 ± 3.2 Gt CO_2-eq year^{-1}), methane (CH_4) ~19% (10.1 ± 3.1 Gt CO_2-eq year^{-1}), and nitrous oxide (N_2O) 6.2% (2.8 ± 0.7 Gt CO_2-eq year^{-1}), and the rest comes from fluorinated gases (IPCC 2019).

The global C flux data (Gt C year^{-1}) averaged over the 1959–2017 period show that 82% of the total CO_2 emissions were caused by fossil CO_2 emissions and the rest (i.e., 18%) resulted from land use and land use changes (Quéré et al., 2018). The emitted C is partitioned among the three reservoirs, namely, the atmosphere (45%), land (30%), and ocean (24%).

The challenge is to curb emissions of CO_2 and other GHGs in order to limit warming to less than 2°C relative to the preindustrial era. Several negative emissions technologies have been suggested (Smith et al., 2015). These include (i) bioenergy with C capture and storage; (ii) direct air capture of CO_2 from ambient air by engineered chemical reactions; (iii) enhanced weathering of minerals to remove CO_2 from the atmosphere and store the products in soils or bury them in land or the deep ocean; (iv) afforestation and reforestation to fix atmospheric C in biomass and soils; (v) increasing ocean C uptake via biological (fertilizing nutrient-limited areas) or chemical (by enhancing alkalinity) manipulations; (vi) altered agricultural practices to increase C storage in soils; and (vii) converting biomass to recalcitrant biochar for use as a soil amendment.

The global C cycle is presented in Figure 3.1. The C cycle consists of the exchange of C fluxes between terrestrial (soil including permafrost, vegetation, atmosphere), fossil, and ocean reservoirs. The exchange of C between various reservoirs occurs primarily through the exchange of CO_2 gas resulting from the formation and decomposition of organic matter and inorganic C. Microbial respiration and fires are the main agents for C fluxes from organic matter, whereas photosynthesis is the main process for the terrestrial organic C sink. Precipitation and dissolution are the main processes for the flow of inorganic C between reservoirs.

The ocean pool, estimated at 38,000 Pg (petagrams = 1×10^{15} g = 1 billion metric tons), is the largest C pool. Among the terrestrial C pools, soil stores the largest amount of C (organic C only), stored primarily in SOM, at between 2400 and 3400 Pg (to 1 m depth), the atmospheric pool at 890 Pg (see Box 3.1), and the vegetation pool at 450–650 Pg.

In addition to the SOC pool, there is a soil inorganic C (SIC) pool that ranges from 695 to 748 Pg of C in mass and is most important in the subsurface horizons of arid and semiarid soils (Batjes, 1996). The sources of the SIC pool are primary (lithogenic) carbonates and pedogenic (secondary) carbonates, the latter being more important in C sequestration. The pedogenic carbonates are formed when H_2CO_3 chemically reacts with Ca^{2+} and/or Mg^{2+} in the soil solution in the upper portion of the profile and is then leached to lower soil horizons via irrigation/precipitation. The rate of SIC sequestration via this mechanism may be 0.25–1 Mg C ha^{-1} year^{-1} (Wilding, 1999). Accordingly, the role that soils, particularly SOM, play in the global C cycle is immense, both in serving as a pool in sequestering C and also as a flux in releasing C.

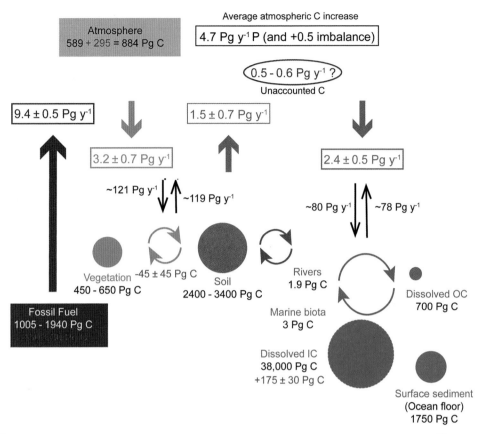

FIGURE 3.1 A schematic representation of the global C cycle showing C stocks (1 Pg C = 1 Gt C = 10^{15} g C) in terrestrial (soil including permafrost, vegetation, atmosphere), fossil, and ocean reservoirs. The changes in fossil fuel, atmosphere, soil/vegetation, and oceanic pool stocks since the Industrial Era (1750) are in red numbers. Annual carbon exchange (in Pg C year^{-1}) between the terrestrial, atmosphere, and ocean pools is indicated by *black arrows*. Large *colored arrows* show annual "anthropogenic" fluxes of C averaged over the 2008–2017 period. The *dark brown arrow* represents emissions from fossil fuels (coal, oil, and gas), cement, and gas flaring; the orange arrow represents CO_2 emissions from land-use change activities, largely from deforestation. About 43% of the anthropogenic C accumulates in the atmosphere and the remaining C is absorbed by terrestrial ecosystems (*green arrow*) and oceans (*blue arrow*), so called "carbon sinks." There is a small budget imbalance (in the *purple box*), which is due to an overestimation of the C emissions and/or underestimation of the C sinks. The data in the figure are based on Ciais et al. (2013), Quere et al. (2018), and Tifafi et al. (2018).

3.4.1 Land Use and SOM

Key points: Land use change affects C stocks in the soil, with about a 50% decrease in SOC stock after land use changes from native forest and pasture to crops and an increase (18% −53%) in SOC stocks from a change in land use from crops to forest or pasture.

Land use and crop and soil management have drastic effects on the SOC stocks and thus soil C sequestration. Both an increase and decrease in SOC stocks have been found depending on land use change and the management practices used. Soil disturbance causing the breakdown of soil aggregates and ambient environmental changes resulting from a land

---

BOX 3.1

How much carbon (by mass) is in the atmosphere?

The atmospheric concentration of CO_2 concentration by volume in 2021 = 417 ppm

The atmospheric concentration of CO_2 concentration by weight in 2021 = 417 × [44.009/28.9647] = 633.59 mg kg^{-1} (ppm)

where molar mass of CO_2 = 44.009 g mole^{-1}, and mean molar mass of air = 28.9647 g mole^{-1}

According to Trenberth and Smith (2005), the total mean mass of the atmosphere is = 5.1480 × 10^{18} kg.

Thus the total weight of CO_2 in preindustrial era = 633.59 × 5.1480 × 10^{18} = 3261.72 × 10^{18} mg = 3261 Pg

Therefore the total mass of CO_2 is 3261 Pg, which is equivalent to 890 Pg of carbon.

---

use change have profound effects on both the input and decomposition rate of organic C. Based on a metaanalysis, Guo and Gifford (2002) showed that soil C stocks decline after land use changes from native forest to crops (about 50% up to a 60 cm depth) and pasture to crops (50% or more) and conversely SOC stocks increased after land use changes from crops to secondary forest (+53%) and crops to pasture or plantation (+18%−19%).

The decline in SOC after land use conversion from native vegetation to cropping is well known and largely contributes to (i) reduced input of organic matter, (ii) increased decomposability of organic residues, and (iii) tillage that decreases the physical protection of organic matter in soil aggregates (Post and Kwon, 2000). In addition to these, SOC can be lost through erosion processes and its leaching into subsoil or groundwater. The rate of SOC loss due to land use change from natural to agricultural ecosystems is more significant and rapid in tropical than in temperate-region soils, is greater from cropland than from pastureland, and is greater from soils with high SOC levels than with low initial levels (Mann, 1986).

Changes in agricultural management practices, such as conservation tillage or no-till and using legumes in crop rotation, can help in increasing SOC stocks in agricultural soils. The use of limited or no-till can dramatically reduce C losses from soils by reducing mineralization and erosion and promoting C sequestration. A global study comprising 67 long-term agricultural experiments estimates that a change from conventional tillage to no-tillage has the potential to sequester 57±14 g C m^{-2} year^{-1}, excluding wheat−fallow systems (West and Post, 2002). This study also found that increased crop rotation complexity (i.e., changing from monoculture to continuous rotation) can sequester 20 ± 12 g C m^{-2} year^{-1}, excluding a rotation from continuous corn to corn−soybean. Under a no-till system, SOC accumulates in the top 15−30 cm soil, which leads to a highly stratified vertical distribution of SOC. The effect of a no-till management system on SOC accumulation is slow and a peak sequestration rate is observed in 5−10 years, and the accumulation is negligible after 15−20 years.

Despite the gains in the soil C pool and thus C sequestration resulting from no-tillage agriculture and cover crops, these must be balanced by considering CO_2 fluxes due to the manufacture of applied inorganic N fertilizers and irrigation of crops (Schlesinger, 1999). Carbon dioxide is produced in N fertilizer production (0.375 moles of C per mole of N produced), and fossil fuels are used in pumping irrigation water. Additionally, the groundwater of arid regions can contain as much as 1% dissolved Ca and CO_2 versus 0.04% in the atmosphere. When such waters are applied to arid soils, CO_2 is released into the atmosphere $\left(Ca^{2+} + 2HCO_3^- \rightarrow CaCO_3\downarrow + H_2O + CO_2\uparrow\right)$.

Afforestation, reforestation, agroforestry, and the management of organic C in mineral soils are suggested to be effective land-based options for increasing C sequestration. The improved management of cropland and grazing lands, using practices such as perennial crops and pastures and cover crops, has good potential to increase SOC without requiring land use change. Similarly, improved and sustainable forest management and the management of fires can increase C sequestration in soils.

Carbon sequestration can also be significantly enhanced by restoring soils degraded by erosion, desertification, salinity, compaction, and mining operations. Agronomic practices, such as increasing SOM, decreasing soil erosion, and improving crop and fertilizer management, have the potential to increase SOC and contribute to climate change adaptation and mitigation in cropland.

3.5 Composition of SOM

Key points: Carbon, O, H and N are the main constituents of SOM. The chemical composition of SOM is generally consistent, and therefore the total concentrations of important nutrient elements (N, P, and S) can be deduced from the organic C concentration and approximate ratio of C:N:P:S in soils. SOM consists of nonhumic substances (e.g., amino acids, carbohydrates, lipids), humic substances (or humified products), and black C.

3.5.1 Composition and Constituents of SOM

The main constituent elements of SOM are C (52%–58%), O (34%–39%), H (3.3%–4.8%), and N (3.7%–4.1%). As shown in Table 3.4, the elemental composition of humic acid (HA)

TABLE 3.4 The elementary composition of humic acids from different soils.[a]

Soil	Percentage				Ratio		
	C	H	N	O	C/N	C/H	O/H
A[b]	52.39	4.82	3.74	39.05	14.0	10.9	8.1
B	57.47	3.38	3.78	35.37	15.2	17.0	10.4
C	58.37	3.26	3.70	34.67	15.7	17.9	10.6
D	58.56	3.40	4.09	33.95	14.3	17.2	10.0

[a] *From Kononova (1966).*
[b] *Soils A, B, C, and D represent soils of varying genesis, taxonomy, and physico-chemical properties.*

from several soils is similar. A comprehensive review of SOM data from the last three centuries shows that the C concentration of SOM is closer to the lower end (median = 50.8% and mean = 51.3%) of the range in Table 3.4.

The C/N ratio (mass) in SOM is between 14 and 16 (Table 3.4). Other prominent elements in SOM are P and S. Stevenson and Cole (1999) suggested an average C:N:P:S stoichiometry equivalent to 108:8:1:1 (rounded values, by mass and only organic forms of P and S used) in soil humus or SOM, excluding plant and animal residues or microbial biomass. Contrastingly, Kirkby et al. (2011) calculated a stoichiometry of 52:5:1:1 (C:N:P:S, on a mass basis) using data for mineral soils from 22 countries.

In a recent and more comprehensive analysis of more than 2000 globally distributed soil samples from all soil horizons Tipping et al. (2016) showed that the stoichiometric relationships of C to N, P, and S in SOM vary between the two end-members, that is, nutrient-poor SOM and nutrient-rich SOM. The elemental ratio of SOM across a range of ecosystems and all horizons can be described using a simple mixing model, with a C:N:P:S stoichiometry of 909:35:1:5 (rounded values, by mass) and 63:8:1:1 for nutrient-poor SOM and nutrient-rich SOM, respectively. The major exception in this model is tropical soils, which generally have very low organic P concentrations. A simple approach like this can serve as a useful guide for SOM characterization, SOM turnover, and nutrient dynamics in soils.

SOM, excluding organic residues and biomass, consists of nonhumic and humic substances (or humified products). Additionally, black C has been recognized as an important and sometimes significant component of OM in the soil. The nonhumic substances have recognizable physical and chemical properties and consist of carbohydrates, proteins, peptides, amino acids, fats, waxes, and low-molecular-weight acids. These compounds are attacked easily by soil microorganisms and persist in the soil only for a brief time.

The amounts of nonhumic and humic substances in soils differ substantially (Table 3.5). The amount of lipids can range from 2% in forest soil humus to 20% in acid peat soils. Similarly, amino acids, including proteins and peptides, may vary from 15% to 45% and carbohydrates from 5% to 25%. Humic substances range from 33% to 75% of the total SOM, with grassland soils having higher quantities of HA and forest soils having higher amounts of FA (Stevenson, 1982). Among the humic compounds, humin contains the highest C concentration, whereas lipids have the highest C concentration among the nonhumic compounds. Most of these SOM constituents are not water soluble, but they are soluble in strong bases.

In some soils a considerable portion of the organic matter can consist of finely distributed charred organic matter (black carbon, charcoal) originating from vegetation fires. The data on black C concentration in soils are limited, but in some soils up to 45% of the organic C could be present as black C. Natural fires lead to an incomplete combustion of organic materials that leaves behind a continuum of charred aromatic structures condensed to different degrees. Both the chemical recalcitrance and interactions of the charred structures with minerals protect them from rapid degradation, and the residence

TABLE 3.5 The composition of soil organic matter in humic, nonhumic substances, and black C and their C concentrations.[a]

Component	Range present in SOM (% by weight)	Range of C in component (% by weight)
Humic substances		
Humic acid	10–20	54–59
Fulvic acid	10–40	41–51
Humin	8–50	62–67
Nonhumic substances		
Amino acids (includes peptide and proteins)	15–45	33–65[b]
Carbohydrates	5–25	40–45
Lipids	2–6[c]	69–79[d]
Black or pyrogenic carbon	0–60 (mean 13.7)[e]	50–90

[a] Pribyl (2010).
[b] Median value is 46% and the range represents the 95th percentile range of the published values.
[c] In acid peats the value may be as high as 20%.
[d] Median value is 78% and the range represents the 95th percentile range of the published values.
[e] The range of black C is in SOC and not SOM (Reisser et al., 2016).
From Stevenson and Cole (1999), with permission.

time of black C in soils can range from a few hundred to more than 10,000 years.

3.5.2 Nonhumic Substances

(a) Carbohydrates

Structural components of plants, such as simple sugars, cellulose, hemicelluloses, pectin, and root exudates, contribute to carbohydrates in soils. Cellulose is the most abundant biopolymer in the soil, being the major structural component of most plants and also produced by some bacteria. Cellulose constitutes up to about 45% of the dry weight of wood. Cellulose is composed of D-glucose monomers, which are joined by a β-1-4 glycosidic bond (Figure 3.2). Cellulose is decomposed by both bacteria and fungi, and it is usually present in trace concentrations in the

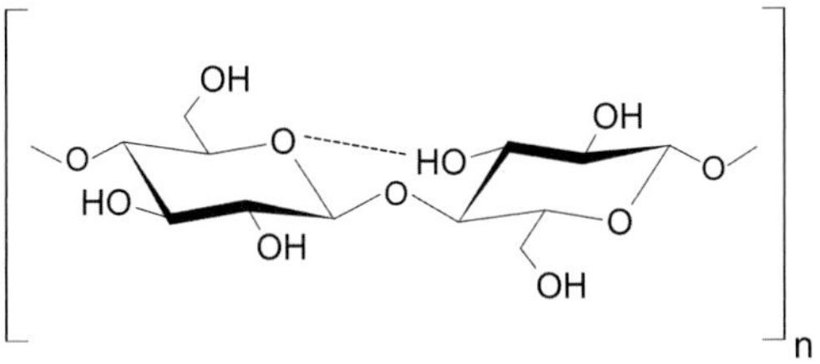

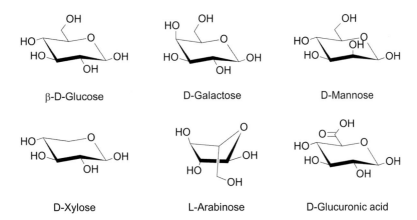

FIGURE 3.2   Basic structural unit of cellulose formed by β-1,4-linked D-glucose units.

soil. The accumulation of cellulose occurs under anaerobic conditions, such as in peat formation. The anaerobic degradation of cellulose is carried out by protozoa and slime molds, producing methane and water.

Hemicelluloses, also known as polyoses, are low-molecular-weight polysaccharides and the second largest chemical constituents of woody (25%–30% of total wood dry weight) and grassy plant cell walls along with lignin and cellulose. These compounds also exist in bacteria, fungi, and algae. Hemicelluloses are heteropolymers with side chains and branching that is composed of different sugar units (Figure 3.3), mainly pentoses (e.g., D-xylose, D-ribose, L-arabinose), hexoses (D-glucose, D-galactose, D-mannose), hexuronic acids (D-glucuronic

acid, D-galacturonic acid), and deoxyhexoses (D-fucose, L-rhamnose). In hemicelluloses the degree of polymerization is lower; that is, they are composed of shorter molecules than cellulose. Deciduous woody trees contain about 75% pentoses and 25% hexose, and the reverse contents occur in coniferous trees.

Pectins are more complex and highly branched polysaccharides, which mainly consist of galactose, arabinose, and hexuronic acids. Pectins are present in herbaceous plants and fruits, and they are abundant in the primary walls of plant cells. Hemicelluloses and pectins are decomposed by several bacteria and fungi under aerobic and anaerobic conditions; their decomposition rate is greater than cellulose.

These plant compounds generally serve as an excellent source of energy for soil microbes; thus they undergo continual degradation, and in this process polysaccharides and other carbohydrates are synthesized. The concentration of simple sugars in the soil is very small and most of the carbohydrates are present as polysaccharides. Carbohydrates constitute between 5% and 25% of the SOM and have a significant role in soil aggregation, metal complexation, acting as a source of N and P, CEC, and biological activity in soils.

β-D-Glucose   D-Galactose   D-Mannose

D-Xylose   L-Arabinose   D-Glucuronic acid

FIGURE 3.3   Structures of major sugar monomers that constitute hemicelluloses in plants.

(b) Amino Acids, Proteins, and Amino Sugars

Proteins are the most abundant polymeric components in living cells. In soils amino acids are predominantly present in protein or peptide-bound forms and the concentration of free amino acids is generally small. The concentrations of amino acids in the soil solution vary and range from 0.1 to 50 µM.

Amino acids are the fundamental structural units of proteins. In most contexts the term "amino acids" refers to the α-amino acids, where both amine ($-NH_2$) and carboxyl ($-COOH$) groups are attached to the same central C atom along with a side chain (R group) specific to each amino acid (see Sidebar 3.3). Another important characteristic of amino acids is that they act as both an acid and a base, that is, zwitterions (see Sidebar 3.4), at a certain pH.

i.e., a carboxylic ($-C(=O)OH$) group. In the example below alanine, an amino acid, has pK_a values of 2.34 and 9.69 for the carboxylic and amine group, respectively.

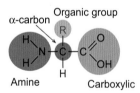

| Positive charge at low pH | Zwitterion at about pH 6 | Negative charge at high pH |

Ionization state of alanine at different pHs.

All proteins are composed of the same set of 20 α-amino acids, except for some bacterial peptides and some peptide antibiotics. The amino acids of peptides and proteins are linked together by peptide bonds that are formed by the condensation of the α-carboxyl group of one amino acid with the α-amino group of another amino acid (Sidebar 3.5). The chemical structures of amino acids found in soils are shown in Figure 3.4.

There is a large variability in the proportion of different α-amino acids present in soils from different regions (Table 3.6). However, neutral compounds form more than 60% of the N compounds. Basic compounds generally exist in greater proportion in the semitropical and tropical soils than in temperate soils, and the reverse is true for acidic compounds, which are present in greater concentrations in temperate soils than in the other two regions. Tropical soils containing noncrystalline or amorphous Al and Fe compounds may retain acidic amino acids, particularly aspartic acid, possibly by forming chemical complexes. In contrast to previous results, Warren and Taranto (2010) extracted large amounts of acidic amino acids using 1 M KCl in temperate grassland soils and suggested that weak extractants in earlier studies yielded lower concentrations.

Sidebar 3.3 — A Generalized Structure of an α-Amino Acid

Sidebar 3.4 — Amino Acids Are Zwitterions!

A zwitterion is a molecule with two or more functional groups, of which at least one has a positive and one has a negative charge, and the net charge of the entire molecule is zero at a specific pH. Amino acids are the best-known examples of zwitterions; they contain at least a basic group, i.e., an amine ($-NH_2$) group and an acidic group,

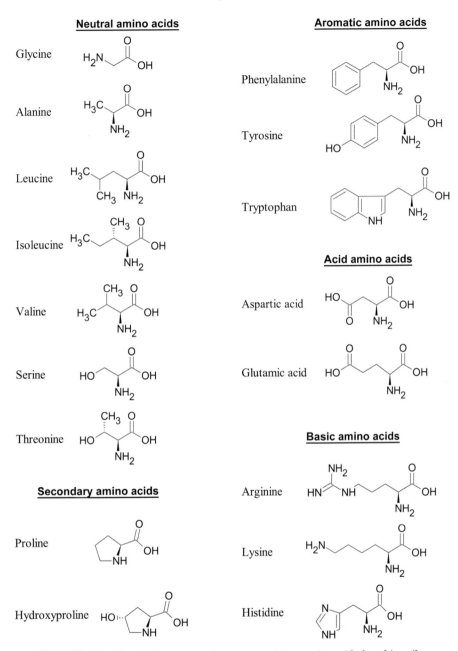

FIGURE 3.4   Chemical structure of common protein α-amino acids found in soils.

3. Chemistry of Soil Organic Matter

TABLE 3.6 Proportion of acid, neutral, and basic compounds of α-amino acid nitrogen in soils from different climatic zones.

Climatic zone	Proportion of α-amino acid-nitrogen compounds (%)		
	Acidic	Neutral	Basic
Temperate	12.6–25.0	66.6–76.2	8.4–9.8
Semitropical	0.8–10.9	61.2–70.8	10.3–35.5
Tropical	1.2–6.4	65.0–85.6	8.0–29.1

Myo-inositol hexakisphosphate.

Inositols exist in various states of phosphorylation (bound to between one and six phosphate groups). In soils four isomeric forms, i.e., myo, d-chiro, scyllo, and neo, exist, and myo-inositol hexakisphosphate is the most dominant form, constituting up to 90% of the inositols (Giles et al., 2011).

Sulfur: Like N, more than 95% of S exists in organic forms in soils. Most common S compounds in soils are S-containing amino acids (cysteine, methionine, cystine), vitamins (e.g., biotin, thiamine), and antibiotics. A significant amount of S is present in ester sulfates with C—O—S linkages (i.e., is not directly bonded to C), e.g., arylsulfate, alkylsulfates, phenolic sulfate, sulfated polysaccharides, choline sulfate, and sulfated lipids. Some organic S could be present as sulfamates (C—N—S) and sulfated thioglycosides (N—O—S).

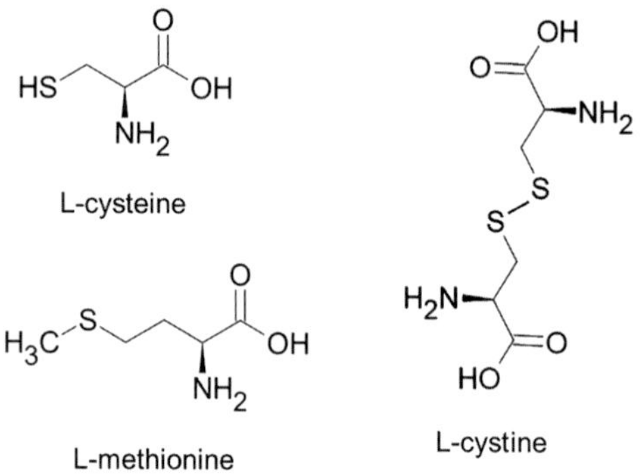

Sidebar 3.5 — Peptide Bond Formation Between Two α-Amino Acids

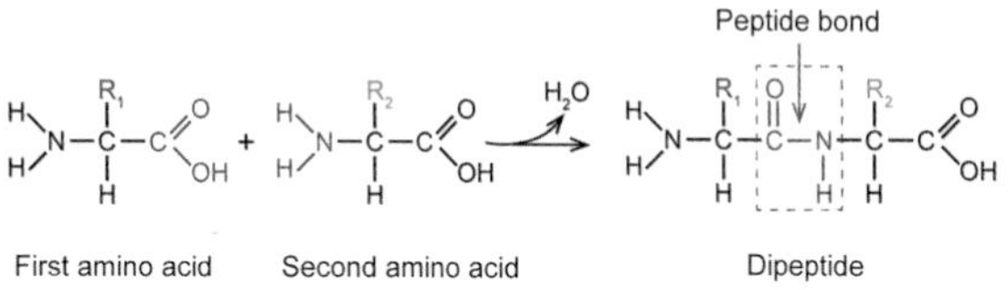

First amino acid   Second amino acid   Dipeptide

Soil microorganisms secrete protease enzymes to hydrolyze peptide bonds in proteins and peptides, which release their constituent amino acids that are utilized by the microorganisms as sources of C and N. The main organic compounds of N, P, and S in SOM are discussed in Sidebar 3.6.

Sidebar 3.6 — Nitrogen, P, and S Compounds in SOM

Nitrogen: About 95%–98% of the total N in soil is present in organic matter. The main N compounds are proteins, peptides, amino acids, amino sugars, peptides, and nucleic acids.

Phosphorus: Organic P constitutes between 30% and 65% of total P in mineral soils. The large variability is associated with soil type. Principal organic compounds of P are inositol phosphates, phospholipids, nucleic acids, and their derivatives.

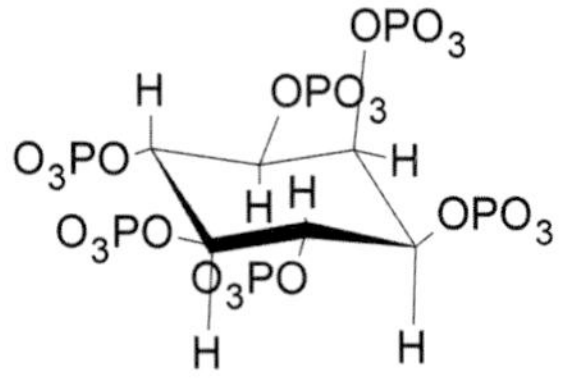

L-cysteine

L-methionine

L-cystine

Common S-containing amino acids in soils.

Based on the dominance of a small number of amino acids (e.g., alanine, aspartic acid, glutamic acid, and glycine) in soils with a composition similar to the cells of microorganisms, it has been suggested that these are derived from microbial cells via a similar biochemical process (Stevenson and Cole, 1999).

The most prominent amino sugars present in soils are glucosamine and galactosamine, with an average contribution of 60% and 30%, respectively, of the total amino sugars in soils (Joergensen, 2018). Muramic acid and mannosamine make up most of the rest of the 10% of soil amino sugars. The structures of common sugars in SOM are presented in Sidebar 3.7. Amino sugars in soils are solely derived from microbial residues as plants do not produce these compounds. Muramic acid is present in the murein layers of the bacterial cell wall, and it is a very specific biomarker for bacterial residues in soils. Similarly, glucosamine is a specific biomarker for fungal residues, provided corrections are made for the bacterial origin of glucosamine. Glucosamine is a constituent of a polymer chitin that is present in the cell walls of higher fungi.

(c) Lignins

Lignins are considered to be the third most abundant, after cellulose and hemicellulose, plant-derived organic compounds that are added into the soil. Lignins are polymers and their basic structural unit (monolignols) is comprised of aromatic rings with sidechains and $-OH$ (hydroxy) and $-OCH_3$ (methoxyl) groups linked by various strong covalent bonds (alkyl-aryl, ether, $C-O-C$, and $C-C$). The three primary structural units of lignins are three cinnamyl alcohols: p-coumaryl alcohol, coniferyl alcohol, and sinapyl alcohol (Figure 3.5). Different plant species synthesize

three-dimensional macromolecules of lignins via the dehydrogenative polymerization of different cinnamyl alcohols. For example, the lignins of gymnosperms are usually made up entirely of guaiacyl propane monomers (derived from coniferyl alcohols), whereas angiosperm lignins consist of nearly equal proportions of guaiacyl propane and syringyl propane (derived from sinapyl alcohol) units. Lignins in grasses have more complex structures with nearly equal proportions of guaiacyl propane, syringyl propane, and p-hydroxyphenyl propane (derived from p-coumaryl alcohol) units.

Sidebar 3.7 — Structure of Amino Sugars in Soils

Glucosamine

Galactosamine

Mannosamine

Muramic acid

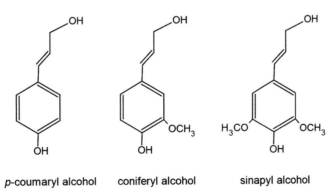

FIGURE 3.5 Structure of three primary building units of lignin, monolignols, found in "natural" plant lignin.

The lignin concentration in plant residues (on dry weight) varies from about 5% in ryegrass to 20% basis in grasses, wheat, sorghum, and corn and up to 40% in wood and barks of beech and spruce. Another important aspect of lignin is that it is synthesized only in vascular plants and is not present in bacteria and fungi.

Lignins are arguably the most resistant component of plant residues in the soil that are selectively preserved due to their complex structures with aromatic components. However, this hypothesis may not be true, as some research shows that lignin degradation by specialized microorganisms (e.g., white-rot fungus, *Phanerochaete chrysosporium*) is relatively rapid. Additionally, it has been observed that plant residues with greater lignin content are more completely decomposed and leave less residual C in the soil compared to lower lignin residues. Thus high lignin residues contribute a proportionally small amount to new SOM.

(d) Lipids

Soil lipids represent a diverse range of compounds that mostly originate from both plants and animals as products of deposition, exudation, and decomposition, as well as various other sources, such as fungi, bacteria, and mesofauna (Table 3.7).

These heterogeneous organic compounds are conveniently grouped as an analytical group that is insoluble in water and soluble in organic solvents, such as benzene, chloroform, hexane, ether, and acetone or a mixture of organic solvents. Lipids in soils range from simple fatty acids and esters to complex molecules, including sterols, aliphatic hydrocarbons, ketones, fats, waxes, and resins.

Cutin and suberin are complex polymers with cross-linked structures constituted by (16- or 18-carbon) hydroxy fatty acids. Cutin, along with associated waxes, is a primary constituent of the cuticle (Sidebar 3.8), which is present on both sides of the outer-epidermal walls of leaves and fruits. Suberin is a hydrophobic polymer that

TABLE 3.7   Different classes of lipid compounds originating from plants and microorganisms in the soil.

Class of compound	Plants	Algae	Fungi	Bacteria
Hydrocarbons				
n-Alkanes	×	×	×	×
Branched alkanes	×	×	×	×
Olefins	×	×		×
Cyclic alkanes	×			
Ketones				
Monoketones	×			×
β-Diketones	×			
Secondary alcohols	×			×
Alkane diols (diesters)	×			
Free fatty acids	×			
Wax esters				
Primary alcohol esters	×	×	×	×
Triesters	×			
Primary alcohols	×	×		×
Aldehydes	×			
Terpenoids	×			

From Dinel et al. (1990), with permission.

accumulates in the apoplastic regions of many noncutinized boundary cell layers. It is present in the wall portion of cork cells, which are found on the underground parts of root vegetables and tubers, woody stems, and healed wounds.

Sidebar 3.8 — Cuticle and Apoplast

Cuticle is an extracellular hydrophobic or waxy layer covering the aerial epidermis of all land plants; it provides protection against desiccation and external environmental stresses.

Apoplast is the continuous space external to the plasma membranes throughout the plant that facilitates the transport of water and solutes across a tissue or organ.

Globally, lipids constitute 4%–8% of the total organic C in the soil. However, quantities as high as 42% have been reported in a cultivated organic soil (Dinel et al., 1990). Soil types and vegetation have a major control on the soil lipid content.

3.5.3 Humic Substances

Humic substances represent the most active fraction of SOM and are extremely common and widely distributed in the earth. According to Szalay (1964), the amount of C in the earth as HAs (60×10^{11} Mg) exceeds that which occurs in living organisms (7×10^{11} Mg). HAs are found in soils, waters, sewage, compost heaps, marine and lake sediments, peat bogs, carbonaceous shales, lignites, and brown coals. While they are not harmful, they are not desirable in potable water (Stevenson, 1982).

In addition to the definition in Table 3.1 in Section 3.1 (Stevenson, 1982), humic substances can be defined as "a general category of naturally occurring, biogenic, heterogeneous organic substances that can generally be characterized as being yellow to black in color, of high molecular weight (MW), and refractory" (Aiken et al., 1985b). They are amorphous, partly aromatic, polyelectrolyte materials that no longer have specific chemical and physical characteristics associated with well-defined organic compounds (Box 3.2; Schnitzer and Schulten, 1995). Humic substances have been described and defined differently in the literature for a long time, and this has become one of the hotly debated topics in soil science over the last decade. The differences in the description of

---

BOX 3.2

Molecular structure of soil organic matter

While we know the elemental and functional group composition of HS, definitive knowledge of the basic "backbone structure" of SOM is still an enigma. Many structures have been proposed, and each of them is characterized by similar functional groups and the presence of aliphatic and aromatic components.

Use of advanced analytical techniques such as ^{13}C NMR spectroscopy has provided spectra of whole soils that show paraffinic C, OCH_3–C, amino acid–C, C in carbohydrates and aliphatic structures containing OH groups, aromatic C, phenolic C, and C in CO_2H groups (Arshad et al., 1988). Based on the spectra, the aromaticity of SOM has been shown to seldom exceed 55%, with the aliphaticity often being higher than aromaticity (Schnitzer and Preston, 1986). Similar to the ^{13}C NMR studies, pyrolysis field ionization mass spectrometry (Py-FIMS) studies on SOM in whole soils show the presence of carbohydrates; phenols; lignin monomers; lignin dimers; alkanes; fatty acids; n-alkyl mono-, di-, and triesters; n-alkylbenzenes; methylnapthalenes; methylphenanthrenes; and N compounds (Schnitzer and Schulten, 1992). The carbohydrates, proteinaceous materials (amino acids, peptides, and proteins), and lipids (alkanes, alkenes, saturated and unsaturated fatty acids, alkyl mono-, di-, and triesters) in SOM are tightly bound by the aromatic SOM compounds (Schnitzer, 2000). Altogether, more than 300 different compounds have been identified as degradative products of SOM, and many of these compounds may be related directly to the structure of SOM. The unresolved and controversial part is how these building blocks are linked together and whether there are unique structure classes of compounds of SOM or not.

Continued

BOX 3.2   (cont'd)

Molecular structure of soil organic matter

Schulten and coworkers (Schulten and Schnitzer, 1997; Schulten et al., 1998) have proposed two- and three-dimensional (averaged) structures of HA, SOM, and whole soil. Using Py-FIMS and Curie-Point pyrolysis GC-MS (gas chromatography—mass spectrometry), Schulten and Schnitzer (1993) proposed a two-dimensional structure for HA (Box Figure 3.1) in which aromatic rings are linked covalently by aliphatic chains. Oxygen is present as carboxyls, phenolic and alcoholic hydroxyls, esters, ethers, and ketones while nitrogen

is present as heterocyclic structures and as nitriles. The elemental composition of the structure in the figure is $C_{308}H_{328}O_{90}N_5$; it has a molecular weight of 5.540 Da and an elemental analysis of 66.8% C, 6.0% H, 26.0% O, and 1.3% N. The C skeleton has high microporosity containing voids of different dimensions that can trap and bind other organics (e.g., carbohydrates and proteinaceous materials) and inorganic components (clay minerals and metal oxides) and water.

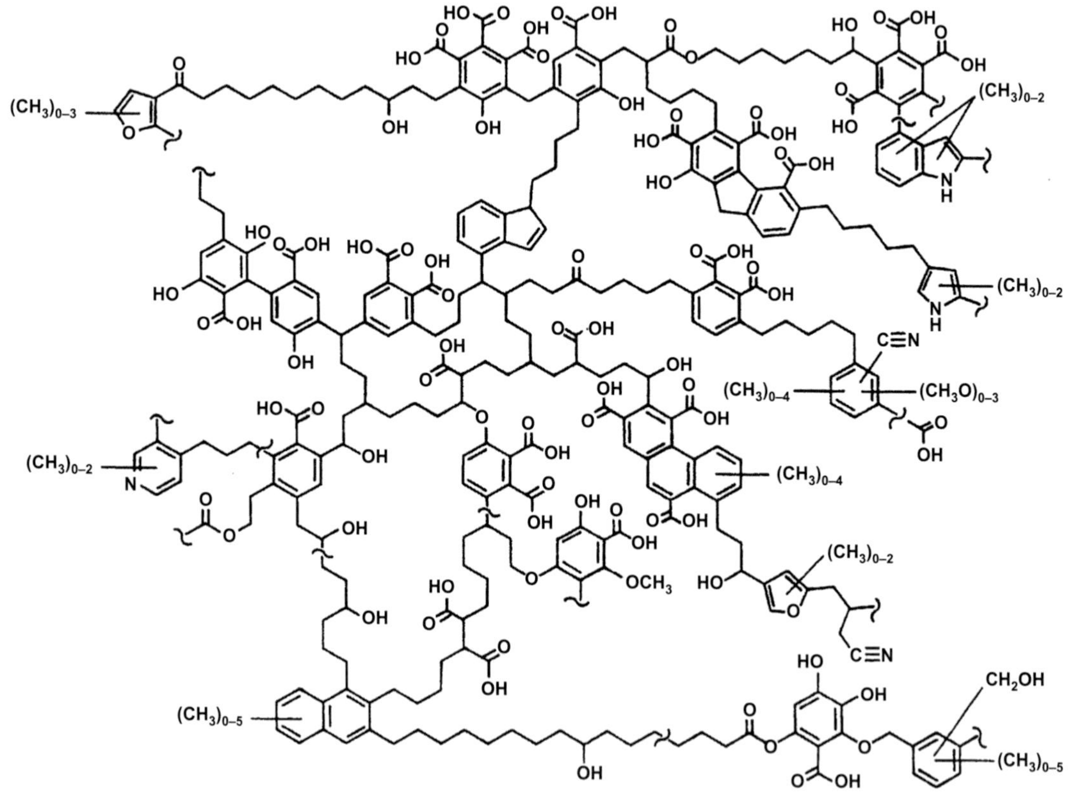

BOX FIGURE 3.1   Two-dimensional humic acid model structure. *From Schulten and Schnitzer (1993), with permission.*

BOX 3.2 *(cont'd)*

Molecular structure of soil organic matter

As discussed in Section 3.6, there are different and contrasting hypotheses for the genesis of SOM. The existence of a unique macromolecular structure is therefore questionable. The conformation of SOM varies with environmental conditions, such as the origin of SOM (soil vs. fluvial), solution chemistry (pH, ionic strength), and mineralogy. There is not a regularly repeating structural unit or set of units that is characteristic of HS (MacCarthy, 2001). Consequently, no two molecules of HS are alike (Dubach and Mehta, 1963). There is large heterogeneity and complexity in SOM; thus it is likely that there is no unique structure of SOM.

---

humic substances are attributed to characteristics: high MW versus low MW, whether it is of a predominant aromatic structure or not, whether it is formed by microbial synthesis or is of plant origin, and whether it is refractory or not. The literature on the evolution of humic substances can be summarized into two different models: (i) the humic polymer model and (ii) the molecular aggregation or supramolecular associations model. These aspects will be discussed in the "Genesis of Humic Substances" section.

There is a multitude of paths that HS can take in the environment (Figure 3.6). Water is obviously the most important medium that affects the transport of HS. A host of environmental conditions affect HS, ranging from oxic to anoxic environments and from particulate to dissolved HS. Additionally, the time range that HS remains in the environment is wide. It can range from weeks and months for HS in surface waters of lakes, streams, and estuaries to hundreds of years in soils and deep aquifers (Aiken et al., 1985b).

(a) Humic Substances – Extraction and Fractionation

Before one can suitably study SOM, it must be separated from the inorganic soil components. The fractionation of SOM lessens the heterogeneity of HS so that physical and chemical techniques can be used to study their structure and molecular properties (Hayes and Swift, 1978).

Archard (1786) was the first scientist to use $0.1-0.5$ M NaOH (or KOH) for this purpose, and since then, dilute NaOH has been established as the most efficient extractant for OM from soils. In the classical fractionation scheme (Figure 3.7) the alkaline extract is usually partitioned into three fractions: HA, FA, and humin. HA is the fraction of alkaline extract that coagulates when the extract is acidified to a very low pH (about 1.0), that is, soluble in alkali (base) and insoluble in acid. FA remains in solution in the acidified extract and is soluble in both alkali and acid. Humin is the fraction that remains in the soil that is not soluble in both alkali and acid.

Alkali extractions can dissolve silica, contaminating the humic fractions, and dissolve protoplasmic and structural components from organic tissues. Milder extractants, such as $Na_4P_2O_7$ and EDTA, dilute acid mixtures with HF, and organic solvents, can also be employed. However, less SOM is extracted and there are other problems, such as the difficulty in removing pyrophosphate from the extracted humic substances (Stevenson, 1982).

In addition to extraction and precipitation procedures, gel permeation chromatography,

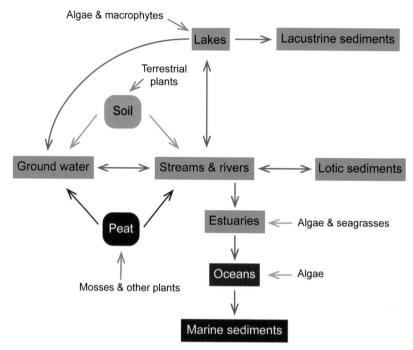

FIGURE 3.6   Flowpaths of humic substances in the terrestrial and water environments. From Aiken et al. (1985a), with permission.

ultrafiltration membranes, adsorption on hydrophobic resins (XAD, nonionic methylmethacrylate polymer), adsorption on ion exchange resins, adsorption on charcoal and Al_2O_3, and centrifugation are also used for SOM fractionation (Buffle, 1984; Thurman, 1985). New electrophoretic methods, including polyacrylamide gel electrophoresis, isoelectric focusing, and isotachophoresis, are also promising techniques for SOM fractionation (Hayes and Swift, 1978). The use of XAD resins is considered by many researchers to be the best method to fractionate or isolate HS (Thurman et al., 1978; Thurman, 1985). The alkali extraction procedure has allowed the isolation of a closely knit family of molecules for the characterization of generalized chemical structures and properties and functions of humic substances, which has advanced our understanding of the role of SOM in many soil processes and ecosystem services.

In recent years some "state-of-the-art techniques" have emerged that allow the characterization of organic matter without any pretreatments. The combined application of techniques, such as nanoscale secondary ion mass spectrometry (NanoSIMS), X-ray computed tomography, Fourier transform ion cyclotron resonance mass spectrometry (FT-ICR-MS), and synchrotron-based scanning transmission X-ray microscopy (STXM), X-ray absorption fine structure (XAFS), and Fourier transform infrared (FTIR) spectroscopy, has provided new insights into the molecular structure and composition of SOM and the association of organic compounds with various soil minerals. Obtaining such information is not possible with bulk analysis of SOM.

(b) Properties of Humic Substances

Humic substances range in diameter from 0.001 to 1 μm. While HAs fit into this size range, some of

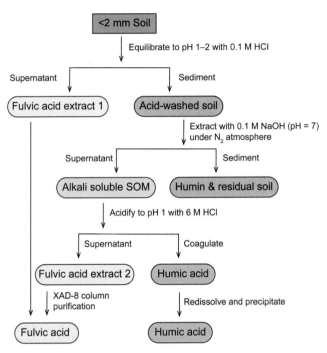

FIGURE 3.7   Extraction and fractionation of humic substances from soil organic matter based on the method developed by the International Humic Substance Society (Swift, 1996).

the lower-molecular-weight FAs are smaller. Humic substances are hydrophilic and consist of globular particles, which in aqueous solution contain hydration water. Stevenson (1982) notes that HSs are thought of as coiled, long-chain molecules or two- or three-dimensional cross-linked macromolecules whose negative charge is primarily derived from the ionization of acidic functional groups such as carboxyls. Pyrolysis field ionization mass spectrometry (Py-FIMS) analysis shows that HA material is rich in carbohydrates, phenols, monomeric and dimeric lignins, n-fatty acids, and N-compounds, and FA is also dominated by carbohydrates, phenols, and lignins. The major components of the humin fraction are carbohydrates, phenols, monomeric lignins, and alkyl esters.

The average MW of HS may range from 500 to 5000 Da for FA to 3000 to 1,000,000 Da for HA (Stevenson, 1982). Soil HAs have higher MWs

than aquatic HAs. The MW measurements depend on pH, concentration, and ionic strength.

One of the problems in studying humin was the lack of a suitable procedure for the isolation of humin from soil mineral components. Carbon-13 (^{13}C) nuclear magnetic resonance (NMR) spectroscopy has greatly assisted in the study of humin since the high content of mineral matter in humin is not a factor. Humin is similar to HA. It is slightly less aromatic (organic compounds that behave like benzene; Aiken et al., 1985b) than HA, but it contains a higher polysaccharide content (Wright and Schnitzer, 1961; Acton et al., 1963; Schnitzer and Khan, 1972). More recent data on humin isolates show that humin composition is dominated by aliphatic hydrocarbons (see Sidebar 3.9 for aliphatic and aromatic compounds) with small amounts of carbohydrates and peptide materials and lignin residues (Simpson et al., 2007; Song et al., 2008, 2011).

Sidebar 3.9 — Aliphatic and Aromatic Compounds

Aliphatic compounds are a broad category of organic carbon compounds containing C and H joined together only in straight chains, branched chains, or nonaromatic rings; the carbon–carbon bonds may be saturated or unsaturated, e.g., long-chain amino acids such as arginine, lysine, and fatty acids.

Arginine

Stearic acid

Aromatic compounds are a large category of unsaturated C compounds characterized by one or more planar rings of C atoms joined by covalent bonds of two different kinds, i.e., sigma bonds between C and H and pi (π) electrons of C atoms, e.g., aromatic amino acids such as phenylalanine, tyrosine, condensed tannins, and phenanthrene.

Phenylalanine   *Phenanthrene*

The lack of reproducibility in analytical methods makes the study of HS difficult and exacerbates the task of deriving a precise elemental composition. Table 3.8 shows the elemental composition of HS isolated from soil, fresh water, and peat environments all over the world (Rice and MacCarthy, 1991). The data show large ranges for all elements and humic compounds; however, standard

TABLE 3.8   Elemental composition (% weight basis) of ash-free humic compounds isolated from soil, fresh water, marine, peat, and coal samples from all over the world.

Sample	C	H	O	N	S
All sources samples[a]					
Humic acid					
Mean ± SD	55.1 ± 5.0	5.0 ± 1.1	35.6 ± 5.8	3.5 ± 1.5	1.8 ± 1.6
Range	37.2 − 75.8	1.6 − 11.7	7.9 − 56.6	0.5 − 10.5	0.1 − 8.3
No. of samples	410	410	410	410	160
Fulvic acid					
Mean ± SD	46.2 ± 5.4	4.9 ± 1.0	45.6 ± 5.5	2.5 ± 1.6	1.2 ± 1.2
Range	35.1 − 75.4	0.43 − 7.2	16.9 − 55.9	0.45 − 8.2	0.10 − 3.6
No. of samples	214	214	214	214	71
Humin					
Mean ± SD	56.1 ± 2.6	5.5 ± 1.0	34.7 ± 3.4	3.7 ± 1.3	0.4 ± 0.3
Range	48.3 − 61.6	4.2 − 7.3	28.8 − 45.1	0.9 − 6.0	0.1 − 0.9
No. of samples	26	26	26	24	16

TABLE 3.8 Elemental composition (% weight basis) of ash-free humic compounds isolated from soil, fresh water, marine, peat, and coal samples from all over the world.—cont'd

Sample	C	H	O	N	S
Humic acid					
Soil					
Mean $\pm$ SD	55.4 $\pm$ 3.8	4.8 $\pm$ 1.0	36.0 $\pm$ 3.7	3.6 $\pm$ 1.3	0.8 $\pm$ 0.6
Range	37.2 − 64.1	1.6 − 8.0	27.1 − 52.0	0.5 − 7.0	0.1 − 4.9
No. of samples	215	215	215	215	67
Fresh water					
Mean $\pm$ SD	51.2 $\pm$ 3.0	4.7 $\pm$ 0.6	40.4 $\pm$ 3.8	2.6 $\pm$ 1.6	1.9 $\pm$ 1.4
Range	43.8 − 56.0	3.5 − 6.5	30.9 − 48.2	0.6 − 8.0	0.4 − 4.3
No. of samples	56	56	56	56	13
Peat					
Mean $\pm$ SD	57.1 $\pm$ 2.5	5.0 $\pm$ 0.8	35.2 $\pm$ 2.7	2.8 $\pm$ 1.0	0.4 $\pm$ 0.2
Range	50.5 − 62.8	3.6 − 6.6	30.7 − 43.2	0.6 − 3.9	0.1 − 0.7
No. of samples	23	23	23	21	12
Fulvic acid					
Soil					
Mean $\pm$ SD	45.3 $\pm$ 5.4	5.0 $\pm$ 1.0	46.2 $\pm$ 5.2	2.6 $\pm$ 1.3	1.3 $\pm$ 1.1
Range	35.1 − 75.4	3.2 − 7.0	16.9 − 55.9	0.45 − 5.9	0.1 − 3.6
No. of samples	127	127	127	127	45
Fresh water					
Mean $\pm$ SD	46.7 $\pm$ 4.3	4.2 $\pm$ 0.7	45.9 $\pm$ 5.1	2.3 $\pm$ 2.1	1.2 $\pm$ 0.9
Range	39.2 − 56.3	0.43 − 5.9	34.7 − 55.8	0.47 − 8.2	0.16 − 3.1
No. of samples	63	63	63	63	14
Peat					
Mean $\pm$ SD	54.2 $\pm$ 4.3	5.3 $\pm$ 1.1	37.8 $\pm$ 3.7	2.0 $\pm$ 0.5	0.8 $\pm$ 0.6
Range	46.9 − 60.8	4.2 − 7.2	31.1 − 44.3	1.2 − 2.6	0.2 − 1.9
No. of samples	12	12	12	12	11

[a] All sources of samples include soil, fresh water, marine, peat, and coal samples.
From Rice and MacCarthy (1991), with permission.

deviation values are generally small. For example, the mean C concentration in HA for samples ($n = 410$) from all sources is 55.1%, 68% of the C concentration values are within $\pm 5.0\%$ (± 1 SD), and 96% of the C concentration values are within $\pm 10.0\%$ (± 2 SD) (Rice and MacCarthy, 1991). The mean C and H concentrations in HA and FA extracted from peat are the highest, and the concentrations are the lowest in water extracted HA and FA, except for the C concentration in FA, that is, the lowest in soil-extracted samples. The unique chemical compositions of HA, FA, and humin that are different from each other may be related to their environmental conditions of formation and/or the biochemical processes by which they are derived. These data also suggest minimal changes in HS during the extraction process. Based on the mean values of soil-extracted HS, the mean formulae of $C_{18}H_{19}O_9N$ for HA and $C_{15}H_{19}O_{11}N$ for FA, disregarding S, could be derived.

Atomic ratios of H/C, O/C, and N/C can be useful in identifying types of HS and in devising structural formulas (Table 3.9). Based on the information in Table 3.9, it appears that the O/C ratio is the best indicator of humic types. Soil HA O/C ratios are about 0.50, while FA O/C ratios are close to 0.75. The H/C ratio of HA is about 1.0 compared to 1.25 for FA. The N/C ratio in HA and FA is more variable, but in general, the ratio is larger in FA than in HA, indicating that, in general, FA will have more N than HA.

Several methods can be used to quantitate the functional groups of SOM, particularly the acidic groups, because of their contribution to the CEC of soils. The main acidic groups are carboxyl (R—C(=O)OH) and acidic OH groups (presumed to be phenolic OH), with carboxyls being the most important group (Table 3.10). Smaller amounts of alcoholic OH, quinonic, ketonic, and methoxylic groups are also found. In fertile soils with near-neutral pH most of the carboxylic groups dissociate to the carboxylate (COO⁻) anion and thus contribute to the soil's cation exchange capacity. A derivation of acid dissociation constants is given in Box 3.3.

The development of negative and positive charge on SOM functional groups is discussed in Sidebar 3.10.

TABLE 3.9   Atomic ratio of elements in humic and fulvic acids isolated from soils from all over the world.

H/C	O/C	N/C	Reference
Soil humic acid			
1.0	0.48	0.04	Schnitzer and Khan (1978)
1.1	0.50	0.02	Ishiwatari (1975)
1.1	0.47	0.06	Hatchet et al. (1994)
0.8	0.46	0.01	Steelink (1985)
0.97	0.57	0.04	Leenheer (1980)
1.16	0.45	0.07	Chen et al. (1978) [a]
1.03	0.49	0.06	Rice and MacCarthy (1991) [b]
1.023	0.489	0.043	Average
Soil fulvic acid			
1.40	0.74	0.04	Schnitzer and Khan (1978)
0.83	0.70	0.06	Ishiwatari (1975)
0.93	0.64	0.03	Malcolm et al. (1981)
1.78	0.80	0.07	Chen et al. (1978) [a]
1.32	0.77	0.07	Rice and MacCarthy (1991) [b]
1.252	0.730	0.054	Average

[a] Atomic ratio was calculated from the mean value of HAs and FAs reported in the study.
[b] Atomic ratio was calculated from the mean value of HAs and FAs extracted from soils reported in the study.

Sidebar 3.10 — Charge on SOM Functional Groups

(i) Negative charge on SOM functional groups.

$$R-CH_2-NH_2 + H^+ \rightarrow R-CH_2-NH_3^+$$

$$R-\overset{\overset{H}{\|}}{C}-NH_2 + H^+ \rightarrow R-CH_2-NH_3^+$$

(ii) Positive charge on SOM functional groups.

Amines and amides are the main basic groups of SOM, and they provide most of the anion exchange capacity of the soils, particularly in temperate soils. Further discussion on the basic groups is provided in Sidebar 3.11.

TABLE 3.10   Important functional groups of soil organic matter.

Functional group	Structure
Acidic groups	
Carboxyl	$R-C(=O)-OH$[a]
Enol	$R-CH=CH-OH$
Phenolic	$Ar-OH$[a]
Quinone	$Ar=O$
Neutral groups	
Alcohol	$R-CH_2-OH$
Ether	$R-CH_2-O-CH_2-R$
Ketone	$R-C=O(-R)$
Aldehyde	$R-C=O(-H)$
Ester	$R-C=O(-OR)$
Basic groups	
Amine	$R-CH_2-NH_2$
Amide	$R-C=O(-NH-R)$

[a] *R is an aliphatic (a broad category of carbon compounds having only a straight, or branched, open-chain arrangement of the constituent carbon atoms; the carbon–carbon bonds may be saturated or unsaturated; Aiken et al., 1985a) backbone and Ar is an aromatic ring.*
Adapted from Stevenson (1982) and Thurman (1985), with permission.

Sidebar 3.11 − Basic Functional Groups of SOM

Amines are considered derivatives of NH_3 wherein one or more hydrogen atoms have been replaced by an alkyl or aryl group. An alkyl group is an alkane lacking a hydrogen atom. An aryl group contains an aromatic ring. In primary amines one of three hydrogen atoms in NH_3 is generally replaced by an alkyl or aryl group, abbreviated as $R-NH_2$, where R is an alkyl or aryl group. Secondary amines have two carbons that are bonded (alkyl, aryl, or both) to the nitrogen together with one hydrogen, abbreviated as $R-NH$. Tertiary amines have three carbons bonded to the nitrogen and are denoted as $R-N$.

Amides are derived from carboxylic acids by the replacement of the hydroxyl group (OH) by an amino group. Amides are qualified as primary $(R'(C=O)-NH_2)$, secondary $(R'C(=O)-NH-R)$, and tertiary $(R'(C=O)-N-R2)$, where R and R' are groups other than hydrogen.

Imine is a functional group that has a carbon–nitrogen double bond ($C=N$).

Amine, amide, and imine are basic functional groups, and they have been identified in humic substances.

Total functional group acidities ranging from 12,000 to 23,100 $mmol_c \ kg^{-1}$ have been reported for soil HS (Table 3.11). The total acidities of FA are greater than those for HA. FAs generally have high quantities of carboxyls and alcoholic groups, while HA contains high quantities of phenolic groups (Table 3.11).

3.5.4 Black Carbon

Black carbon in the soil is derived from the incomplete burning of vegetation, either through natural processes such as wildfires or human-induced burning of vegetation for land clearing or other purposes. A continuum of charred

BOX 3.3

Calculation of acid dissociation constants (K_a) and derivation of the Henderson–Hasselbalch equation

(I) Calculation of K_a and pK_a values

Suppose that a 0.1 M CH_3COOH (acetic acid) solution is ionized 1.3% at 298 K. What is the K_a for CH_3COOH?

$$CH_3COOH + H_2O \overset{K_a}{\leftrightarrow} H_3O^+ + CH_3COO^-$$
$$\text{(B3.3a)}$$

acid   base

or

$$CH_3COOH \overset{K_a}{\leftrightarrow} H_3O^+ + CH_3COO^- \quad \text{(B3.3b)}$$

acid   base

(*Remember that free H^+ does not exist in aqueous solutions and that a proton is transferred to H_2O in all acid ionization reactions to form hydronium ions, H_3O^+*).

Brönsted and Lowry defined an acid as a proton donor and a base as a proton acceptor. According to Lewis, an acid is an electron pair acceptor and a base is an electron pair donor. The definitions of Lewis are general and also include species that do not have a reactive H^+ (Harris, 1987). In Equation (B3.3a) CH_3COOH is an acid and CH_3COO^- is a base because the latter can accept a proton to become CH_3COOH. Acetic acid and CH_3COO^- are a conjugate acid–base pair. The K_a for the second reaction in Equation (B3.3b) is

$$K_a = \frac{[H^+][CH_3COO^-]}{[CH_3COOH]} \quad \text{(B3.3c)}$$

where brackets indicate concentration in mol liter^{-1}, $[H^+] = 0.1$ mol liter^{-1} $\times$ 0.013 = 0.0013 mol liter^{-1}, $[CH_3COO^-] = 0.1$ mol liter^{-1} $\times$ 0.013 = 0.0013 mol liter^{-1}, and $[CH_3COOH] = (0.1$ mol liter$^{-1}) - (0.0013$ mol liter$^{-1}) = 0.0987$ mol liter^{-1}. Substituting these values in Equation (B3.3c),

$$K_a = \frac{\left[0.0013 \text{ mol liter}^{-1}\right]\left[0.0013 \text{ mol liter}^{-1}\right]}{\left[0.0987 \text{ mol liter}^{-1}\right]}$$

$$= 1.0712 \times 10^{-5} \text{mol liter}^{-1}$$
$$\text{(B3.3d)}$$

The higher the K_a, the more dissociation of the acid into products and the stronger the acid. Acetic acid has a low K_a and thus is slightly dissociated. It is a weak acid. One can also calculate a pK_a for CH_3COOH where

$$pK_a = -\log K_a = -\log 1.712 \times 10^{-5} \text{ mol liter}^{-1}$$
$$\text{(B3.3e)}$$

$$pK_a = 4.77 \quad \text{(B3.3f)}$$

The lower the pKa, the stronger the acid. For example, hydrochloric acid (HCl), a strong acid, has a $pK_a = -3$.

(II) Derivation of Henderson–Hasselbalch Equation

One can derive a relationship, known as the Henderson–Hasselbalch equation, between pK_a and pH that is very useful in studying weak acids and in preparing buffers. A buffer is a solution whose pH is relatively constant when a small amount of acid or base is added. For example, zwitterions, as previously mentioned, are naturally occurring buffers in soils, and one of the main reasons that SOM is effective at buffering soil pH. Since soils behave as weak acids, the Henderson–Hasselbalch equation is useful in understanding how, if pH and pK_a are known, the protonation and deprotonation of functional groups (formally defined in Chapter 5; an example would be the carboxyl [R–C=O–OH, where R is an aliphatic group] functional group of soil organic matter in soils) can be assessed:

BOX 3.3   (cont'd)

Calculation of acid dissociation constants (K_a) and derivation of the Henderson —Hasselbalch equation

$$-\log K_a = -\log\left[\frac{-\log[H^+][CH_3COO^-]}{[CH_3COOH]}\right] \tag{B3.3g}$$

Rearranging,

$$-\log K_a = -\log[H^+] - \log\frac{[CH_3COO^-]}{[CH_3COOH]} \tag{B3.3h}$$

Simplifying,

$$pK_a = pH - \log\frac{[CH_3COO^-]}{[CH_3COOH]} \tag{B3.3i}$$

Rearrangement yields

$$pH = pK_a + \log\frac{[CH_3COO^-]}{[CH_3COOH]} \tag{B3.3j}$$

or

$$pH = pK_a + \log\frac{[\text{conjugate base}]}{[\text{acid}]} \tag{B3.3k}$$

If $[CH_3COO^-] = [CH_3COOH]$, then

$$pH = pK_a + \log 1 \tag{B3.3l}$$

Since $\log 1 = 0$, Eq. (B3.3l) becomes

$$pH = pK_a \tag{B3.3m}$$

Therefore when the $pH = pK_a$, 50% of the acid is dissociated (CH_3COO^-) and 50% is undissociated (CH_3COOH). If one is studying the dissociation of a carboxyl functional group associated with soil organic matter, and pK_a is 5, at pH 5, 50% would be in the undissociated carboxyl ($R-C=O-OH$) form and 50% would be in the dissociated carboxylate ($R-C=O-O^-$) form.

TABLE 3.11   Total acidity, i.e., charge (in $mmol_c\ kg^{-1}$) and acidity associated with different functional groups ($mmol_c\ kg^{-1}$) of soil humic and fulvic acids.

Functional group	Humic acids		Fulvic acids	
	Mean	Range	Mean	Range
Total acidity	6700	5600–8900	10,300	6400–14,200
Carboxyl ($-C=O(-OH)$)	3600	1500–5700	8200	5200–11,200
Phenolic ($Ar-OH$)[a]	3900	2100–5700	3000	300–5700
Alcoholic ($R-OH$)[a]	2600	200–4900	6100	2600–9500
Total carbonyl ($C=O$)	2900	100–5600	2700	1200–4200
Methoxyl (OCH_3)	600	300–800	800	300–1200

[a] *Ar is an aromatic ring and R is an aliphatic carbon.*
Adapted from Stevenson (1982) and Thurman (1985), with permission.

aromatic structures, such as char, coal, and soot, can result from the uncontrolled burning of biomass. Due to the predominant aromatic structures of black C and interactions of the charred structures with clay minerals, the residence time of the bulk of black C in the soil is rather long, ranging from a few hundred years to over 1000 years.

Black carbon can produce positive effects in certain soil types and environmental conditions. For example, it can increase pH in acidic soils, enhance water retention in sandy soils, increase nutrient concentration and availability, and adsorb pollutants (Kookana et al., 2011). Due to these beneficial effects, black C has been intentionally produced, commonly called biochar, for soil and other environmental applications in the last few decades. Biochar is produced by burning biomass in oxygen-deficient environments, and it has been applied to soils for long-term C storage and other benefits on forest and agricultural land. Because of its long-term residence time in the soil, black C (including biochar) is an important sink for C in the soil (Fang et al., 2014). With prolonged aging in the soil, there is some oxidation of black C, and carboxyl, phenolic hydroxyls, and quinone-type carbonyl groups develop on its surfaces, which contribute to sorption reactions in the soil (Singh et al., 2014).

Black C concentration in SOC ranges from 0% to 60%, with an average of 13.7%, of the total OC (Reisser et al., 2016). The highly variable concentration of black C has been attributed to different environmental factors, including fire and land use, climatic conditions, and soil properties. The accuracy of the existing black C data in soils has been questioned because of the analytical difficulties and variable and inconsistent techniques used in the analyses.

3.6 Genesis of Humic Substances in the Soil

Key points: Various pathways have been proposed for the evolution of humic substances in the soil, and there is no conclusive evidence or consensus at this point yet.

Plants deposit organic C constituents in the soil both above- and belowground. Plant residues include structural litter from dead shoot and root tissues, dissolved organic C from root exudates, and leaf litter leachate. In addition to plant residues, dead faunal tissues are also added to the soil. Organic residues in the soil can undergo different degradation reactions in natural systems, which include biotic (enzymatically catalyzed), pyrolytic, and abiotic reactions exclusive of pyrolytic reactions. However, the microbial degradation of organic residues is the main degradation mechanism in most natural systems, with sporadic fires contributing in some conditions. Under suitable conditions, soil microbes degrade complex organic compounds via transformation, decomposition, and mineralization processes (Sidebar 3.12). In a stable ecosystem, such as a forest or grassland, equilibrium exists between the formation and decomposition of humic substances.

Sidebar 3.12 — Degradation Mechanisms of SOM

Mineralization: Microbial conversion of organic substances into soluble inorganic constituents, including carbon dioxide, water, and nutrients.

Transformation: Transfer of organic carbon within a distinct chemical structure to another chemical structure caused by enzymatic attack or chemical reactions.

Decomposition: Breakdown of organic macromolecules into smaller organic molecules and inorganic constituents of organic matter. Decomposition is usually mediated by microorganisms and includes depolymerization and oxidation reactions; however, it could be due to physical, chemical, or biological processes.

Several pathways have been proposed to explain the genesis of humus substances during the decay of plant and animal residues in soil. These can be considered under two broad groups: (i) the humic polymer model and (ii) the molecular aggregation model.

3.6.1 Humic Polymer Model

The key processes for the origin of humic substances in this model are the polycondensation and polymerization of organic compounds formed from the decomposition of plant and microbial precursors. Polymers are formed by the condensation of monomers with the release of a low-molecular-weight byproduct or by the successive addition of low-molecular-weight monomer molecules to the active sites at the end of the growing chain. Humic substances in this model are thus considered large polymers with unique chemical structures that are different from those of the plant and microbial precursors. The process of humification is defined as the progressive transformation and conversion of organic residues to humic substances via biochemical and abiotic processes.

Several mechanisms have been proposed for the formation of soil HS under this model (Figure 3.8). Selman Waksman's classical theory, the so-called lignin theory, was that HSs are modified lignins that remain after microbial attack (pathway 4 of Figure 3.8). The modified lignins are characterized by a loss of methoxyl (OCH$_3$) groups and the presence of (ortho)-hydroxyphenols and oxidation of aliphatic side chains to form COOH groups. These lignins undergo more modifications and then result first in HA and then FA. Pathway 1, which is not considered significant, assumes that HSs form from sugars (Stevenson, 1982).

The polyphenol theory of HS genesis via pathways 2 and 3 (in Figure 3.8) involving quinones has found a wider acceptance. In pathway 3 lignin is an important component of HS creation, but phenolic aldehydes and acids released from lignin during microbial attack enzymatically are altered to quinones, which polymerize in the absence or presence of amino compounds to form humic-like macromolecules. Pathway 2 (Figure 3.8) is analogous to pathway 3, except the polyphenols are microbially synthesized from nonlignin C sources (e.g., cellulose) and oxidized by enzymes to quinones and then to HS (Stevenson, 1982).

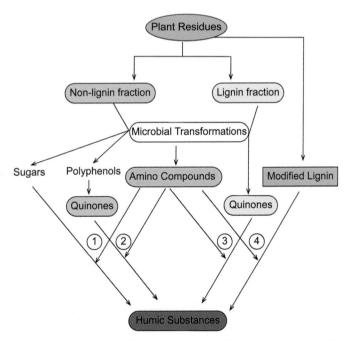

FIGURE 3.8 Mechanisms for the formation of soil humic substances. Amino compounds synthesized by microorganisms are seen to react with modified lignin (pathway 4), quinones (pathways 2 and 3), and reducing sugars (pathway 1) to form complex dark-colored polymers. From Stevenson (1982), with permission.

While the polyphenol theory pathways are generally more acceptable than the other pathways in the humic polymer model, all four pathways may occur in all soils. However, one pathway is usually prominent in a certain soil environment. For example, pathway 4, the lignin pathway, may be primary in poorly drained soils while the polyphenol pathways (2 and 3) may predominate in forest soils (Stevenson, 1982).

The humification polymer models have been criticized for several reasons. Various pathways of this model have been developed based on the chemical and physical data of soil HA isolates or group of isolates or model molecules. Similarly, polymerization processes for the model have been carried out under controlled conditions in the laboratory. During extraction, oxidation and modification of organic compounds can occur and thus may not represent the actual structural and chemical composition that exists under natural conditions. Also, protoplasmic and structural components of fresh organic tissues are dissolved in the extraction process and mixed with humic extracts.

The recalcitrance or selective preservation of humic substances has also been questioned based on the shorter residence time of complex organic compounds such as lignin in the soil. Soil HA isolates obtained with 0.1 M NaOH extraction of soils are composed of mixtures of compounds, including aliphatic acids, ethers, esters, and alcohols; aromatic lignin-derived fragments; and polysaccharides and polypeptides, and thus do not represent a single identifiable entity. Schnitzer and Monreal (2011) proposed a new pathway for HS synthesis rather than the humic polymer model, in which the synthesis and excretion processes of soil microbial polyketides (PKs) may constitute the central unit structure for HS synthesis. Due to their bioactivity, strong and rapid adsorption to clay particles, and the high energy content of their bonds, the microbial PKs (Sidebar 3.13) are very stable against biological decomposition and thus represent passive C pools in the soil. Moieties of the PK adsorb carbohydrates,

proteins, lipids, and N-heterocyclic molecular associations, which can be further modified and polymerized by catalytic mineral surfaces to form large soil humified macromolecules of diverse chemical compositions. These processes create large structures and confer additional stability to the trapped organic molecules.

Sidebar 3.13 — Soil Microbial Polyketides

Polyketides are structurally diverse and biologically active secondary metabolites derived from bacteria, fungi, plants, and animals. In their simplest form polyketides contain alternating carbonyl ($-C=O$) groups and methylene groups $-CO-CH_2-$) and include several compounds that have antihelmintic, insecticidal, antibiotic, and antienzymic properties even at low concentrations.

Metabolites are the intermediates and products of metabolism; the term metabolite is usually restricted to small molecules.

Primary metabolites are present in all microorganisms and are involved in the growth, development, and reproduction of the organism.

Secondary metabolites do not play a role in growth, development, and reproduction. They are produced through the modification of primary metabolite syntheses, typically formed during the end or near the stationary phase of growth.

Biosynthesis is an enzyme-catalyzed, multistep process where substrates are converted into more complex products in living organisms.

There are about 100,000 secondary plant, animal, and microbial metabolites with a MW of less than 2500 Da, and out of these, about 50,000 have origin in soil microorganisms. PKs consist of aromatic, polyaromatic, alkylaromatic, and alkylated molecular structures such as polyphenols, macrolides, polyenes, enediynes, and polyethers. Plants, animals, and soil microorganisms synthesize polyphenols, which consist of compounds having a moderate solubility in water,

MW between 500 and 4000 Da, >12 phenolic hydroxyl groups, and 5—7 aromatic rings per 1000 Da (Quideau et al., 2011). Using Py-FIMS, Monreal et al. (2010) showed that putative PKs represented by alkylaromatic, aromatic, phenolic, and lipid structures made up the great majority of SOM in the clay and nanosized soil fractions, and the age of this carbon was >1000 years.

3.6.2 Molecular Aggregation Model

In contrast to the humic polymer model, the molecular aggregation model or supramolecular associations is based on the aggregation of several relatively small and chemically diverse biomolecules, which are derived from the degradation and decomposition of dead biological material (Wershaw, 1986, 2004; Piccolo, 2001; Sutton and Sposito, 2005). The molecular masses of humic substances, according to this model, are much lower (500—6000 Da) than those reported in the polymerization model. Further, these biomolecules are linked together by hydrogen bonds and other hydrophobic dispersive forces, such as van der Waals, $\pi-\pi$, and CH$-\pi$ interactions (Sidebar 3.14).

Sidebar 3.14 — $\pi-\pi$ and CH$-\pi$ Interactions

Aromatic $\pi-\pi$ interactions are attractive and noncovalent bonds between organic compounds containing aromatic components at the intermolecular overlap of orbitals in conjugated systems.

The CH$-\pi$ hydrogen bond is an attractive molecular force between the C—H orbital and π —system. This is the weakest of the hydrogen bonds.

Wershaw (1986) initially proposed the model, so called the "membrane model," in which humic substances exist as micelles in solution and as bilayer membrane coatings on mineral grains or bound to charged clay surfaces in soils and sediments (Figure 3.9; Sidebar 3.15). The sizes of molecular aggregates are dependent on the solution pH and concentration of humic substances.

(A)   **(B)**

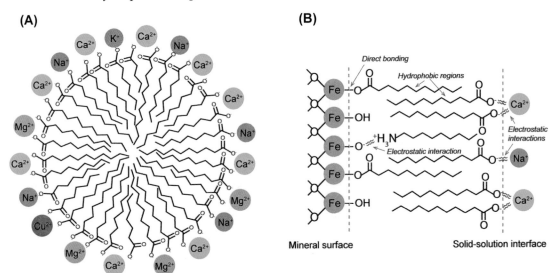

FIGURE 3.9 Schematic representation of amphiphilic humic compounds: (a) a micelle structure in solution formed by the self-aggregation of compounds with a hydrophobic interior and a hydrophilic exterior surface where cations can be adsorbed by electrostatic interaction and (b) a membrane-like structure formed on the surface of an iron oxyhydroxide that has both positive and negative charges; humic compounds are directly bound to or electrostatically attracted to the oxide surface; cations are attached to the negative charge on humic substances via electrostatic interaction at the water interface based on Wershaw (1986).

The aggregates of humic substances were found to mainly consist of lignin, carbohydrate, and lipid fragments derived from the enzymatic degradation of plant tissues. It was also established that in aggregates positively charged amide or amino groups of amino acids (or proteins) were electrostatically bound to clay surfaces or negatively charged carboxylate groups of HAs. Polyvalent cations, such as Fe^{3+}, Al^{3+}, and Mg^{2+}, also contribute to aggregate humic compounds. The incorporation of inorganic and organic molecules in the humic aggregates is also considered.

In the molecular aggregation model FAs are considered associations of small hydrophilic molecules that carry enough acidic functional groups to keep the FA clusters dispersed in solution at any solution pH. Contrary to this, the HAs are composed of associations of predominantly hydrophobic compounds, such as polymethylenic chains, fatty acids, and steroid compounds, which are stabilized by hydrophobic dispersive forces at neutral pH. At lower pH values, supramolecular associations grow progressively in size when intermolecular hydrogen bonds are increasingly formed until they flocculate.

There is an ongoing debate about the accurate representation of SOM with experimental evidence in support of and against both models; neither of the models completely and unambiguously accounts for the range of complex properties observed for SOM. Although, the current interpretations are more in favor of the molecular aggregation model. Additional theories have been proposed in relation to SOM. It has been suggested that, for a better understanding, SOM be separated into compartments: (i) the partially degraded plant tissues, (ii) the microbial biomass, (iii) the organic coatings on mineral particles, (iv) the black (pyrolytic) carbon, (v) the organic precipitates, and (vi) dissolved organic matter in soil solution. The drawback of this theory is that in the separation of different compartments, using physical or chemical procedures, there may be alteration in the structure and composition of humic substances. Additionally, the compartments are not homogeneous and unique compounds, with each compartment being composed of a set of chemical substances, perhaps containing both inorganic and organic substances. These compartments are expected to interact with each other and the inorganic soil components. Therefore even after the detailed characterization of SOM in each compartment, it may not be possible to model the entire SOM in different natural environments.

3.6.3 Soil Continuum Model

Lehmann and Kleber (2015) proposed the "soil continuum model," which considers SOM as a continuum of organic fragments that are continuously processed by the decomposer community toward a smaller molecular size. The breakdown of large molecules in plant material leads to a decrease in the size with concurrent increases in polar and ionizable groups on resulting compounds and thus their increased solubility in water. During the decomposition process in the soil, the opportunity for chemical interactions with minerals and incorporation into aggregates increases the protection of OM against further decomposition. The model does not consider the products resulting from microbial transformation, which have been shown to have a greater degree of inherent persistence because of their composition than the original plant materials from which they are formed (Hayes and Swift, 2020). The continuum model of SOM is an overly simplified view of the different microbial, chemical, and other processes that occur in the soil. While it is well accepted that all forms of SOM eventually convert to carbon dioxide and water, pathways (such as polymerization and molecular aggregation) other than incorporation into aggregates and interactions with minerals cannot be disregarded.

3.6.4 The Way Forward

In a complex system such as soils SOM is likely to be evolved via multiple pathways depending on various factors, such as the nature of organic residues (both above- and belowground), black C, a range of microbes and other organisms, varying soil properties including clay mineralogy, and climatic conditions. At present, there is no conclusive evidence for the existence of a single pathway for the evolution of SOM. A generalized model of the SOM cycle representing different pathways and processes is presented in Figure 3.10.

Soil humic substances are evolved by intertwined microbiological, chemical, and physical processes, which have a variety of complex chemical structures and a specific structure of humic substances may not be universally present. Some chemical compounds are known to have recalcitrant chemical structures; these compounds may be present in plant tissues (e.g., n-alkanes, n-fatty acids, and lignin) or produced by soil microbes and fauna (e.g., chitin, murein, melanin, polyphenols, and polyquinones), and pyrolyzed C results from natural fires and anthropogenic activities. Such resistant structures of both microbial and plant origin may

serve as a template or backbone of different families of humic substances that are extracted with the procedure described in Figure 3.7.

3.7 Organo-mineral Associations in the Soil

Key points: Organic matter is preserved in association with minerals in the soil. The stabilization of SOM is via (i) restricted accessibility to microbes due to the occlusion of OM in soil aggregates and/or (ii) chemical interactions with mineral surfaces involving ligand exchange, polyvalent cation bridges, and weak interactions.

Physico-chemical associations between soil minerals and plant or microbial-derived organic matter can result in the protection of theoretically thermodynamically unstable (or labile) compounds within soil for decades. Soil minerals often constitute a mixture of phyllosilicates, oxides, and hydroxides, as well as short-range order and noncrystalline minerals in some soil types. These particles are generally found in the clay (<2 μm) size fraction of soils.

The accessibility of SOM to microorganisms can be hindered through mineral interactions by adsorbing or occluding organic compounds and immobilizing exoenzymes responsible for the depolymerization reactions.

SOM interacts with soil minerals, particularly with clay minerals (including metal oxides), and these interactions have profound effects on both organic and inorganic materials. Useful reviews on this topic can be found in a number of sources (Mortland, 1970; Greenland, 1971; Mortland et al., 1986; Schnitzer, 1986; Kögel-Knabner et al., 2008; Kleber et al., 2015). Clays tend to stabilize SOM (see Sidebar 3.16), and a correlation is often observed between the clay content and SOM. It has been estimated that from 52% to 98% of all C in soils occurs in mineral-organic matter associations (Stevenson, 1982).

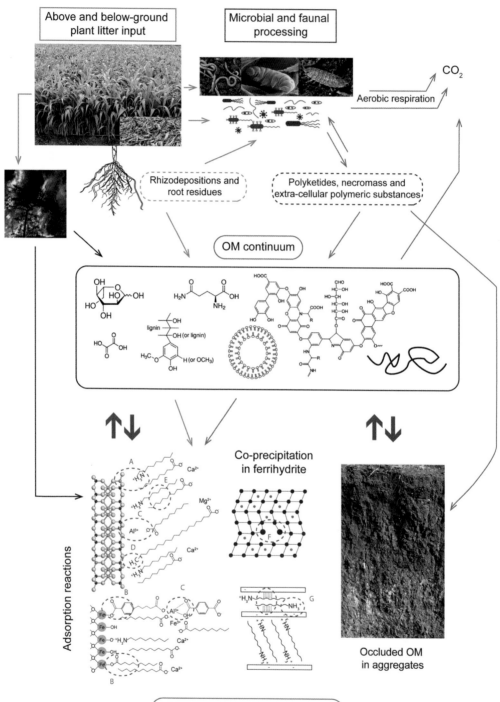

Above and below-ground plant litter input

Microbial and faunal processing

CO_2

Aerobic respiration

Rhizodepositions and root residues

Polyketides, necromass and extra-cellular polymeric substances

OM continuum

Co-precipitation in ferrihydrite

Adsorption reactions

Occluded OM in aggregates

Organic matter-Minerals interactions

FIGURE 3.10 A generalized model of soil organic matter representing major pathways and components. In the organic matter continuum organic substances can range from simple compounds (e.g., organic acids, simple sugars) to complex and large substances derived from the microbial processing of organic residues and organic substances resulting from pyrolytic and abiotic processes. Mineral-associated (involving phyllosilicates, metal oxides including poorly crystalline and short-range order minerals) OM involving electrostatic interactions (encircled area A), inner-sphere complexation (encircled area B), cation bridges (encircled area C), van der Waals forces (encircled area D), hydrophobic interactions (encircled area E), coprecipitation (encircled area F), and charged organic cations in the interlayers of an expansible layer silicate (encircled area G) are shown in the figure. Occluded OM in aggregates is protected against decomposition due to inaccessibility to soil microbes and decreased diffusion of enzymes and oxygen. The OM in aggregates includes cementing agents of biological origins, including microbial plants and faunal.

Sidebar 3.16 — SOM Stabilization

SOM stabilization is a broad term that defines the protection of SOM from mineralization. The rate of turnover of OM is decreased (and mean residence time increased) by association with minerals; SOM stability is the integrated effect of organo-mineral interactions, accessibility, and recalcitrance.

3.7.1 Occluded Organic Matter

Some organic substances glue soil particles together, resulting in stable aggregates of different sizes: microaggregates (53–250 μm) and macroaggregates (>250 μm). Various compounds of plant, microbial, and faunal origins, such as microbial cells, secretions, extracellular polymeric substances, fungal hyphae, root exudates, and faunal mucus, act as a cementing agent in binding together soil particles. The importance of polysaccharides produced by both plants and microbes as glues in soil aggregates is well recognized.

Glomalin, an extracellular protein produced by arbuscular mycorrhizal fungi, may also promote soil aggregation. There is widespread evidence for the protection of OM from degradation in soil aggregates. Organic matter occluded in soil aggregates may not be accessible to microorganisms and their enzymes. Additionally, restricted oxygen supply in the pore space may reduce the microbial mineralization of SOM.

3.7.2 Complexation of Organic Matter With Minerals

The complexation of SOM with minerals is a primary stabilization mechanism. The types of interactions involved in SOM–clay complexes include (i) electrostatic interactions (cation and anion exchange processes), (ii) cation and anion bridges (whereby the polyvalent metal forms a bridge between a functional group of an organic compound and the inorganic surface to which it is bound), (iii) inner-sphere complexation (or specific adsorption), (iv) coprecipitation, (vi) interactions via van der Waals forces, (vii) hydrophobic interactions, and (viii) hydrogen bonding. Two or more of these mechanisms may occur simultaneously, depending on the type of organic material, the nature of the exchangeable ion on the clay surface, the surface acidity, and the moisture content of the system (Schnitzer, 1986).

Electrostatic bonding (outer-sphere complexation) can occur via cation or anion exchange or protonation (Figure 3.10). The cationic property of a weakly basic organic molecule is strongly pH dependent. Thus this mechanism is dependent on the basic character of the organic molecule, the pH of the system, the properties of the organic cation or chain length, and the type of cation on the clay surface. The sorption of organic compounds on negatively charged clay minerals occurs only when polyvalent metal cations such as Al^{3+}, Fe^{3+}, Ca^{2+}, and Mg^{2+} are present on the exchange sites since they can neutralize the negative charge on both the clay minerals and the deprotonated organic functional groups, for example, COO^- (Figure 3.10). Calcium is weakly held as a cation bridge and can be easily displaced, whereas Al^{3+} and Fe^{3+} are usually bound nonelectrostatically and are difficult to remove (Schnitzer, 1986).

Physical adsorption or retention of organic substances via van der Waals forces is weak and results from changes in the electric charge density of individual atoms. An electrically positive fluctuation in one atom causes an electrically negative fluctuation in a neighboring atom, resulting in a net attractive force. Adsorption due to physical forces occurs with neutral polar and nonpolar molecules, particularly those with high MW (Schnitzer, 1986).

Hydrogen bonding results from the linkage between two electronegative atoms through bonding with a single H^+ ion. The H^+ ion is a bare nucleus with a 1+ charge and a propensity to share electrons with those atoms that contain an unshared electron pair such as O. The hydrogen bond is weaker than ionic or covalent bonds, but stronger than van der Waals attractive forces (Stevenson, 1982). The association of hydrophobic (nonpolar) organic molecules at the mineral−water interface is referred to as a hydrophobic interaction (Sidebar 3.17), which does not include intermolecular forces. The hydrophobic interactions are caused by the net effects of attractive and repulsive forces that occur between mineral surfaces, dissolved organic matter, and water.

Sidebar 3.17 − Hydrophobic and Hydrophilic Functional Groups of SOM

Soil organic matter possesses both hydrophobic and hydrophilic groups, also known as polar molecules and nonpolar groups, respectively.

Hydrophobic (Greek "water fearing or hating") is a molecule or part of a molecule with low polarity, usually characterized by few (if any) polar bonds and/or hydrogen bond acceptors and/or hydrogen bond donors. The hydrophobic characteristic results from the uncharged nature of chemical groups that don't combine with water molecules. Alkyl chain and aromatic rings in organic matter substances are hydrophobic; these groups tend to avoid interaction with water and hydrophilic surfaces.

Water is a polar molecule, which means it carries a partial charge between its atoms. Oxygen, being an electronegative atom, draws the electrons of each bond closer to its core, which gives oxygen a partial negative charge and each hydrogen a partial positive charge.

Hydrophilic ("water bonding or loving") is a molecule or a part of a molecule with high polarity, generally characterized by the presence of a significant number of polar bonds and/or hydrogen bond acceptors and/or hydrogen bond donors. Chemical groups with hydrophilic tendency include ionic (charged) groups and groups that contain oxygen or nitrogen atoms. Carboxyl, hydroxyl, amine, and phenoxy are common hydrophilic groups of SOM. Some molecules, such as phospholipids and fats have a hydrophilic head and a hydrophobic tail.

In addition to microbially processed OM, other organic acids, organic amine cations, and amino acids can also interact with clay minerals. Organic acids are negatively charged in the pH range of most soils, but they are pH-dependently charged; some adsorption can occur through H bonding and van der Waals forces when the pH is below the pK_a of the acidic group and when the organic acid is in the undissociated form. Organic materials such as proteins and charged organic cations can also be adsorbed in the interlayers of expansible layer silicates like montmorillonite.

The properties of both mineral (surface charge and specific surface area) and organic acids (pK_a value and functional groups), along with the solution pH, have a large influence on the adsorption of organic acids at low concentrations that might exist in natural environments (Yeasmin et al., 2014). It was observed that iron oxides (including goethite and ferrihydrite) adsorbed much greater

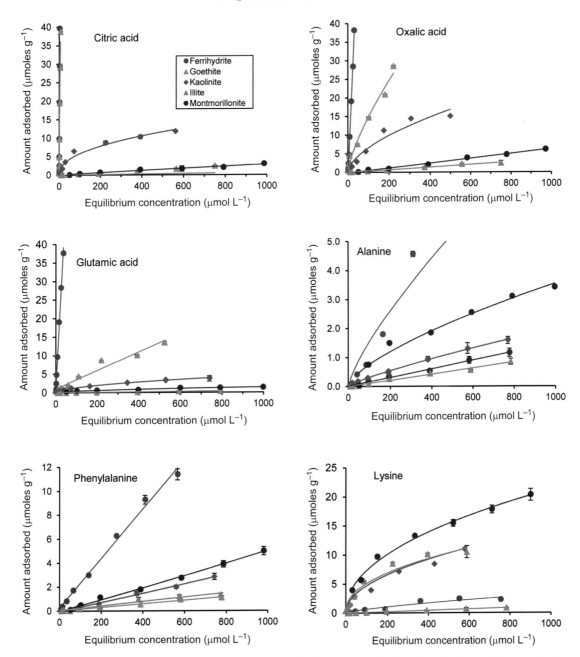

FIGURE 3.11   Adsorption isotherms of citric acid, oxalic acid, glutamic acid, alanine, phenylalanine, and lysine onto ferri-hydrite, goethite, kaolinite, illite, and montmorillonite. Symbols denote experimental data points (mean ± standard error, $n =$ 3), while solid lines represent Freundlich isotherm fitted data from Yeasmin et al. (2014).

amounts of organic acids (e.g., citric acid, oxalic acid, and glutamic acid) than phyllosilicates (montmorillonite, illite, and kaolinite) at the equilibrium solution pH (5.8–8.8), where the organic acids were present in deprotonated states (negatively charged) (Figure 3.11). At these equilibrium solution pH values, Fe oxides would carry a net positive charge and phyllosilicates would predominately possess a negative charge on their surfaces (Yeasmin et al., 2014). Conversely, greater adsorption of lysine was found on phyllosilicates than on Fe oxides, which is expected as predominantly the cationic form of lysine is adsorbed on negatively charged phyllosilicates. Montmorillonite with a greater layer charge adsorbed more lysine than illite and kaolinite. The adsorption of two other amino acids on minerals was relatively small and related to both the specific surface area and charge of minerals with greater adsorption onto ferrihydrite and montmorillonite than the other minerals.

The adsorption of organic ligands at mineral—water interfaces depends on many factors, including solution pH, solution concentration and ionic composition, ligand structure, and residence time (Kleber et al., 2015). Inner- and outer-sphere complexation reactions generally demonstrate a strong pH dependency, with greater adsorption at low pH values. Also, inner-sphere complexation reactions are observed at low pH values, while at neutral and alkaline pH, adsorption via an outer-sphere complexation mechanism is predominant. The influence of pH on adsorption reactions involving van der Waals forces and H-bonding is generally minor or not observed.

At low solution concentrations, organic ligands form inner-sphere surface complexes, and conversely, at high ligand concentrations, outer-sphere complexes are formed. Anions, such as PO_4^{3-} and Cl^- competing for adsorption sites on minerals, can markedly decrease the amount of organic ligand adsorbed. In contrast, polyvalent cations, that is, Fe^{3+} and Al^{3+} at low pH values and Ca^{2+} and Mg^{2+} at high pH values, increase the adsorption of organic ligands via a cation bridging mechanism.

As noted earlier, advanced analytical techniques have greatly enhanced our understanding of SOM in terms of how functional groups are distributed and assemble at mineral surfaces and the mechanisms of SOM—mineral complexation. Chen et al. (2014a) used a combination of STXM and carbon near-edge X-ray absorption fine structure (NEXAFS) spectroscopy to determine the distribution of C in the clay fraction of a pasture soil, the association of C with different elements, and C functional group distribution. Chemical imaging and NEXAFS spectra were obtained using image sequence scans (stacks) over a range of photon energies at a specific element edge. The image sequences were converted to optical density (OD) values. The latter are directly proportional to the thickness of the individual compound in the X-ray beam path. The C distribution map was determined by subtracting the below-edge OD map from the above-edge map. To determine the different regions of C, principal component analyses (PCAs) and cluster analyses (CAs) were carried out on the image sequence data for each individual element (Chen et al., 2014a). Figure 3.12 shows STXM maps of the distribution of C (A and B) and regions of C (C and D) along with C NEXAFS data indicating the functional groups present in the different C regions (E and F) in the clay fractions of a pasture soil located at the summit and footslope, respectively. The brighter white areas in the distribution maps (A and B) indicate higher OD values that can be translated to higher C areas. From Figures C and D, one notes that there are four major regions in the summit soil and three regions in the footslope soil. The predominant C functional groups in the two soils were aromatic C, carboxyl C, and polysaccharides. Regressing elemental OD versus C OD, Chen et al. (2014a) observed strong correlation with even stronger with Ca (Fe and Al).

In the past the correlation with Ca had been associated with the formation of bridging complexes with SOM and metal oxides (Mikutta et al., 2007). Sowers et al. (2018a) observed greater DOM adsorption in the presence of Ca,

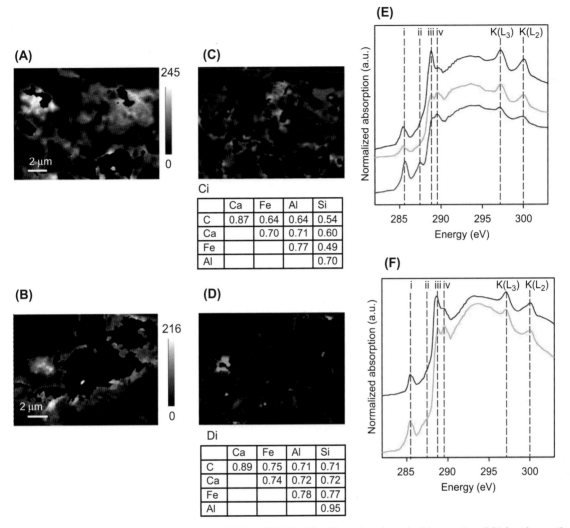

FIGURE 3.12 Carbon image difference maps (OD289-OD282) of the thin regions from the (a) summit and (b) footslope soil clay particles. The gray scale indicates thickness in nanometers. C 1s cluster indices map showing the distribution of C functional groups in the thin regions from the (c) summit soil clay particles, with three distinct regions (*red, green*, and *blue*) and (d) footslope soil clay particles, with two distinct regions (*green* and *blue*). The C 1s NEXAFS spectra were extracted from the regions in the cluster indices maps for the (e) summit and (f) footslope soil clay particles, respectively. Note that the color of the spectra and the color of the region from which they were extracted are the same. Spectra features identified by the *vertical dashed lines* correspond to (i) aromatic (C=C) (285.5 eV), (ii) aliphatic C—H (287.4 eV), (iii) carboxylic (COOH) (288.6 eV), and (iv) polysaccharide C—OH (289.5 eV) functional groups. The peaks at higher energies are from K^+ and correspond to its L_3 and L_2 edges. Correlation coefficients from pairs of thickness values of different elements within the distribution maps of the (Ci) Summit and (Di) footslope soil clay particles. *From Chen et al. (2014a), with permission.*

and as pH increased, the amount of DOM adsorbed decreased (Figure 3.13). The latter is attributed to the increasing negative charge of the ferrihydrite as pH increased, causing less sorption of the negatively charged DOM. The

greater adsorption of DOM with Ca addition suggests that Ca enhances the sequestration of DOM to the ferrihydrite.

To help elucidate the potential mechanism for the role Ca plays in enhancing the sequestration,

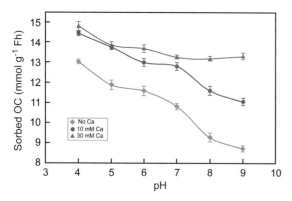

FIGURE 3.13　The adsorption of dissolved organic matter (DOM) at different pH values (4–9) onto ferrihydrite (Fh) – (*green*) with no added Ca, (*blue*) in the presence of 10 mM Ca, and (*orange*) in the presence of 30 mM Ca. An initial C/Fe ratio of 4.7 was used for all samples and experiments were performed in triplicate. *From Sowers et al. (2018a), with permission.*

Sowers et al. (2018a, b) employed a number of molecular-scale spectroscopic and imaging tools. Using attenuated total reflectance Fourier transform infrared (ATR-FTIR) spectroscopy, Sowers et al. (2018a) concluded that Ca formed a ternary complex with OC and Fe. Calcium addition increased the symmetric COO^- peak intensity compared to samples without Ca addition, suggesting a greater association of carboxylic moieties of the DOM with ferrihydrite reacted with Ca. A shift in the symmetric COO^- band suggested that a ligand exchange mechanism was operational between the Fe and COO^-.

In another study Sowers et al. (2018b) employed STXM and CNEXAFS to further investigate the mechanism of Ca complexation with OC and iron-bearing minerals. Figure 3.14 (Sowers et al., 2018b) shows OD maps derived from STXM-NEXAFS data for C, Ca, and Fe. In Figure 3.14a, where no Ca is added, one sees a strong coassociation between C and Fe as represented by the purple region. When Ca is added, the purple regions are mainly gone, and there is a prevalence of white, gray, and yellow regions with hot spots of pink. The white and gray regions show coassociation between Ca, Fe, and C. The yellow regions suggest coassociation of C and Ca. The OD correlation plots of Fe, Ca, and C are shown in Figure 3.15 (Sowers et al., 2018b). Strong correlations occur between Ca and Fe with a C thickness up to 135 nm and with Ca and C. The high correlation between Ca and C relative to the much lower correlation between C and Fe suggests that C is more closely associated with Ca in ternary systems containing Fe, Ca, and C, perhaps indicating that Ca serves as a bridge between C and Fe to promote complexation. The important role of Ca in promoting C-mineral complexation suggests that agricultural practices such as liming soils could promote the sequestration of C.

The accumulation of OM on mineral surfaces may not be continuous, as preferential adsorption in patches at reactive surface sites (e.g., at singly

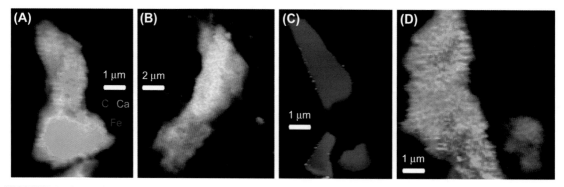

FIGURE 3.14　Color-coded composite RGB optical density maps created from STXM-NEXAFS data. *Red, green,* and *blue* represent carbon, calcium, and iron, respectively. RGB maps are shown for ferrihydrite–DOM (a), ferrihydrite–Ca–DOM (b), ferrihydrite–citric acid (c), and ferrihydrite–Ca–citric acid samples (d). Cyan regions in (a) and (d) were identified as regions too thick for transmission. *From Sowers et al. (2018b), with permission.*

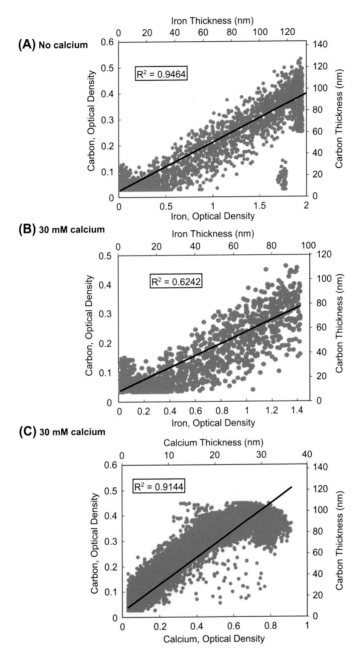

FIGURE 3.15   Elemental optical density correlation plots between C, Ca, and Fe for DOM-bearing ferrihydrite (a) in the absence of Ca and (b) and (c) in the presence of 30 mM Ca. Linear correlation coeffcients are provided for each plot. Thickness values for each element are also provided. *From Sowers et al., 2018b, with permission.*

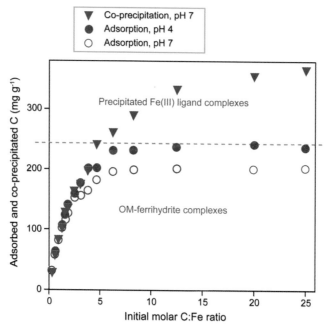

FIGURE 3.16   Adsorption and coprecipitation of dissolved organic matter (DOM) at varying (0.3–25.0) molar C/Fe ratios (Chen et al., 2014b). The dotted red line indicates the maximum adsorption value of DOM onto ferrihydrite.

coordinated hydroxyls on the surfaces of Fe oxides) and rough surfaces have been observed. In addition to adsorption, the coprecipitation of organic matter in poorly crystalline phases of Fe and Al, such as ferrihydrite and $Al(OH_3)$, can occur under highly acidic (pH ~4.5) soil conditions. Coprecipitation is also a common process in natural environments caused by changes in pH or redox potential. Chen et al. (2014b) synthesized coprecipitates and adsorption complexes of Fe−OM using dissolved organic C (DOC) extracted from litter from the forest floor of a Ultisol. Figure 3.16 (Chen et al., 2014b) shows that in both cases as the C:Fe ratio (0.3–25.0) increases, the amount of C sorption increases but is greater in the coprecipitated system than in the adsorption system.

By conducting Fe EXAFS analyses, Chen et al. (2014b) showed that in a coprecipitation system, as the C:Fe ratio increased, precipitated insoluble Fe(III)-organic complexes formed, which explained the higher sorption of C with coprecipitation. To better assess the stability of the C, they

conducted desorption studies using several desorbing agents (Figure 3.17). With all desorption solutions, the C was more stable under coprecipitation compared to adsorption, which suggests a greater biological stability with coprecipitation and greater retention of C in micropores in coprecipitated systems. One sees that even with the strongest desorption solution, sodium pyrophosphate $(Na_4P_2O_7)$, considerable C was retained under both adsorption and coprecipitation, showing the resiliency of SOM when complexed to Fe-bearing minerals. The C seemed to be more stable at lower loadings for both adsorption and coprecipitation mechanisms.

Chen et al. (2014b) conducted STXM studies under lower and higher C loadings (Figure 3.18). The images show that at lower loadings, the C was more homogeneously distributed on the surface of ferrihydrite, which could make it more strongly bound to the ferrihydrite surface compared to the higher loadings where the C appeared to be more heterogeneously

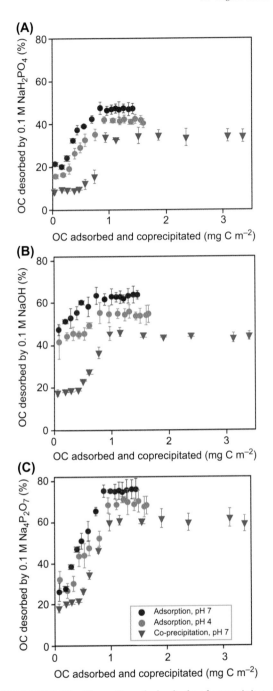

FIGURE 3.17 Desorption of adsorbed and coprecipitated dissolved organic matter with ferrihydrite after a single 24-h extraction with (a) 0.1 M NaH$_2$PO$_4$; (b) 0.1 M NaOH, and (c) 0.01 M Na$_4$P$_2$O$_7$. From Chen et al. (2014b).

distributed and perhaps layered, making it less resistant to desorption.

The adsorption and coprecipitation of OM with clay minerals (including oxides) decrease its susceptibility toward biological and chemical degradation and result in increased mean residence times in the soil. Often the OM in the finer fraction (silt and clay) is older and has a longer turnover time than OM in the coarse (sand) size fraction of the soil. The interaction of OM with Al and Fe oxides is the main reason for its stability in Podzols.

Yeasmin et al. (2017) used a sequential density fractionation scheme to isolate organic matter associated with discrete minerals in four contrasting soils. The separated density fractions were subjected to HF and NaOCl treatments, which confirmed three types of mineral−OM associations: (i) Fe and Al oxides, (ii) phyllosilicates, and (iii) quartz and feldspars. The OM associated with Fe and Al oxide surfaces mostly consisted of aromatic and carboxylate C, whereas the OM associated with phyllosilicates was predominantly protonated amide N and aromatic C, and polysaccharides and protonated amide N were associated with quartz and feldspars. The OM associated with metal oxides was protected more against oxidative degradation compared to the OM in association with phyllosilicates, and quartz and feldspar phases contributed the least to the OM preservation.

3.7.3 Modelling the Assembly of Molecular Compounds on Mineral Surfaces

While a number of models have been proposed regarding C sorption onto mineral surfaces, the molecular assembly of organic compounds at the mineral surface is still not well understood (Coward et al., 2019). One model promoted by Kleber et al. (2007), referred to as the zonal model, proposes that OM compounds assemble in the following sequence: adsorption of polar organic functional groups or hydrophobic interactions, which promote

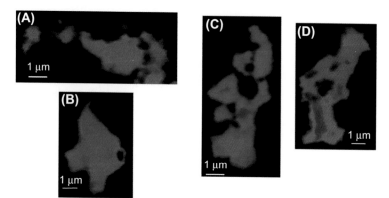

FIGURE 3.18   Color-coded composite maps of carbon (in *blue*) and iron (in *red*) for the (a) adsorption complexes with a molar C/Fe ratio of 1.5, (b) coprecipitates with a molar C/Fe ratio of 1.5, (c) the adsorption complexes with a molar C/Fe ratio of 4.5, and (d) coprecipitates with a molar C/Fe ratio of 5.6 From Chen et al. (2014b).

the formation of a membrane-like bilayer containing a hydrophobic zone and an outer kinetic zone, which is weakly held by cation bridging, hydrogen bonding, and other weak interactions. The onion model involves the formation of an initial layer of CO_2H- and NH_2-containing compounds and hydrophobic zones in succeeding layers (Sollins et al., 2006; Jones and Singh, 2014; Chasse et al., 2015). Another model proposes the formation of patchy structures where primary hydrophobic acid compounds, in patches, react directly with the surface via strong specific interactions and form sorption sites for molecules that can be adsorbed through weak interactions such as H-bonding or van der Waals interactions (Vogel et al., 2014).

Recently, Coward et al. (2019) employed kinetic assays and ultrahigh-resolution Fourier transform ion cyclotron resonance mass spectrometry (FT-ICR-MS) with precise time resolution to determine the molecular association of dissolved organic matter sorbed on goethite. There were three sequential sorption events: retention of aromatic and polycyclic aromatic compounds in 0–15 min, sorption of lignin-like and highly aliphatic compounds in a period of 30–120 min, and simple aliphatic compounds and nitrogenated protein residues and monomers sorbing in 120–240 min (Figure 3.19, Coward et al., 2019). This model supports the concept of a "zonal" assembly model.

3.8 Functional Groups and Charge Characteristics

Key points: Surface functional groups of SOM contribute to a significant portion of CEC of the surface soil, particularly in lighter-textured soils. SOM makes strong complexes with environmentally important trace elements and has a strong affinity for hydrophobic compounds such as pesticides.

3.8.1 Contribution to Soil Cation Exchange Capacity

The surface areas and CEC of SOM, as given earlier, are higher than those of clay minerals. The role that SOM plays in the retention of ions is indeed significant, even in soils where the SOM content is very low. It has been estimated that up to 80% of the CEC of soils is due to organic matter (Stevenson, 1982).

Organic matter is a variable charge soil component. Since its point of zero charge (PZC), defined as the pH at which the colloid particle has no net charge (see Chapter 2 for discussions on points of zero charge), is low, about 3, SOM is overall negatively charged at pH values greater than 3. As pH increases, the degree of negative charge increases due to the deprotonation or dissociation of H^+ from functional groups.

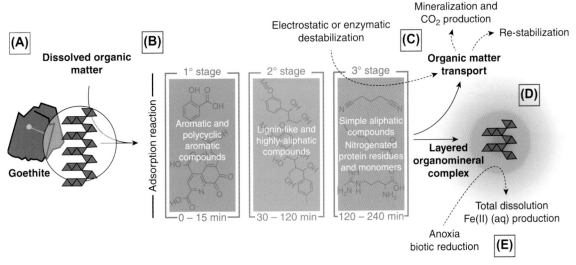

FIGURE 3.19 Conceptual model of sequential adsorptive structuring of organic matter (OM) onto goethite. Exposure of goethite crystal surfaces to (a) dissolved OM catalyzes a tripartite adsorption process (b), driven by temporal molecular fractionation occurring at the solid–solution interface. Compounds of distinct chemical composition and stability self-assemble sequentially via Fe–OM and OM–OM interactions. (c) Adsorption results in fractionated transport of nonsorbed OM, which may lead to CO_2 production or downstream restabilization, and (d) the formation of a layered, organo-mineral complex. (e) The latter may be destabilized via electrostatic or enzymatic mobilization of outer-sphere, loosely bound OM, resulting in increased contributions of aliphatic and proteinaceous compounds or reduction-mediated total dissolution of the mineral itself and all bound OM. From Coward et al. (2019), with permission.

The major acidic functional groups are carboxyl groups ($pK_a < 5$), quinones, and enols (Table 3.10). Quinones can dissociate as readily as carboxyl groups, and they have as low a pK_a as phenolic OH groups. Since carboxyl and phenolic groups can deprotonate at pH values common in many soils, they are major contributors to the negative charge of soils. It has been estimated that up to 55% of the CEC from SOM is due to carboxyl groups (Broadbent and Bradfield, 1952), while about 30% of the CEC of SOM up to pH 7 is due to the quinonic, phenolic, and enolic groups. Neutral and basic functional groups important in SOM are also given in Table 3.10.

Two seminal papers (Helling et al., 1964; Yuan et al., 1967) in the scientific literature clearly established the importance of SOM in contributing to the CEC of a soil. Helling et al. (1964) measured the CEC of 60 Wisconsin soils at pH values between 2.5 and 8.0 and, through multiple correlation analysis, determined the contributions of OM and clay at each pH and

the variation in CEC of each as the pH changed (Table 3.12).

The authors concluded that the CEC of the clay fraction changed much less as pH increased compared to the CEC of the SOM fraction. They attributed this to the dominance of the permanent charge of the clay fraction, primarily composed of montmorillonite and vermiculite, which would increase little with the increase in pH. The small increases in the CEC of the clay fraction were attributed to the dissociation of edge functional groups on the kaolinite surfaces as pH increased.

However, the CEC of the organic fraction of the soils increased dramatically with pH: from 360 mmol$_c$ kg^{-1} at pH 2.5 to 2150 mmol$_c$ kg^{-1} at pH 8.0. This was due to the dissociation of H^+ from the functional groups of the SOM. At pH 2.5, 19% of the CEC was due to SOM, while at pH 8.0, SOM accounted for 45% of the total CEC.

The importance of SOM in sandy soils to the cation exchange capacity is shown in Table 3.13.

TABLE 3.12 Contribution of organic matter and clay fractions to soil CEC as influenced by pH.

Buffer pH	Clay fraction (mmol$_c$ kg^{-1} clay)	Organic fraction (mmol$_c$ kg^{-1} clay)	% of CEC due to SOM
2.5	380	360	19
3.5	450	730	28
5.0	540	1270	37
6.0	560	1310	36
7.0	600	1630	40
8.0	640	2150	45

From Helling et al. (1964), with permission.

TABLE 3.13 Contribution of organic matter and clay to the cation exchange capacities of sandy soils.

Soil group	Average CEC (mmol$_c$ kg^{-1} clay)	Relative contribution (%) SOM	Clay
Entisols			
Psamments	52.6	74.9	25.1
Aquipsamments	38.4	86.8	15.2
Quartipsamments	56.3	75.7	24.3
Acid family	38.3	78.7	21.3
Nonacid family	42.1	95.4	4.6
Phosphatic family	105.8	77.4	22.6
Inceptisols			
Aquepts and Umbrepts	81.7	69.2	30.8
Mollisols			
Aqualls	129.3	66.4	33.6
Spodosols			
Aquods	55.3	96.5	3.5
All soils	67.7	76.1	23.9

Adapted from Yuan et al. (1967).

From 66.4 to 96.5% of the CEC of these soils was due to SOM. In a large study comprising over 2100 soil samples from different soil types and climatic regions of sub-Saharan Africa it was found that OM contributed about 60% of the CEC (Asadu et al., 1997).

SOM is a major contributor to the buffering capacity of soils. A typical titration curve of peat HA and soil HA is shown in Figure 3.20. As base is added, pH increases, illustrating the large buffering capacity of HS that is apparent over a wide pH range and the weak acid characteristics of HS. Most buffer curves of HS from acid soils have a pK$_a$ of about 6. The titration curve (Figure 3.20) can be divided into three zones. Zone I, the most acid region, represents the dissociation of carboxyl groups, while Zone III represents the dissociation of phenolic OH and other very weak acid groups. Zone II

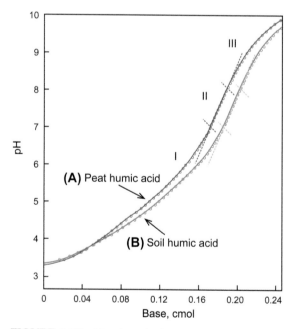

FIGURE 3.20 Titration of a (a) peat and (b) soil humic acid. The small wavy lines on the curves indicate endpoints for the dissociation of weak acid groups having different but overlapping dissociation constants. *From Stevenson (1982), with permission.*

is an intermediate range attributed to a combination of dissociation of weak carboxyl and that of very weak acid groups (Stevenson, 1982).

The weak acid character of SOM is ascribed to complexation with free metals, such as Al^{3+}, Fe^{3+}, and Cu^{2+}, and hydroxy-Al and hydroxy-Fe materials. Thus important functional groups such as carboxyl groups are not always found as free groups in soils but are complexed with metals (Martin and Reeve, 1958).

3.8.2 Humic Substance–Metal Interactions

The complexation of metal ions by SOM is extremely important in affecting the retention and mobility of metal contaminants in soils and waters. Several different types of SOM–metal reactions can occur (Figure 3.21).

These include reactions between dissolved organic carbon (DOC), which is the organic C passing through a 0.2-μm filter, and metal ions, complexation reactions between suspended (colloidal) organic matter (SOC retained on a 0.2-μm silver filter) and metal ions, and bottom sediment and metal ions. The functional groups of SOM have different affinities for metal ions (Charberek and Martell, 1959), given as follows:

$$-O^- > -NH_2 > -N=N- > N/ \\ \text{(enolate)} \quad \text{(amine)} \quad \text{(azo compounds)} \quad \text{(ring N)}$$

$$> -COO^- > -O^- - > /C=O \\ \text{(carboxlate)} \quad \text{(ether)} \quad \text{(carbonyl)}$$

If two or more organic functional groups (e.g., carboxylate) coordinate the metal ion, forming an internal ring structure, chelation, a form of complexation, occurs (Figure 3.22). The total binding capacity of HA for metal ions is about $200-600$ μmol g^{-1}. About 33% of this total is due to retention on cation complexing sites. The major complexing sites are carboxyl and phenolic groups.

(a) Factors Affecting Metal–Ligand Interactions

The types of interactions between metal ions and ligands, such as inorganic ions (anions)

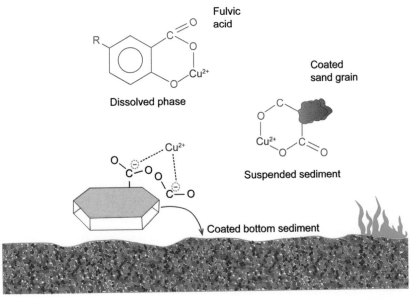

FIGURE 3.21 Complexation of metal ions by organic matter in suspended sediment, bottom sediment, colloidal, and dissolved phases. *From Thurman (1985), with permission.*

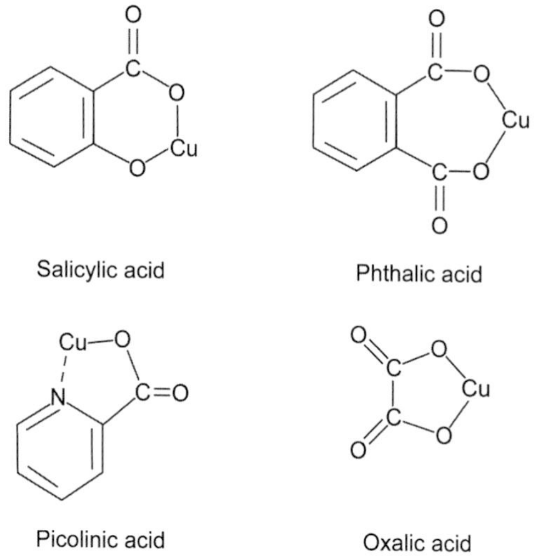

Salicylic acid   Phthalic acid

Picolinic acid   Oxalic acid

FIGURE 3.22   Chelation of metal ions by organic compounds. *From Thurman (1985), with permission.*

Sidebar 3.18 — The Ligand Is an Atom, Functional Group, or Molecule Attached to the Central Atom of a Coordination Compound.

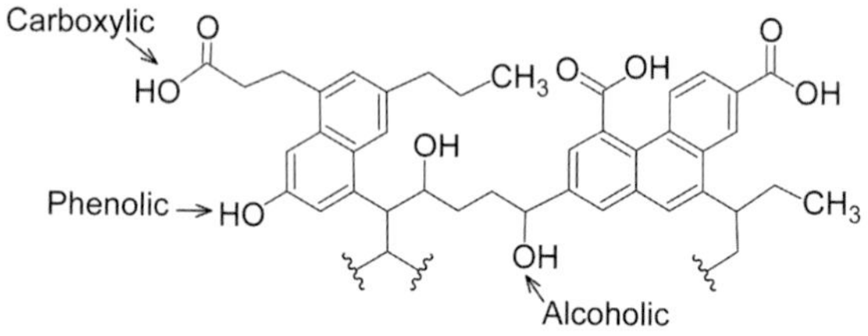

Ligands of SOM shown in a hypothetical structure of a humic substance.

and organic (e.g., carboxyl and phenolic groups of SOM) molecules (Sidebar 3.18), can be predicted based on: (i) the hydrolysis properties of elements and (ii) the concept of "hardness" and "softness" of metals and electron-donor atoms of complexing sites (Buffle and Stumm, 1994).

(a.1.) Hydrolysis Properties

Inorganic elements in the periodic table can be divided into three groups based on their reactions with OH^- or O^{2-} (Figure 3.23). Group 1 elements form nondissociated oxo-complexes (e.g., SO_4^{2-}, PO_4^{3-}) and oxyacids (e.g., $As(OH)_3$). Group 2 elements are highly hydrolyzed but can also occur as hydrated cations (e.g., Fe(III)),

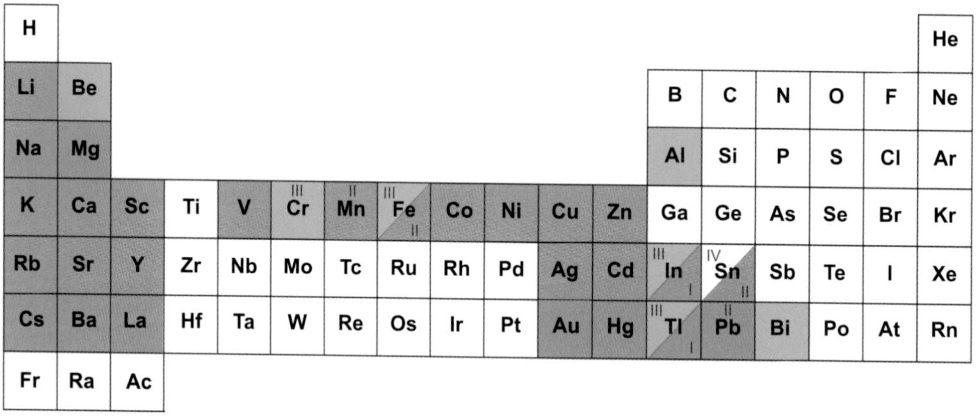

FIGURE 3.23   Group 1 elements (white squares), Group 2 elements (yellow-brown-shaded squares), and Group 3 elements (green-shaded squares); the relevant valence states of the elements are given. Group 1, 2, and 3 elements are defined in text. Redrawn based on Buffle and Stumm (1994).

and Group 3 elements do not have very stable hydroxo-complexes even at high pH (e.g., Ca^{2+}, Zn^{2+}). The major ligand in water is OH^-; ligands other than OH^- will combine only with Group 3 elements and to some extent with Group 2, but not at all with Group 1 elements.

(a.2.) Hard and Soft Characteristics

Elements can also be classified based on their hard and soft characteristics (see Box 3.4 for a discussion of hard and soft elements). Hard cations (Group 1) such as alkali metals (Na^+ and K^+) and alkaline earth metals (Mg^{2+} and Ca^{2+})

BOX 3.4

Lewis acids and bases and "The Principles of Hard and Soft Acids and Bases" (HSAB Principle)[a]

Definitions and characteristics

$$A + : B \rightarrow A : B$$

A = Lewis acid

: B = Lewis base   (B3.4a)

$A : B$ = acid-base complex

Lewis Acid. An atom, molecule, or ion in which at least one atom has a vacant orbital in which a pair of electrons can be accommodated; thus a Lewis acid is an electron pair acceptor. All metal atoms or ions are Lewis acids. Most cations are Lewis acids. In Equation (B3.4a) above, when A is a metal ion, B is referred to as a ligand. Lewis acids are coordinated to Lewis bases or ligands.

Lewis Base. An atom, molecule, or ion that has at least one pair of valence electrons not being shared in a covalent bond; thus a Lewis base is an electron pair donor. Most anions are Lewis bases.

Hard Acid. The acceptor atom is of high positive charge and small size and does not have easily excited outer electrons. A hard acid is not polarizable and associates with hard bases through ionic bonds.

Soft Acid. The acceptor atom is of low positive charge and large size and has several easily excited outer electrons. A soft acid is polarizable and prefers soft bases through covalent bonds.

Hard Base. The donor atom is of low polarizability and high electronegativity, is hard to

reduce, and is associated with empty orbitals of high energy and thus is accessible.

Soft Base. The donor atom is of high polarizability and low electronegativity, is easily oxidized, and is associated with empty, low-lying orbitals.

Classification of Lewis acids and bases

Lewis acids	Lewis bases
Hard acids: Group 1 metals[b]	Hard bases
H^+, Li^+, Na^+, K^+, Be^{2+}, Mg^{2+}, Ca^{2+}, Sr^{2+}, Fe^{3+}, Al^{3+}, Se^{3+}	H_2O, OH^-, F^-, PO_4^{3-}, SO_4^{2-}, Cl^-, CO_3^{2-}, ClO_4^-, NO_3^-
Transition acids: Group 2 metals	Transition bases
Cr^{2+}, Mn^{2+}, Fe^{2+}, Co^{2+}, Ni^{2+}, Cu^{2+}	Br^-, NO_2^-, N_2
Soft acids: Group 3 metals	Soft bases
Ag^+, Au^+, Tl^+, Cu^+, Zn^{2+}, Cd^{2+} Hg^{2+}, Pb^{2+}, Sn^{2+}	I^-, CN^-, CO

[a] *Adapted from Pearson (1963, 1968) and Buffle and Stumm (1994).*
[b] *Refers to metal groups in the text discussion of hard and soft characteristics of elements.*

interact via electrostatic, ionic reactions, while soft cations (Group 3) such as Cu^{2+}, Zn^{2+}, and Cd^{2+} react to form covalent bonds. Transition metals (Group 2) form complexes of intermediate strength.

The degree of hardness can be determined from the term Z^2/r, where Z and r are the charge and radius of the cation, respectively. The preference of hard metals (Group 1) for ligand atoms decreases in the following order:

$$F > O > N \sim Cl \qquad > \qquad Br > I > S$$

Hard donor atoms   Soft donor atoms

This order is reversed for soft metal ions (Group 3). Thus the hard donor atoms such as F and O prefer hard metal ions, while the soft donor atoms such as I and S prefer soft metal ions (Box 3.4).

Ligands can be classified as follows (Figure 3.24) (Buffle, 1984; Buffle and Stumm, 1994): (1) simple inorganic ligands, X, which are major anions — their donor atom is oxygen and they prefer hard metals; (2) "hard" sites of natural organic matter (NOM), referred to as L_H — they are mainly carboxyl and phenolic sites; and (3) "soft" sites of SOM, denoted as L_S, which are N- and S-containing sites. Based on the concentration of ligands and metals in aqueous systems, one can make predictions about metal–ligand interactions.

Group 1 metals (mainly alkali metal and alkaline earth metal cations) prefer hard ligands (Figure 3.24) but form weak bonds (complexes) with them. Thus complexation would occur when the concentrations of metals or ligands are high and the predominant complex would be with inorganic ligands (anions), X.

Group 2 metals, especially divalent transition metals, have an affinity for both hard and soft sites and can react with all three groups of ligands (Figure 3.24). These metals will compete for L_H sites with metals from Group 1, which are less strongly bound but at higher concentrations, and for L_S sites with Group 3 metals, which are at lower concentrations but are more strongly bound. Group 3 metals have a greater affinity for soft sites (L_S) than for hard sites (L_H) or X ligands.

(b) Determination of Stability Constants of Metal–HS Complexes

The determination of stability constants for HS–metal complexes provides information on the affinity of the metal for the organic ligand, and they also provide valuable insights into the fate of heavy metals in the environment. Box 3.5 outlines how stability constants are calculated. Stability constants can be determined using potentiometric titration, chromatography, ultrafiltration, equilibrium dialysis, ion-specific electrodes (ISEs), differential pulse anodic stripping voltammetry (DPASV), fluorescence spectrometry, and modeling (Scatchard method). For more details on these methods, the reader should consult Buffle (1984) and Thurman (1985).

Stability constants are affected by the source of the HS and extraction or isolation procedure employed, concentration of HS, ionic strength (see Chapter 4) of the solution, temperature, pH, method of analysis of the complex, and

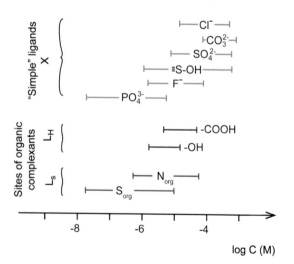

FIGURE 3.24 Ranges of concentrations of ligands or complexing sites in natural fresh waters. $\equiv S\!-\!OH$ refers to inorganic solid surface sites. $-COOH$ and $-OH$ refer to total concentrations of carboxyl and phenolic sites in natural organic matter. N_{org} and S_{org} refer to total concentrations of organic nitrogen and sulfur, respectively. *From Buffle (1984), with permission.*

BOX 3.5

Determination of stability constants for metal–HS complexes

The complexation reaction between a metal ion M and the ith deprotonated ligand (L_i) or binding site in a multiligand mixture at constant pH can be written as (Perdue and Lytle, 1983)

$$M + L_i \rightleftharpoons ML_i \qquad (B3.5a)$$

The stability constant for this reaction can be expressed as

$$K_i^{cond} = \frac{[ML_i]}{[M][L_i]} \qquad (B3.5b)$$

where brackets indicate the concentration, that is, $[ML_i]$ is the concentration of metal bound to ligands (sites) of the ith class, $[M]$ is the concentration of free metal, and $[L_i]$ is the unbound ligand concentration.

It is usually not possible to analytically determine L_i because of H^+ competition and the possibility that the ith metal binding ligand may bind several protons. A more realistic expression of Equation (B3.5b) is

$$K_i'^{cond} = \frac{[ML_i]}{[M][H_{x_i}L_i]} \qquad (B3.5c)$$

where $[H_{x_i}L_i]$ is the concentration of all forms of the ith ligand not bound to M. Both K_i^{cond} and $K_i'^{cond}$ are referred to as conditional stability constants. Conditional stability constants are valid only for the conditions (i.e., pH, neutral salt concentration) stated for a particular system. By definition, the value of K_i^{cond} and $K_i'^{cond}$ will change with changes in pH or neutral salt concentration. Note also that $K_i'^{cond}$ is related to K_i^{cond} through the expression

$$K_i'^{cond} = \frac{[ML_i]}{[M][L_i]} \times \frac{[L_i]}{[H_{x_i}L_i]} = K_i^{cond} \times \frac{[L_i]}{[H_{x_i}L_i]} \qquad (B3.5d)$$

The conditional stability constant $(K_i'^{cond}) = K_i^{cond}$ times the fraction of $[H_{x_i}L_i]$ not protonated. This fraction is constant at constant pH and neutral salt concentration.

It is not possible to determine individual values of K_i^{cond} or $K_i'^{cond}$ in complex mixtures of ligands, such as often occurs with natural systems. Rather, average K_i^{cond} $\left(\overline{K}^{cond}\right)$ or $K_i'^{cond}$ $\left(\overline{K}'^{cond}\right)$ values are determined from experimental data:

$$K_i'^{cond} = \frac{\sum_i[ML_i]}{[M]\sum_i[L_i]} = \frac{\sum_i K_i^{cond}[L_i]}{\sum_i[L_i]} \qquad (B3.5e)$$

$$K_i'^{cond} = \frac{\sum_i[ML_i]}{[M]\sum_i[H_{x_i}L_i]} = \frac{\sum_i K_i'^{cond}[H_{x_i}L_i]}{\sum_i[H_{x_i}L_i]}$$
$$= \frac{(C_M - [M])}{[M]((C_L - C_M) + [M])} \qquad (B3.5f)$$

where C_M and C_L are stoichiometric concentrations of metal and total ligand, respectively. Note that as before $K_i'^{cond}$ is related to K_i^{cond} through an expression similar to Equation (B3.5d):

$$\overline{K}' = \overline{K}^{cond} \times \frac{\sum_i[L_i]}{\sum_i[H_{x_i}L_i]} \qquad (B3.5g)$$

The values of $\overline{K}'^{cond}$ stability constants are the easiest to determine experimentally and are the most common values published in the scientific literature. Their dependence on the total concentration of metal and ligand present in a given system (Equation B3.5f), however, reinforces the fact that these are conditional stability constants and are only valid for the environmental conditions under which they were determined. The application of the concept of stability constants for metal–HS complexes requires using average stability constants determined under similar environmental conditions.

method of data manipulation and stability constant calculation (Thurman, 1985).

Schnitzer and Hansen (1970) calculated conditional stability constants (K_i^{cond}) for metal–FA complexes, based on continuous variations and ion exchange equilibrium methods, and found that the order of stability was:

$$Fe^{3+} > Al^{3+} > Cu^{2+} > Ni^{2+} > Co^{2+} > Pb^{2+}$$
$$> Ca^{2+} > Zn^{2+} > Mn^{2+} > Mg^{2+}$$

The stability constants were slightly higher at pH 5.0 than at pH 3.5, which is due to the higher dissociation of functional groups, particularly carboxyl groups, at pH 5.0. Also, H^+ and the metal ions compete for binding sites on the ligand, and less metal is bound at the lower pH.

3.8.3 Effect of HS–Metal Complexation on Metal Transport

The complexation of HS with metals can beneficially as well as deleteriously affect the fate of metals in soils and waters. The speciation (see Chapter 4) of metals can be affected by these complexes as well as oxidation–reduction reactions (see Chapter 8). For example, HA can act as a reducing agent and reduce Cr(VI), the more toxic form of Cr, to Cr(III). As a Lewis hard acid, Cr(III) will form a stable complex with the carboxyl groups on the HA, further limiting its availability for plant or animal uptake.

Humic substances can serve as carriers of toxic metals, forming complexes that are stable and enhance the transport of toxic metals in waters. In water treatment scenarios these stable, soluble complexes can diminish the removal effectiveness of heavy metals from leachate waters of hazardous waste sites when precipitation techniques are employed as a remediation technique (Manahan, 1991). Additionally, the efficiency of membrane processes (reverse osmosis) and resin processes (ion exchange) used to treat the leachate waters is diminished. On the other hand, the binding of heavy metals

to HS can enhance the biological treatment of inorganic contaminants by reducing the toxicity of the heavy metals to the microbes. Additionally, inorganic anions such as phosphate and cyanide can be removed from water as well through mixed ligand complexation.

3.8.4 Effect of HS–Al^{3+} Complexes on Plant Growth

The importance of HS–Al^{3+} complexes on plant growth is illustrated in studies that have shown that crops often grow well on soils with a pH of <4 in the surface horizon (Evans and Kamprath, 1970; Thomas, 1975). Organic matter contents of the surface horizons, which were enhanced due to no-tillage practices, were fairly high (4%–5%). Exchangeable Al^{3+} (Al^{3+} bound electrostatically on soil colloids such as clay minerals and SOM) and the activity of Al^{3+} (see Chapter 4 for a discussion of ion activity) in the soil solution was low.

Thomas (1975) studied HS–Al^{3+} complexation effects on plant growth with a Maury silt loam soil from Kentucky with an average pH of 6.1 and an organic matter content ranging from 5.11% in the 0- to 7.5-cm layer to 0.80% in the 37.5- to 45-cm layer. Various amounts of 0.1 M HNO_3 were added such that the pH of the soils was 3.10–3.30. No exchangeable H^+ was found in the soils. The relationship between organic matter concentration and exchangeable Al^{3+} at several pH values is shown in Figure 3.25.

One sees that the influence of SOM on exchangeable Al^{3+} at any given pH is greatest up to 2.5%, with a smaller effect at higher SOM contents. The effect of organic matter on exchangeable Al^{3+} was greater at the lower pH values. At pH 3.5, an increase from 1 to 2% organic matter lowered exchangeable Al from 6 to 4.2 $mmol_c$ kg^{-1}. These results show that even small increases in SOM result in a significant reduction in exchangeable Al^{3+} and the activity of Al^{3+} in the soil solution. Therefore the lack of any deleterious effect on plant growth

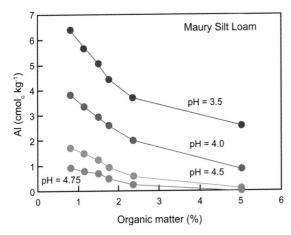

FIGURE 3.25   The relation between organic matter concentration and exchangeable Al at acidic pH values. *From Thomas (1975), with permission.*

even at a low pH may be due to the complexation of Al^{3+} with SOM and the removal of Al^{3+} and perhaps other toxic metals from the soil solution.

3.8.5 Effect of HS on Mineral Dissolution

Humic substances can also cause mineral dissolution. Lichens, bacteria, and fungi, which often grow on mineral surfaces, enhance this breakdown by producing organic complexing materials. At pH > 6.5, HA and FA can attack and degrade minerals and form water-soluble and water-insoluble metal complexes (Schnitzer, 1986). HAs can extract metals from galena (PbS), pyrolusite (MnO_2), calcite ($CaCO_3$), and malachite ($Cu_2(OH)_2CO_3$). HA can also extract metals from silicates (Baker, 1973), but to a lesser degree than from previously mentioned minerals. Various metal sulfides and carbonates can be solubilized by HA, including Pb(II), Zn(II), Cu(II), Ni, Fe(III), and Mn(IV). Solubilization ranges from 2100 μg for PbS to 95 μg for ZnS.

3.9 Retention of Pesticides and Other Organic Substances by Humic Substances

Most pesticides (hydrophobic organic compounds) have a strong affinity for SOM. SOM has an important effect on the bioactivity, persistence, biodegradability, leachability, and volatility of pesticides. In fact, perhaps SOM is the soil component most important in the retention of nonionic pesticides. Because of this assumption, soil organic carbon sorption coefficient (K_{OC}, details in Chapter 5) values of pesticides are frequently used as measures of their relative potential mobility in soils and in "fugacity" models to describe the partitioning of pesticides in the terrestrial environment, namely, soil/water/atmosphere systems (Wauchope et al., 2002). Large variabilities, between 40 and 60%, have been found in the K_{OC} values of the same pesticide in different studies (Wauchope et al., 2002). The authors considered about half of the variation due to experimental errors including SOC estimation, and the rest was attributed to the structural composition of SOM. Thus the amount of a pesticide that must be added to soils is affected by both the quantity and characteristics (e.g., aromaticity, H/O ratio) of SOM and mineral—OM interactions in the soil (Ahmad et al., 2001; Wauchope et al., 2002; Kookana et al., 2014). Olk et al., (2019) have made several observations in relation to pesticide sorption on SOM: (i) a greater affinity for HA than FA, (ii) a greater affinity for HS richer in aromatic C than in aliphatic C, (iii) a greater affinity for HS derived from terrestrial than from aquatic environments, and (iv) pesticides are sorbed more strongly by HS of larger molecular sizes than HS of smaller molecular sizes.

Factors that affect the retention of pesticides by SOM are the number, type, and accessibility of HS functional groups, nature of the pesticides (e.g., cationic, ionizable (i.e., weakly acidic, weakly basic, or zwitterionic), or nonionic molecules), properties of the soil including types and quantities of clay minerals, pH, exchangeable cations, moisture, and temperature (Stevenson, 1982; Kookana et al., 2014). Bailey and White (1970) noted several properties of pesticides, particularly ionic pesticides, that affect the retention and type of bonding on soil components: (i) type of the functional group, that is, carboxyl, alcoholic hydroxyl, or amine; (ii) nature of the substituting group, which alters the functional group, and the position of the substituting groups in relation to the functional group, which may affect bonding; and (iii) the presence and magnitude of unsaturation in the molecule, which affects the lyophilic and lyophobic balance. These factors determine important chemical properties such as the acidity or basicity of the compound, as suggested by the pK_a of the compound.

The mechanisms by which pesticides are retained by SOM are not clearly understood. However, both *ab*sorption/partitioning and *ad*sorption processes are often involved and hence the nonspecific term sorption is often used to describe their retention to SOM. The pesticide molecule may be retained in the internal voids of humus molecules that are sieve like (Khan, 1973). The hole or pore-filling mechanism has been shown to be an important process in the retention of organic compounds, where both flexible (e.g., humic substances) and fixed (e.g., black carbon) pores may be involved in the retention process (Xia and Pignatello, 2001).

The adsorption of pesticides on HS occurs via ion exchange and protonation, H bonding, van der Waals forces, or ligand exchange (where an organic functional group, such as a carboxyl or hydroxyl, displaces an inorganic hydroxyl or water molecule of a metal ion at the surface of a soil mineral, such as a metal oxide; this results in an inner-sphere complex; see discussions on ligand exchange and inner-sphere complexes in Chapter 5), and cation and water bridging. These mechanisms are discussed in detail in Koskinen and Harper (1990).

For nonionic pesticides, hydrophobic bonding (partitioning) is the most common adsorption mechanism, as indicated earlier. This partitioning on hydrophobic SOM surfaces results from a weak solute—solvent interaction or the low solubility or hydrophobic nature of the solute. Important hydrophobic sites on HS include fats, waxes, and resins and aliphatic side chains. Since humus has an aromatic framework and contains polar groups, it may have both hydrophobic and hydrophilic sites. More details on pesticide and organic chemical reaction mechanisms with SOM are given in Chapter 5.

Soluble humic substances can enhance the transport of pesticides in soils and into groundwaters. FAs, which have a low MW and high acidities and are more soluble than HA, can transport pesticides and other organic materials quite effectively. For example, the downward movement of the highly hydrophobic insecticide dichlorodiphenyltrichloroethane (DDT) in the organic layers of some forest soils has been ascribed to water-soluble, humic substances (Ballard, 1971).

Humic substances can also serve as reducing agents and chemically alter pesticides. The alteration is enhanced by the presence of reactive groups such as phenolic, carboxyl, enolic, heterocyclic, aliphatic-OH, and semiquinone, like those contained in FA and HA. The presence of stable free radicals in HS would also indicate that they can cause chemical alterations of pesticides. The hydroxylation of the chloro-s-triazines is an example of the nonbiological transformation of a pesticide by HS (Stevenson, 1982).

Suggested Reading

Aiken, G.R., McKnight, D.M., Wershaw, R.L. (Eds.), 1985. Humic Substances in Soil, Sediments, and Water. Wiley—Interscience, New York, NY.

Buffle, J., 1984. Natural organic matter and metal-organic interactions in aquatic systems. In: Sigel, H. (Ed.), Metal Ions in Biological Systems. Dekker, New York, NY, pp. 165—221.

Buffle, J., Stumm, W., 1994. General chemistry of aquatic systems. In: Buffle, J., DeVitre, R.R. (Eds.), Chemical and Biological Regulation of Aquatic Systems. CRC Press, Boca Raton, FL, pp. 1—42.

Christman, R.F., Gjessing, E.T. (Eds.), 1983. Aquatic and Terrestrial Humic Materials. Ann Arbor Science, Ann Arbor, MI.

Dinel, H., Schnitzer, M., Mehuys, G.R., 1990. Soil lipids: origin, nature, content, decomposition, and effect on soil physical properties. In: Bollag, J.M. (Ed.), Soil Biochemistry. vol. 6. Marcel Dekker, New York, NY, pp. 397–429.

Hayes, M.H.B., MacCarthy, P., Malcolm, R.L., Swift, R.S. (Eds.), 1983. Humic Substances. II: In Search of Structure. Wiley, New York, NY.

Huang, P.M., Schnitzer, M. (Eds.), 1996. Interactions of Soil Minerals With Natural Organics and Microbes. SSSA Special Publications. vol. 17. Soil Science Society of America, Inc., Madison, WI.

Schnitzer, M., 2000. A lifetime perspective on the chemistry of soil organic matter. Adv. Agron 68, 1–58.

Schnitzer, M., Khan, S.U. (Eds.), 1972. Humic Substances in the Environment. Dekker, New York, NY.

Stevenson, F.J., 1982. Humus Chemistry. Wiley, New York, NY.

Suffet, I.H., MacCarthy, P. (Eds.), 1989. Aquatic Humic Substances. Influence of Fate and Treatment of Pollutants. Advances in Chemistry Series. vol. 219. American Chemical Society, Washington, DC.

Thurman, E.M., 1985. Organic Geochemistry of Natural Waters. Kluwer Academic, Hingham, MA.

Problem Set

Q1: A soil contains 3.5% organic carbon in the 0–10 cm layer. Calculate the total amount (stock) of organic carbon and nitrogen in the top 10 cm soil layer of a 1-ha field, assuming the bulk density of the 0–10 cm soil layer is 1.27 g cm^{-3} and C:N ratio in the organic carbon is 12:1.

Q2: In the table below typical elemental concentrations (g kg^{-1}) of soil organic matter, bacterial, wheat straw, and wheat straw biochar are given.

Element	Soil organic matter	Bacteria	Wheat straw	Wheat straw biochar
Carbon	565	500	460	676
Nitrogen	52	140	5	10.5
Phosphorus	10	30	0.8	1.8
Sulfur	8	10	0.6	1.1

(i) Calculate and compare the C:N ratio for each material.

(ii) If bacteria are using wheat straw as a source of energy and assimilating 60% carbon and respiring the rest, what percentage of their N can be derived from the straw. Assume all N in the straw is available to bacteria.

(iii) Using the principles, calculate the percentage of P and S in the bacteria that can be derived from wheat straw.

(iv) From what other sources might the additional N, P and S be obtained by soil bacteria?

Q3: Following the chemical oxidation procedure of Walkley and Black for organic carbon measurement, 0.75 g soil was digested with 10 mL 0.1667 M $K_2Cr_2O_7$ solution and 20 mL concentrated H_2SO_4 in a 500-mL Erlenmeyer flask. After the digestion had completed, 3–4 drops of o-phenanthroline ferrous sulfate indicator were added into the digest, and then it was titrated with 0.5 M $FeSO_4$. The end point (dirty brown to purple via green) was obtained after adding 8.6 mL of $FeSO_4$ solution. A blank sample (i.e., without soil) was processed similarly to the soil sample; in the titration 19.8 mL of $FeSO_4$ solution was used. Calculate the concentration of organic carbon in the soil sample.

Q4: **(a).** Define the term amphoteric. Based on your definition, explain why soil organic matter (SOM) is or is not amphoteric.

(b). Considering the charge characteristics of clay minerals and metal oxides (Chapter 2), is SOM more like smectite or birnessite? Explain.

(c). Given the descriptions of Lewis acid and Lewis base, which better describes SOM? Explain.

Q5: In order to understand the affinity of copper (Cu) for soils you collected from a contaminated site, you want to determine the stability constant.

(a). First of all, you must extract the OM from the soil. How would you go about doing this?

(b). From the extraction, you find that the ratio of acid extractable to base extractable material is 2. You make a $1 \, g \, L^{-1}$ suspension of the soil-extracted SOM and add 2 mM Cu solution to obtain a final $[Cu] = 0.8$ mM. You will conduct the experiment at pH 6.5.

(i) Write the stoichiometric chemical equation that represents the reaction between the metal and OM. Considering this equation, what must be done to conduct the experiment under the specified conditions?

(ii) Calculate the average stability constant, K'^{cond}, for the Cu and SOM (with the composition mentioned in part b), which corresponds to the chemical equation. (Use the minimum average molecular weights of FA and HA).

(iii) If you conducted the experiment under more acidic conditions, how would the K'^{cond} for the CuL_i system change?

(iv) How do you think K'^{cond} would be different if the metal contaminant in question was lead?

Q6: In light of the concern about global warming and the potential problems associated with increased temperatures, researchers propose using soils to sequester carbon. Explain what carbon sequestration is and list the major pools and their contribution to the global carbon cycle. What management practices might be used to limit the loss of carbon from soils?

Q7: Using the hard and soft acid case theory, comment on the reactivity of the following pairs of ion:

a) $H^+ + Br^-$

b) $Pb^{2+} + SO_4^{2-}$

c) $Fe^{3+} + Cl^-$

d) $Mn^{2+} + OH^-$

e) $Cd^{2+} + N_{org}$

f) $Mg^{2+} + -COOH$.

Q8: A soil solution with pH 5.2 was extracted from the rhizosphere soil. The solution contains 0.005 M of acetic acid. The dissociation reaction and pK_a value of acetic acid are given below:

$$CH_3COOH \rightarrow CH_3COO^- + H^+ \, (pK_a = 4.77)$$

Calculate the concentration of CH_3COOH and CH_3COO^- in the soil solution.

Q9: Using the following data (% w/w) of dry, ash-free soil humic compounds that were obtained from the International Humic Substances Society, answer the questions given below.

Sample	C	H	O	N	S
Summit Hill Soil_humic acid	51.03	4.29	43.96	2.71	1.19
Elliott Soil IV_fulvic acid	54.0	4.84	37.90	5.13	0.64

(i) Calculate the maximum potential cation exchange capacity (in $mmol_c \, kg^{-1}$) that can be contributed by the two acids, assigning all oxygen to carboxyl groups. Compare the results for the two acids and also compare them with common phyllosilicates, namely, smectite, vermiculite, illite, and kaolinite.

(ii) Calculate the maximum potential anion exchange capacity (in $mmol_c$ kg^{-1}) that can be contributed by assigning all nitrogen to the amine group.

(iii) If a surface soil contains 3% organic carbon, assuming that OC consists of 50% of humic and FA each, calculate the maximum contribution of OC to the soil cation exchange capacity (in $mmol_c$ kg^{-1}). The total CEC of the soil including both mineral and organic fractions is 650 $mmol_c$ kg^{-1}; comment on the contribution of the two fractions to the soil CEC.

(iv) How much anion exchange capacity can be potentially contributed by OC?

References

Acton, C.J., Paul, E.A., Rennie, D.A., 1963. Measurements of the polysaccharide content of soils. Can. J. Soil Sci. 43, 141–150.

Ahmad, R., Kookana, R.S., Alston, A.M., Skjemstad, J.O., 2001. The nature of soil organic matter affects sorption of pesticides. 1. Relationship with carbon chemistry as determined by ^{13}C CPMAS NMR spectroscopy. Environ. Sci. Technol. 35, 878–884.

Aiken, G.R., McKnight, D.M., Wershaw, R.L. (Eds.), 1985a. Humic Substances in Soil, Sediments, and Water. John Wiley & Sons (Interscience), New York, NY.

Aiken, G.R., McKnight, D.M., Wershaw, R.L., MacCarthy, P., 1985b. An introduction to humic substances in soil, sediment, and water. In: Aiken, G.R., McKnight, D.M., Wershaw, R.L. (Eds.), Humic Substances in Soil, Sediments, and Water. John Wiley & Sons (Interscience), New York, NY, pp. 1–9.

Archard, F.K., 1786. Chemische untertersuchung des Torfs. Crells Chemical Annalen. 2, 391–403.

Arshad, M.A., Ripmeester, J.A., Schnitzer, M., 1988. Attempts to improve solid-state 13C NMR spectra of whole mineral soils. Can. J. Soil Sci. 68, 593–602.

Asadu, C.L.A., Diels, J., Vanlauwe, B., 1997. A comparison of the contributions of clay, silt and organic matter to the effective CEC of soils of Sub-Saharan Africa. Soil Sci. 162, 785–794.

Bailey, G.W., White, J.L., 1970. Factors influencing the adsorption, desorption, and movement of pesticides in soil. Residue Rev. 32, 29–92.

Baker, W.E., 1973. The role of humic acids from Tasmania podzolic soils in mineral degradation and metal mobilization. Geochim. Cosmochim. Acta. 37, 269–281.

Baldock, J.A., Broos, K., 2011. Soil organic matter. In: Huang, P.M., Li, Y., Sumner, M.E. (Eds.), Handbook of Soil Sciences: Properties and Processes Second edition. CRC Press/Taylor Francis Group, Boca Raton, FL, pp. 11.1–11.52.

Ballard, T.M., 1971. Role of humic carrier substances in DDT movement through forest soil. Soil Sci. Soc. Am. Proc. 35, 145–147.

Batjes, N.H., 1996. Total C and N in the soils of the world. Eur. J. Soil Sci. 47, 151–163.

Bohn, H.L., McNeal, B.L., O'Conner, G.A., 1985. Soil Chemistry Second edition. John Wiley & Sons, New York, NY.

Broadbent, F.E., Bradford, G.R., 1952. Cation exchange groupings in soil organic fraction. Soil Sci. 74, 447–457.

Buffle, J., 1984. Natural organic matter and metal-organic interactions in aquatic systems. In: Sigel, H. (Ed.), Metal Ions in Biological Systems. Dekker, New York, NY, pp. 165–221.

Buffle, J., Stumm, W., 1994. General chemistry of aquatic systems. In: Buffle, J., DeVitre, R.R. (Eds.), Chemical and Biological Regulation of Aquatic Systems. CRC Press, Boca Raton, FL, pp. 1–42.

Charberek, S., Martell, A.E., 1959. Organic Sequestering Agents. John Wiley & Sons, New York, NY.

Chasse, A.W., Ohno, T., Higgins, S.R., Amirbahman, A., Yildirim, N., Parr, T.B., 2015. Chemical force spectroscopy evidence supporting the layer-by-layer model of organic matter binding to iron (oxy) hydroxide mineral surfaces. Environ. Sci. Technol. 49, 9733–9741.

Chen, C., Dynes, J.J., Wang, J., Karunakaran, C., Sparks, D.L., 2014a. Soft X-ray spectromicroscopy study of mineral-organic matter associations in pasture soil clay fractions. Environ. Sci. Technol. 48, 6678–6686.

Chen, C., Dynes, J.J., Wang, J., Sparks, D.L., 2014b. Properties of Fe-organic matter associations via coprecipitation versus adsorption. Environ. Sci. Technol. 48, 13751–13759.

Chen, Y., Senesi, N., Schnitzer, M., 1978. Chemical and physical characteristics of humic and fulvic acids extracted from soils of the Mediterranean region. Geoderma. 20, 87–104.

Ciais, P., Sabine, C., Govindasamy, B., Bopp, L., Brovkin, V., Canadell, J., et al., 2013. Carbon and other biogeochemical cycles. In: Stocker, T., Qin, D., Platner, G.-K., et al. (Eds.), Climate Change 2013: The Physical Science Basis. Cambridge University Press, Cambridge, pp. 465–570.

Cornwell, W.K., Cornelissen, J.H.C., Amatangelo, K., Dorrepaal, E., Eviner, V.T., Godoy, O., et al., 2008. Plant species traits are the predominant control on litter decomposition rates within biomes worldwide. Ecol. Lett. 11, 1065–1071.

Coward, E.K., Ohno, T., Sparks, D.L., 2019. Direct evidence for temporal molecular fractionation of dissolved organic matter at the iron oxyhydroxide interface. Environ. Sci. Technol. 53, 642–650.

Dinel, H., Schnitzer, M., Mehuys, G.R., 1990. Soil lipids: origin, nature, content, decomposition, and effect on soil physical properties. In: Bollag, J.-M. (Ed.), Soil Biochemistry. Marcel Kekker Inc., New York, NY, pp. 397–429.

Drake, T.W., Wickland, K.P., Spencer, R.G.M., McKnight, D.M., Striegl, R.G., 2015. Ancient low-molecular-weight organic acids in permafrost fuel rapid carbon dioxide production upon thaw. Proc. Natl. Acad. Sci. U.S.A. 112, 13946–13951.

Dubach, P., Mehta, N.C., 1963. The chemistry of soil humic substances. Soils Fert. 26, 293–300.

Evans, C.E., Kamprath, E.J., 1970. Lime response as related to percent Al saturation, solution Al and organic matter response. Soil Sci. Soc. Am. Proc. 34, 893–896.

Fang, Y., Singh, B., Singh, B.P., Krull, E., 2014. Biochar carbon stability in four contrasting soils. Eur. J. Soil Sci. 65, 60–71.

Garcia-Pausas, J., Romanyà, J., Montané, F., Rios, A.I., Taull, M., Rovira, P., et al., 2017. Are soil carbon stocks in mountain grasslands compromised by land-use changes? In: Catalan, J., Ninot, J.M., Aniz, M.M. (Eds.), High Mountain Conservation in a Changing World. Advances in Global Change Research, 62. Springer, Cham, pp. 207–230.

Giles, C.D., Cade-Menun, B.J., Hill, J.E., 2011. The inositol phosphates in soils and manures: Abundance, cycling, and measurement. Can. J. Soil Sci. 91, 397–416.

Gollany, H.T., Rickman, R.W., Liang, Y., Albrecht, S.L., Machado, S., Kang, S., 2011. Predicting agricultural management influence on long-term soil. Agron. J. 103, 234–246.

Greenland, D.J., 1971. Interactions between humic and fulvic acids and clays. Soil Sci. 111, 34–41.

Guo, L.B., Gifford, R.M., 2002. Soil carbon stocks and land use change: a meta analysis. Glob. Change Biol. 8, 345–360.

Harris, D.C., 1987. Quantitative Chemical Analysis. Second edition. Freeman, New York, NY.

Hatcher, P.G., Bortiatynski, J.M., Knicker, H., 1994. NMR techniques (C, H, and N) in soil chemistry in 15th World Congress of Soil Science. Commission II Transactions 3a, 23–44. Acapulco, Mexico.

Hayes, M.H.B., Swift, R.S., 1978. The chemistry of soil organic colloids. In: Greenland, D.J., Hayes, M.H.B. (Eds.), The Chemistry of Soil Constituents. John Wiley & Sons (Interscience), New York, NY, pp. 179–230.

Hayes, M.H.B., Swift, R.S., 2018. An appreciation of the contribution of Frank Stevenson to the advancement of studies of soil organic matter and humic substances. J. Soils Sediments 18, 1212–1231.

Hayes, M.H.B., Swift, R.S., 2020. Vindication of humic substances as a key component of organic matter in soil and water. Adv. Agron. 163, 1–37.

Helling, C.S., Chester, G., Corey, R.B., 1964. Contribution of organic matter and clay to soil cation-exchange capacity as affected by the pH of the saturation solution. Soil Sci. Soc. Am. Proc. 28, 517–520.

IPCC, 2019. 2019 Refinement to the 2006 IPCC Guidelines for National Greenhouse Gas Inventories. https://www.ipcc.ch/report/2019-refinement-to-the-2006-ipcc-guidelines-for-national-greenhouse-gas-inventories/.

Ishiwatari, R., 1975. Chemical nature of sedimentary humic acid. In: Povoledo, D., Golterman, H.L. (Eds.), Humic Substances, Their Structure and Function in the Biosphere. Centre for Agricultural Publications and Documentation, Wageningen, pp. 87–107.

Jenny, H., 1941. Factors of Soil Formation. McGraw-Hill, New York, NY.

Jenny, H., Bingham, F., Padilla-Saravia, B., 1948. Nitrogen and organic matter contents of equatorial soils of Columbia, South America. Soil Sci 66, 173–186.

Joergensen, R.G., 2018. Amino sugars as specific indices for fungal and bacterial residues in soil. Biol. Fertil. Soils. 54, 559–568.

Jones, E., Singh, B., 2014. Organo-mineral interactions in contrasting soils under natural vegetation. Front. Environ. Sci. 2, 2.

Khan, S.U., 1973. Equilibrium and kinetic studies on the adsorption of 2,4-D and picloram on humic acid. Can. J. Soil Sci. 53, 429–434.

Kirkby, C.A., Kirkegaard, J.A., Richardson, A.E., Wade, L.J., Blanchard, C., Batten, G., 2011. Stable soil organic matter: a comparison of C:N:P:S ratios in Australian and other world soils. Geoderma. 163, 197–208.

Kleber, M., Eusterhues, K., Keiluweit, M., et al., 2015. Mineral-organic associations: formation, properties, and relevance in soil environments. Adv. Agron. 130, 1–140.

Kleber, M., Sollins, P., Sutton, R., 2007. A conceptual model of organo-mineral interactions in soils: Self-assembly of organic molecular fragments into zonal structures on mineral surfaces. Biogeochemistry. 85, 9–24.

Kögel-Knabner, I., Guggenberger, G., Kleber, M., Kandeler, E., Kalbitz, K., Scheu, S., et al., 2008. Organo-mineral associations in temperate soils: integrating biology, mineralogy, and organic matter chemistry. J. Plant Nutr. Soil Sci. 171, 61–82.

Kononova, M.M., 1966. Soil Organic Matter. Pergamon, New York, NY.

Kookana, R.S., Ahmad, R., Farenhorst, A., 2014. Sorption of pesticides and its dependence on soil properties: chemometrics approach for estimating sorption. In: Chen, W., Sabljic, A., Cryer, S., Kookana, R. (Eds.), Non-first Order Degradation and Time-dependent Sorption of Organic Chemicals in Soil, ACS Symposium Series, Vol. 1174. Amer. Chem. Soc., Washington, D.C, pp. 221–240. Chapter 12.

Kookana, R.S., Sarmah, A.K., Van Zwieten, L., Krull, E., Singh, B., 2011. Biochar application to soil: agronomic and environmental benefits and unintended consequences. Adv. Agron. 112, 103−143.

Koskinen, W.C., Harper, S.S., 1990. The retention process: mechanisms. In: Cheng, H.H. (Ed.), Pesticides in the Soil Environment: Processes, Impacts, and Modeling. SSSA Book Series 2. Soil Science Society of America, Madison, WI, pp. 51−77.

Leenheer, J.A., 1980. Origin and nature of humic substances in the water of the Amazon River basin. Acta Amazon. 10, 513−526.

Lehmann, J., Kleber, M., 2015. The contentious nature of soil organic matter. Nature. 528, 60−68.

MacCarthy, P., 2001. The principles of humic substances. Soil Sci. 166, 738−751.

Malcolm, R.L., Wershaw, R.L., Thurman, E.M., Aiken, G.R., Pickney, D.J., Kaakinen, J., 1981. Reconnaisance sampling and characterization of aquatic humic substances at the Yuma Desalting Test Facility, Arizona. In: U.S. Geol. Surv. Water Resour. Invest. Series No. 81−42. U.S. Geological Survey, Reston, VA.

Manahan, S.E., 1991. Environmental Chemistry, Fifth edition. Lewis Publishers, Chelsea, MI.

Mann, L.K., 1986. Changes in soil carbon storage after cultivation. Soil Sci. 142, 279−288.

Martin, A.E., Reeve, R., 1958. Chemical studies of podzolic illuvial horizons: III. Titration curves of organic matter suspensions. J. Soil Sci. 9, 89−100.

Mikutta, R., Mikutta, C., Kalbitz, K., Scheel, T., Kaiser, K., Jahn, R., 2007. Biodegradation of forest floor organic matter bound to minerals via different binding mechanisms. Geochim. Cosmochim. Acta. 71, 2569−2590.

Monreal, C.M., Sultan, Y., Schnitzer, M., 2010. Soil organic matter in nano-scale structures of a cultivated Black Chernozem. Geoderma. 159, 237−242.

Mortland, M.M., 1970. Clay-organic complexes and interactions. Adv. Agron. 22, 75−117.

Mortland, M.M., Shaobai, S., Boyd, S.A., 1986. Clay-organic complexes as adsorbents for phenol and chlorophenols. Clays Clay Miner. 34, 581−585.

Nelson, D.W., Sommers, L.E., 1996. Total carbon, organic carbon, and organic matter. In: Sparks, D.L. (Ed.), Methods of Soil Analysis. Part 3. Chemical Methods. Soil Sci. Soc. Amer. and Amer. Soc. Agro., Madison, WI, pp. 961−1010.

Olk, D.C., Bloom, P.R., Perdue, E.M., McKnight, D.M., Chen, Y., Farenhorst, A., et al., 2019. Environmental and agricultural relevance of humic fractions extracted by alkali from soils and natural waters. J. Environ. Qual. 48 (2), 217−232.

Pearson, R.G., 1963. Hard and soft acids and bases. J. Am. Chem. Soc 85, 3533−3539.

Pearson, R.G., 1968. Hard and soft acids and bases, HSAB. Part 1. Fundamental principles. J. Chem. Educ 45, 581−587.

Perdue, E.M., Lytle, C.R., 1983. Distribution model for binding of protons and metal ions by humic substances. Environ. Sci. Technol 17, 654−660.

Piccolo, A., 2001. The supramolecular structure of humic substances. Soil Sci. 166, 810−832.

Post, W.M., Kwon, K.C., 2000. Soil carbon sequestration and land-use change: processes and potential. Glob. Change Biol. 6, 317−327.

Pribyl, D.W., 2010. A critical review of the conventional SOC to SOM conversion factor. Geoderma 156, 75−83.

Quéré, C.L., Andrew, R.M., Friedlingstein, P., Sitch, S., et al., 2018. Global carbon budget. Earth Syst. Sci. Data. 10, 2141−2194.

Quideau, S.A., Chadwick, O.A., Benesi, A., Grahama, R.C., Anderson, M.A., 2001. A direct link between forest vegetation type and soil organic matter composition. Geoderma. 104, 41−60.

Quideau, S., Deffieux, D., Douat-Casassus, C., Pouységu, L., 2011. Plant polyphenols: chemical properties, biological activities, and synthesis. Angew. Chem. Int. Ed. 50, 586−621.

Reisser, M., Purves, R.S., Schmidt, M.W.I., Abiven, S., 2016. Pyrogenic carbon in soils: a literature-based inventory and a global estimation of its content in soil organic carbon and stocks. Front. Earth Sci. 4, 80.

Rice, J.A., MacCarthy, P., 1991. Statistical evaluation of the elemental composition of humic substances. Org. Geochem. 17, 635−648.

Saxton, K.E., Rawls, W.J., Romberger, J.S., Papendick, R.I., 1986. Estimating generalized soil-water characteristics from texture. Soil Sci. Soc. Am. J. 50, 1031−1036.

Schlesinger, W.H., 1999. Carbon sequestration in soils. Science. 284, 2095.

Schnitzer, M., 2000. A lifetime perspective on the chemistry of soil organic matter. Adv. Agron. 68, 1−58.

Schnitzer, M., Hansen, E.H., 1970. Organo-metallic interactions in soils: 8. An evaluation of methods for the determination of stability constants of metal-fulvic acid complexes. Soil Sci. 109, 333−340.

Schnitzer, M., Khan, S.U. (Eds.), 1972. Humic Substances in the Environment. Dekker, New York, NY.

Schnitzer, M., Khan, S.U. (Eds.), 1978. Soil Organic Matter. Elsevier, New York, NY.

Schnitzer, M., Monreal, C.M., 2011. Quo vadis soil organic matter research?: a biological link to the chemistry of humification. Adv. Agron. 113, 139−213.

Schnitzer, M., Preston, C.M., 1986. Analysis of humic acids by solution−and solid-state carbon-13 nuclear magnetic resonance. Soil Sci. Soc. Am. J. 50, 326−331.

Schnitzer, M., Schulten, H.-R., 1992. The analysis of soil organic matter by pyrolysis-field ionization mass spectrometry. Soil Sci. Soc. Am. J. 56, 1811−1817.

Schnitzer, M., Schulten, H.-R., 1995. Analysis of organic matter in soil extracts and whole soils by pyrolysis-mass spectrometry. Adv. Agron. 55, 168–218.

Schulten, H.-R., Leinmeher, P., Schnitzer, M., 1998. Analytical pyrolysis and computer modeling of humic and soil particles. In: Huang, P.M., Senesi, N., Buffle, J. (Eds.), Structure and Surface Reactions of Soil Particles. John Wiley & Sons, New York, NY, pp. 282–324.

Schulten, H.-R., Schnitzer, M., 1993. A state of the art structural concept for humic substances. Naturwissenschaften. 80, 29–30.

Schulten, H.-R., Schnitzer, M., 1997. Chemical model structures for soil organic matter and soils. Soil Sci. 162, 115–130.

Schuur, E.A.G., Bockheim, J., Canadell, J.G., Euskirchen, E., Field, C.B., Goryachkin, S.V., et al., 2008. Vulnerability of permafrost carbon to climate change: Implications for the global carbon cycle. BioScience. 58, 7–14.

Simpson, A.J., Song, G., Smith, E., Lam, B., Novotny, E.H., Hayes, M.H.B., 2007. Unravelling the structural components of soil humin by use of solution-state nuclear magnetic resonance spectroscopy. Environ. Sci. Technol. 41, 876–883.

Singh, B., Fang, Y., Cowie, B.C.C., Thomsen, L., 2014. NEXAFS and XPS characterisation of carbon functional groups of fresh and aged biochars. Org. Geochem. 77, 1–10.

Smith, P., Davis, S.J., Creutzig, F., et al., 2015. Biophysical and economic limits to negative CO_2 emissions. Nat. Clim. Change. 6 (1), 42–50.

Sollins, P., Swanston, C., Kleber, M., Filley, T., Kramer, M., Crow, S., et al., 2006. Organic C and N stabilization in a forest soil: evidence from sequential density fractionation. Soil Biol. Biochem. 38 (11), 3313–3324.

Song, G., Hayes, M.H.B., Novotny, E.H., Simpson, A.J., 2011. Isolation and fractionation of soil humin using alkaline urea and dimethylsulphoxide plus sulphuric acid. Naturwissenschaften. 98, 7–13.

Song, G., Novotny, E.H., Simpson, A.J., Clapp, C.E., Hayes, M.H.B., 2008. Sequential exhaustive extraction of a Mollisol soil, and characterizations of humic components, including humin, by solid and solution state NMR. Eur. J. Soil Sci. 59, 505–516.

Sowers, T.D., Stuckey, J.W., Sparks, D.L., 2018a. The synergistic effect of calcium on organic carbon sequestration to ferrihydrite. Geochem. Trans. 19, 1–11.

Sowers, T.D., Adhikari, D., Wang, J., Yang, Y., Sparks, D.L., 2018b. Spatial associations and chemical composition of organic carbon sequestered in Fe, Ca, and organic carbon ternary systems. Environ. Sci. Technol. 52, 6936–6944.

Sowers, T.D., Wani, R.P., Coward, E.K., Fischel, M.H.H., Betts, A.R., Douglas, T.A., et al., 2020. Spatially resolved organo-mineral interactions across a permafrost chronosequence. Environ. Sci. Technol. 54, 2951–2960.

Steelink, C., 1985. Implications of elemental characteristics of humic substances. In: Aiken, G.R., McKnight, D.M., Wershaw, R.L. (Eds.), Humic Substances in Soil, Sediments, and Water. John Wiley & Sons (Interscience), New York, NY, pp. 457–476.

Stevenson, F.J., 1982. Humus Chemistry. John Wiley & Sons, New York, NY.

Stevenson, F.J., Cole, M.A., 1999. Cycles of Soil: Carbon, Nitrogen, Phosphorus, Sulfur, Micronutrients. Second edition. Wiley, New York, NY.

Sutton, R., Sposito, G., 2005. Molecular structure in soil humic substances: the new view. Environ. Sci. Technol. 39, 9009–9015.

Swift, R.S., 1996. Organic matter characterization. In: Sparks, D.L., Page, A.L., Helmke, P.A., Loeppert, R.H., Soltanpour, P.N., Tabatabai, M.A., et al. (Eds.), Methods of Soil Analysis. Part 3. Chemical Methods. Book Series no. 5. Soil Sci. Soc. Amer. and Amer. Soc. Agro., Madison, WI, pp. 1011–1069.

Szalay, A., 1964. Cation exchange properties of humic acids and their importance in the geochemical enrichment of UO^{++} and other cations. Geochim. Cosmochim. Acta. 22, 1605–1614.

Thomas, G.W., 1975. Relationship between organic matter content and exchangeable aluminum in acid soil. Soil Sci. Soc. Am. Proc. 39, 591.

Thurman, E.M., 1985. Organic Geochemistry of Natural Waters. Kluwer Academic Publishers, Hingham, MA.

Thurman, E.M., Malcom, R.L., Aiken, G.R., 1978. Prediction of capacity factors for aqueous organic solutes on a porous acrylic resin. Anal. Chem. 50, 775–779.

Tifafi, M., Guenet, B., Hatté, C., 2018. Large differences in global and regional total soil carbon stock estimates based on SoilGrids, HWSD, and NCSCD: intercomparison and evaluation based on field data from USA, England, Wales, and France. Glob. Biogeochem. Cycles. 32, 42–56.

Tipping, E., Somerville, C.J., Luster, J., 2016. The C:N:P:S stoichiometry of soil organic matter. Biogeochemistry. 130, 117–131.

Vogel, C., Mueller, C.W., Höschen, C., Buegger, F., Heister, K., Schulz, S., et al., 2014. Submicron structures provide preferential spots for carbon and nitrogen sequestration in soils. Nature Commun. 5, 2947.

von Lützow, M., Kögel-Knabner, I., Ekschmitt, K., Matzner, E., Guggenberger, G., Marschner, B., et al., 2006. Stabilization of organic matter in temperate soils: mechanisms and their relevance under different soil conditions – a review. Eur. J. Soil Sc. 57, 426–445.

Warren, C.R., Taranto, M.T., 2010. Temporal variation in pools of amino acids, inorganic and microbial N in a temperate grassland soil. Soil Biol. Biochem. 42, 353–359.

Wauchope, R.D., Yeh, S., Linders, J.B.H.J., Kloskowski, R., Tanaka, K., Rubin, B., et al., 2002. Pesticide soil sorption parameters: theory, measurement, uses, limitations and reliability. Pest Manag. Sci. 58, 419–445.

Wershaw, R.L., 1986. A new model for humic materials and their interactions with hydrophobic chemicals in soilwater or sediment-water systems. J. Contam. Hydrol. 1, 29–45.

Wershaw, R.L., 2004. Evaluation of Conceptual Models of Natural Organic Matter (Humus) From a Consideration of the Chemical and Biochemical Processes of Humification, Scientific Investigations Report No. 2004-5121. U.S. Geological Survey, Reston, VA.

West, T.O., Post, W.M., 2002. Soil organic carbon sequestration rates by tillage and crop rotation. Soil Sci. Soc. Am. J. 66 (6), 1930–1946.

Wilding, L.P., 1999. Comments on paper by R. Lal, H.M. Hassan and J. Dumanski. In: Rosenberg, N.J., Izaurralde, R.C., Malone, E.L. (Eds.), Carbon Sequestration in Soils: Science, Monitoring and Beyond. Batelle Press, Columbus, OH, pp. 146–149.

Wright, J.R., Schnitzer, M., 1961. An estimate of the aromaticity of the organic matter of a Podzol soil. Nature (London). 190, 703–704.

Xia, G., Pignatello, J.J., 2001. Detailed sorption isotherms of polar and apolar compounds in a high-organic soil. Environ. Sci. Technol. 35, 84–94.

Yeasmin, S., Singh, B., Johnston, C.T., Sparks, D.L., 2017. Organic carbon characteristics in density fractions of soils with contrasting mineralogies. Geochim. Cosmochim. Acta. 218, 215–236.

Yeasmin, S., Singh, B., Kookana, R.S., Farrell, M., Sparks, D.L., Johnston, C.T., 2014. Influence of mineral characteristics on the retention of low molecular weight organic compounds: a batch sorption–desorption and ATR-FTIR study. J. Colloid Interface Sci. 432, 246–257.

Yuan, T.L., Gammon Jr., N., Leighty, R.G., 1967. Relative contribution of organic and clay fractions to cation-exchange capacity of sandy soils from several groups. Soil Sci. 104, 123–128.

Soil Solution — Solid Phase Equilibria

4.1 Introduction

Key Points: Interactions between solutes in the soil solution, such as metals and ligands, affect their mobility and transport. Inner-sphere and outer-sphere complexes are two types of interactions in which ions in solution can bind with each other or to the solid phase components of the soil, such as clay minerals and organic matter.

There are numerous complex and dynamic chemical reactions that can occur in soils. These reactions occur and interact with each other through the soil solution (Figure 4.1). The soil solution is the aqueous liquid phase of the soil and its solutes (Glossary of Soil Science Terms, 2008). Most solutes in the soil solution are ions, which occur either as free hydrated ions (e.g., Al^{3+}, which is expressed as $Al(H_2O)_6^{3+}$) or as various complexes with organic functional groups (e.g., carboxyl) or inorganic ligands (e.g., Cl^-, NO_3^-, or SO_4^{2-}). A ligand is an ion or molecule that can exist independently (Atkins et al., 2006), and it attaches and forms bonds with other ions, molecules, or mineral surfaces. Sidebar 4.1 shows two examples of potential ligands that can be found in the soil solution. When metal ions (such as Cd^{2+} and Pb^{2+}) and ligands (phosphate and arsenate) directly interact in the soil solution, they form what are known as inner-sphere complexes, where no water molecules are present between the metal ion and the ligand. The same term is applied when ions bond in this fashion to a mineral surface.

Sidebar 4.1 — Example chelator molecules

Two potential chelator molecules are shown below. The negative charge on chelators allows them to associate and complex with cations in the soil solution.

Deprotonated oxalic acid

Sulfate ion

An outer-sphere complex is formed when a water molecule is positioned between the metal ion (e.g., Ca^{2+} and Na^+) and the ligand (Cl^- and NO_3^-). Outer-sphere complexes are not as tightly bound as inner-sphere complexes (further descriptions of inner- and outer-sphere complexes are given in Chapter 5). An uncharged outer-sphere complex is often referred to as an ion pair (e.g., Ca^{2+} plus SO_4^{2-} ions are known to form ion pairs, e.g., $CaSO_4^0$). Ion pairs can enhance the leaching of solutes from the soil

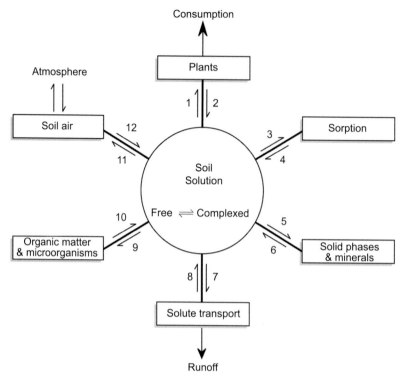

FIGURE 4.1   Dynamic equilibria reactions in soils. There are many reactions that take place in the soil solution, which are identified in reactions 1 through 12. *From Lindsay (1979), with permission.*

profile during soil weathering. Soil solutions therefore are composed of a variety of ion species, either complexed or noncomplexed. Determining the quantity and distribution of different ion complexes in an aqueous system such as the soil solution is a process called speciation.

The soil solution is the medium from which plants take up ions, such as nitrate and phosphate ions (1 in Figure 4.1), and in which plant exudates reside (2). The uptake of nutrients by plants removes nutrients from the soil solution. Ions in the soil solution can be sorbed (bound) to organic and inorganic components of the soil (3) and sorbed ions can be desorbed (released) into the soil solution (4). If the soil solution is supersaturated with respect to a mineral, that mineral can precipitate (5) until equilibrium is reached. This process forms secondary minerals (solid phases)

in soils and naturally occurs as the soil weathers. If the soil solution becomes undersaturated with any mineral in the soil, the mineral can dissolve (6) until equilibrium is reached. Ions in the soil solution can be transported from the soil and (7) into groundwater or removed through surface runoff processes. Transport can also carry nanoparticles, colloids, and nutrients and contaminants associated with them (Sidebar 4.2). Through evaporation and drying, the upward movement of ions can also occur (8). This process is caused by the capillary action in soil micropores and is very common in arid and semiarid regions. Microorganisms can remove ions from the soil solution (9), and when the organisms die and organic matter is decomposed (mineralized), ions are released into the soil solution (10). Gases (e.g., O_2) diffuse from the atmosphere and into the soil air in the pore spaces and dissolve into the soil solution (11), or gases (e.g., CO_2) can diffuse from the soil (e.g., CO_2 produced via microbial respiration) into the atmosphere (12). It is important to note that many of these processes are equilibrium based as well as kinetically controlled (time dependent).

Sidebar 4.2 — Mineral Nanoparticles

An example of a mineral nanoparticle (hematite) with a size of 1 and 5 nm. The nanoparticle is only several octahedral layers thick. Nanoparticles such as these can enhance the transport of contaminants bound to their surfaces. *From Hochella et al. (2008), with permission.*

4.2 Soil Air Composition

Key Point: The amount of carbon dioxide and oxygen in the soil solution varies drastically with field conditions.

Nitrogen and oxygen comprise the majority of our atmosphere, with carbon dioxide composing a smaller portion. By volume, there is 78.08% nitrogen (N_2), 20.95% oxygen (O_2), and 0.934%

argon in the atmosphere (dry air). The current amount of carbon dioxide (CO_2) is approximately 417 ppm (0.0417%), which has been steadily increasing since the 1950s.

However, in the soil pore space that is not filled with water (the soil air) the amounts of O_2 and CO_2 are very different from those in the atmosphere above the soil. This variation is due to respiration during microbial activity in the soil and the respiration of plant roots. In soil air the amount of CO_2 can be 10 times higher than in the atmosphere, and the level can reach as high as several percent. In a compilation of data sets from around the world soil CO_2 concentrations ranged from 0.04% to 13.0% in the upper several meters of the soil; this variability is due to several factors including organic matter content, temperature, and soil moisture (Amundson and Davidson, 1990). The saturation of a soil with water will greatly impact the quantity of oxygen in the soil pore space. However, the relative concentration of nitrogen only varies slightly in the soil profile.

4.3 Soil Solution Composition

Key Points: The soil solution is composed of water, dissolved gases, dissolved ions, dissolved organic matter (DOM), and nano- to microsized particles. The soil solution can be sampled in the field or obtained from a soil brought into the laboratory. Various methods and procedures used to obtain the soil solution have their benefits and drawbacks and affect the resulting sample differently.

4.3.1 Soil Solution

Sidebar 4.3 — Soil Pore Space
There exist a wide range of soil pore sizes, from nanometer-sized pores to macroscopic pores created by soil aggregation, soil biota including microorganisms, soil animals, and plant roots. In general, pore size can be divided into macropores and micropores, with micropores being smaller than 30 μm.

The soil solution is found in both macropores and micropores between the solid phases present in soil (Sidebar 4.3). The concentrations of cations and anions in the soil solution are directly related to the parent material and the extent of soil weathering. Table 4.1 shows the average ion composition of soil solutions for soils from around the world. Cation and anion concentrations can vary depending on the specific ecosystem and use of the soil. One will note that Ca is the most prevalent metal cation in the soil solution, which is typical for most soils. Nitrate (NO_3^-), chloride (Cl^-), and sulfate (SO_4^{2-}) are the common anions (ligands) in more acidic soils, while carbonate (CO_3^{2-}) can be important in some alkaline soils, for example, in arid or semiarid regions.

4.3.2 Obtaining Soil Solution Samples

The concentration of ions in the soil solution and the ability of solid components in the soil to resupply a depleted ion to the soil solution are both important properties. Measuring ion concentration in the soil solution is critical in predicting and understanding reactions in the soil environment, such as weathering, bioavailability,

mobility, and geochemical cycling of nutrients and inorganic and organic chemicals (Wolt, 1994). However, measuring the components of the soil solution in situ is very difficult and, technically, not possible for most ion species. In addition, the actual concentration of the ion species in the soil solution changes with changes in soil moisture content (Wolt, 1994). Sidebar 4.4 discusses some of the effects of sample treatment on soil solution chemistry.

Sidebar 4.4 − Changes to the Soil Solution

The composition of an extracted soil solution is affected by many different variables. For example, soil moisture, the choice of displacement or extraction technique, soil pretreatment, and storage conditions can all impact the concentrations of cations and anions. Extraction and analysis of the soil solution immediately following sampling is ideal; however, this is often impractical. Air-dried and sieved soil is commonly used for soil storage, but this can also increase proton donation from water molecules to the surface of dry clays, likely increasing surface acidity upon rewetting. Freezing field moist soils can inhibit changes to the extracted solution over short periods of time

TABLE 4.1   Total ion composition of soil solutions from world soils.

| | | Total ion composition of soil solution (mmol liter^{-1}) | | | | | | | | | | |
| | | Metals/Cations | | | | | | | Ligands | | | |
Location of soils	pH	Ca	Mg	K	Na	NH$_4$	Al	Si	HCO$_3$	SO$_4$	Cl	NO$_3$
California, US[a]	7.43	12.86	2.48	2.29	3.84	0.86	−	3.26	3.26	5.01	3.07	21.36
Georgia, US[b]	6.15	0.89	0.29	0.07	0.16	−	0.0004	−	−	1.18	0.21	0.1
United Kingdom[c]	7.09	1.46	0.12	0.49	0.31	−	0.01	−	−	0.32	0.75	0.69
Australia[d]	5.75	0.27	0.4	0.38	0.41	2.6	−	0.85	0.85	0.87	1.67	0.32

[a] From Eaton et al. (1960); soil solution obtained via pressure plate extraction at 10 kPa moisture. The pH and total composition for each ion represent the mean for seven soils.
[b] From Gillman and Sumner (1987); soil solution obtained via centrifugal displacement at 7.5 kPa moisture. The pH and total composition for each ion represent the mean for four Ap (surface) horizons.
[c] From Kinniburgh and Miles (1983); soil solution obtained via centrifugal displacement at field moisture contents. The pH and total composition for each ion represent the mean for ten surface soils.
[d] From Gillman and Bell (1978); soil solution obtained via centrifugal displacement at 10 kPa moisture. The pH and total composition for each ion represent the mean value for six soils.
From Wolt (1994), with permission.

(e.g., 4 days), but the long-term storage of frozen soils can promote soil desiccation (Wolt, 1994). Given these potential changes in the soil solution, the experimental methods should state how the soils were treated and stored prior to analysis.

A number of laboratory methods are used to obtain samples of the soil solution. These are described in detail by Adams (1971) and Wolt (1994) and will be discussed only briefly here. Further discussion on these techniques and variations of them can also be found in Schwab (2012) and Nieminen et al. (2013). The methods can be classified as displacement techniques, including (1) column displacement (pressure or tension displacement, with or without a displacing solution), (2) centrifugation (e.g., with an immiscible liquid), and (3) saturation extracts (e.g., saturation pastes).

In the column displacement method (1) a field moist soil is packed into a glass column, and then soil solution is forced out of the soil using a miscible or immiscible displacing solution. The displaced soil solution is collected at the bottom of the column and analyzed for its ion contents. The column displacement method is probably the most reliable method for obtaining a sample of the soil solution. However, the procedure is tedious, requiring an analyst who is experienced in packing the column with soil. Furthermore, the procedure is very time consuming.

The centrifugation method (2) uses an immiscible liquid, such as carbon tetrachloride or ethyl benzoylacetate, to physically remove the soil solution (which is essentially water) from the solid matrix of the moist soil. During centrifugation (>1 MPa for from 30 min to 2 h), the immiscible liquid passes through the soil and the soil solution is displaced upward. The displaced soil solution is then decanted and filtered through a phase-separating filter paper (to remove traces of the immiscible liquid) and then through a 0.2 μm membrane filter to remove suspended clay particles (Wolt, 1994). Centrifugation methods are the most widely used because they are relatively simple and the equipment to carry out the procedure is usually available. The centrifugation method can be more efficient than positive pressure or vacuum methods for soils with moisture contents below saturation (Schwab, 2012).

In the saturation extract method (3) a saturated soil paste is prepared by adding distilled water to a soil sample in a beaker, stirring with a spatula, and tapping the beaker occasionally to consolidate the soil—water mixture. At saturation, the soil paste should glisten. Prior to extraction, the saturated soil paste should be allowed to stand for 4—16 h. Then the soil paste is transferred to a funnel containing a filter paper. The funnel is connected to a flask, vacuum is applied, and the filtrate is collected.

Field methods for obtaining the soil solution include the use of different types of lysimeters. Lysimeters can be divided into two general categories, namely, those that employ nondestructive and "semidestructive" methods; nondestructive methods include tension lysimeters, while semidestructive zero-tension lysimeters can cause long-term changes to the soil (Nieminen et al., 2013). Lysimeters have many different configurations, including block or monolithic, zero tension or pan, and porous cup vacuum systems, and the particular method will influence the soil solution that is obtained (Schwab, 2012).

4.4 Speciation of the Soil Solution

Key Points: Determining whether a complex is inner-sphere or outer-sphere and to what each ion is complexed is a process called speciation. The speciation of the soil solution refers to determining the distribution of ions in their various chemical forms. This is an important process because ion complexes in the soil solution can influence the availability of nutrient elements and leaching of nutrient ions and contaminants from the soil profile and help predict the precipitation−dissolution processes in the soil system.

While total ion concentration (i.e., the sum of free and complexed ions) of the soil solution, as measured by analytical techniques such as spectrometry, chromatography, and colorimetry, provides important information on the quantities of ions available for plant uptake and movement through the soil profile, it is important that the speciation or chemical forms of the free and complexed ions be known. The speciation of ions influences their reactivity with the solid phases in soils. Ions in the soil solution can form a number of species due to hydrolysis, complexation, and redox (see Chapter 8) reactions. Hydrolysis reactions involve the splitting of a hydrogen ion (H^+) ion from a water molecule. The remaining hydroxyl group (OH^-) can then form an inner-sphere complex with a metal ion (Baes and Mesmer, 1976). The general hydrolysis reaction for a hydrated metal ($[M(H_2O)_x]^{n+}$) in solution is:

$$[M(H_2O)_x]^{n+} \leftrightarrow \left[M(OH)_y(H_2O)_{x-y}\right]^{(n-y)+} + yH^+$$

$$(4.1)$$

where n^+ refers to the charge on the metal ion, x is the coordination number, and y is the number of H^+ ions released into solution. The degree of hydrolysis for a solvated metal ion (i.e., extent of the reaction in Equation [4.1]) is a function of pH. This can be noted because H^+ is a product of hydrolysis. Thus hydrolysis releases protons into solution. The hydrolysis of metal cations increases with pH. For example, as pH increases, the concentration of H^+ decreases, driving the reaction in Equation (4.3) to the right. Additionally, at a low pH (i.e., acidic conditions), there is

a high concentration of protons, which decreases the dissociation of water molecules that occupy a metal's sphere of hydration (Essington, 2015). Metal cations in solution are generally hydrated in a coordination sphere of hydration, and the rate at which metal ions react with ligands is partly dependent on the rate at which water enters and exits this sphere of hydration (Stumm and Morgan, 1996). This rate is called the water exchange rate and the values for cations commonly found in soils (both contaminated and noncontaminated) can be found in Table 4.2.

TABLE 4.2   Rate constants for water exchange.

Metal ion	Hydrated notation (metal aquo complex)	k $(s^{-1})^a$
Pb^{2+}	$Pb(H_2O)_6^{2+}$	7×10^9
Hg^{2+}	$Hg(H_2O)_4^{2+}$	2×10^9
Cu^{2+}	$Cu(H_2O)_6^{2+}$	1×10^9 to 4.4×10^9
K^+	$K(H_2O)_6^+$	1×10^9
Na^+	$Na(H_2O)_6^+$	5×10^8
Ca^{2+}	$Ca(H_2O)_6^{2+}$	3.2×10^8 to 6×10^8
Cd^{2+}	$Cd(H_2O)_6^{2+}$	3×10^8
Zn^{2+}	$Zn(H_2O)_6^{2+}$	7×10^7
Mn^{2+}	$Mn(H_2O)_6^{2+}$	2.1×10^7 to 3×10^7
Fe^{2+}	$Fe(H_2O)_6^{2+}$	4×10^6 to 4.4×10^6
U^{4+}	$U(H_2O)_{10}^{4+}$	5×10^6
Co^{2+}	$Co(H_2O)_6^{2+}$	2×10^6 to 3.2×10^6
Mg^{2+}	$Mg(H_2O)_6^{2+}$	3×10^5 to 6.7×10^5
Ni^{2+}	$Ni(H_2O)_6^{2+}$	3×10^4 to 3.2×10^4
Fe^{3+}	$Fe(H_2O)_6^{3+}$	1.6×10^2 to 2.0×10^2
Al^{3+}	$Al(H_2O)_6^{3+}$	1 to 1.29

[a] Select values are tabulated in Luther (2016) and were retrieved from Dunand et al. (2003); Eigen and Mass (1966); Helm and Merbach (2005); Helm et al. (2005); Margerum et al. (1978); Richens (1997). Additional values retrieved from Morel and Hering (1993) and Stumm and Morgan (1996).

A conceptual model for water exchange is presented in Sidebar 4.5.

Sidebar 4.5 — Conceptual Model of Water Exchange

Ions do not exist in solution as isolated atoms but are hydrated with water molecules. Water exchange and the rate at which water exchanges around metal ions plays a critical role in ligand exchange. Here, a hexaaqua Fe(III) atom (in blue) $[Fe(H_2O)_6^{3+}]$ is shown.

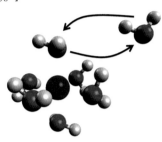

One water molecule from the bulk water is depicted exchanging with one of the six water molecules in the coordination sphere of hydration. The sphere of hydration is the six water molecules surrounding Fe(III). Additionally, Fe(III) is generally only soluble at low pH values because of its rapid hydrolysis. The iron atom is orange, oxygen atoms are red, and hydrogen atoms are gray.

The rate of complex formation, which is related to the rate of water exchange, can be used to gauge whether a complex is inert or labile. The rate of the reaction is important because inert complexes can have kinetic stability (Luther, 2016). For example, complexes that form in 1 min or less can be considered labile (Taube, 1952). Additionally, most of the chemical reactions of interest in soil chemistry are related to either ligand exchange reactions or electron exchange reactions. While both reactions are critical to understanding the fate and transport of metals in soils, a prerequisite to understanding the redox mechanisms (Chapter 8) is an understanding of ligand replacement (Casey and Swaddle, 2003). This is because electron transfer will occur via inner-sphere or outer-sphere electron transfer, and the formation of inner-sphere and outer-sphere complexes depends on ligand exchange.

A general complexation reaction between a hydrated metal $[(M(H_2O)_x]n^+$ and a ligand, L^{-1}, to form an outer-sphere complex is given in Equation (4.2), where a and b are stoichiometric coefficients:

$$a\left[M(H_2O)_x\right]^{n+} + bL^{1-} \leftrightarrow \left[M(H_2O)_x\right]_a L_b \quad (4.2)$$

While this reaction may seem straightforward, analytically, it is not possible to determine all the individual ion species that can occur in soil solutions. This is because they are too numerous. To speciate the soil solution, one must apply ion association or speciation models using the total concentration data for each metal and ligand in the soil solution. A mass balance equation for the total concentration of a metal, M_T^{n+}, in the solution phase can be written as:

$$M_T^{n+} = M^{n+} + M_{ML} \quad (4.3)$$

where M^{n+} represents the concentration of free, hydrated metal ion species and M_{ML} is the concentration of metal ion associated with the remaining metal—ligand complexes. It is important to realize that there can be many different types of ligands that complex with metals in solution. The ligands in the metal—ligand complexes could be inorganic anions or humic substances such as humic and fulvic acids. Similar mass balance equations can be written for the total concentration of each metal and ligand in the soil solution.

The formation of each metal—ligand complex in the soil solution can be described using an expression similar to Equation (4.5) and a conditional stability or conditional equilibrium constant, K^{cond}. As mentioned in the previous chapter, conditional stability constants vary with the composition and total electrolyte concentration of the soil solution. Conditional stability constants for inorganic complexes, however, can be related to thermodynamic stability or thermodynamic equilibrium constants (K°), which are independent of the chemical composition of the soil solution at a particular temperature and pressure. Box 4.1 illustrates the relationship between thermodynamic and conditional equilibrium constants.

To illustrate how mass balance equations like Equation (4.3) and conditional equilibrium constants can be used to speciate a metal ion in a soil solution, let us calculate the concentration of the different Ca species present in a typical soil solution from a basic (alkaline) soil. Assume that the pH of a soil solution is 7.9 and the total concentration of Ca (Ca_T) is 3.715×10^{-3} mol liter^{-1}. For simplicity's sake, and realizing that other complexes occur, let us assume that Ca exists only as free hydrated Ca^{2+} ions and as complexes with the inorganic ligands CO_3, SO_4, and Cl (i.e., $[CaCO_3^0]$, $[CaHCO_3^+]$, $[CaSO_4^0]$, and $[CaCl^+]$). The Ca_T can be expressed as the sum of free and complexed forms as follows:

$$Ca_T = [Ca^{2+}] + [CaCO_3^0] + [CaHCO_3^+]$$
$$+ [CaSO_4^0] + [CaCl^+] \tag{4.4}$$

BOX 4.1

Thermodynamic and conditional equilibrium constants

For any reaction (Lindsay, 1979)

$$aA + bB \rightleftharpoons cC + dD \tag{B4.1a}$$

A thermodynamic equilibrium constant, K°, can be written as

$$K^\circ = \frac{(C)^c(D)^d}{(A)^a(B)^b} = \frac{Products}{Reactants} \tag{B4.1b}$$

where the small superscript letters refer to stoichiometric coefficients and A, B, C, and D are chemical species. One strategy to help keep this clear, which is also useful for other equilibrium constants, is to remember that it is the products divided by the reactants. The parentheses in Equation (B4.1b) refer to the chemical activity of each of the chemical species.

Thermodynamic equilibrium constants are independent of changes in the electrolyte composition of soil solutions because they are expressed in terms of activities, not concentrations. Unfortunately, the activities of ions in solutions generally cannot be measured directly as opposed to their concentration. The activity and concentration of an ion in solution can be related, however, through the expression

$$(B) = [B]\,\gamma_B \tag{B4.1c}$$

where [B] is the concentration of species B in mol liter^{-1} and γ_B is a single-ion activity coefficient for species B.

Using the relationship in Equation (B4.1c), one can rewrite Equation (B4.1b) in terms of concentrations rather than activities:

$$K^\circ = \frac{[C]^c\gamma_C^c[D]^d\gamma_D^d}{[A]^a\gamma_A^a[B]^b\gamma_B^b} = \frac{[C]^c[D]^d}{[A]^a[B]^b}\left(\frac{\gamma_C^c\gamma_D^d}{\gamma_A^a\gamma_B^b}\right)$$
$$= K^{cond}\frac{\gamma_C^c\gamma_D^d}{\gamma_A^a\gamma_B^b} \tag{B4.1d}$$

where K^{cond} is the conditional equilibrium constant. The ratio of the activity coefficients is the correct term that relates K^{cond} to K°. The value of the activity coefficients, and therefore the value of the ratio of the activity coefficients, changes with changes in the electrolyte composition of the soil solution. In very dilute solutions the value of the activity coefficients approaches 1, and the conditional equilibrium constant equals the thermodynamic equilibrium constant.

where the square brackets indicate species' concentrations in mol liter^{-1}. Using Equation (4.3), we can write the following expression for the complex $CaCO_3^0$ (as well as similar expressions for each complex in Equation [4.4]) as follows:

$$[Ca^{2+}] + [CO_3^{2-}] \rightleftharpoons [CaCO_3^0] \qquad (4.5)$$

Assuming dilute conditions as discussed in Box 4.1, where ion activity coefficients are 1, each of the complexes in Equation (4.4) can be described by a conditional stability constant:

$$K_1^{cond} = \frac{Products}{Reactants} = \frac{[CaCO_3^0]}{[Ca^{2+}][CO_3^{2-}]} \qquad (4.6)$$

$$K_2^{cond} = \frac{[CaHCO_3^+]}{[Ca^{2+}][HCO_3^-]} \qquad (4.7)$$

$$K_3^{cond} = \frac{[CaSO_4^0]}{[Ca^{2+}][SO_4^{2-}]} \qquad (4.8)$$

$$K_4^{cond} = \frac{[CaCl^+]}{[Ca^{2+}][Cl^-]} \qquad (4.9)$$

Since Equations (4.6)–(4.9) have $[Ca^{2+}]$ as a common term, the original Equation (4.4) can be rewritten in terms of the concentrations of $[Ca^{2+}]$ and each of the inorganic ligands (Sposito, 2008). This is done algebraically where $[Ca^{2+}]$ becomes a common factor removed from each complex. Note that while each complex in Equation (4.7) does not have $[Ca^{2+}]$ specifically written as a term, it can be removed by simply placing it into the denominator position:

$$Ca_T = [Ca^{2+}]\left\{ 1 + \frac{[CaCO_3^0]}{[Ca^{2+}]} + \frac{[CaHCO_3^+]}{[Ca^{2+}]} \right.$$
$$\left. + \frac{[CaSO_4^0]}{[Ca^{2+}]} + \frac{[CaCl^+]}{[Ca^{2+}]} \right\} \qquad (4.10)$$

Thus Equations (4.4) and (4.10) are the same. To relate this expression in terms of conditional stability constants, we rearrange Equation (4.6) as:

$$\frac{[CaCO_3^0]}{[Ca^{2+}]} = K_1^{cond} \times [CO_3^{2-}] \qquad (4.11)$$

Repeating this process for the other complexes allows us to view the total Ca speciation with respect to conditional stability constants, many of which are known and tabulated. The tabulated stability constants can be used to calculate the aqueous speciation using a software program as described further below.

Sidebar 4.6 – Parameters Required for Equilibrium Speciation Calculations

When performing calculations using a software to determine the equilibrium speciation of a solution, there are many factors to consider, many of which are required inputs by the user into the software itself. Thus consideration for multiple parameters is needed to produce a realistic model of the system of interest. Types of complexes, species, and factors that can be incorporated into equilibrium speciation models can include:

- Inorganic molecules (cations, anions)
- Organic ligands (organic acids, dissolved organic matter)
- Stability constants
- Mineral surfaces
- Solubility product constants
- Precipitated solids or solids already present in the system. These can include solids such as:
  - Oxides, hydroxides, and oxyhydroxides
  - Mixed metal hydroxides
  - Carbonates
  - Sulfides
  - Silicates
- Adsorbed cations and anions
- Surface complexation models
- pH
- Redox potential
- Ionic strength
- Temperature

Rearranging Equations (4.6)–(4.9) and replacing each term in Equation (4.10) with the result yields:

$$Ca_T = [Ca^{2+}]\left\{1 + \left(K_1^{cond} \times [CO_3^{2-}]\right)\right.$$

$$+ \left(K_2^{cond} \times [HCO_3^-]\right) + \left(K_3^{cond} \times [SO_4^{2-}]\right)$$

$$\left. + \left(K_4^{cond} \times [Cl^-]\right)\right\} \qquad (4.12)$$

Thus Equations (4.4) and (4.12) are equivalent. However, Equation (4.12) uses the conditional stability constants and individual ion concentrations to account for total Ca. This facilitates our ability to speciate the solution because those values are either known or determined analytically, respectively.

As noted earlier, mass balance equations like Equation (4.3) can be expressed for total metal and ligand concentration, M_T^{n+} and L_T^{1-}, respectively, for each metal and ligand species. The mass balance expressions are transformed into coupled algebraic equations like Equations (4.4), (4.10), and (4.12). In order to solve this equation and calculate the distribution of dissolved chemical species, the total concentrations of metals and ligands must be known and are input into the calculation. Additionally, the conditional stability constants for all possible complexes between each ligand and metal must be known. These constants are incorporated into databases and equilibrium modeling software. The complexes are also pH dependent, so the pH of the soil solution must be known. With that information, mass balance expressions, such as Equation (4.12), can be written for each chemical species (Sposito, 2008).

The free metal cation concentrations and the metal–ligand complex concentrations are unknown, but the metal–ligand conditional stability constants are known, as are the total concentrations of all important metals and ligands. The conditional stability constants and ligand concentrations are substituted into the reaction equation and replace their respective metal–ligand complex and free cation concentrations, yielding Equation (4.12).

The algebraic Equation (4.12) can be solved iteratively using successive approximation to obtain the free ion concentrations (i.e., the free cation and free ligand concentrations) by using a number of possible computer equilibrium models (see Table 4.3 for examples of programs often used in soil and geochemical research). A common iterative approach is the Newton–Raphson algorithm. The computer equilibrium models contain thermodynamic databases (e.g., Martell and Smith, 1976) and computational algorithms. Thorough discussions of these models can be found in Jenne (1979), Melchior and Bassetteds (1990) and Mattigod and Zachara (1996). The thermodynamic databases contain equilibrium constants for aqueous complex species as well as other equilibrium parameters. See Sidebar 4.6 for additional factors that can be included in these models.

The free ion concentrations obtained from the iterative approach can then be used to calculate the concentrations of the metal–ligand complexes. For example, the complex $CaHCO_3^+$ forms based on the following reaction:

$$Ca^{2+} + H^+ + CO_3^{2-} \rightarrow CaHCO_3^+ \qquad (4.13)$$

The conditional equilibrium constant (K_2^{cond}) for calcium bicarbonate ($CaHCO_3^+$) is 9.33×10^{10} at 298 K. Substituting this value into Equation (4.9) yields

$$9.33 \times 10^{10} = \frac{[CaHCO_3^+]}{[Ca^{2+}][H^+][CO_3^{2-}]} \qquad (4.14)$$

and solving for $[CaHCO_3^+]$ yields

$$[CaHCO_3^+] = 9.33 \times 10^{10}[Ca^{2+}][H^+][CO_3^{2-}] \qquad (4.15)$$

TABLE 4.3   Software used for aqueous soil and geochemical analyses.[a]

Program	Reference or website
C-SALT	Smith et al. (1995)
Chem_Transport	Selim (2016); http://www.spess.lsu.edu/chem_transport
CHEAQS Next	Verweij and Simonin (2020) https://www.cheaqs.eu
CHESS	Van der Lee and De Windt (2002)
ECOSAT	Keizer and Van Riemsdijk (2002)
FITEQL	Herbelin and Westall (1999)
The Geochemist's Workbench	Bethke and Yeakel (2010); Bethke, C. M. (2022); https://www.gwb.com
GEOCHEM	Sposito and Mattigod (1980)
GEOCHEM-PC	Parker et al. (1995)
GEOSURF	Sahai and Sverjensky (1998)
Hydra/Medusa	Puigdomenech (2006); https://www.kth.se/che/medusa
JESS	May et al. (2019)
MICROQL	Westall (1979)
MINEQL	Westall et al. (1976)
MINEQL+	Schecher and McAvoy (2007)
MINFIT	Xie et al. (2016)
MINSORB	Bradbury and Baeyens (1997)
MINTEQ	Felmy et al. (1984)
ORCHESTRA	Meeussen (2003)
PHREEQC	Parkhurst and Appelo (2013)
REDEQL	Morgan (1972)
REDEQL2	McDuff et al. (1973)
REDEQL.EPA	Ingle et al. (1978)
REDEQL.UMD.	Harris (1981)
SOILCHEM	Sposito and Coves (1988)
SOLMNEQ	Kharaka and Barnes (1973)
TICKET-UWM	Farley et al. (2011); http://www.unitworldmodel.net
TITRATOR	Cabaniss (1987)
Visual MINTEQ	Gustafsson (2011); https://vminteq.lwr.kth.se
WATCHEM	Barnes and Clarke (1969)
WATEQ	Truesdall and Jones (1974)

(Continued)

TABLE 4.3   Software used for aqueous soil and geochemical analyses.[a]—cont'd

Program	Reference or website
WATEQ2	Ball et al. (1979)
WATEQ3	Ball et al. (1981)
WHAM	Tipping (1994)

[a] *Examples of computer software used in equilibrium, transport, or batch and kinetic modeling for soil and aquatic geochemical research. Geochemical models can use an equilibrium constant approach for describing ion speciation and computing ion activities. Other models use the Gibbs free energy minimization approach. Some of the uses for these models include calculations for acid–base equilibria, solubility of elements, complexation reactions for metals and dissolved (natural) organic matter, oxidation–reduction equilibrium reactions, and surface complexation modeling. Others can be used for reaction-transport modeling and the determination of equilibrium constants. See, Baham (1984), Mattigod and Zachara (1996), Brezonik and Arnold (2012), Xie et al., (2016) and references for additional sources.*

Once the free concentrations of Ca^{2+}, H^+, and CO_3^{2-} are determined, the concentration of the complex, $[CaHCO_3^+]$, can be calculated using Equation (4.15).

The calculated concentrations for the various species can be checked by inserting them into the mass balance equation (Equation 4.4) and determining if the sum is equal to the known total concentration. This can be illustrated for Ca using the data in Table 4.4. Table 4.4 shows the calculated concentrations of seven metals, six ligands, and metal–ligand complexes (99 total complexes were assumed to be possible) of a representative example of a soil solution. To compare the total measured Ca concentration (Ca_T) to the sum of the calculated Ca species concentrations, one would use the following mass balance equation:

$$
\begin{aligned}
Ca_T = &\left[Ca^{2+}\right] + [\text{Ca-CO}_3 \text{ complexes}] \\
&+ [\text{Ca-SO}_4 \text{ complexes}] \\
&+ [\text{Ca-Cl complexes}] \\
&+ [\text{Ca-PO}_4 \text{ complexes}] \\
&+ [\text{Ca-NO}_3 \text{ complexes}] \\
&+ [\text{Ca-OH complexes}]
\end{aligned}
\tag{4.16}
$$

Using the data in Table 4.4, which are values calculated from the GEOCHEM-PC chemical speciation model:

$$
\begin{aligned}
Ca_T = 3.715 \times 10^{-3} = &\left[3.395 \times 10^{-3}\right] \\
&+ \left[8.396 \times 10^{-5}\right] \\
&+ \left[1.871 \times 10^{-4}\right] \\
&+ \left[1.607 \times 10^{-8}\right] \\
&+ \left[9.050 \times 10^{-6}\right] \\
&+ \left[2.539 \times 10^{-5}\right] \\
&+ \left[3.600 \times 10^{-7}\right]
\end{aligned}
\tag{4.17}
$$

where all concentrations are expressed in mol liter^{-1} and are the concentrations of free Ca and Ca complexed with the following anions found in Table 4.4: Ca^{2+}, CO_3^{2-}, SO_4^{2-}, Cl^-, PO_4^{2-}, NO_3^-, and OH^-, respectively. When all the individual species are added together, one sees that Ca_T compares well to the sum of free and complexed Ca species:

$$
Ca_T = 3.715 \times 10^{-3} \approx 3.717 \times 10^{-3} \tag{4.18}
$$

Table 4.5 shows the primary distribution of free metals and ligands and metal–ligand complexes for the soil solution data in Table 4.4. One sees

TABLE 4.4 Speciation of a soil solution. Concentrations of free metals, free ligands, total metal complexed by each ligand, and total concentrations of metals and ligands for a soil solution.[a,b] The pH of the solution is set at 7.9.

Free ligands	CO$_3$	SO$_4$	Cl	PO$_4$	NO$_3$	OH	
	1.297×10^{-5}	7.607×10^{-4}	1.971×10^{-3}	6.375×10^{-10}	2.473×10^{-3}	8.910×10^{-7}	

Free metals		Total metal complexed by each ligand[c]						Total metals
Ca	3.395×10^{-3}	8.396×10^{-5}	1.871×10^{-4}	1.607×10^{-8}	9.050×10^{-6}	2.539×10^{-5}	3.600×10^{-7}	3.715×10^{-3}
Mg	5.541×10^{-4}	9.772×10^{-6}	2.399×10^{-5}	2.090×10^{-6}	1.995×10^{-6}	1.047×10^{-5}	9.500×10^{-8}	6.026×10^{-4}
K	9.929×10^{-4}	1.412×10^{-6}	3.630×10^{-6}	7.580×10^{-7}	8.300×10^{-8}	1.202×10^{-6}	2.188×10^{-10}	1.000×10^{-3}
Na	4.126×10^{-3}	1.148×10^{-5}	2.398×10^{-5}	5.011×10^{-6}	4.360×10^{-7}	1.995×10^{-6}	1.000×10^{-9}	4.169×10^{-3}
Cu	3.224×10^{-8}	1.659×10^{-6}	2.000×10^{-9}	1.047×10^{-10}	2.900×10^{-8}	1.202×10^{-10}	1.800×10^{-8}	1.738×10^{-6}
Zn	5.335×10^{-6}	6.165×10^{-6}	3.710×10^{-7}	7.000×10^{-9}	1.090×10^{-8}	1.900×10^{-8}	5.490×10^{-7}	1.259×10^{-5}
H$^+$	1.400×10^{-7}	2.512×10^{-3}	6.610×10^{-10}	8.710×10^{-20}	1.230×10^{-5}	1.097×10^{-12}		
Total ligands		2.570×10^{-3}	1.000×10^{-3}	1.995×10^{-3}	2.188×10^{-5}	2.512×10^{-3}		

[a] Adapted from Mattigod and Zachara (1996), with permission, using the GEOCHEM-PC (Parker et al., 1995) chemical speciation model.
[b] All concentrations are expressed in mol liter^{-1}.
[c] The concentrations of the metals complexed with the ligands are those data within the area marked by lines and designated as total metal complexed by each ligand. For example, 8.396×10^{-5} mol liter^{-1} Ca is complexed with CO$_3$.

that Ca, Mg, K, and Na exist predominantly as free metals while Cu mainly occurs complexed with CO$_3$ and Zn is primarily speciated as the free metal (42.38%) and complexed with CO$_3$ (49.25%). The ligands Cl and NO$_3$ (>98%) occur as free ions, while about 20% of SO$_4$ and 10% of PO$_4$ are complexed with Ca. These findings are consistent with the discussion in Chapter 3 (see Box 3.1) on hard and soft acids and bases. Hard acids such as Ca, Mg, K, and Na and hard bases such as Cl and NO$_3$ would not be expected to form significant complexes except at high concentrations of the metal or ligand (Cl or NO$_3$). However, soft acids such as Cu and Zn can form complexes with hard bases such as CO$_3$.

4.4.1 Performing Equilibrium Speciation Calculations

One of the most effective methods to understand equilibrium speciation modeling and the software used to perform these calculations is to actually perform them oneself. In this section and beyond several examples and considerations are presented for equilibrium speciation calculations. While there are many software programs available to perform this work (Table 4.3), Visual MINTEQ is used here because it is a commonly used software and freely available. When setting up an equilibrium speciation calculation, there are many variables to consider (Sidebar 4.6). However, in this first example a limited number of variables will be used in order to illustrate the concepts of cation and anion complexation (i.e., the formation of an ion pair in aqueous solution). Additional examples will discuss precipitation reactions.

In this first example (Box 4.2) we consider Table 4.1, where values of typical soil solutions are given. The average Ca concentration in the examples of soil solutions from California soils is 12.86 mM. To simplify the calculation, we will start with 12.86 mM of both Ca and SO$_4^{2-}$. The average pH is stated as 7.43. Thus in this first example we will consider three input components: calcium, sulfate, and pH. The software itself has equilibrium constants (sometimes termed as stability constants or association constants in other literature) and solubility product constants incorporated into its database. Thus the resulting calculation utilizes those three input variables to compute what the resulting solution is predicted to be.

TABLE 4.5   Primary distribution of free metals and ligands and metal—ligand complexes for a soil solution.[a]

Metal	Free metal	Metal—ligand complexes (%)					
		CO_3	SO_4	Cl	PO_4	NO_3	OH
Ca	91.36[b]	2.26	5.03	0.43	0.23	0.68	–
Mg	91.95	1.6	4.02	0.35	0.33	1.72	0.03
K	99.29	0.14	0.36	0.08	–	0.12	–
Na	98.97	0.27	0.58	0.12	0.01	0.05	–
Cu	1.86	95.26	0.13	–	1.68	–	1.06
Zn	42.38	49.25	2.94	0.06	0.87	0.16	4.34

Ligand	Free ligand	Metal—ligand complexes (%)						
		Ca	Mg	K	Na	Cu	Zn	H
CO_3	0.50[b]	3.26	0.38	0.06	0.44	0.06	0.24	95.05
SO_4	76.07	18.71	2.43	0.36	2.4	–	0.04	–
Cl	98.8	0.8	0.1	0.04	0.25	–	–	–
PO_4	38.22	9.19	0.38	0.38	1.98	0.13	0.5	49.59
NO_3	98.45	1.01	0.41	0.05	0.08	–	–	–

[a] Distribution for soil solution data in Table 4.5, from Mattigod and Zachara (1996), with permission.
[b] Percentages represent the amount of metal or ligand as a free or complexed species. For example, 91.36% of Ca occurs as the free metal ion while 2.26% occurs as a Ca—CO_3 metal—ligand complex, and 5.03, 0.43, 0.23, and 0.68% occur as Ca—SO_4, Ca—Cl, Ca—PO_4, and Ca—NO_3 metal—ligand complexes.

4.5 Ion Activity and Activity Coefficients

Key Points: The concentration of an ion in a solution, measured through analytical techniques, is related to but is different from the ion's activity. The ion's activity is affected by the presence of other ions (i.e., ionic strength) and valence number.

The activity coefficient is a correction term used to account for "nonideal effects," or the effects of other solutes on the solute of interest in a solution. The activity of an ion in solution is a measure of its "reactivity" toward other solution components, and this "reactivity" can be decreased or increased via interactions with other ions in the solution. These interactions can range from covalent bonding and dipole (van der Waals) interactions to long-range coulombic repulsion and attraction due to similar or opposing ionic charges. Additionally, volume exclusion effects, where a portion of a molecule is not accessible due to the presence of other molecules, can also affect ion activity. The interactions between solutes accounted for by using the activity coefficient are in addition to those interactions that take place between the solute and solvent (e.g., H_2O) molecules and the solvation effect, which is the electrostatic interaction between ions and, for example, polar water molecules. The concentration of solutes affects the intensity of all these interactions and thus also the ion activity coefficient value (Morel and Hering, 1993). In an "ideal" solution the composition of the rest of the system has no effect on the solute of interest; thus in "ideal solutions" the activity is equal to concentration (Stumm and Morgan, 1996). The activity of an ion in solution is a dimensionless quantity, but it is derived by multiplying the activity coefficient by molarity. Another way of thinking about activity is to consider it as an "effective concentration" and necessary to consider because ions in solution do not behave in an ideal fashion. To describe chemical kinetics and equilibrium, chemists account for the nonideal behavior of solutes by replacing molar abundance (e.g., molarity or molality) with activity (Bickmore and Wander, 2018).

As illustrated in Box 4.1, the activity of an ionic species in solution is related to its concentration in solution through its single-ion activity coefficients (Equation B4.1c). To account for the effective size of hydrated ions, single-ion activity coefficients are typically calculated using the extended Debye—Hückel equation (Stumm and Morgan, 1996):

$$\log \gamma_i = - A\, Z_i^2 \left(\frac{I^{0.5}}{1 + Ba_i\, I^{0.5}} \right) \qquad (4.19)$$

BOX 4.2

An equilibrium speciation calculation for calcium and sulfate

This example illustrates how to set up a simple calculation in Visual MINTEQ (Gustafsson, 2011). To illustrate the basic steps involved with performing a calculation, it does not include additional components offered by Visual MINTEQ, such as adsorption, surface complexation, precipitation of solid phases, or gas equilibrium. Some of those components will be illustrated later. First, consider the species we wish to speciate, their concentrations, and the pH (Ca 12.86 mM, 12.86 mM SO_4^{2-}, pH 7.43). In the software (image below) set or "fix" the pH to 7.43 using the dropdown menu. Then, allow ionic strength to be calculated. Ionic strength will be calculated from the input of components (i.e., Ca^{2+} and SO_4^{2-}). With respect to

the concentrations of components, note that the software works within the context of molality. Even if mg L^{-1} is selected, the software internally converts the values into molality. See Chapter 1 for a discussion on molality versus molarity, which describes that for most of our systems of interest at low ionic strength, where molality and molarity are approximately equal. From the list of components, select "Ca^{+2}" and enter 12.86 into the text box. Select "Millimolal" as a concentration unit and select "Add to list." This step ensures that each component is entered before running the calculation. Repeat those steps for sulfate, listed as "$SO4^{-2}$." Check that all components are present by selecting the "View/edit list." Hydrogen will

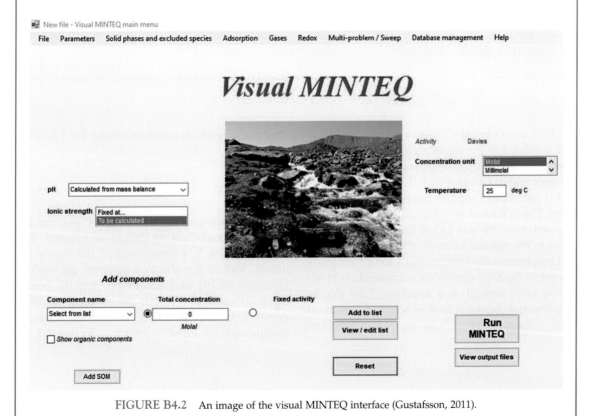

FIGURE B4.2 An image of the visual MINTEQ interface (Gustafsson, 2011).

Continued

BOX 4.2 (cont'd)

An equilibrium speciation calculation for calcium and sulfate

also be present by default, so there will be three components. Then select "Run MINTEQ," and the output values will appear. Additionally, by selecting "View species distribution" on the output screen, one can view the species that are dissolved in the aqueous media as a percentage (similar to the information presented in Table 4.5). The following is an image of the lcisual MINTEQ interface (Figure B4.2) and Table 4.6 contains some of the resulting values of the calculation:

A summation of the concentrations of the two major Ca species (Ca^{2+} at 8.6059×10^{-3} M and 4.2537×10^{-3} M $CaSO_4(aq)$) yields a concentration of 12.8596 mM, which is essentially equal to the starting concentration of 12.86 mM. Additionally, the calculated percentage can be determined by using the ratio of the component to the starting concentration. For example, (8.6059E-03)/(1.286E-02)(100) = 66.9%.

TABLE 4.6 Equilibrium speciation data of Ca^{2+} and SO_4^{2-} at pH 7.43.

Component	Concentration (M)	% of total concentration
Ca^{2+}	8.6059E-03	66.922
$CaOH^+$	2.7771E-08	—
$CaSO_4(aq)$	4.2537E-03	33.078
HSO_4^-	1.8654E-08	—
SO_4^{2-}	8.6059E-03	66.922

The A term typically has a value of 0.5 at 298 K and is related to the dielectric constant of water. The B term is also related to the dielectric constant of water and has a value of ≈ 0.33 at 298 K. The A and B parameters are also dependent on temperature. The a_i term is an adjustable parameter corresponding to the size (Å) of the ion. The Z_i term refers to the valence (charge) of the ion in solution and I represents the ionic strength. Ionic strength is related to the total electrolyte concentration in solution. Ionic strength is a measure of the degree of interaction between ions in solution, which, in turn, influences the activity of ions in solution.

The ionic strength is calculated using the following expression:

$$I = \frac{1}{2}\sum_i C_i\, Z_i^2 \qquad (4.20)$$

where C_i is the concentration in mol liter^{-1} of the ith species and Z_i is its charge or valence. The capital epsilon $\sum$ extends over all the ions in solution. Values for the a_i parameter (in nm) in Equation (4.19) and single-ion activity coefficients at various I values for an array of inorganic ions are listed in Table 4.7. Sample calculations for I and γ_i are given in Box 4.3.

TABLE 4.7 Single-ion activity coefficients at various ionic strengths at 298 K calculated using Equation (4.19).

Ion	a_i (nm)[a]	Ionic strength, I (mol liter^{-1})				
		0.001	0.005	0.01	0.05	0.1
Inorganic ions of charge 1+ or 1−						
H^+	0.90	0.967	0.933	0.914	0.860	0.830
Li^+	0.60	0.965	0.929	0.907	0.835	0.800
$Na^+, HCO_3^-, H_2PO_4^-, H_2AsO_4^-, IO_3^-$	0.40−0.45	0.964	0.928	0.902	0.820	0.775
$OH^-, F^-, ClO_4^-, MnO_4^-, HS^-, IO_4^-$	0.35	0.964	0.926	0.900	0.810	0.760
$K^+, Cl^-, Br^-, I^-, CN^-, NO_2^-, NO_3^-$	0.30	0.964	0.925	0.899	0.805	0.755
$Rb^+, Cs^+, NH_4^+, Tl^+, Ag^+$	0.25	0.964	0.924	0.898	0.800	0.750
Inorganic ions of charge 2+ or 2−						
Mg^{2+}, Be^{2+}	0.80	0.872	0.755	0.690	0.520	0.450
$Ca^{2+}, Cu^{2+}, Zn^{2+}, Mn^{2+}, Fe^{2+}, Ni^{2+}, Co^{2+}$	0.60	0.870	0.749	0.675	0.485	0.405
$Sr^{2+}, Ba^{2+}, Ra^{2+}, Cd^{2+}, Hg^{2+}, S^{2-}$	0.50	0.868	0.744	0.670	0.465	0.380
$Pb^{2+}, CO_3^{2-}, MoO_4^{2-}$	0.45	0.868	0.742	0.665	0.455	0.370
$Hg_2^{2+}, SO_4^{2-}, SeO_4^{2-}, CrO_4^{2-}, HPO_4^{2-}$	0.40	0.867	0.740	0.660	0.455	0.355
Inorganic ions of charge 3+ or 3−						
$Al^{3+}, Fe^{3+}, Cr^{3+}, Se^{3+}, Y^{3+}$	0.90	0.738	0.540	0.445	0.245	0.180
PO_4^{3-}	0.40	0.725	0.505	0.395	0.160	0.095

[a]Adjustable parameter corresponding to size of ion.
Adapted from Kielland (1937), with permission.

The extended Debye–Hückel equation (Equation 4.21) is one of several equations that have been derived to account for ion–ion interactions in aqueous solutions (Stumm and Morgan, 1996). Single-ion activity coefficients calculated using the extended Debye–Hückel equation begin to deviate from actual measurements, however, when the ionic strength of the soil solution is >0.2 mol liter^{-1}. For more concentrated solutions (I < 0.5 mol liter^{-1}), the Davies (1962) equation is often used:

$$\log \gamma_i = -A\, Z_i^2 \left(\frac{I^{0.5}}{1 + I^{0.5}} - 0.3I \right) \quad (4.21)$$

Special equations for calculating single-ion activity coefficients in more concentrated solutions (I > 0.5 mol liter^{-1}) also exist but are beyond the scope of this chapter. Such concentrated solutions are seldom, if ever, encountered in soil solutions.

The ionic strength of the aqueous medium (i.e., the ionic strength of water, soil pore water, or a batch experiment) is an important variable to consider not only when performing equilibrium speciation calculations but also during adsorption experiments (Chapter 5). Often, this is termed the "background electrolyte," and typically experiments are done under a constant ionic strength or constant background electrolyte concentration. The background

---

BOX 4.3

Calculation of ionic strength and single-ion activity coefficients

1. Calculate the ionic strength (I) of 0.10 mol liter^{-1} KNO$_3$ and 0.10 mol liter^{-1} K$_2$SO$_4$ solutions.

I for 0.10 mol liter^{-1} KNO$_3$:

$$I = \frac{1}{2}\Sigma_i C_i Z_i^2$$

$$I = \frac{1}{2}\left\{(0.10)(+1)^2 + (0.10)(-1)^2\right\}$$

$$= 0.10 \text{ mol liter}^{-1}$$

I for 0.10 mol liter^{-1} K$_2$SO$_4$:

$$I = \frac{1}{2}\Sigma_i C_i Z_i^2$$

$$I = \frac{1}{2}\left\{2(0.10)(+1)^2 + (0.10)(-2)^2\right\}$$

$$= 0.30 \text{ mol liter}^{-1}$$

KNO$_3$ is a 1:1 electrolyte since the cation and anion both have a valence of 1. With these simple electrolytes, I equals the molarity of the solution. For other stoichiometries (e.g., K$_2$SO$_4$, a 2:1 electrolyte), I is greater than the molarity of the solution (Harris, 1987).

2. Calculate γ_i at 298 K for Cd^{2+} when I = 0.01 mol liter^{-1}. Using Table 4.5, one sees that $\gamma_{Cd^{2+}}$ is 0.67. One can also calculate $\gamma_{Cd^{2+}}$ using Equation (4.21) as

$$\log\gamma_{Cd^{2+}} = \frac{(-0.5)\,(2)^2(0.01)^{0.5}}{1 + (0.33)(0.5 \times 10)(0.01)^{0.5}}$$

$$\log\gamma_{Cd^{2+}} = -0.1717$$

$$\gamma_{Cd^{2+}} = 0.67$$

The value of a_i in Equation (4.21) is reported in Å. To convert from nm (unit used for a_i in Table 4.5) to Å, multiply by 10.

---

electrolyte can be considered as a way to "swamp" or inundate the system in order to maintain a constant ionic strength throughout the reaction (Stumm and Morgan, 1996). Often a salt such as sodium nitrate is used. By employing a background electrolyte, any small changes in ionic strength due to adsorption are minimal. Thus a constant ionic strength is maintained throughout the reaction. For example, using a background electrolyte approximately 10 or more times more concentrated than the species of interest helps maintain constant ion activity coefficients; additionally, if the concentrations of solutes are much smaller than that of the background electrolyte, the activity coefficients can be assumed to be one with respect to using the constant ionic medium as a reference state (Stumm and Morgan, 1996).

4.6 Dissolution and Solubility Processes

Key Points: Dissolution is the dissolving of a solid in the soil, for example, a primary mineral. Precipitation is the process of forming a new solid from the soil solution, for example, a secondary mineral. These are two important chemical reactions that determine the fate of inorganic mineral components in soils.

Soil formation, weathering processes, and contaminant mobility are all affected by the dissolution–precipitation equilibria of the solid phase. The precipitation of minerals occurs only when supersaturation conditions exist in the soil solution (the soil solution contains more solute than should be present at equilibrium), while dissolution occurs only when the soil solution is undersaturated with respect to the inorganic mineral components in the soil. This means that soils are in a continuous state of dissolution and precipitation.

Calcareous minerals, such as gypsum ($CaSO_4 \cdot 2H_2O$), dissolve relatively rapidly and congruently; that is, after dissolution, the stoichiometric proportions of the elements in gypsum (Ca and SO_4) in the solution are the same as those in the dissolving mineral. However, many soil weathering reactions are incongruent, meaning dissolution is nonstoichiometric, and the concentration of dissolved ions is different from the ratio found in the dissolving soil mineral. This incongruency is often caused by the precipitation of a new solid phase (Stumm, 1992), either separately or on the surface of the original mineral that is undergoing weathering.

Silicate minerals dissolve to form various secondary minerals that are important in soils.

example, anorthite, a primary mineral in the feldspar group, weathers into kaolinite and other byproducts. Albite weathers into smectite, and biotite weathers to form the aluminum and iron oxides gibbsite and goethite, respectively:

These reactions illustrate that water and carbonic acid are the major reactants, and through dissolution, Ca^{2+}, Mg^{2+}, K^+, Na^+, and HCO_3^- are released into the soil solution. Carbonic acid is present in the soil porewater through the equilibrium reaction with carbon dioxide, and carbon dioxide is a natural product of plant roots and soil microorganisms:

$$CO_2(g) + H_2O \rightleftharpoons H_2CO_3 \ (carbonic \ acid) \quad (4.25)$$

$$H_2CO_3 \rightleftharpoons HCO_3^- + H^+ \quad (4.26)$$

$$HCO_3^- \rightleftharpoons CO_3^{2-} + H^+ \quad (4.27)$$

Through Equations 4.25–4.27, the production of protons will occur with an increase in dissolved carbon dioxide gas in water. It is because of the presence of carbon dioxide in the atmosphere and its equilibrium reaction

$$CaAl_2Si_2O_8 (anorthite) + 2CO_2 + 3H_2O \rightarrow Al_2Si_2O_5(OH)_4 (kaolinite) + Ca^{2+} + 2HCO_3^-$$

$$(4.22)$$

$$3NaAlSi_3O_8 (albite) + 2H_2O + Mg^{2+} \rightarrow 2Na_{0.5}(Al_{1.5}Mg_{0.5})Si_4O_{10}(OH)_2 (smectite)$$

$$+ 2Na^+ + H_4SiO_4 \quad (4.23)$$

$$K(MgFe^{2+})[AlSi_3]O_{10}(OH)_2 \ (biotite) + \frac{1}{2}O_2 + 3CO_2 + 9H_2O \rightarrow Al(OH)_3 \ (gibbsite)$$

$$+ 2FeOOH \ (goethite) + K^+ + Mg^{2+} + 3HCO_3 + 3H_4SiO_4 \quad (4.24)$$

Overall weathering reactions that can result in the formation of secondary minerals are shown below (Stumm and Wollast, 1990). For

in water, such as in natural rainfall water, that the pH of pristine rainwater is approximately 5.65 and not a neutral pH of 7 (Stumm and

Morgan, 1996). The extent, however, to which these weathering reactions occur in actual soils depends upon the soil-forming factors (parent materials, climate, biology, topography, time, and to a large extent, humans in urban areas).

Thermodynamic dissolution constants (K_{dis}^0) are used to calculate or predict the dissolution and precipitation reactions that will occur in soils and soil pore waters. For example, the dissolution of gibbsite can be expressed as—

$$Al(OH)_3 (gibbsite) \rightleftharpoons Al^{3+} + 3OH^- \quad (4.28)$$

Often the solubility product constants (K_{so}) for hydroxide minerals such as gibbsite are given with respect to protons (H^+) and not the hydroxyl group (OH^-). This is the equilibrium constant for the dissolution of the hydroxide mineral but written with respect to H^+. One of the differences between reporting K_{so} values for hydroxide minerals, such as gibbsite, and reporting those of other nonhydroxide minerals is that in the dissolution reaction, OH^- is replaced by H^+ using the reaction for the formation of liquid water ($OH^- + H^+ = H_2O$) (Sposito, 2008). This is commonly done because the equilibrium concentration of hydroxyl ions (OH^-) will be much smaller than protons (H^+) for pH < 7. In these cases the formation reaction for water ($3OH^- + 3H^+ = 3H_2O$) is used to write the expression with respect to the proton. By adding water to both sides of the Equation (4.28):

$$Al(OH)_3 + 3OH^- + 3H^+ \rightleftharpoons Al^{3+} + 3OH^- + 3H_2O$$
$$(4.29)$$

And now subtracting $3OH^-$ from both sides:

$$Al(OH)_3 + 3H^+ \rightleftharpoons Al^{3+} + 3H_2O \quad (4.30)$$

Considering the format for thermodynamic equilibrium constants presented above in Equation B4.2, the dissolution constant K_{dis}^0 for this reaction can be expressed as

$$K_{dis}^0 = \frac{(Al^{3+})(H_2O)^3}{(Al(OH)_3)(H^+)^3} = \frac{Products}{Reactants} \quad (4.31)$$

There are, however, several simplifications to this equation for dissolution constants that can be considered. The first one is that the activity of gibbsite [$Al(OH)_3$] or any other solid phase is defined as 1 if it exists in a pure form (has no structural imperfections) and is at standard temperature (T = 298.15 K) and pressure (0.101 MPa). If the solid phase is not pure, as is often the case in soils, one cannot equate activity to 1. The activity of water is also assumed to be 1 since soil solutions are essentially infinitely dilute compared to the concentration of water molecules. Assuming an activity of 1 for both the solid phase and water, Equation (4.31) can be simplified as

$$K_{dis}^0 = \frac{(Al^{3+})}{(H^+)^3} = K_{so}^0 \quad (4.32)$$

The K_{so}^0 term is the thermodynamic solubility product constant (Sposito, 1989) and is numerically equal to K_{dis}^0 when the solid phase is pure (there are no structural imperfections) and the aqueous solution phase is infinitely dilute (activity of water = 1). The K_{so}^0 values for a number of inorganic mineral components found in soils are known and have been tabulated (Lindsay, 1979). Those are the values that are incorporated into the software databases for equilibrium speciation modeling, as discussed in Box 4.2.

The right side of Equation (4.32) ([i.e., (Al^{3+})/$(H^+)^3$) is often referred to as the ion activity product (IAP) and, together with the K_{so}^0, is used to determine whether the soil solution is in equilibrium with a given inorganic mineral component. If the ratio of the $IAP/K_{so}^0 = 1$, the soil solution is said to be in equilibrium with a given soil solid phase. If the ratio of $IAP/K_{so}^0 > 1$, the solution is supersaturated with respect to the solid phase. During supersaturation, precipitation will occur. A ratio of $IAP/K_{so}^0 < 1$ indicates that the soil solution is undersaturated. During undersaturation, dissolution will occur. The ratio of the activities of the products to the activities of the reactants can also often be termed a reaction quotient (Q).

Software tools, such as Visual MINTEQ, are also capable of calculating the IAPs and comparing them with the solubility product constants to determine if a mineral is under- or oversaturated with respect to the ion composition of the solution (Box 4.4). This is carried out by determining the Saturation Index (SI), where $SI = \log IAP - \log K_{so}$. The dissolution of a generic solid $M_aL_{b(s)}$, where M is the metal and L is the ligand, can be written as follows (Sposito, 2008):

$$M_aL_{b(s)} = M^{m+} + L^{l-} \tag{4.33}$$

And the IAP can be defined by

$$IAP \equiv (M^{m+})(L^{l-}) \tag{4.34}$$

BOX 4.4

Using equilibrium speciation modeling to calculate saturation indices

Building off the example in Box 4.2, we will predict under what conditions any solids may form in the system with calcium and sulfate using Visual MINTEQ. The initial conditions to use for this calculation will be the same as before (Ca^{2+} 12.86 mM, SO_4^{2-} 12.86 mM, and pH 7.43) and can be entered into the software and the calculation run. On the output menu, to see what minerals may precipitate, select the "Display saturation indices" option and the following results are displayed in Table 4.8 (an additional column has been added to show the mineral formula). Saturation Index = $\log IAP - \log K_{so}$. Also, $\log K_{so}$ is not displayed in the Visual MINTEQ result.

As can be noted in the "Display saturation indices" option, the software recognizes what elements are input into the calculation to provide a list of all the potential solids that may precipitate out from solution based on that chemical composition. In this case the Saturation Index shows that all values are negative (and also appear as blue text to indicate that saturation

TABLE 4.8   Saturation index of Ca^{2+} and SO_4^{2-} at pH 7.43.

Mineral	Formula	Log IAP	Log K_{so}	Saturation Index
Anhydrite	$CaSO_4$	−4.728	−4.36	−0.368
Gypsum	$CaSO_4 \cdot 2H_2O$	−4.729	−4.61	−0.119
Lime	CaO	12.496	32.6993	−20.204
Portlandite	$Ca(OH)_2$	12.495	22.704	−10.209

has not been achieved for that particular mineral). Thus under these particular reaction conditions, we can expect all Ca^{2+} and SO_4^{2-} to remain in solution. This is also evident by selecting the "Equilibrated mass distribution" option on the output page, where both Ca^{2+} and SO_4^{2-} are listed as 100% dissolved and 0% precipitated. Other output options are also presented, including "Total sorbed" and "% sorbed." However, those options were not included in the calculation.

Continued

BOX 4.4   *(cont'd)*

Using equilibrium speciation modeling to calculate saturation indices

However, it is often useful to predict at what pH a solid may precipitate, particularly when planning experimental reactions in the laboratory. For example, if the goal of an experiment is to study the adsorption of dissolved ions onto a mineral surface over a range of pH values, the upper pH limit of the experiment can be calculated to avoid saturating the system with respect to that ion. This can be illustrated with a continuation of the Ca^{2+} and SO_4^{2-} example. Starting with the same experimental parameters, on the main input page, select "Solid phases and excluded species" from the menu bar at the top. Select "Specify possible solid phases." This option instructs the software to consider the precipitation of solids that can potentially form if the solution becomes oversaturated with respect to that solid. These solids will not influence the equilibrium calculation unless the solution becomes oversaturated with respect to that solid (Gustafsson, 2011). After selecting solids, which in this case would be anhydrite, gypsum, lime, and portlandite, go back to the main input menu. Then, select "Multi-problem/Sweep" from the menu bar. This will allow the user to select the parameter to "sweep"

or vary throughout the desired change in pH. For example, to sweep from pH 7 to 12 at 0.1 pH intervals, select "Sweep: one parameter is varied," choose the pH, set the start value to 7, and input 0.1 into the "Increments between values" textbox. The value of the "State number of problems" textbox should be 71, and this will set up the calculation from 7 to 14 at 0.1 pH intervals. In order to monitor the dissolved concentrations of Ca^{2+} and SO_4^{2-}, select these components from the dropdown menu and add them as "Total dissolved" components. A multitude of other components can also be selected if desired. Select "Save and Back" to go back to the main input menu and run the calculation. The output displays that the starting pH is 7.0 and there are 71 output files that can be selected from a dropdown menu. However, a new option is now available, which is the "Selected sweep results." Select that option to view how the solution ion concentration varies with pH. The resulting graph can be obtained.

In the case of Ca^{2+} and SO_4^{2-} the formation of gypsum is not expected throughout this pH range. Gypsum is a readily soluble mineral and is typically found on soil surfaces in arid environments, where evaporation rates are high. Hydroxide minerals, however, are a potent source for the retention and precipitation of cations. Portlandite (calcium hydroxide) is predicted to start to precipitate at a pH of 12.7, and at this pH, the blue line in the above graph begins to drop because Ca is decreasing in solution concentration and being incorporated into portlandite. While this pH is much higher than that in soils, it illustrates that calcium is a soluble ion, which is one reason why it can be leached easily from the soil in humid regions. Values in Table 4.9 (below) are dissolved ion concentrations at pH 13 and can be compared to those

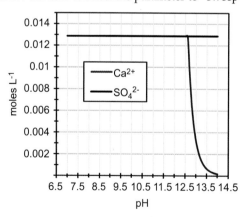

FIGURE B4.3   Solubility of Ca^{2+} and SO_4^{2-} over pH 7–14.

BOX 4.4 *(cont'd)*

Using equilibrium speciation modeling to calculate saturation indices

TABLE 4.9   Equilibrium speciation data of Ca^{2+} and SO_4^{2-} at pH 13.

Component	Concentration (M) in solution	% of total concentration in solution	% dissolved	% precipitated
Ca^{2+}	1.3355E-03	42.141	24.644	75.356
$CaOH^+$	1.2975E-03	40.940	—	—
$CaSO_4(aq)$	5.3623E-04	16.920 for Ca^{2+} 4.170 for SO_4^{2-}	—	—
HSO_4^-	5.8418E-14	—	—	—
SO_4^{2-}	1.2324E-02	95.830	100	0

values in Table 4.6. The $CaSO_4(aq)$ component is comprised of both Ca^{2+} and SO_4^{2-}, and because portlandite is precipitating out of solution at pH 13 while sulfate remains in solution, a portion of Ca is no longer available for complexation with sulfate.

The % of dissolved and % of precipitated calcium and sulfate can also be calculated to illustrate that about 75.4% of calcium is expected to precipitate at pH 13. However, no sulfate is expected to precipitate. The changes in the Saturation Index for each mineral are shown in Table 4.10, where it can be seen that portlandite is now in equilibrium with the solution because it is now saturated and thus will precipitate out.

While Visual MINTEQ and other equilibrium speciation modeling software will give a thermodynamic understanding and prediction as to whether a solid will form or not, it is important to remember that thermodynamic controls on mineral formation are not the only factor to consider. Additional considerations for the kinetics (i.e., the rate of reactions) of a reaction also need to be recognized. Reaction kinetics are discussed in Chapter 7, and their importance can be highlighted here. For example, if a particular mineral takes a long time to form (i.e., it has a kinetically slow reaction), then it is unlikely that a precipitate will form through the duration of a short-term experiment.

TABLE 4.10   Saturation Index of Ca^{2+} and SO_4^{2-} at pH 13.

Mineral	Formula	Log IAP	Saturation Index
Anhydrite	$CaSO_4$	−5.621	−1.261
Gypsum	$CaSO_4 \cdot 2H_2O$	−5.624	−1.014
Lime	CaO	22.705	−9.994
Portlandite	$Ca(OH)_2$	22.704	0.000

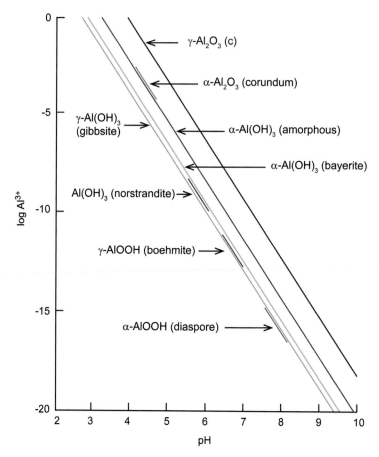

FIGURE 4.2   Solubility diagram for various aluminum oxides and hydroxides. *From Lindsay (1979), with permission.*

Thus the IAP for gypsum and gibbsite are defined by the products of their activities (Sposito, 2008):

$$\text{Gypsum IAP} = (Ca^{2+})(SO_4^{2-}) \qquad (4.35)$$

$$\text{Gibbsite IAP} = (Al^{3+})(OH^-)^3$$

$$\text{or } (Al^{3+})(H^+)^{-3} \qquad (4.36)$$

4.6.1 Stability Diagrams

A more useful and informative way of determining whether a solid phase controls the concentration of an element in the soil solution, and if it does, what the solid is, is to use a stability diagram (Figure 4.2). Stability diagrams are constructed by converting solubility relationships such as that in Equation (4.34) into log terms and rearranging terms to form a straight-line (linear) relationship. The log K_{dis}^0 value can be calculated from the standard free energy accompanying the reaction (ΔG_r^0), which is equal to:

$$\Delta G_r^0 = \Sigma \Delta G_f^0 \text{ products} - \Sigma \Delta G_f^0 \text{ reactants}$$

$$(4.37)$$

The sum of all the standard free energies of formation for products and reactants is denoted

TABLE 4.11 Equilibrium reactions of aluminum oxides and hydroxides at 298 K with corresponding $\log K_{dis}^0$.

Reaction number	Equilibrium reaction	$\log K_{dis}^0$
1	$0.5\gamma\text{-Al}_2O_3(c) + 3H^+ \rightleftharpoons Al^{3+} + 1.5H_2O$	11.49
2	$0.5\alpha\text{-Al}_2O_3(corundum) + 3H^+ \rightleftharpoons Al^{3+} + 1.5H_2O$	9.73
3	$Al(OH)_3(amorphous) + 3H^+ \rightleftharpoons Al^{3+} + 3H_2O$	9.66
4	$\alpha\text{-Al}(OH)_3(bayerite) + 3H^+ \rightleftharpoons Al^{3+} + 3H_2O$	8.51
5	$\gamma\text{-AlOOH}(boehmite) + 3H^+ \rightleftharpoons Al^{3+} + 2H_2O$	8.13
6	$Al(OH)_3(norstrandite) + 3H^+ \rightleftharpoons Al^{3+} + 3H_2O$	8.13
7	$\gamma\text{-Al}(OH)_3(gibbsite) + 3H^+ \rightleftharpoons Al^{3+} + 3H_2O$	8.04
8	$\alpha\text{-AlOOH}(diaspore) + 3H^+ \rightleftharpoons Al^{3+} + 2H_2O$	7.92

From Lindsay (1979), with permission.

as $\Sigma\Delta G_f^0$. The standard free energy of formation is a value typically expressed in kilojoules per mole (kJ mol^{-1}), though it can also be given in units such as joules mole^{-1} or kilocalories mole^{-1}. The standard free energy of formation (or Gibbs free energy of formation) describes the amount of energy necessary to form or break down a mineral from its individual components. The free energy of formations for many minerals is known and tabulated. Those values are determined in part through temperature-based studies, such as calorimetry. Calorimetric experiments often involve the dissolution of a mineral into a solvent and then measuring the change in the temperature

An example of this process for gibbsite dissolution is shown below. The components used to calculate ΔG_r^0 for gibbsite are $\gamma\text{-Al}(OH)_3$, Al^{3+}, H_2O, and H^+. Gibbsite is one of many different aluminum minerals commonly referred to as aluminum oxides or aluminum oxyhydroxides. While Reaction 7 for gibbsite in Table 4.11 is illustrated below, the same process can be carried out for the other reactions in Table 4.11. The ΔG_r^0 values used to calculate ΔG_r^0 are published values from:

$$Al(OH)_3(gibbsite) + 3H^+ \rightleftharpoons Al^{3+} + 3H_2O \quad (4.38)$$

$$\Delta G_r^0 = \left[\left(-491.61 \text{ kJ mol}^{-1}\right) + 3\left(-237.52 \text{ kJ mol}^{-1}\right)\right] - \left[\left(-1158.24 \text{ kJ mol}^{-1}\right) + 3\left(0 \text{ kJ mol}^{-1}\right)\right]$$

$$= [-1204.16 \text{ kJ mol}^{-1}] - [1158.24 \text{ kJ mol}^{-1}] = -45.92 \text{ kJ mol}^{-1} \quad (4.39)$$

(either an increase or decrease) during mineral dissolution. By performing those experiments on numerous minerals, the results have been used to determine many ΔG_f^0 values for numerous minerals.

Based on Equation (4.39) and because the resulting standard Gibbs free energy of the reaction is negative, the formation of gibbsite is thermodynamically favorable from those reactants. Moreover, by calculating the thermodynamic values, one can also relate ΔG_r^0 to K_{dis}^0 using the following relationship:

$$\Delta G_r^0 = RT \ln K_{dis}^0 \qquad (4.40)$$

In Equation (4.40), where R is the universal gas constant and T is absolute temperature and utilizing the relationship $\ln(x) = 2.303 \log(x)$, at 298 K:

$$\Delta G_r^0 = -\left(0.008314 \text{ kJ K}^{-1}\text{mol}^{-1}\right)$$

$$(298.15 \text{ K})(2.303)\log K_{dis}^0 \qquad (4.41)$$

The constants in Equation (4.41) can be simplified to 5.71 to yield the following expression:

$$\log K_{dis}^0 = \frac{-\Delta G_r^0}{5.71} \qquad (4.42)$$

Thus for Reaction 7 in Table 4.11:

$$\log K_{dis}^0 = \frac{-(-45.92)}{5.71} = 8.04 \qquad (4.43)$$

Based on Equation (4.43), the concentrations of dissolved Al^{3+} and H^+ are related to K_{dis}^0. Thus Equation (4.32) can be combined with Equation (4.43) to produce the solubility line in Figure 4.2 for Reaction 7 as follows:

$$K_{dis}^0 = 10^{8.04} = \frac{(Al^{3+})}{(H^+)^3} \qquad (4.44)$$

By taking the logarithm of both sides and recalling that $pH = -\log(H^+)$, the following equation is produced:

$$\log(10^{8.04}) = \log\left(\frac{(Al^{3+})}{(H^+)^3}\right)$$

$$= \log Al^{3+} - 3 \log H^+ = \log Al^{3+} + 3 \text{ pH} \qquad (4.45)$$

And rearranging Equation (4.45) yields the following in the linearlized form $(y = mx + b)$:

$$\log Al^{3+} = -3pH + 8.04 \qquad (4.46)$$

where $\log Al^{3+}$ is plotted on the y-axis, pH is plotted on the x-axis, and -3 and 8.04 are the slope (m) and y-intercept, respectively.

The advantage of a stability diagram is that the solubility of several different solid phases can be compared at one time. Figure 4.2 is an example of a stability diagram constructed for different aluminum oxides and hydroxides listed in Table 4.11. The positions of the different lines correspond to the solubility of each solid, with the line nearest the axes being the most insoluble. That is, the line nearest the origin of the axes is diaspore; thus diaspore in this example is the least soluble mineral.

The other solubility lines in Figure 4.2 can be similarly developed. The anhydrous aluminum oxides γ-Al_2O_3(c) and α-Al_2O_3 (corundum) depicted in Table 4.11 and Figure 4.2 are high-temperature minerals that are not usually formed in soils. The order of decreasing solubility among the remaining minerals is α-$Al(OH)_3$ (bayerite), γ-AlOOH (boehmite), $Al(OH)_3$ (norstrandite), γ-$Al(OH)_3$ (gibbsite), and α-AlOOH (diaspore). However, the differences in solubility between the last four minerals are very slight. It is also evident from Figure 4.2 that the activity of Al^{3+} in equilibrium with any of the minerals is dependent on pH. The activity of Al^{3+} decreases 1000-fold for each unit increase in pH (slope = 3).

The solubility diagram for several phyllosilicate and oxide minerals found in soils is shown in Figure 4.3. Note that the x-axis is plotted as a function of $\log H_4SiO_4^0$ as opposed to pH in Figure 4.2 and the y-axis is expressed as $\log Al^{3+} + 3pH$ as opposed to $\log Al^{3+}$. This illustrates that solubility diagrams can be constructed in a number of different ways as long as the terms used for each axis are consistent between the different mineral phases of interest.

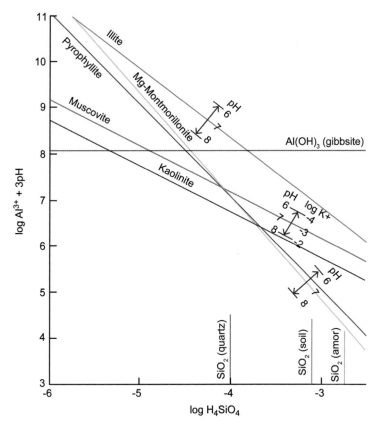

FIGURE 4.3 Solubility diagram of several primary and secondary minerals in equilibrium with 10^{-3} M K$^+$, 10^{-3} M Mg^{2+}, and soil-Fe with indicated changes for pH and K$^+$. *From Lindsay (1979), with permission.*

The stability line for kaolinite in Figure 4.3 can be obtained as follows, assuming the following dissolution reaction:

$$Al_2Si_2O_5(OH)_4 (kaolinite) + 6H^+ \rightleftharpoons 2Al^{3+}$$
$$+2H_4SiO_4^0 + H_2O$$

$$\log K_{dis}^0 = 5.45 \qquad (4.47)$$

Rearranging terms and assuming an activity of 1 for kaolinite and water yields

$$\frac{\left(Al^{3+}\right)^2 \left(H_4SiO_4^0\right)^2}{\left(H^+\right)^6} = 10^{5.45} \qquad (4.48)$$

Taking the log of both sides of Equation (4.50) and solving for the terms on the axes in Figure 4.3 yields the desired linear relationship:

$$2 \log Al^{3+} + 6 \, pH = 5.45 - 2 \log H_4SiO_4^0$$
$$(4.49)$$

Dividing both sides by 2 yields

$$\log Al^{3+} + 3pH = 2.73 - \log H_4SiO_4^0 \qquad (4.50)$$

Stability diagrams such as Figure 4.3 are very useful in predicting the presence of stable solid phases in different soil systems. Remembering that the portion of straight-line relationships closest to the x and y axes represents the most stable mineral phase, it is evident that when the dissolved concentration of silicic acid (H_4SiO_4) increases (i.e., moving to the right on the x-axis), Mg-montmorillonite is the most stable of the phyllosilicates (e.g., at 10^{-3} M H_4SiO_4). Additionally, as pH increases, the stability line for Mg-montmorillonite moves closer to the x- and y-axes, indicating that stability increases. The shorter parallel lines indicate the stability with changes in pH. For muscovite, an additional comparison is given for changes in the concentration of potassium. The vertical lines for quartz and the other sources of SiO_2 are straight lines in Figure 4.3 because there is no Al^{3+} in the crystalline structure, and the dissolution reaction is essentially independent of pH. At lower concentrations in the soil solution (log $H_4SiO_4^0 < -3.7$), the 1:1 clay mineral kaolinite becomes the most stable solid phase. This is consistent with the observation that kaolinite is the dominant phyllosilicate mineral in acid soils that receive substantial amounts of rainfall (i.e., soils that have extensive soil leaching). Under more extreme leaching conditions (log $H_4SiO_4^0 < -5.3$), gibbsite becomes the most stable solid phase.

Because gibbsite is one of the most stable minerals in soils, it is often used as the reference mineral that limits the activity of Al^{3+} in soil solutions. The y-axis in Figure 4.3 can be interpreted as a combination of the concentration of the free metal ion Al^{3+} along with pH. Recall that the dissolve concentration of Al^{3+} is also dependent on pH (Equation 4.46). According to Equation (4.46), if pH = 6, the log concentration of Al^{3+} will be -9.96.

4.7 Understanding Trace Metal Solubility Using Equilibrium Speciation Calculations and Metal Hydroxide Solids

Key Points: The formation of metal hydroxides is a key driver for the precipitation of Fe, Al, and

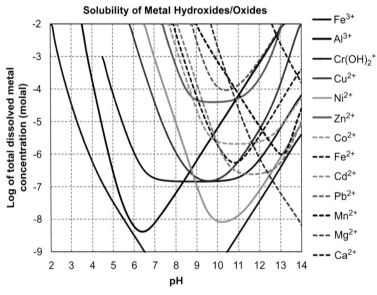

FIGURE 4.4 The solubility of aqueous metal cations with respect to the solubility product constants of their metal hydroxide/oxide precipitates. In these models, initially, 10 mM of each metal is present in solution, except for Cr, which is present at 1 mM. The figure is modeled after Dyer et al. (1998) and was produced using Visual MINTEQ. Units are given in molality, which, at low concentrations, is equivalent to molarity.

other metal hydroxides in soils. The solubility of these hydroxides illustrates why some elements, such as Fe and Al, are much less mobile than other elements, such as Ca and Mg.

An analogy of metal solubility in soils and mineral systems can be made by comparing the equilibrium concentration of each metal with respect to the formation of metal hydroxide/oxide. In Figure 4.4, which is modeled after Dyer et al. (1998), the aqueous concentrations of 13 different cationic species are presented over the pH range of 2—14. The aqueous concentrations were modeled using Visual MINTEQ, typically with the amorphous or crystalline metal hydroxide/oxide being the sole solid phase present. To determine the change in metal solubility when the metal hydroxide/oxide is allowed to precipitate, a sweep function was applied over pH 2—14. Hydrolysis is one of the first reactions that metal cations undertake when the pH of a solution changes. This can be analogous to changes in soil pH or changes in pH in laboratory batch reactions. The presence of organic ligands in soils will modify metal solubility.

In Figure 4.4 the amorphous or crystalline forms of the metal hydroxide or other common hydroxides were used, except for Pb, where PbO was used (Dyer et al., 1998). For example, for Fe and Al, ferrihydrite and gibbsite were used as the solids. There are several important trends to point out in Figure 4.4. The variability of metal hydroxide solubility is large, with some metals such as Fe and Al being much more insoluble at lower pH values than the other metals. Iron and Al are two of the first cations to hydrolyze at low pH values. This is one reason why Fe and Al oxides are more persistent in soils, because their hydroxides are very insoluble over a wide pH range, particularly even at slightly acidic pH values. Iron and Al hydrolyze very quickly and are thus generally very low in concentration in soil porewater (except in acidic soils). Magnesium and Ca, on the other hand, have much higher solubilities. The precipitation of Ca and Mg hydroxides will only occur when pH values are exceedingly high. The curves

for Mg and Ca appear at around pH 9.8 and pH 12.5, respectively, which are much higher than almost all soils. Often, Mg and Ca that are present in soils and primary minerals are the first cations to be leached and weathered out of the soil profile. One reason for this is that they do not readily form hydroxides. This is true for soils that receive significant rainfall, such as Oxisols, where Fe and Al oxides can persist. Soils in arid climates can become enriched in Ca due to the formation of calcite, which is not a hydroxide mineral.

Over pH 2—14, many metals exhibit a rapid decrease in solubility. As the metal hydroxide solids begin to precipitate out, they pull metal cations out of the solution and into the solid phase, creating newly formed hydroxide precipitates. As pH increases, the solubility of metals also tends to increase. For example, Al is most insoluble at approximately pH 6.4 with respect to gibbsite formation. Above pH 6.4, however, the aqueous species of Al begin to increase. While Al is still in the +3 valence state, the dissolved species is predominantly the negatively charged $Al(OH)_4^-$, which increases Al solubility. The addition of more hydroxyl groups gives a negative charge to the molecule and increases its solubility. The hydrolysis of Al is very rapid and can incorporate other coions adjacent to the mineral—water interface as it hydrolyzes to form new minerals such as layered double hydroxides (Chapter 5).

While different elements hydrolyze at different pH values, solubility also depends heavily on the redox state of an element. Many of the trace metals of interest in Figure 4.4 are in the +2 valence states, which is their most common state in soils. There are several important exceptions, such as Cr, Fe, and Mn. For Fe, the aqueous concentration of Fe(III) immediately begins to decrease as pH increases because of the formation of $Fe(OH)_3$, which is very insoluble. The precipitation of $Fe(OH)_3$ drives the aqueous concentration of Fe(III) very low. This is one reason that Fe can be a limiting micronutrient for some plants and in aquatic systems. The reduced form of Fe [Fe(II)] is much more soluble over a larger pH range. It is only at alkaline pH values above pH 8 when $Fe(OH)_2$ begins to precipitate, which decreases the total dissolved Fe(II) concentration. While these types of comparisons provide an important basic understanding of metal solubility, it is essential to consider the precipitation of other mineral phases, including mixed metal phases (Chapter 5), when modeling environmental systems. Mixed metal phases can have lower solubility product constants that are not typically found in thermodynamic databases. In addition to thermodynamics, the rates of mineral precipitate formation (i.e., the kinetics of a reaction) should also be considered. Soils are often not at equilibrium and the continual flux of elements within porewater will affect mineral dissolution and precipitation.

4.7.1 Precipitation Terminology

There are several mechanisms (i.e., ways) that new solid phase precipitates in a solution or mineral suspension can form. These mechanisms are categorically different: (1) homogeneous precipitation, (2) heterogeneous precipitation, and (3) surface precipitation (Siebecker et al., 2018). In order for homogeneous precipitation to occur, there must be a "supersaturation" of the aqueous ion. The newly formed solid phase will not form until the solute concentration exceeds the solubility product with respect to that solid phase. In addition, an energy barrier must be overcome (Stumm and Morgan, 1996). In homogeneous precipitation no preexisting mineral surface is necessary for a new mineral to form. However, the role of the mineral surface is often key in catalyzing the formation of new precipitates in soils because it helps overcome that energy barrier. Heterogeneous precipitation will occur on an existing solid surface (Atkins, et al., 2006), such as a clay mineral present in soil or a mineral suspension. In heterogeneous nucleation the mineral surface is a fixed solid where there is no mixing of sorptive ions with ions that have dissolved from the sorbent. This is quite different

from surface precipitation, where the mineral–water interface is a "mixing zone" for all sorptive ions, including ions that have dissolved from the adsorbent (Stumm and Morgan, 1996). Because we now view many clay minerals in soils as actively undergoing dissolution, reprecipitation, and atom exchange, the surface precipitation mechanism is the most common formation route of new precipitates.

Suggested Reading

Adams, F., 1971. Ionic concentrations and activities in the soil solution. Soil. Sci. Soc. Am. Proc. 35, 420–426.

Drever, J.I. (Ed.), 1985. The Chemistry of Weathering, Vol. 149. NATO ASI Ser. C, Reidel, Dordrecht, The Netherlands.

Garrels, R.M., Christ, C.L., 1965. Solutions, Minerals, and Equilibria. Freeman, Cooper, San Francisco, CA.

Lindsay, W.L., 1979. Chemical Equilibria in Soils. Wiley, New York, NY.

Lindsay, W.L., 1981. Solid-phase solution equilibria in soils. ASA Spec. Publ. 40, 183–202.

Mattigod, S.V., Zachara, J.M., 1996. Equilibrium modeling in soil chemistry. In: Sparks, D.L. (Ed.). Society of America Book Series 5. Soil Science Society of America, Madison, WI, pp. 1309–1358.

Stumm, W., Wollast, R., 1990. Coordination chemistry of weathering. Kinetics of the surface-controlled dissolution of oxide minerals. Rev. Geophys. 28, 53–69.

Wolt, J., 1994. Soil Solution Chemistry. Wiley, New York, NY.

Problem Set

Q1. When the supersaturation of the soil solution occurs with respect to a cation, would you expect mineral precipitation or dissolution to occur? Why?

Q2. What is hydrolysis?

Q3. How is K^{cond} related to K°?

Q4. When would the conditional equilibrium constant be equal to the thermodynamic equilibrium constant?

Q5. Calculate the single-ion activity coefficient for calcium at 0.01 M.

Q6. When electrolytes have a valence charge of 2, will the ionic strength equal the molarity of the solution? Why or why not?.

Q7. Calculate the ionic strength of 0.001 M $Ni(NO_3)_2$.

Q8. What is the single-ion activity coefficient for cadmium at 0.01 M?

Q9. Based on the equilibrium speciation calculations of a basic soil (Table 4.4), which metal would you expect to occur as the most complexed with ligands in a soil solution?

Q10. Based on the equilibrium speciation calculations of a basic soil (Table 4.4), sulfate is predicted to complex the most with which metal?

Q11. When are molarity and molality indistinguishable?

Q12. The following question deals with the use of Visual MINTEQ. In a solution of 3 mM Ca^{2+} and 3 mM CO_3^{-2} at what pH is calcite predicted to start to precipitate? What percentage of the Ca^{2+} component is predicted to be precipitated at pH 8.1? Parameters/Hints: Assume 25°C and let the ionic strength be calculated. Select "Specify possible solid phases" to add calcite. Use the sweep mode in 0.1 step intervals.

Q13. When considering thermodynamic dissolution constants, if a solid has no structural imperfections or elemental substitutions, what is its activity?

Q14. Rank the Al oxides/hydroxide minerals in order of decreasing solubility (most to least soluble): bayerite, boehmite, corundum, gibbsite, diaspore, and gamma-alumina.

References

Amundson, R.G., Davidson, E.A., 1990. Carbon-dioxide and nitrogenous gases in the soil atmosphere. J. Geochem. Explor. 38, 13–41.

Atkins, P., Overton, T., Rourke, J., Weller, M., Armstrong, F., 2006. Shriver and Atkins' Inorganic Chemistry, Fourth Edition. Oxford University Press, Oxford.

Baes Jr., C.F., Mesmer, R.E., 1976. The Hydrolysis of Cations. John Wiley & Sons, New York.

Ball, J.W., Jenne, E.A., Cantrell, M.W., 1981. WATEQ3: A geochemical model with uranium added. US Geol. Surv. Open-File Rep. 81–1183.

Ball, J.W., Jenne, E.A., Nordstrom, D.K., 1979. WATEQ2 – a computerized chemical model for trace and major element speciation and mineral equilibria of natural waters. In: Jenne, E.A. (Ed.), Chemical Modeling in Aqueous Systems, American Chemical Society Symposium. Series No, 93. American Chemical Society, Washington, D.C, pp. 815–835.

Baham, J., 1984. Prediction of ion activities in soil solutions: Computer equilibrium modeling. Soil Sci. Soc. Am. J 48, 523–531.

Barnes, I., Clarke, F.E., 1969. Chemical properties of ground water and their corrosion and encrustation effects on wells. Geological Survey professional paper, 498–D.

Bethke, C.M., 2022. In: Geochemical and Biogeochemical Reaction Modeling, 3rd Ed. Cambridge University Press, New York, NY, USA.

Bethke, C.M., Yeakel, S., 2010. The Geochemist's Workbench, Release 8.0 GWB Essentials Guide, Hydrogeology Program. University of Illinois. https://www.gwb.com.

Bradbury, M.H., Baeyens, B., 1997. A mechanistic description of Ni and Zn sorption on Na-montmorillonite Part II: Modelling. J. Contam. Hydrol. 27 (3), 223–248.

Brezonik, P.L., Arnold, W.A., 2012. Water chemistry: fifty years of change and progress. Environ. Sci. Technol. 46, 5650–5657.

Bickmore, B.R., Wander, M.C.F., 2018. Activity and activity coefficients. In: White, W.M. (Ed.), Encyclopedia of Geochemistry: A Comprehensive Reference Source on the Chemistry of the Earth. Springer International Publishing, Cham, pp. 21–23.

Cabaniss, S.E., 1987. TITRATOR: an interactive program for aquatic equilibrium calculations. Environ. Sci. Technol. 21, 209–210.

Casey, W.H., Swaddle, T.W., 2003. Why small? The use of small inorganic clusters to understand mineral surface and dissolution reactions in geochemistry. Rev. Geophy. 41 (2), 1008.

Dyer, J.A., Scrivner, Noel, C., Dentel, S.K., 1998. A practical guide for determining the solubility of metal hydroxides and oxides in water. Environ. Prog. 17, 1–8.

Dunand, F.A., Helm, L., Merbach, A.E., 2003. Solvent exchange on metal ions. Adv. Inorg. Chem. 54, 1–69.

Eaton, F.M., Harding, R.B., Ganje, T.J., 1960. Soil solution extractions at tenth-bar moisture percentages. Soil Sci 90, 253−258.

Eigen, M., Mass, G., 1966. Über die kinetic der metallkomplexbildung der alkali- und erdalkaliionen in wässrigenlösungen. Zeitschrift für Physikalische Chemie Neue Folge 49, 1633−1770.

Essington, M.E., 2015. Soil and Water Chemistry: An Integrative Approach, Second Edition. CRC Press, Boca Raton, FL.

Farley, K.J., Carbonaro, R.F., Fanelli, C.J., Costanzo, R., Rader, K.J., Di Toro, D.M., 2011. Ticket-UWM: a coupled kinetic, equilibrium, and transport screening model for metals in lakes. Environ. Toxicol. Chem. 30, 1278−1287.

Felmy, A.R., Girvin, D.C., Jenne, E.A., 1984. MINTEQ. A computer program for calculating aqueous geochemical equilibria. NTIS PB84-157418. EPA-600/3-84-032. National Technical Information Service.

Glossary of Soil Science Terms, 2008. Soil Science Glossary Terms Committee and Soil Science Society of America. Soil Sci. Soc. Am., Madison, WI.

Gillman, G.P., Bell, L.C., 1978. Soil solution studies on weathered soils from tropical North Queensland. Aust. J. Soil Res 16, 67−77.

Gillman, G.P., Sumner, M.E., 1987. Surface charge characterization and soil solution composition of four soils from the southern piedmont of Georgia. Soil Sci. Soc. Am. J 51, 589−594.

Gustafsson, J.P., 2011. Visual MINTEQ 3.0 User Guide. Royal Institute of Technology, Stockholm, Sweden.

Harris, D.C., 1987. Quantitative Chemical Analysis, 2nd ed. Freeman, New York.

Harris, D.K., Ingle, S.E., Magnunson, V.R., Taylor, D.K. (1981). REDEQL.UMD, Department of Chemistry, University of Minnesota, Duluth.

Helm, L., Merbach, A.E., 2005. Inorganic and bioinorganic solvent exchange mechanisms. Chem. Rev. 105, 1923−1959.

Helm, L., Nicole, G.M., Merbach, A.E., 2005. Water and proton exchange processes on metal ions. Adv. Inorg. Chem. 57, 327−379.

Herbelin, A.L., Westall, J.C., 1999. FITEQL 4.0: A computer Program for the Determination of Chemical Equilibrium Constants from Experimental Data. Oregon State University, Corvallis, OR.

Hochella Jr., M.F., et al., 2008. Nanominerals, mineral nanoparticles, and earth systems. Science 319, 1631−1635.

Ingle, S.E., Schultt, M.D., Shults, D.W., 1978. A User's Guide for REDEQL.EPA. Environmental Protection Agency, Corvallis, OR.

Jenne, E.A., 1979. Chemical modeling-goals, problems, approaches, and priorities. In: Jenne, E.A. (Ed.), Chemical Modeling in Aqueous Systems. American Chemical Society Symposium Series No. 93, Washington, DC, pp. 3−21.

Keizer, M.G., Van Riemsdijk, W.H., 2002. ECOSAT version 4.7. A Computer Program for the Calculation of Speciation and Transport in Soil-Water Systems. Department of Environmental Sciences, Soil Quality, Wageningen University, Wageningen, The Netherlands.

Kharaka, Y.K., Barnes, J., 1973. SOLMNEQ: Solution-Mineral Equilibrium Computations. 73-002, NTIS PB-215899. U.S. Geological Survey Techniques of Water-Resources Investigations.

Kielland, J., 1937. Individual activity coefficients of ions in aqueous solutions. J. Am. Chem. Soc. 59, 1675−1678.

Kinniburgh, D.G., Miles, D.L., 1983. Extraction and chemical analysis of interstitial water from soils and rocks. Environ. Sci. Technol 17, 372−368.

Luther III, G.W., 2016. Inorganic Chemistry for Geochemistry and Environmental Sciences: Fundamentals and Applications. Wiley, Chennai.

Mattigod, S.V., Zachara, J.M., 1996. Equilibrium modeling in soil chemistry. In: Sparks, D.L. (Ed.), Methods of Soil Analysis: Part 3-Chemical Methods. SSSA Book Ser. 5, Soil Sci. Soc. Am., Madison, WI, pp. 1309−1358.

Martell, A.E., Smith, R.N., 1976. Critical Stability Constants. Plenum, New York, NY.

May, P.M., Rowland, D., Murray, K., May, E.F., 2019. JESS. http://jess.murdo ch.edu.au/jess_about .shtml.

Margerum, D.W., Cayley, G.R., Weatherburn, D.C., Pagenkopf, G.K., 1978. Kinetics and mechanisms of complex formation and ligand exchange. In: Martell, A.E. (Ed.), Coordination Chemistry, 2. ACS Monograph 174 ACS, Washington, DC, pp. 1−221.

McDuff, R.E., Morel, F.M.M., Morgan, J.J., 1973. Description and use of the chemical equilibrium program REDEQL2. California Institute of Technology, Pasadena, CA.

Meeussen, J.C.L., 2003. ORCHESTRA: an object-oriented framework for implementing chemical equilibrium models. Environ. Sci. Technol 37 (6), 1175−1182.

Melchior, D.C., Bassetteds, R.L. (Eds.), 1990. Chemical modeling of aqueous systems II. ACS Symposium Series No. 416. American Chemical Society, Washington, DC.

Morel, F.M.M., Hering, J.G., 1993. Principles and Applications of Aquatic Chemistry. John Wiley and Sons, Inc., New York.

Morel, F.M.M., Morgan, J.J., 1972. A numerical method for computing equilibrium in aqueous chemical systems. Environ. Sci. Technol. 6, 58−67.

Nieminen, T.M., Derome, K., Meesenburg, H., De Vos, B., 2013. Soil solution: sampling and chemical analyses. In: Ferretti, M., Fischer, R. (Eds.), Forest Monitoring : Methods for Terrestrial Investigations in Europe With an Overview of North America and Asia. Elsevier, London, pp. 301−315.

Parker, D.R., Chaney, R.L., Norvell, W.A., 1995. Chemical equilibrium models: Applications to plant nutrition research. In: Loeppert, R.H., Schwab, A.P., Goldberg, S.

(Eds.), Chemical Equilibrium and Reaction Models. Soil Science Society of America, Madison, WI, pp. 163–200.

Parkhurst, D.L., Appelo, C.A.J., 2013. Description of input and examples for PHREEQC version 3 – A computer program for speciation, batch-reaction, one-dimensional transport, and inverse geochemical calculations. In: U.S. Geological Survey Techniques and Methods, Book 6, Modeling Techniques, (Chapter 43). U.S. Geological Survey, Reston, VA.

Puigdomenech, I., 2006. HYDRA (Hydrochemical Equilibrium-Constant Database) and MEDUSA (Make Equilibrium Diagrams Using Sophisticated Algorithms) Programs. Royal Institute of Technology, Stockholm. http://www.kemi.kth.se/medusa/.

Richens, D.T., 1997. The Chemistry of Aqua Ions. John Wiley & Sons, Chichester.

Sahai, N., Sverjensky, D.A., 1998. GEOSURF: a computer program for modeling adsorption on mineral surfaces from aqueous solution. Comput. Geosci. 24 (9), 853–873.

Selim, M., 2016. Chem_Transport Software Models for Chemical Kinetic Retention and Transport in Soils and Geological Media User's Manual. School of Plant, Environmental and Soil Science, LSU-Agcenter, Baton Rouge. http://www.spess.lsu.edu/chem_transport/. (Accessed July 2022).

Schecher, W.D., McAvoy, D.C., 2007. MINEQL+: A Chemical Equilibrium Modeling System, Version 4.6. Environmental Research Software, Hallowell, ME.

Schwab, P., 2012. Soil solution. In: Huang, P.M., Li, Y., Sumner, M.E. (Eds.), Handbook of Soil Sciences. CRC, Taylor & Francis, Boca Raton, FL; London, pp. 12-1–12-23.

Siebecker, M.G., Li, W., Sparks, D.L., 2018. The important role of layered double hydroxides in soil chemical processes and remediation: what we have learned over the past 20 years. In: Sparks, D.L. (Ed.), Advances in Agronomy, vol. 147. Academic Press, Cambridge, MA, pp. 1–59.

Smith, G.R., Tanji, K.K., Jurinak, J.J., Burau, R.G., 1995. C-SALT – A chemical equilibrium model for multicomponent solutions. In: Loeppert, R.H., et al. (Eds.), Chemical Equilibrium and Reaction Models, SSSA Special Publications 42. ASA and SSSA, Madison, WI, pp. 289–324.

Sposito, G., 1989. The Chemistry of Soils. Oxford Univ. Press, New York.

Sposito, G., 2008. The Chemistry of Soils, Second Edition. Oxford University Press, New York, NY.

Sposito, G., Coves, J., 1988. SOILCHEM: A Computer Program for the Calculation of Chemical Equilibria in Soil Solutions and Other Natural Water Systems. Kearney Foundation of Soil Science, University of California, Riverside.

Sposito, G., Mattigod, S.V., 1980. GEOCHEM: A Computer Program for the Calculation of Chemical Equilibria in Soil Solutions and Other Natural Water Systems. Kearney Foundation of Soil Science, University of California, Riverside.

Stumm, W. (Ed.), 1992. Chemistry of the solid-water interface. John Wiley & Sons, New York, NY.

Stumm, W., Morgan, J.J., 1996. Aquatic Chemistry: Chemical Equilibria and Rates in Natural Waters, Third Edition. John Wiley and Sons, Inc., New York, NY.

Taube, H., 1952. Rates and mechanisms of substitution in inorganic complexes in solution. Chem. Rev. 50, 69–126.

Tipping, E., 1994. WHAM – A chemical-equilibrium model and computer code for waters, sediments, and soils incorporating a discrete site electrostatic model of ion-binding by humic substances. Comput. Geosci. 20, 973–1023.

Truesdall, A.H., Jones, B.F., 1974. WATEQ, a computer program for calculating chemical equilibria of natural waters. J. Res. U.S. Geol. Surv. 2, 223.

Van der Lee, J., De Windt, L., 2002. CHESS Tutorial and Cookbook. Users Manual Nr. LHM/RD/02/13. Ecole des Mines de Paris, Fontainebleau.

Verweij, W., Simonin, J.-P., 2020. Implementing the Mean Spherical Approximation model in the speciation code CHEAQS next at high salt concentrations. J. Solution. Chem. 49, 1319–1327.

Westall, J.C., 1979. MICROQL – A Chemical Equilibrium Program in. BASIC, Swiss Federal Institute, Duebendorf, Switzerland.

Westall, J.C., Zachary, J.L., Morel, F.M.M., 1976. MINEQL – A Computer Program for the Calculation of Chemical Equilibrium Composition of Aqueous Systems. In: Technical Note No. 18. Massachusetts Institute of Technology, Cambridge, MA.

Xie, X., Giammar, D.E., Wang, Z., 2016. MINFIT: a spreadsheet-based tool for parameter estimation in an equilibrium speciation software program. Environ. Sci. Technol. 50, 11112–11120.

Zhu, M.Q., Paul, K.W., Kubicki, J.D., Sparks, D.L., 2009. Quantum chemical study of arsenic (III, V) adsorption on Mn-oxides: Implications for arsenic(III) oxidation. Environ. Sci. Technol. 43, 6655–6661.

Sorption Phenomena on Soils

5.1 Introduction and Terminology

Key points: Sorption is the binding of ions and molecules to soil colloids. It is one of the most important soil chemical reactions in soils because it has a profound effect on the availability and mobility of nutrient elements and contaminants.

Adsorption can be defined as the accumulation of a substance or material at an interface between the solid surface and the bathing solution. Examples of substances or materials include cations, anions, or organic molecules. Adsorption includes the removal of solute molecules (dissolved substances) from the solution and/or solvent and their attachment to solid surfaces (Stumm, 1992). The solvent is the continuous phase of a solution in which the solute is dissolved. Adsorption does not include surface precipitation (which will be discussed later in this chapter) or polymerization (formation of small multinuclear inorganic species such as dimers or trimers) processes. However, adsorption, surface precipitation, and polymerization are all examples of "sorption," which is a general term used when the retention mechanism at a surface is unknown. There are various sorption mechanisms involving both physical and chemical processes that could occur at soil mineral surfaces (Figure 5.1a). These will be discussed in detail later in this chapter and in other chapters.

It would be useful before proceeding any further to define a number of terms pertaining to the retention (adsorption/sorption) of ions and molecules in soils. The adsorbate is the material that accumulates at an interface, the solid surface on which the adsorbate accumulates is referred to as the adsorbent, and the molecule or ion in solution that has the potential of being adsorbed is the adsorptive (Figure 5.1b). If the general term sorption is used, the material that accumulates at the surface, the solid surface, and the molecule or ion in solution that can be sorbed are referred to as sorbate, sorbent, and sorptive, respectively (Stumm, 1992).

Adsorption is one of the most important chemical processes in soils. It determines the quantity of plant nutrients, metals, pesticides, and other organic chemicals retained on soil surfaces and therefore is one of the primary processes that affect the transport of nutrients and contaminants in soils. Adsorption also affects the electrostatic properties, for example, coagulation and settling, of suspended particles and colloids (Stumm, 1992). More recently, the role of mineral–organic interactions involving sorption reactions has been recognized in relation to the preservation of organic matter in the soil.

Both physical and chemical forces are involved in the adsorption of solutes from solution. Physical forces include van der Waals forces (e.g., partitioning) and electrostatic outer-sphere complexes (e.g., ion exchange). Chemical forces resulting from short-range interactions include inner-sphere complexation that involves a ligand

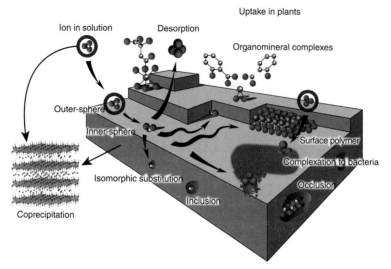

FIGURE 5.1A   Various mechanisms of sorption of an ion at the mineral–water interface. The following processes are illustrated: an aqueous ion in solution, hydrated by water (*pink circle*); adsorption of an ion to a mineral surface (*blue blocks*) via the formation of an outer-sphere complex; loss of hydration water and formation of an inner-sphere complex; lattice diffusion and isomorphic substitution within the mineral lattice; rapid lateral diffusion and either formation of a surface polymer or adsorption on a ledge (which maximizes the number of bonds to the atom). Upon particle growth, surface polymers end up embedded in the lattice structure. The adsorbed ion can diffuse back in solution, either because of dynamic equilibrium or as a product of surface redox reactions. Additionally, the mineral surface itself can dissolve, and those solutes combine with other ions in soil pore water to form new precipitates (coprecipitation). *From Manceau et al. (2002), with permission.*

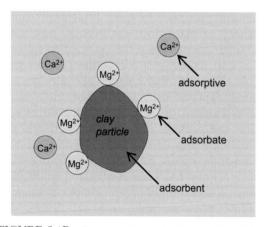

FIGURE 5.1B   Terms pertaining to the retention of ions and molecules in soils. These three terms are not interchangeable and have specific assignments for each part of an adsorption reaction. The adsorptive is in solution and has the potential to be bound to the surface — but it is not bound. The adsorbate is what is bound at the surface, and the adsorbent is the solid itself, which in many cases is a clay mineral.

exchange mechanism, covalent bonding, and hydrogen bonding (Stumm and Morgan, 1981). The physical and chemical forces involved in adsorption are discussed in the sections that follow.

5.2 Surface Functional Groups

Key points: On a mineral surface, the atoms that are in contact with the water are typically oxygen and hydroxyl atoms. Some of these surface functional groups can be protonated or deprotonated depending on the solution conditions such as pH. Surface functional groups are the reactive sites where sorption processes occur.

Surface functional groups in soils play a significant role in adsorption processes. A surface functional group is "a chemically reactive molecular unit bound into the structure of a solid at its periphery such that the reactive components of the unit can be bathed by a fluid" (Sposito, 1989). Surface functional groups can be organic (e.g., carboxyl, carbonyl, phenolic) or inorganic molecular units. The major inorganic surface functional groups in soils are the siloxane surface groups associated with the plane of oxygen atoms bound to the silica tetrahedral layer of a phyllosilicate and hydroxyl groups associated with the edges of clay minerals such as kaolinite, amorphous materials, and metal oxides, oxyhydroxides, and hydroxides. A cross-section of the surface layer of a metal oxide is shown in Figure 5.2. In Figure 5.2a the surface is unhydrated and has metal ions that are Lewis acids and that have a reduced coordination number. The oxide anions are Lewis bases. In Figure 5.2b the surface metal ions coordinate to H_2O molecules forming a Lewis acid site, and then a dissociative chemisorption (chemical bonding to the surface) leads to a hydroxylated surface (Figure 5.2c) with surface OH groups (Stumm, 1987, 1992).

The surface functional groups can be protonated or deprotonated by adsorption of H^+ and OH^-, respectively, as shown below:

$$\text{S-OH} + \text{H}^+ \rightleftharpoons \text{S-OH}_2^+ \qquad (5.1)$$

$$\text{S-OH} \rightleftharpoons \text{S-O}^- + \text{H}^+ \qquad (5.2)$$

Here the Lewis acids are denoted by S and the deprotonated surface hydroxyls are Lewis bases. The water molecule is unstable and can be exchanged for an inorganic or organic anion (Lewis base or ligand) in the solution, which then bonds to the metal cation. This process is called ligand exchange (Stumm, 1987, 1992).

The Lewis acid sites are present not only on metal oxides, such as on the edges of gibbsite or goethite, but also on the edges of clay minerals such as kaolinite. There are also singly coordinated OH groups on the edges of clay minerals. At the edge of the octahedral sheet, OH groups are singly coordinated to Al^{3+}, and at the edge of the tetrahedral sheet, they are singly coordinated to Si^{4+}. The OH groups coordinated to Si^{4+} dissociate only protons; however, the OH coordinated to Al^{3+} dissociate and bind protons. These edge OH groups are called silanol (SiOH) and aluminol (AlOH), respectively (Sposito, 1989; Stumm, 1992).

Spectroscopic analyses of the crystal structures of oxides and phyllosilicates show that different types of hydroxyl groups have different reactivities. Goethite (α-FeOOH) has four types of surface hydroxyls whose reactivities are a function of the coordination environment of the O in the FeOH group (Figure 5.3). The FeOH groups are A-, B-, or C-type sites, depending on whether the O is coordinated with 1, 3, or 2 adjacent Fe(III) ions. The fourth type of site is a Lewis acid-type site, which results from the chemisorption of a water molecule on a bare Fe(III) ion. Sposito (1984) has noted that only A-type sites are basic; that is, they can form a complex with H^+, and A-type and Lewis acid sites can release a proton. The B- and C-type sites

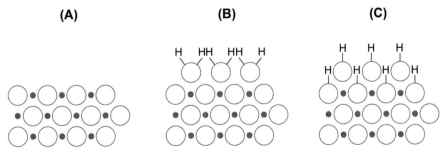

(A)   **(B)**   **(C)**

FIGURE 5.2   Cross-section of the surface layer of a metal oxide. (•) Metal atoms (e.g., Fe, Al, Mn), (O) oxygen atoms. (a) The metal ions in the surface layer have a reduced coordination number and exhibit Lewis acidity. (b) In the presence of water the surface metal ions may coordinate H_2O molecules. (c) Dissociative chemisorption leads to a hydroxylated surface. *From Schindler (1981), with permission.*

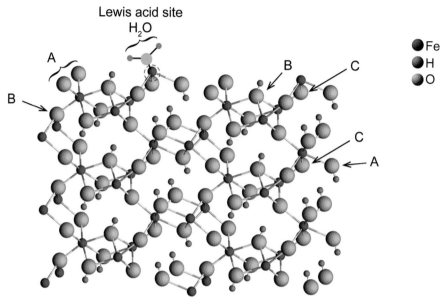

FIGURE 5.3   Types of surface hydroxyl groups on goethite: singly (A-type), triply (B-type), and doubly (C-type) coordinated hydroxyl groups or oxygen atoms, coordinated to Fe(III) atoms; and a Lewis acid site, which includes the Fe(III) atom coordinated to a water molecule. It is important to distinguish that the actual Lewis acid is the Fe(III) atom, while the Lewis base is the oxygen atom in the water molecule, which donates electrons to bond with Fe(III).

are considered unreactive. Thus A-type sites can be either a proton acceptor or a proton donor (i.e., they are amphoteric). The water coordinated with Lewis acid sites may be a proton donor site, that is, an acidic site.

Clay minerals have both aluminol and silanol groups. Kaolinite has three types of surface hydroxyl groups: aluminol, silanol, and Lewis acid sites (Figure 5.4).

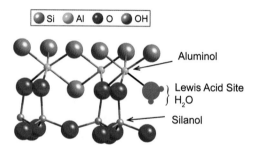

FIGURE 5.4   Surface hydroxyl groups on kaolinite. Besides the OH groups on the basal plane, there are aluminol groups, Lewis acid sites (at which H_2O is adsorbed), and silanol groups, all associated with ruptured bonds along the edges of the kaolinite.

5.3 Surface Complexes

Key points: When the interaction of a surface functional group (e.g., O^-) with an ion or molecule present in the soil solution (e.g., Cu^{2+}) creates a stable bond (e.g., O—Cu), it is called a surface complex. Through this process, the ion is bound to the mineral surface, and the overall reaction is referred to as surface complexation.

There are two types of surface complexes that can form: outer-sphere and inner-sphere. Figure 5.5 shows surface complexes between metal cations and siloxane ditrigonal cavities on 2:1 clay minerals. Such complexes can also occur on the edges of clay minerals. If a water molecule is present between the surface functional group

(A)   **(B)**

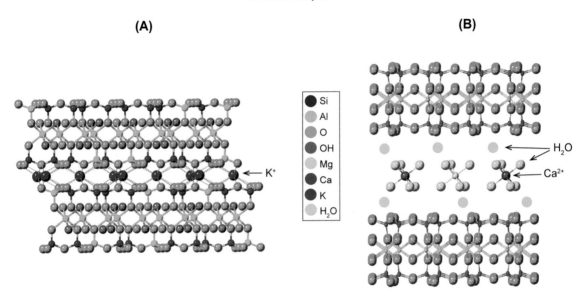

FIGURE 5.5 Examples of (a) inner- and (b) outer-sphere complexes formed between metal cations and siloxane ditrigonal cavities on 2:1 clay minerals. *Redrawn from Sposito (1984), with permission.*

and the bound ion or molecule, the surface complex is termed outer-sphere (Sposito, 1989).

If there is not a water molecule present between the ion or molecule and the surface functional group to which it is bound, this is an inner-sphere complex. Inner-sphere complexes can be monodentate (metal is bonded to only one oxygen) and bidentate (metal is bonded to two oxygens) and mononuclear and binuclear (Figure 5.6).

A polyhedral approach can be used to determine molecular configurations of ions sorbed on mineral surfaces. Using this approach, one can interpret metal–metal distances derived from molecular-scale studies (e.g., XAFS) and octahedral linkages in minerals. Possible configurations (Figure 5.7) include: (i) a single corner (SC) monodentate mononuclear complex in which a given octahedron shares one oxygen with another octahedron; (ii) a double corner (DC) bidentate binuclear complex in which a given octahedron shares two nearest oxygens with two different octahedra; (iii) an edge (E) bidentate mononuclear complex in which an octahedron shares two nearest oxygens with another octahedron; and (iv) a face

(F) tridentate mononuclear complex in which an octahedron shares three nearest neighbors with another octahedron (Charlet and Manceau, 1992). A polyhedral approach can be applied, with molecular-scale data (e.g., EXAFS), to determine the possible molecular configurations of ions sorbed on mineral surfaces. An example of this can be illustrated for Pb(II) sorption on Al oxides (Bargar et al., 1997). An introduction to EXAFS is in Section 5.11.

There are a finite number of ways that Pb(II) can be linked to Al_2O_3 surfaces, with each linkage resulting in a characteristic Pb–Al distance. These configurations are shown in Figure 5.7. Pb(II) ions could adsorb in a monodentate, bidentate, or tridentate fashion. Using the averaged EXAFS-derived Pb–O bond distance of 2.25 Å, the known Al–O bond distances for AlO_6 octahedra (1.85–1.97 Å), and the AlO_6 octahedron edge lengths (i.e., O–O separations of 2.52–2.86 Å), the range of Pb–Al interatomic distances for Pb(II) sorbed to AlO_6 octahedra can be determined. Specifically, for monodentate Pb sorption to corners of AlO_6 octahedra, $R_{Pb–Al}$ ≈ 4.10–4.22 Å. For bridging bidentate sorption of Pb to corners of neighboring AlO_6 octahedra,

(A)   **(B)**

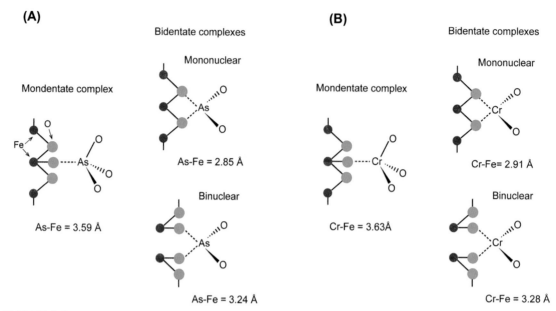

FIGURE 5.6   Schematic illustration of the surface structure of (A) As(V) and (B) Cr(VI) on goethite based on the local coordination environment determined with EXAFS spectroscopy. Red circles indicate the metal(loid) and blue circles indicate oxygen. *Redrawn from Fendorf et al. (1997), with permission.*

$R_{Pb-Al} \approx 3.87-3.99$ Å). And lastly, for bidentate sorption of Pb to edges of AlO_6 octahedra, $R_{Pb-Al} \approx 2.91-3.38$ Å. The less favorable configuration of face-sharing mononuclear tridentate would have a Pb–Al distance of 2.4–3.1 Å. Based on the EXAFS data, the Pb–Al distances for Pb sorbed on the Al oxides were between 3.20 and 3.32 Å, which are consistent with edge-sharing mononuclear bidentate inner-sphere complexation (Figure 5.7).

The type of surface complexes, based on molecular-scale investigations, that occur with metals and metalloids sorbed on an array of mineral surfaces is given in Table 5.1. Environmental factors such as pH, surface loading, ionic strength, type of sorbent, and time all affect the type of sorption complex or product. An example of this is shown for Pb sorption on montmorillonite over an I range of 0.006–0.1 and a pH range of 4.48–6.77 (Table 5.2). Employing XAFS analysis, at pH 4.48 and I = 0.006, outer-sphere complexation on basal

planes in the interlayer regions of the montmorillonite predominated. At pH 6.77 and I = 0.1, inner-sphere complexation on the edge sites of montmorillonite was most prominent, and at pH 6.76, I = 0.006 and pH 6.31, I = 0.1, both inner- and outer-sphere complexation occurred. These data are consistent with other findings that inner-sphere complexation is favored at higher pH and ionic strength (Elzinga and Sparks, 1999). Clearly, there is a continuum of adsorption complexes that can exist in soils.

Outer-sphere complexes involve electrostatic coulombic interactions that are weak compared to inner-sphere complexes in which the binding is covalent or ionic. Outer-sphere complexation is usually a rapid process that is reversible, and adsorption occurs only on surfaces of opposite charge to the adsorbate.

Inner-sphere complexation is usually slower than outer-sphere complexation; it is often not reversible; and it can increase, reduce, neutralize, or reverse the charge on the sorptive regardless

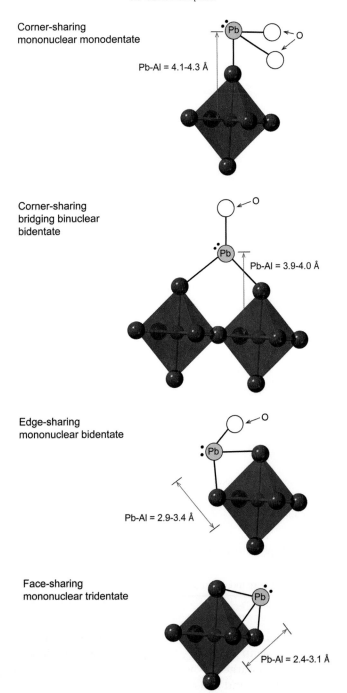

Corner-sharing
mononuclear monodentate

Pb-Al = 4.1-4.3 Å

Corner-sharing
bridging binuclear
bidentate

Pb-Al = 3.9-4.0 Å

Edge-sharing
mononuclear bidentate

Pb-Al = 2.9-3.4 Å

Face-sharing
mononuclear tridentate

Pb-Al = 2.4-3.1 Å

FIGURE 5.7   Characteristic Pb—Al separations for Pb(II) adsorbed to AlO_6 octahedra. In order to be consistent with the EXAFS and XANES data, the Pb(II) ions are depicted as having trigonal pyramidal coordination geometries. *Redrawn from Bargar et al. (1997), with permission.*

TABLE 5.1   Early studies on sorption mechanisms for metals (cations) and oxyanions on soil minerals.

Adsorbate (cations)	Sorbent	Sorption mechanism	Molecular probe	Reference
Cd(II)	Manganite	Inner-sphere	XAFS	Bochatay et al. (2000)
Co(II)	Silica Kaolinite	Cohydroxide precipitates Co multinuclear complexes	XAFS XAFS	O'Day et al. (1996) O'Day et al. (1994a,b)
Cr(III)	Goethite, hydrous ferric oxide	Inner-sphere and Cr-hydroxide	XAFS	Charlet and Manceau (1992)
Cu(II)	Silica	Cu-hydroxide clusters	XAFS, EPR	Xia et al. (1997)
Ni(II)	Pyrophyllite γ-Al_2O_3	Mixed Ni-Al hydroxide (LDH) surface precipitates NiAl hydrotalcite-type coprecipitates	XAFS	Scheidegger et al. (1996) d'Espinose de la Caillerie (1995)
Pb(II)	γ-Al_2O_3	Inner-sphere monodentate mononuclear	XAFS	Chisholm-Brause et al. (1990b)
Sr(II)	Ferrihydrite	Outer-sphere	XAFS	Axe et al. (1998)
Zn(II)	Alumina powders	Inner-sphere bidentate and mixed metal-Al hydroxide precipitates	XAFS	Trainor et al. (2000)
Arsenite (As(III))	Goethite	Inner-sphere bidentate binuclear	ATR-FTIR	Sun and Doner (1996)
Arsenate (As(V))	Goethite	Inner-sphere bidentate binuclear, inner-sphere monodentate	XAFS	Waychunas et al. (1993)
Boron	Amorphous $Al(OH)_3$	Inner-sphere	ATR-FTIR DRIFT-FTIR	Su and Suarez (1995)
Carbonate	Gibbsite	Inner-sphere monodentate	ATR-FTIR	Su and Suarez (1997)
Chromate (Cr(VI))	Goethite	Inner-sphere	XAFS	Fendorf et al. (1997)
Phosphate	Goethite	Inner-sphere bidentate and monodentate	ATR-FTIR	Tejedor-Tejedor and Anderson (1990)
Selenate (Se(VI)	Goethite	Outer-sphere	XAFS	Hayes et al. (1987)
Selenite (Se(IV))	Goethite	Inner-sphere bidentate	XAFS	Hayes et al. (1987)
Sulfate	Hematite	Inner-sphere monodentate	ATR-FTIR	Hug (1997)

of the original charge. Adsorption of ions via inner-sphere complexation can occur on a surface regardless of the surface charge. It is important to remember that outer- and inner-sphere complexations can, and often do, occur simultaneously.

Ionic strength effects on sorption are often used as indirect evidence for whether an outer-sphere or inner-sphere complex forms (Hayes and Leckie, 1986). For example, strontium (Sr(II)) sorption on γ-Al_2O_3 is highly dependent on the ionic strength of the background

TABLE 5.2 Effect of solution ionic strength (I) and pH on the type of Pb(II) adsorption complexes on montmorillonite.

I (M)	pH	Removal from solution (%)	Adsorbed Pb(II) (mmol kg^{-1})	Primary adsorption complex[a]
0.1	6.77	86.7	171	Inner-sphere
0.1	6.31	71.2	140	Mixed
0.006	6.76	99.0	201	Mixed
0.006	6.40	98.5	200	Outer-sphere
0.006	5.83	98.0	199	Outer-sphere
0.006	4.48	96.8	197	Outer-sphere

[a] *Based on results from XAFS data analysis.*
From Strawn and Sparks (1999), with permission.

electrolyte, NaNO$_3$, while Co(II) sorption is unaffected by changes in I (Figure 5.8). The lack of ionic strength effects on Co(II) sorption would suggest the formation of an inner-sphere complex, which is consistent with findings from molecular scale spectroscopic analyses (Hayes and Katz, 1996; Towle et al., 1997). The strong dependence of Sr(II) sorption on ionic strength, suggesting outer-sphere complexation, is also consistent with spectroscopic findings (Katz and Boyle-Wight, 2001).

While the effects of ionic strength can provide important indirect evidence for the formation and strength of an inner-sphere bond, spectroscopic information is needed to determine whether a bond is monodentate, bidentate, or even tridentate. The distinction of denticity is important because it can reveal why a certain ion appears to be more tightly bound to a mineral or solid surface than other ions. For example, the strength of an inner-sphere bond will depend on denticity; when denticity increases, the stability of the complexed ion at the surface also tends to increase. A monodentate mononuclear bond is an ion bound to a single surface functional group, and the interatomic distance between the functional group and the metal atom tends to be the longest in this configuration. This type of bond will also tend to be weaker than if the metal were bound via a bidentate fashion. The same is true for bidentate or tridentate surface complexes, which are attached to a surface (e.g., vacancy site) via two or three bonds, respectively; in general, it will be more difficult to remove (desorb) bidentate and tridentate surface complexes versus monodentate complexes, with all other factors, for example, ionic strength, pH, and the temperature being equal (Brown et al., 1999).

5.4 Adsorption Isotherms

Key points: Adsorption isotherms are carried out in the laboratory to describe the affinity of an ion to a mineral surface over a range of concentrations and/or pH values under a constant temperature. They provide important insight as to how much sorbate a sorbent can retain on its surface and how the solution pH affects the sorption behavior.

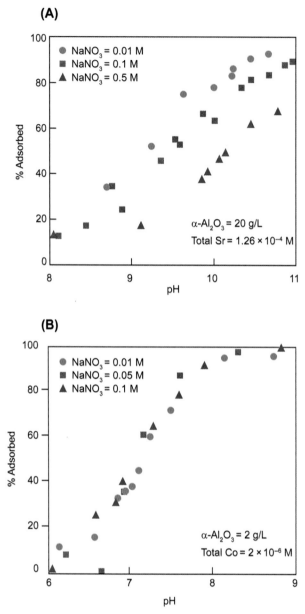

FIGURE 5.8   Effect of increasing ionic strength on pH adsorption edges for (A) a weakly sorbing divalent metal, Sr(II), and (B) a strongly sorbing divalent metal ion, Co(II). *From Katz and Boyle-Wight (2001), with permission.*

One can conduct an adsorption experiment as explained in Box 5.1. The quantity of adsorbate can then be used to determine an adsorption isotherm.

An adsorption isotherm, which describes the relation between the activity or equilibrium concentration of the adsorptive and the quantity of adsorbate on the surface at a constant

BOX 5.1

Conducting an adsorption experiment

Adsorption experiments are carried out by equilibrating (shaking, stirring) an adsorptive solution of a known composition and volume with a known amount of adsorbent at a constant temperature and pressure for a period of time such that an equilibrium (adsorption reaches a steady state or no longer changes after a period of time) is attained. The pH and ionic strength are also controlled in most adsorption experiments.

After equilibrium is reached (it must be realized that true equilibrium is seldom reached, especially with soils), the adsorptive solution is separated from the adsorbent by using centrifugation, settling, or filtering, and then analyzed.

It is very important to equilibrate the adsorbent and adsorptive long enough to ensure that steady state has been reached. However, one should be careful that the equilibration process is not so lengthy that precipitation or dissolution reactions occur (Sposito, 1984). Additionally, the degree of agitation used in the equilibration process should be forceful enough to effect good mixing but not so vigorous that adsorbent modification occurs (Sparks, 1989). The method that

one uses for the adsorption experiment, e.g., batch or flow, is also important. While batch techniques are simpler, one should be aware of their pitfalls, including the possibility of secondary precipitation and alterations in equilibrium states. More details on these techniques are given in Chapter 7.

One can determine the degree of adsorption by using the following mass balance equation:

$$\frac{(C_0 V_0) - (C_f V_f)}{m} = q \qquad \text{(B5.1a)}$$

where q is the amount of adsorption (adsorbate per unit mass of adsorbent) in mol kg^{-1}, C_0 and C_f are the initial and final adsorptive concentrations, respectively, in mol L^{-1}, V_0 and V_f are the initial and final adsorptive volumes, respectively, in liters, and m is the mass of the adsorbent in kilograms. Adsorption could then be described graphically by plotting C_f or C (where C is referred to as the equilibrium or final adsorptive concentration) on the x-axis versus q on the y-axis.

---

temperature, is usually employed to describe adsorption. One of the first solute adsorption isotherms was described by van Bemmelen (1888), and he later described experimental data using an adsorption isotherm.

Adsorption can be described by using four general types of isotherms (S, L, H, and C), which are shown in Figure 5.9. With an S-type isotherm, the slope initially increases with adsorptive concentration but eventually decreases and becomes zero as vacant adsorbent sites are filled. This type of isotherm indicates that at low concentrations, the surface has a low affinity for the adsorptive, which increases

at higher concentrations. The L-shaped (Langmuir) isotherm is characterized by a decreasing slope as concentration increases since vacant adsorption sites decrease as the adsorbent becomes covered. Such adsorption behavior could be explained by the high affinity of the adsorbent for the adsorptive at low concentrations, which then decreases as concentration increases. The H-type (high-affinity) isotherm is indicative of strong adsorbate—adsorptive interactions such as inner-sphere complexes.

The C-type isotherms are indicative of a partitioning mechanism whereby adsorptive ions or molecules are distributed or partitioned between

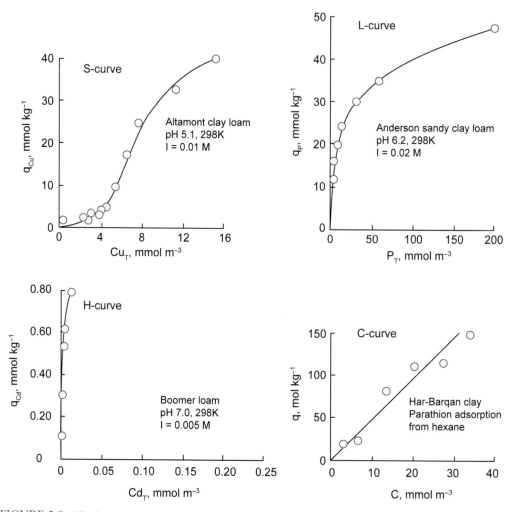

FIGURE 5.9   The four general categories of adsorption isotherms. *Redrawn from Sposito (1984), with permission.*

the interfacial phase (i.e., the solid) and the bulk solution phase without any specific bonding between the adsorbent and adsorbate. The C-type isotherm often describes hydrophobic sorption or the sorption of organic contaminants to SOM. SOM content is generally correlated with the amount of contaminant that a soil can sorb because hydrophobic compounds tend to dissolve into SOM. The phrase "likes-dissolve-likes" can be used to help remember this. Essentially, this means that molecules, such as

nonpolar organic contaminants, will tend to associate with other molecules in soils that also have large amounts of nonpolar regions, such as SOM. The organic molecules are retained in the SOM via weak van der Waals forces. Dissolved organic carbon (DOC) can also impact the mobility of organic contaminants, where DOC can compete with solid-phase OC for the sorption of the contaminant. This can increase the contaminant concentration in soil pore-waters, potentially increasing the mobility of the contaminant. Organic contaminant association with DOC can occur for DOC molecules that are similar chemically and structurally to solid-phase OC, such as having hydrophobic regions where nonpolar contaminants can sorb. Box 5.2 contains further discussion of partition coefficients.

One should realize that adsorption isotherms are purely descriptions of macroscopic data and do not definitively prove a reaction mechanism. Mechanisms must be gleaned from molecular investigations, for example, the use of spectroscopic techniques. Thus the conformity of experimental adsorption data to a particular isotherm does not indicate that this is a unique description of the experimental data and that only adsorption is operational. Thus one cannot differentiate between adsorption and precipitation using an adsorption isotherm, even though this has been done in the soil chemistry literature. For example, some researchers have described data using the Langmuir adsorption isotherm and have suggested that one slope at lower adsorptive concentrations represents adsorption and a second slope observed at higher solution concentrations represents precipitation. This is an incorrect use of an adsorption isotherm since molecular conclusions are being made and, moreover, depending on experimental conditions, precipitation and adsorption can occur simultaneously.

5.5 Equilibrium-Based Adsorption Models

Key point: While sorption isotherms provide important empirical evidence for ion binding on mineral surfaces, utilizing the isotherm data

---

BOX 5.2

Partitioning coefficients

A partitioning mechanism is usually suggested from linear adsorption isotherms (C-type isotherm, Figure 5.9), reversible adsorption/desorption, a small temperature effect on adsorption, and the absence of competition when other adsorptives are added; i.e., adsorption of one of the adsorptives is not affected by the inclusion of a second adsorptive.

A partition coefficient, K_p, can be obtained from the slope of a linear adsorption isotherm using the equation

$$q = K_p C \qquad \text{(B5.2a)}$$

where q was defined earlier, and C is the equilibrium concentration of the adsorptive. The K_p provides a measure of the ratio of the amount of a material adsorbed to the amount in solution.

Partition mechanisms have been invoked for a number of organic compounds, particularly for nonionic organic compounds and some pesticides (Chiou et al., 1977, 1979, 1983). The partition coefficient will generally increase when SOM content increases. If a contaminant has very low solubility in water, then it will generally have a higher K_p value, meaning that it will prefer to sorb to SOM versus remain in solution.

A convenient relationship between K_p and the fraction of organic carbon (f_{oc}) in the soil is the organic carbon–normalized partition coefficient, K_{oc}. This relationship utilizes the partition coefficient to help predict and calculate contaminant partitioning with respect to the amount (%) of OC in a soil, and it can be expressed as

$$K_{OC} = K_P / f_{OC} \qquad \text{(B5.2b)}$$

and applying it to complex soil systems via the models described below allow for the extrapolation of the empirical data to explain and predict the fate and transport of dissolved ions under different soil conditions and in different soil types.

There is an array of equilibrium-based models that have been used to describe adsorption on soil surfaces. These include the widely used Freundlich equation, a purely empirical model, the Langmuir equation, and double-layer models, including the diffuse double-layer (DDL), Stern, and surface complexation models, which are discussed in the following sections.

5.5.1 Freundlich Equation

The Freundlich equation, which was first used to describe gas phase adsorption and solute adsorption, is an empirical adsorption model that has been widely used in environmental soil chemistry. It can be expressed as

$$q = K_d C^{1/n} \qquad (5.3)$$

where q and C were defined earlier, K_d is the distribution coefficient, and n is a correction factor. By plotting the linear form of Equation (5.3), $\log q = 1/n \log C + \log K_d$, the slope is the value of $1/n$, and the intercept is equal to $\log K_d$. If $1/n = 1$, Equation (5.3) becomes equal to Equation (5.2a) (Box 5.2), and K_d is a partition coefficient, K_p. One of the major disadvantages of the Freundlich equation is that it does not predict an adsorption maximum. The single K_d term in the Freundlich equation implies that the energy of adsorption on a homogeneous surface is independent of surface coverage. While researchers have often used the K_d and $1/n$ parameters to make conclusions concerning mechanisms of adsorption and have interpreted multiple slopes from Freundlich isotherms (Figure 5.10) as evidence of different binding sites, such interpretations are speculative. Plots such as those of Figure 5.10 cannot be used for

delineating adsorption mechanisms at soil surfaces.

5.5.2 Langmuir Equation

Another widely used adsorption model is the Langmuir equation. It was developed by Irving Langmuir (1918) to describe the adsorption of gas molecules on a planar surface. It was first applied to soils by Fried and Shapiro (1956) and Olsen and Watanabe (1957) to describe phosphate sorption on soils. Since that time, it has been heavily employed in many fields to describe sorption on colloidal surfaces. As with the Freundlich equation, it best describes sorption at low sorptive concentrations. However, even here, failure occurs. Beginning in the late 1970s, researchers began to question the validity of its original assumptions and consequently its use in describing sorption on heterogeneous surfaces such as soils and even soil components (see references in Harter and Smith, 1981).

To understand why concerns have been raised about the use of the Langmuir equation, it would be instructive to review the original assumptions that Langmuir (1918) made in the development of the equation. They are (Harter and Smith, 1981): (i) Adsorption occurs on planar surfaces that have a fixed number of sites that are identical, and the sites can hold only one molecule. Thus only monolayer coverage is permitted, which represents maximum adsorption. (ii) Adsorption is reversible. (iii) There is no lateral movement of molecules on the surface. (iv) The adsorption energy is the same for all sites and independent of surface coverage (i.e., the surface is homogeneous), and there is no interaction between adsorbate molecules (i.e., the adsorbate behaves ideally).

Most of these assumptions are not valid for the heterogeneous surfaces found in soils. As a result, the Langmuir equation should only be used for purely qualitative and descriptive purposes.

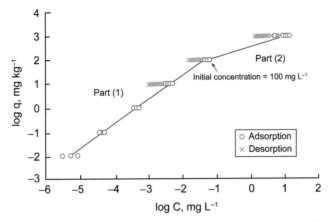

FIGURE 5.10 Use of the Freundlich equation to describe zinc adsorption ($\bigcirc$)/desorption ($\times$) on soils. Part 1 refers to the linear portion of the isotherm (initial Zn concentration <100 mg L^{-1}) while Part 2 refers to the nonlinear portion of the isotherm. *From Elrashidi and O'Connor (1982), with permission.*

The Langmuir adsorption equation can be expressed as

$$q = kCb/(1 + kC) \qquad (5.4)$$

where q and C were defined previously, k is a constant related to the binding strength, and b is the maximum amount of adsorptive that can be adsorbed (monolayer coverage). In some of the literature x/m, the weight of the adsorbate/unit weight of adsorbent, is plotted in lieu of q. Rearranging to a linear form, Equation (5.4) becomes

$$C/q = 1/kb + C/b \qquad (5.5)$$

Plotting C/q versus C, the slope is 1/b and the intercept is 1/kb. An application of the Langmuir equation to adsorption of zinc on a soil is shown in Figure 5.11a. One will note that the data were described well by the Langmuir equation when the plots were resolved into two linear portions.

A number of other investigators have also shown that sorption data applied to the Langmuir equation can be described by multiple, linear portions. Some researchers have ascribed these to sorption on different binding sites. Some investigators have also concluded that if sorption data conform to the Langmuir equation, this indicates an adsorption mechanism, while

deviations would suggest precipitation or some other mechanism. However, it has been clearly shown that the Langmuir equation can equally well describe both adsorption and precipitation (Veith and Sposito, 1977). Thus mechanistic information cannot be derived from a purely macroscopic model like the Langmuir equation. While it is admissible to calculate maximum sorption (b) values for different soils and to compare them in a qualitative sense, the calculation of binding strength (k) values seems questionable. A better approach for calculating these parameters is to determine energies of activation from kinetic studies (see Chapter 7).

Some investigators have also employed a two-site or two-surface Langmuir equation to describe sorption data for an adsorbent with two sites of different affinities. This equation can be expressed as

$$q = \frac{b_1 k_1 C}{1 + k_1 C} + \frac{b_2 k_2 C}{1 + k_2 C} \qquad (5.6)$$

where the subscripts refer to sites 1 and 2, for example, adsorption on high- and low-energy sites. Equation (5.6) has been successfully used to describe sorption on soils of different physicochemical and mineralogical properties. However, the conformity of data to Equation (5.6)

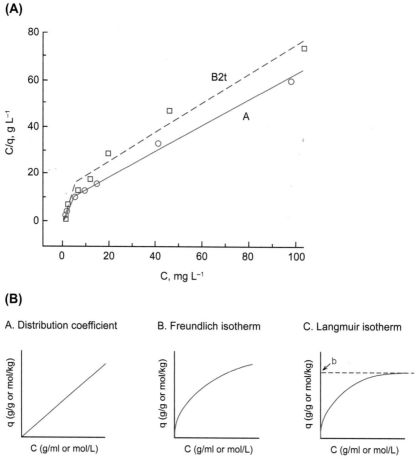

FIGURE 5.11 (a) Zinc adsorption on the A and B2t horizons of a Cecil soil as described by the Langmuir equation. The plots were resolved into two linear portions. *From Shuman (1975), with permission.* (b) Diagrams illustrating the shape and form of different adsorption isotherm equations: (A) distribution coefficient, (B) Freundlich adsorption isotherm, (C) Langmuir adsorption isotherm, where S in g g^{-1} is the concentration of adsorbate on the solid and C in g mL^{-1} is the concentration of the solute (sorptive) in solution. *From Goldberg et al. (2007), with permission.*

does not prove that multiple sites with different binding affinities exist.

Figure 5.11b shows the general shape of linear, Freundlich, and Langmuir adsorption isotherms.

5.5.3 Double-Layer Theory and Models

Hermann von Helmholtz in 1879 was the first to describe the separation of charge on colloids that results in an electrostatic potential difference and the surface charge of colloids in aqueous suspension is balanced by counter ions (Figure 5.12a). Some of the most widely used models for describing sorption behavior are based on the electric double-layer theory developed in the early part of the 20th century. Louis Georges Gouy (1910) and David Leonard Chapman (1913) derived separately an equation describing the ionic distribution in the diffuse layer formed adjacent to a charged surface. The

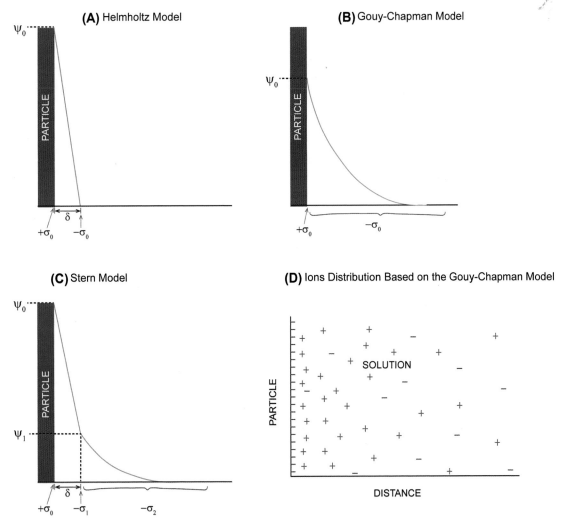

(A) Helmholtz Model

(B) Gouy-Chapman Model

(C) Stern Model

(D) Ions Distribution Based on the Gouy-Chapman Model

FIGURE 5.12   The schematic presentation of colloidal charge and the presence of counter ions balancing the charge according to the (a) von Helmholtz (1879) model, (b, d) diffuse double layer model according to Gouy (1910) and Chapman (1913), and (c) model by Stern (1924). The potentials (ψ) are presented in relation to the distance from the colloid surface. The charge (σ) location is presented on the horizontal axis, which is σ_0 at the surface, σ_1 in the Helmholtz plane, and σ_2 in the diffuse double layer. To retain charge neutrality, $\sigma_0 - \sigma_1 - \sigma_2 = 0$. (a), (b), and (c) are based on the schematic figures published by Stern (1924). Note that in Stern's original representation σ_0 is positive, while in contemporary texts σ_0 is negative. See Section 5.5.4 for additional details on σ_0. Panel (d) is *redrawn from van Olphen (1977), with permission.*

countercharge (charge of opposite sign to the surface charge) can be a diffuse atmosphere of charge or a compact layer of bound charge together with a diffuse atmosphere of charge. The surface charge and the sublayers of compact and diffuse counterions (ions of opposite charge to the surface charge) constitute what is commonly called the diffuse double layer

(Figure 5.12b, d). In 1924 Stern made corrections to the theory accounting for the layer of counterions nearest the surface (Figure 5.12c). When quantitative colloid chemistry came into existence, the "Kruyt" school (Verwey and Overbeek, 1948) routinely employed the Gouy–Chapman and Stern theories to describe the diffuse layer of counterions adjacent to charged particles. Schofield (1947) was among the first persons in soil science to apply the DDL theory to study the thickness of water films on mica surfaces. He used the theory to calculate the negative adsorption of anions (exclusion of anions from the area adjacent to a negatively charged surface) in a bentonite (montmorillonite) suspension.

The historical development of the electrical double-layer theory can be found in several sources (Verwey, 1935; Grahame, 1947; Overbeek, 1952). Excellent discussions of DDL theory and applications to soil colloidal systems can be found in van Olphen (1977), Bolt (1979), and Singh and Uehara (1986).

(a) Gouy–Chapman Model

The Gouy–Chapman model (Gouy, 1910; Chapman, 1913) makes the following assumptions: the distance between the charges on the colloid and the counterions in the liquid exceeds molecular dimensions; the counterions, since they are mobile, do not exist as a dense homoionic layer next to the colloidal surface but as a diffuse cloud, with this cloud containing both ions of the same sign as the surface, or coions, and counterions; the colloid is negatively charged; the ions in solution have no size; that is, they behave as point charges; the solvent adjacent to the charged surface is continuous (same dielectric constant, Sidebar 5.1) and has properties like the bulk solution; the electrical potential is a maximum at the charged surface and drops to zero in the bulk solution (Sidebar 5.2); the change in ion concentration from the charged surface to the bulk solution is nonlinear; and only electrostatic interactions with the surface are assumed (Singh and Uehara, 1986).

Sidebar 5.1 Dielectric constant

The dielectric constant of a solvent is an index of how well the solvent can separate oppositely charged ions. The higher the dielectric constant, the smaller the attraction between ions. It is a dimensionless quantity (Harris, 1987). A dielectric is an insulator.

Sidebar 5.2 Surface charge

Surface charge is the electrical charge (positive or negative) at a mineral's surface; it is a property of the mineral and is impacted by inner- and outer-sphere adsorption. Surface charge causes attraction of counterions. Surface potential is a measure of how much energy (i.e., potential energy, also known as electrical potential energy) is required to bring charged ions to a mineral surface (i.e., to move ions in the solution toward the mineral surface). Surface potential is measured in volts or more commonly millivolts. Surface potential is a function of the distance between the mineral surface and increases closer to the mineral surface.

Figures 5.12b and d shows the Gouy–Chapman model of the DDL, illustrating the charged surface and distribution of cations and anions with distance from the colloidal surface to the bulk solution. Assuming the surface is negatively charged, the counterions are most concentrated near the surface and decrease (exponentially) with distance from the surface until the distribution of coions is equal to that of the counterions (in the bulk solution). The excess positive ions near the surface should equal the negative charge in the fixed layer; that is, an electrically neutral system should exist. Coions are repelled by the negative surface, forcing them to move in the opposite direction so that there is a deficit of anions close to the surface (van Olphen, 1977; Stumm, 1992).

A complete and easy-to-follow derivation of the Gouy–Chapman theory is found in Singh and Uehara (1986) and will not be given here. There are a number of important relationships and parameters that can be derived from the Gouy–Chapman theory to describe the distribution of ions near the charged surface and to predict the stability of the charged particles in soils. These include:

1. The relationship between potential (ψ) and distance (x) from the surface,

$$\tanh[Ze\psi/4kT] = \tanh[Ze\psi_0/4kT]e^{-\kappa x} \quad (5.7)$$

where Z is the valence of the counterion, e is the electronic charge (1.602×10^{-19} C, where C refers to Coulombs), ψ is the electric potential in V, k is Boltzmann's constant (1.38×10^{-23} J K^{-1}), T is the absolute temperature in degrees Kelvin, tanh is the hyperbolic tangent, ψ_0 is the potential at the surface in V, κ is the reciprocal of the double-layer thickness in m^{-1}, and x is the distance from the surface in m.

2. The relationship between the number of ions (n_i) and distance from the charged surface (x):

$$n_i = n_{i_0}\left[\frac{1 - \tanh(-Ze\psi_0/4kT)e^{-\kappa x}}{1 + \tanh(-Ze\psi_0/4kT)e^{-\kappa x}}\right]^2 \quad (5.8)$$

where n_i is the concentration of the ith ions (ions m^{-3}) at a point where the potential is ψ, and n_{i_0} is the concentration of ions (ions m^{-3}) in the bulk solution.

3. The thickness of the double layer is the reciprocal of κ ($1/\kappa$):

$$\kappa = \left(\frac{1000 \text{ dm}^3\text{m}^{-3}e^2N_A\Sigma_iZ_i^2M_i}{\varepsilon kT}\right)^{1/2} \quad (5.9)$$

where N_A is Avogadro's number, Z_i is the valence of ion i, M_i is the molar concentration

of ion i, and ε is the dielectric constant. It should be noted that when SI units are used, $\varepsilon = \varepsilon_r \varepsilon_o$, where $\varepsilon_o = 8.85 \times 10^{-12}$ C^2 J^{-1} m^{-1} and ε_r is the dielectric constant of the medium. For water at 298 K, $\varepsilon_r = 78.54$. Thus in Equation (5.9) $\varepsilon = (78.54)(8.85\times10^{-12}C^2J^{-1}m^{-1})$.

The Gouy–Chapman theory predicts that double-layer thickness ($1/\kappa$) is inversely proportional to the square root of the sum of the product of ion concentration and the square of the valency of the electrolyte in the external solution and directly proportional to the square root of the dielectric constant. This is illustrated in Table 5.3. The actual thickness of the electrical double layer cannot be measured, but it is defined mathematically as the distance of a point from the surface where $d\psi/dx = 0$.

If a surface has an electrical charge, it will generate an electrical potential (ψ) that extends some distance away from the surface. Box 5.3 provides solutions to problems illustrating the relationship between potential and distance from the surface and the effect of concentration and electrolyte valence on double-layer thickness.

The type of colloid (i.e., variable charge or constant charge) affects various double-layer

TABLE 5.3 Approximate thickness of the electric double layer as a function of electrolyte concentration at a constant surface potential.[a]

Concentration of ions of opposite charge to that of the particle (mmol dm^{-3})	Thickness of the double layer (nm)	
	Monovalent ions	Divalent ions
0.01	100	50
1.0	10	5
100	1	0.5

[a] From H. van Olphen, with permission.

BOX 5.3

Electrical double-layer calculations

Problem 1.

Plot the relationship between electrical potential (ψ) and distance from the surface (x) for the following values of x: $x = 0$, 5×10^{-9}, 1×10^{-8}, and 2×10^{-8} m according to the Gouy–Chapman theory. Given: $\psi_0 = 1 \times 10^{-1}$ J C^{-1}, $M_i = 0.001$ mol dm^{-3} NaCl, $e = 1.602 \times 10^{-19}$ C, $\varepsilon = \varepsilon_r \varepsilon_0$, $\varepsilon_r = 78.54$, $\varepsilon_0 = 8.85 \times 10^{-12}$ C^2 J^{-1} m^{-1}, N_A (Avogadro's constant) $= 6.02 \times 10^{23}$ ions mol^{-1}, $k = 1.381 \times 10^{-23}$ J K^{-1}, $R = 8.314$ J K^{-1} mol^{-1}, $T = 298$ K.

First calculate κ, using Equation (5.9):

$$\kappa = \left(\frac{1000 e^2 N_A \Sigma_i Z_i^2 M_i}{\varepsilon kT} \right)^{1/2} \quad \text{(B5.3a)}$$

Substituting values,

$$k = \left(\left(1000 \text{ dm}^3\text{m}^{-3} \right) \left(1.602 \times 10^{-19} \text{ C} \right)^2 \right.$$

$$\left(6.02 \times 10^{23} \text{ ions mol}^{-1} \right) \times$$

$$\left[(1)^2 \left(0.001 \text{ mol dm}^{-3} \right) \right.$$

$$\left. \frac{+(-1)^2 \left(0.001 \text{ mol dm}^{-3} \right) \right]}{(78.54)\left(8.85 \times 10^{-12} \text{ C}^2\text{J}^{-1}\text{m}^{-1} \right)}$$

$$\left. \left(1.38 \times 10^{-23} \text{ JK}^{-1} \right) (298 \text{ K}) \right)^{1/2} \quad \text{(B5.3b)}$$

$$k = \left(1.08 \times 10^{16} \text{m}^{-2} \right)^{1/2} = 1.04 \times 10^8 \text{m}^{-1} \quad \text{(B5.3c)}$$

Therefore $1/\kappa$, or the double-layer thickness, would equal 9.62×10^{-9} m.

To solve for ψ as a function of x, one can use Equation (5.7). For $x = 0$:

$$\tanh\left(\frac{Ze\psi}{4kT} \right) = \tanh\left(\frac{(1)\left(1.602 \times 10^{-19}\text{C}\right)\left(0.1 \text{ J C}^{-1}\right)}{4\left(1.381 \times 10^{-23} \text{ J K}^{-1}\right)(298 \text{ K})} \right)$$

$$\times \left(e^{-\left(1.04 \times 10^8 \text{m}^{-1}\right)(0 \text{ m})} \right) \quad \text{(B5.3d)}$$

$$\tanh\left(\frac{Ze\psi}{4kT} \right) = \tanh\left(\frac{1.60 \times 10^{-20}}{1.64 \times 10^{-10}} \right) e^0 \quad \text{(B5.3e)}$$

$$\tanh\left(\frac{Ze\psi}{4kT} \right) = \tanh\left(9.76 \times 10^{-1} \right)(1) \quad \text{(B5.3f)}$$

$$\tanh\left(\frac{Ze\psi}{4kT} \right) = 0.75 \quad \text{(B5.3g)}$$

The inverse tanh ($\tanh^{-1}$) of 0.75 is 0.97. Therefore

$$\left(\frac{Ze\psi}{4kT} \right) = 0.97 \quad \text{(B5.3h)}$$

Substituting in Equation (5.3h),

$$\frac{(1)\left(1.602 \times 10^{-19} \text{ C}\right)(\psi)}{4\left(1.38 \times 10^{-23} \text{ J K}^{-1}\right)(298 \text{ K})} = 0.97 \quad \text{(B5.3i)}$$

Rearranging, and solving for ψ,

$$\psi = 9.96 \times 10^{-2} \text{ J C}^{-1} \quad \text{(B5.3j)}$$

One can solve for ψ at the other distances, using the approach above.

The ψ values for the other x values are $\psi = 4.58 \times 10^{-2}$ J C^{-1} for $x = 5 \times 10^{-9}$ m, $\psi = 2.72 \times 10^{-2}$ J C^{-1} for $x = 1 \times 10^{-8}$ m, and $\psi = 9.62 \times 10^{-3}$ J C^{-1} for $x = 2 \times 10^{-8}$ m. One can then plot the relationship between ψ and x as shown in Figure B5.3.

BOX 5.3 *(cont'd)*

Electrical double-layer calculations

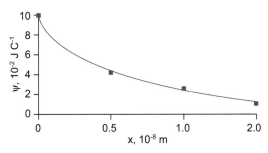

ψ, 10^{-2} J C^{-1}

x, 10^{-8} m

FIGURE B5.3 The relationship between potential and distance from the surface.

Problem 2.

Compare the "thickness" of the double layer $(1/\kappa)$ for 0.001 M (mol dm^{-3}) NaCl, 0.01 M NaCl, and 0.001 M CaCl$_2$.

The $1/\kappa$ for 0.001 mol dm^{-3} NaCl was earlier found to be 9.62×10^{-9} m. For 0.01 M NaCl,

$$k = \left(\left(1000 \text{ dm}^3\text{m}^{-3} \right) \left(1.602 \times 10^{-19}\text{C} \right)^2 \right.$$

$$\left(6.02 \times 10^{23} \text{ ions mol}^{-1} \right)$$

$$\times \left[(1)^2 \left(0.01 \text{ mol dm}^{-3} \right) \right.$$

$$\left. \frac{+(-1)^2 \left(0.01 \text{ mol dm}^{-3} \right) \right]}{(78.54)\left(8.85 \times 10^{-12} \text{ C}^2 \text{ J}^{-1} \text{ m}^{-1} \right)}$$

$$\left(1.38 \times 10^{-23}\text{J K}^{-1} \right) \left(298 \text{ K} \right) \Big)^{1/2} \quad \text{(B5.3k)}$$

$$\kappa = \left(1.08 \times 10^{17} \text{ m}^{-2} \right)^{1/2} = 3.29 \times 10^8 \text{ m}^{-1}$$
$$\text{(B5.3l)}$$

$$\frac{1}{\kappa} = 3.04 \times 10^{-9} \text{ m} \quad \text{(B5.3m)}$$

The $1/\kappa$ for 0.001 M CaCl$_2$ is

$$k = \left(\left(1000 \text{ dm}^3\text{m}^{-3} \right) \left(1.602 \times 10^{-19}\text{C} \right)^2 \right.$$

$$\left(6.02 \times 10^{23} \text{ ions mol}^{-1} \right)$$

$$\times \left[(2)^2 \left(0.001 \text{ mol dm}^{-3} \right) \right.$$

$$\left. \frac{+(-1)^2 (2)\left(0.001 \text{ mol dm}^{-3} \right) \right]}{(78.54)\left(8.85 \times 10^{-12}\text{C}^2\text{J}^{-1}\text{m}^{-1} \right)}$$

$$\left(1.38 \times 10^{-23} \text{ J K}^{-1} \right) \left(298 \text{ K} \right) \Big)^{1/2} \quad \text{(B5.3n)}$$

$$\kappa = \left(3.24 \times 10^{16} \text{ m}^{-2} \right)^{1/2} = 1.80 \times 10^8 \text{ m}^{-1}$$
$$\text{(B5.3o)}$$

parameters including surface charge, surface potential, and double-layer thickness (Figure 5.13). With a variable charge surface (Figure 5.13a), the overall diffuse layer charge is increased at a higher electrolyte concentration (n'). That is, the diffuse charge is concentrated in a region closer

(A)   **(B)**

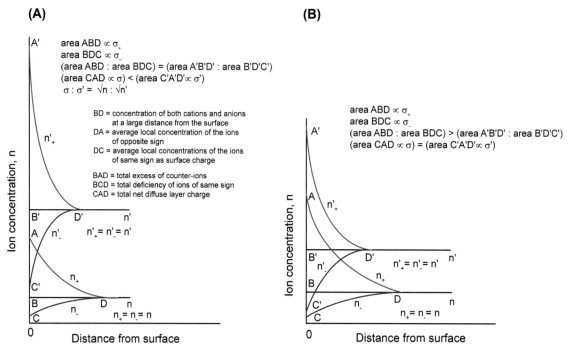

FIGURE 5.13   Charge distribution in the diffuse double layer of a negatively charged particle surface at two electrolyte concentrations, n (lower) and n' (higher). (a) Variable surface charge mineral. (b) Constant surface charge mineral. *From van Olphen (1977), with permission.*

to the surface when electrolyte is added and the total net diffuse charge, $C'A'D'$, which is the new surface charge, is greater than the surface charge at the lower electrolyte concentration, CAD. The surface potential remains the same (Figure 5.13a) but since $1/\kappa$ is less, ψ decays more rapidly with increasing distance from the surface.

In variable charge systems the surface potential is dependent on the activity of PDIs (potential determining ions, e.g., H^+ and OH^-) in the solution phase. The ψ_0 is not affected by the addition of an indifferent electrolyte solution (e.g., NaCl; the electrolyte ions do not react nonelectrostatically with the surface) if the electrolyte solution does not contain PDI and if the activity or concentration of PDI is not affected by the indifferent electrolyte.

In variable charge systems the surface charge (σ_v) is

$$\sigma_v = \left(\frac{2n\varepsilon kT}{\pi}\right)^{1/2} \sinh\left[\left(\frac{Ze}{2kT}\right) \text{constant } \psi_0\right]$$

(5.10)

where sinh is the hyperbolic sin. If the PDIs are H^+ and OH^-, the constant surface potential is related to H^+ by the Nernst equation:

$$\psi_0 = \left(\frac{kT}{e}\right)\ln H^+/H_0^+ \quad \text{or}$$

$$\psi_0 = \left(\frac{kT}{e}\right)2.303(\text{pzc} - \text{pH})$$

(5.11)

where Z is the valence of the PDI, H^+ is the activity of the H^+ ion, H_0^+ is the hydrogen ion activity when ψ_0 is equal to 0, and PZC is the point of zero charge (suspension pH at which a surface has a net charge of 0). Substituting Equation (5.11) for ψ_0 in Equation (5.10) results in

$$\sigma_v = (2n\epsilon kT/\pi)^{1/2} \sinh 1.15Z(pzc - pH) \tag{5.12}$$

Thus σ_v is affected by the valence, dielectric constant, temperature, electrolyte concentration, pH of the bulk solution, and the PZC of the surface.

There are a number of problems with using the Gouy–Chapman double-layer model to describe sorption reactions at soil surfaces. It often fails to describe the distribution of ions adjacent to soil colloidal surfaces. For example, the assumption that the ions behave as point charges results in too high a concentration of counterions calculated to be adjacent to the charged surface. This in turn results in calculated surface potentials being much too high and unrealistic for soil systems. It is also known that the dielectric constant of water next to a charged surface is about 6 versus a value of 80 in the bulk solution. This means there is much more interaction between ions adjacent to the charged surface than assumed present with simple electrostatic attraction of the ions. There is also no provision in the Gouy–Chapman model for surface complexation, and the charged surface in the original model was assumed to be plate shaped and very large. Soil particles have a much more complex shape and vary in size.

(b) Stern Theory

A diagram of the Stern model for describing the electric double layer is shown in Figures 5.14a and b. Stern (1924) modified the Gouy–Chapman model as follows. The first layer of ions (Stern layer) is not immediately at the surface but at a distance away from it. The counterion charge is separated from the surface charge by a layer of thickness δ in which no

charge exists. As a result, the concentration and potential in the diffuse part of the double layer drop to values low enough to warrant the approximation of ions as point charges. Stern further considered the possibility of the specific adsorption of ions and assumed that these ions were located in the Stern layer. The Stern model then assumes that the charge that exists at the surface (σ_0) is balanced by charge in solution, which is distributed between the Stern plane (σ_1) and a diffuse layer (σ_2). The surface charge, σ_0, in C m^{-2}, is

$$\sigma_0 = -(\sigma_1 + \sigma_2) \tag{5.13}$$

where σ_1 is the charge at the Stern plane in C m^{-2} and σ_2 is the diffuse layer charge in C m^{-2}. The charge at the Stern plane is

$$\sigma_1 = N_i Ze/ \{1 + (N_A w/Mn)\exp[-(Ze\psi_\delta + \varphi)/kT]\} \tag{5.14}$$

where N_i is the number of available sites m^{-2} for absorption, M is the molecular weight of the solvent in g mol^{-1}, w is the solvent density in kg m^{-3}, n is the electrolyte concentration in ions m^{-3}, ψ_δ is the Stern potential or electrical potential at the boundary between the Stern layer and the diffuse layer in V, and φ is the specific adsorption potential in the Stern layer (in J), and the other terms have already been defined.

The charge in the diffuse layer is given by the Gouy–Chapman theory, except the reference is now the Stern potential instead of the surface potential,

$$\sigma_2 = (2n\epsilon kT/\pi)^{1/2} \sinh(Ze\psi_\delta/2kT) \tag{5.15}$$

In Equations (5.14) and (5.15) σ_1 and σ_2 are valid for both constant and variable charge systems. Thus the Stern model considers that the distance of the closest approach of a counterion to the charged surface is limited by the size of the ion. With this model, the addition of electrolyte results in not only a compression of the

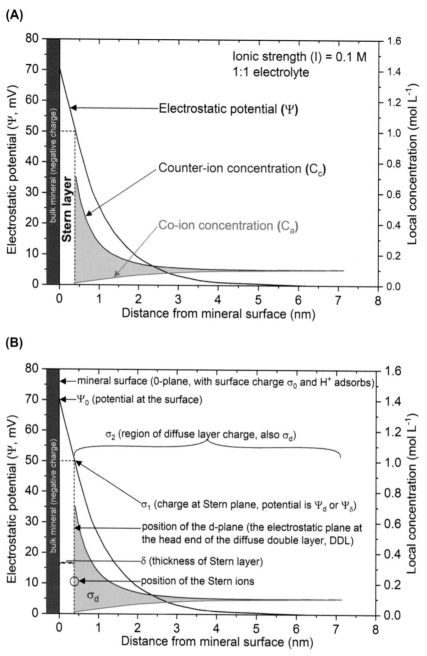

FIGURE 5.14 The calculated charge distribution at the surface of a metal oxide via the basic Stern (BS) model. The d-plane is depicted. Surface charge (σ_0) is neutralized by the charge in the diffuse layer (σ_d). The double layer is composed of the Stern layer (a) and a diffuse double layer. From this model, the concept of a capacitor becomes clear, with a negative charge on the left side of the Stern layer (for a negatively charged particle) and a positive charge on the right side of the Stern layer at the head end of the diffuse double layer. Counterion (C_c) and coion (C_a) concentrations are also depicted. Protons are absorbed at the 0-plane (also called the "naught" plane or surface plane) and sit in the Stern layer. The Stern layer is charge-free. As pointed out by Hiemstra and van Riemsdijk (1996), the surface charge σ_0 is located at a single position (b) on the x-axis (in a "plane" and not in a "layer"). Also, note that even at low values of surface potential, concentrations of 1 M or higher can be found at the head end of the DDL. *Redrawn from Stern (1924); van Olphen (1977); and Hiemstra and van Riemsdijk (1996), with permission.*

diffuse part of the double layer but also in a shift of the counterions from the diffuse layer to the Stern layer and hence in a decrease of the Stern potential, ψ_δ.

Few studies to compare the prediction of surface charge behavior on soil colloids using the Gouy–Chapman and Stern theories have been conducted. In a classic study van Raji and Peech (1972) compared the Gouy–Chapman and Stern models for describing surface charge on variably charged soils from the tropics (Figure 5.15). The relationship between the net negative surface charge as a function of the surface potential for the two models at several electrolyte strengths is shown. The surface potential, ψ_0, was calculated using the Nernst equation, Equation

(5.11). The net surface charge was determined by experimentally measuring point of zero salt effect (PZSE) values. (See the section on points of zero charge in Chapter 2 for the experimental approach for determining PZSE.) The surface charge was calculated using the Gouy–Chapman theory, Equation (5.10), after introducing the value for ψ_0 from the Nernst equation. Surface charge from the Stern model was calculated using Equation (5.14) for σ_1, Equation (5.15) for σ_2, and Equation (5.13). The agreement between the net surface charge, calculated from theory and experimentally measured values, was better when the Stern theory was used than when the Gouy–Chapman theory was employed, except at the 0.001 M NaCl

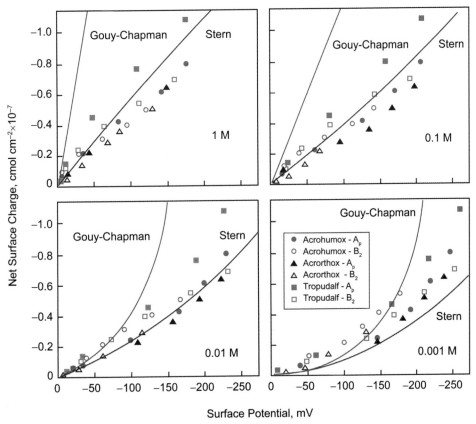

FIGURE 5.15 Comparison of the net negative surface charge of soils as determined by potentiometric titration with that calculated by Gouy–Chapman and Stern theories. The ratio δ/ε', where δ is the thickness of the Stern layer and ε' is the average dielectric constant, used in calculating the net surface charge was 0.015. *From van Raji and Peech (1972), with permission.*

concentration (Figure 5.15). The net surface charge values were lower than those predicted by the Gouy–Chapman theory, although the agreement improved with lower electrolyte concentration and with decreasing surface potential.

(c) Practical Uses of the DDL

There are several ways that surface charge can be manipulated in variably charged field soils. One could increase CEC by lowering the PZC. For example, this could be done by adding an anion that would be adsorbed on the surface and impart more negative charge; Wann and Uehara, (1978) added phosphate to an Oxisol, which resulted in the PZC decreasing with a concomitant increase in net negative charge (Figure 5.16).

The CEC of a soil containing variably charged components increases as pH increases. Liming the soil would increase the pH and the CEC. However, it is difficult to raise the pH of a variable charge soil above 6.5, particularly if the soil has a high buffering capacity along with a high surface area (Uehara and Gillman, 1981; Singh and Uehara, 1986). This is because the OH$^-$ ion, which is produced from the hydrolysis of the CO$_3^{2-}$ ion when CaCO$_3$ is added, can raise the soil pH.

However, in a variably charged system containing hydroxylated surfaces H$^+$ ions, which neutralize the OH$^-$, are released. This results in significant resistance to pH change or buffering.

With a constant surface charge mineral, for example, vermiculite, the total net surface charge, CAD, is not affected by higher electrolyte concentration (n'), but $1/\kappa$ is lower and ψ_0 decreases. In fact, double-layer potential, dielectric constant, temperature, and counterion valence do not influence the sign or magnitude of the surface charge, but any change in these parameters would be offset by a reduction or increase in ψ_0 (Figure 5.13). This is effected by the compression of the double layer, which depends on electrolyte concentration and valence of the counterions.

The Gouy–Chapman model of the electric double layer can be used to predict the effect of electrolyte valence on colloidal stability. The valence of the electrolyte significantly affects the stability (a disperse system) or flocculation status of a suspension. For example, AlCl$_3$ is more effective in flocculation than NaCl. This can be explained by using the Schulze–Hardy Rule: "the coagulative power of a salt is determined by the valency of one of the ions. This prepotent ion is either the negative or the positive

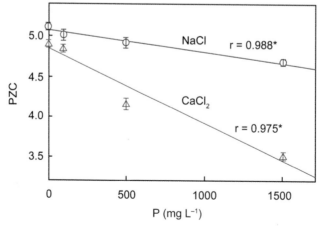

FIGURE 5.16   The effect of applied phosphorus on the point of zero charge (PZC) of an Oxisol soil sample suspended in NaCl and CaCl$_2$ solutions. *From Wann and Uehara (1978), with permission.*
* indicates the correlation coefficient

ion, according to whether the colloidal particles move down or up the potential gradient. The coagulating ion is always of the opposite electrical sign to the particle." In other words, flocculation is mainly determined by the valence of the ion in the electrolyte that is of opposite charge to the surface; that is, with a negatively charged surface, the valence of the cation in the electrolyte is important. Thus the higher the valence, the greater the flocculating power of the electrolyte, and hence the lower the electrolyte concentration needed to cause flocculation.

For example, the following concentration ranges are needed to cause flocculation of suspensions: 25–150 mmol L^{-1} for monovalent ions, 0.5–2.0 mmol L^{-1} for divalent ions, and 0.01–0.1 mmol L^{-1} for trivalent ions. This rule is operational only for indifferent electrolytes that do not undergo any specific (nonelectrostatic such as outer-sphere complexation) adsorption with the surface. Thus the electrolyte cannot contain any PDI or other specifically adsorbed ions; no chemical interactions with the sorptive can occur. The flocculating power of group I cations in the periodic table for negative suspensions decreases slightly in the order $Cs^+ > Rb^+ > NH_4^+ > K^+ > Na^+ > Li^+$. Flocculation can play an important role in the inhibition of contaminants bound to small particles as well as improving soil physical structure.

5.5.4 Surface Complexation Models

Due to the deficiencies inherent in the Gouy–Chapman and Stern models, the retention of ions on soil surfaces has been described using surface complexation models. These models have been widely used to describe an array of chemical reactions including proton dissociation, metal cation and anion adsorption reactions on oxides, clays, and soils, organic ligand adsorption on oxides, and competitive adsorption reactions on oxides. Applications and theoretical aspects of surface complexation models are extensively

reviewed in Goldberg (1992). Surface complexation models (SCMs) often are chemical models based on the molecular descriptions of the electric double layer using equilibrium-derived adsorption data (Goldberg, 1992). SCMs described here include the constant capacitance (CCM), triple-layer (TLM), and modified triple-layer, Stern variable surface charge–variable surface potential (VSC–VSP), generalized two-layer, and the one-pK models. Differences in the surface complexation models lie in the descriptions of the electrical double layer, that is, in the definition and assignment of ions to the planes or layers of adsorption and in differences in the electrostatic equations and the relations between the surface potential and surface charge. These models provide some information on the physical description of the electric double layer, including the capacitance (Sidebar 5.3) and location of adsorbed ions, and they can describe data over a broad range of experimental conditions such as varying pH and I. Two kinds of data can be derived from these models: material balance data, that is, the quantity of a material adsorbed, and information that can be used to describe electrokinetic phenomena (Westall and Hohl, 1980). Common characteristics of surface complexation models are consideration of surface charge balance, electrostatic potential terms, equilibrium constants, capacitances, and surface charge density. A summary of the characteristics of the different surface complexation models is given in Table 5.4.

Sidebar 5.3 Capacitance

Capacitance can be thought of as the ability of a mineral surface to store and hold electrical charge. In a broader sense a capacitor is a device that stores electric energy. This is different from a traditional battery, which stores energy via the chemicals within the battery. The negative charge of the mineral and the positive charges at different planes moving away from the mineral surface are what make this system similar to a parallel plate capacitor.

TABLE 5.4   Characteristics of surface complexation models.[a]

Model	Complexation	Reference state	Relationship between surface charge and surface potential	Surface charge expression
Constant capacitance model	Inner-sphere	Constant ionic medium reference state determines the activity coefficients of the aqueous species in the conditional equilibrium constants.	$\sigma = (CAa/F)\psi_0$ [b] (linear relationship)	$\sigma_p = \sigma_0$
Original triple-layer model	Outer-sphere (metal ions, ligands, cations, and anions); inner-sphere (H^+ and OH^-)	Infinite dilution reference state determines the activity coefficients of the aqueous species in the conditional equilibrium constants.	$\psi_0 - \psi_\beta = \sigma_0/C_1$ $\psi_\beta - \psi_d = (\sigma_0 + \sigma_\beta)/C_2$ $\sigma_{ddl} = -(8RTC\varepsilon_o\varepsilon_r)^{1/2}$ $\sinh(F\psi_d/2RT)$ $\sigma_{dh} = -\sigma_0 - \sigma_\beta - \sigma_{ddl}$	$\sigma_p = \sigma_0 + \sigma_\beta + \sigma_{dh}$
Modified triple-layer model	Inner-sphere in 0-plane; outer-sphere in β-plane	Infinite dilution and zero surface charge.	Same as for the original triple-layer model	$\sigma_p = \sigma_0 + \sigma_\beta + \sigma_{dh}$
Stern variable surface charge–variable surface potential model (four-layer model)	Inner-sphere (H^+, OH^-, and strongly adsorbed oxyanions and metals); outer-sphere (cations and anions)	Defines no surface reactions, no equilibrium constant expressions, and specific surface species.	$\psi_0 - \psi_a = \sigma_0/C_{oa}$ $\psi_a - \psi_\beta = (\sigma_0 + \sigma_a)/C_{a\beta}$ $\psi_\beta - \psi_d = -\sigma_d/C_{\beta d}$	$\sigma_p = \sigma_0 + \sigma_a + \sigma_\beta + \sigma_d$
Generalized two-layer model	Inner-sphere	Infinite dilution for the solution and a reference state of zero charge and potential for the surface.	$\sigma_0 = (CAa/F)\psi_0$ [b] $\sigma_{ddl} = -(8RTC\varepsilon_o\varepsilon_r)^{1/2}$ $\sinh(F\psi_d/2RT)$	$\sigma_p = \sigma_0$
One-pK Stern	Inner-sphere (H^+ and OH^-, strongly adsorbed oxyanions and metals); outer-sphere (metal ions, ligands, cations and anions)	Infinite dilution for the solution, net zero charge and potential, pH = PZC where (e.g., for ferrihydrite and goethite): $[SOH_2^{1/2+}]$ + $[S_3OH^{1/2+}]$ = $[SOH^{1/2-}]$ + $[S_3OH^{1/2-}]$.	$\sigma_0 = C(\psi_0 - \psi_d)$ $\sigma_{ddl} = -(8RTC\varepsilon_o\varepsilon_r)^{1/2}$ $\sinh(F\psi_d/2RT)$ $\sigma_{dh} = -\sigma_0 - \sigma_{ddl}$	$\sigma_p = \sigma_0 + \sigma_{dh}$
CD-MUSIC model (based on the three-plane model)	Inner-sphere (H^+ and OH^-, strongly adsorbed oxyanions and metals); outer-sphere (metal ions, ligands, cations and anions)	Infinite dilution for the solution, net zero charge and potential, pH = PZC where (e.g., for ferrihydrite and goethite): $[SOH_2^{1/2+}]$ + $[S_3OH^{1/2+}]$ = $[SOH^{1/2-}]$ + $[S_3OH^{1/2-}]$.	Same as for the original triple-layer model	$\sigma_p = \sigma_0 + \sigma_\beta + \sigma_{dh}$

[a] Information to prepare the table adapted from Goldberg (1992), Goldberg (2007), and Goldberg (2013).
[b] C refers to capacitance density.

The general balance of surface charge can be written as:

$$\sigma_0 + \sigma_H + \sigma_{is} + \sigma_{os} + \sigma_d = 0 \qquad (5.16)$$

where σ_0 has often been defined as the constant charge or permanent charge in minerals due to ionic or isomorphic substitution; however, in this text σ_0 is considered to be charge at the surface 0-plane. In this case, as has been adopted here, permanent charge in the mineral structure is defined as σ_{sub} (Equation 5.17b). Additionally, σ_H is the net proton charge, which is equal to $\Gamma_H - \Gamma_{OH}$, where Γ is the surface excess concentration (σ_H is equivalent to the dissociation of H^+ in the diffuse layer), σ_{is} is the inner-sphere complex charge as a result of inner-sphere complex formation, σ_{os} is the outer-sphere complex charge as a result of outer-sphere complexes, and σ_d is the charge in the bulk solution that balances the surface charge (these ions do not form any complex with the surface). The σ_0 is negative while σ_H, σ_{is}, and σ_{os} can be positive, negative, or neutral. Although conceptually correct, the terms in Equation (5.16) are not always easy to apply directly in surface complexation models, as, for example, the charge of inner-sphere complexes can be placed in different electrostatic planes depending on the model, or spread out over two or more planes, as is the case in the charge distribution–multisite complexation (CD-MUSIC) model. Another way to describe the net total particle charge (σ_p) on a colloid is to define the surface charge balance from the exact position of the charge along the potential gradient seen in Figure 5.14. In this case the net total particle charge is

$$\sigma_p = \sigma_0 + \sigma_\beta + \sigma_{dh} \qquad (5.17a)$$

In this equation the term σ_0 should be understood as the charge at the surface 0-plane (Figure 5.14). The charge in this plane may have contributions from permanent charge

(σ_{sub}) and proton charge (σ_H), as well as from inner-sphere and even outer-sphere complexes (σ_{is} and σ_{os}). Though, outer-sphere complexes are not included in Equation (5.17b). Further, σ_{dh} is the charge in the head end of the diffuse layer. Charge in the plane separating the inner Stern from the outer Stern layer is σ_β. Thus charge in the 0-plane can be expressed to include

$$\sigma_0 = \sigma_{sub} + \sigma_H + \sigma_{is} \qquad (5.17b)$$

There are differences in how these terms are designated in the literature, which can be confusing for the reader, and therefore we comment on these below:

1. In much of the previous literature the d-plane charge σ_d (Figure 5.14) is treated as the diffuse layer charge. However, as was pointed out by Hiemstra and van Riemsdijk (1996), in Stern layer models such as TLM any charge accumulated in the d-plane itself, that is, in the head end of the diffuse layer, would interact more strongly with the surface than ions further out in the diffuse layer. Therefore in the current literature, including in this chapter, σ_d is often divided into two contributions according to

$$\sigma_d = \sigma_{dh} + \sigma_{ddl} \qquad (5.17c)$$

where dh is diffuse head and ddl is diffuse double layer.

2. In many papers (e.g., Hiemstra and van Riemsdijk, 1996; Venema et al., 1996) the terms σ_β and σ_{dh} are referred to as σ_1 and σ_2, respectively.

In the more advanced interface models such as TLM and TPM (see next section) the Stern layer is split into two, where σ_β is the surface charge in the plane separating the inner Stern from the outer Stern layer. The total particle charge σ_p can be positive or negative but must be balanced by opposite charge in the diffuse layer σ_{ddl}, that is, $\sigma_p = -\sigma_{ddl}$. As pointed out earlier, all surface complexation models contain a balance of surface charge and general surface complexation reactions (Hohl et al., 1980; Goldberg, 1992):

$$SOH + H^+ \equiv SOH_2^+ \tag{5.18}$$

$$SOH \equiv SO^- + H^+ \tag{5.19}$$

$$SOH + M^{n+} \equiv SOM^{(n-1)} + H^+ \tag{5.20}$$

$$2SOH + M^{n+} \equiv (SO)_2 M^{(n-2)} + 2H^+ \tag{5.21}$$

$$SOH + L^{l-} \equiv SL^{(l-1)-} + OH^- \tag{5.22}$$

$$2SOH + L^{l-} \equiv S_2L^{(l-2)-} + 2OH^- \tag{5.23}$$

where SOH is the surface functional group and S represents the metal bound to the surface functional group, for example, OH of an oxide surface or of the aluminol or silanol group of a clay mineral, M is the metal ion, n^+ is the charge on the metal ion, L is a ligand, and l^- is the ligand charge.

The intrinsic equilibrium constants (see Box 5.4 for a discussion of intrinsic and conditional equilibrium constants) for reactions in Equations (5.18)–(5.23) are (Hohl et al., 1980; Goldberg, 1992)

$$K_+^{int} = \frac{(SOH_2^+)}{[SOH](H^+)} \exp(F\psi_i/RT) \tag{5.24}$$

$$K_-^{int} = \frac{[SO^-](H^+)}{[SOH]} \exp(-F\psi_i/RT) \tag{5.25}$$

$$K_{M1}^{int} = \frac{[SOM^{(n-1)}](H^+)}{[SOH](M^{n+})} \exp((n-1)F\psi_i/RT) \tag{5.26}$$

$$K_{M2}^{int} = \frac{[(SO)_2M^{(n-2)}](H^+)^2}{[SOH]^2(M^{n+})} \exp((n-2)F\psi_i/RT) \tag{5.27}$$

$$K_{L1}^{int} = \frac{[SL^{(l-1)-}](OH^-)}{[SOH](L^{l-})} \exp(-(l-1)F\psi_i/RT) \tag{5.28}$$

$$K_{L2}^{int} = \frac{[S_2L^{(l-2)-}](OH^-)^2}{[SOH]^2(L^{l-})} \exp(-(l-2)F\psi_i/RT) \tag{5.29}$$

BOX 5.4

Calculation of conditional and intrinsic equilibrium constants

Consider the following ionization reactions illustrating the variable charge behavior of a surface (Dzombak and Morel, 1990):

$$SOH_2^+ \rightleftharpoons SOH + H^+ \tag{B5.4a}$$

$$SOH \rightleftharpoons SO^- + H^+ \tag{B5.4b}$$

where SOH_2^+, SOH, and SO^- are positively charged, neutral, and negatively charged surface hydroxyl groups, respectively. One can write conditional acidity constants, K_{a1}^{cond} for Equation (5.4a) and K_{a2}^{cond} for Equation (5.4b), as

$$K_{a1}^{cond} = \frac{[SOH](H^+)}{[SOH_2^+]} \tag{B5.4c}$$

$$K_{a2}^{cond} = \frac{[SO^-](H^+)}{[SOH]} \tag{B5.4d}$$

The activity coefficients of the surface species are assumed to be equal. The K_{a1}^{cond} and K_{a2}^{cond} are conditional equilibrium constants because they include effects of surface charge and are thus dependent on the degree of surface ionization,

---

BOX 5.4 *(cont'd)*

Calculation of conditional and intrinsic equilibrium constants

which is dependent on pH. Conditional equilibrium constants thus hold under given experimental conditions, for example, a given pH.

The total free energy of sorption (ΔG^{o}_{tot}) can be separated into component parts:

$$\Delta G^{o}_{tot} = \Delta G^{o}_{int} + \Delta G^{o}_{coul} \qquad (B5.4e)$$

where ΔG^{o}_{int} is the chemical or "intrinsic" free-energy term and ΔG^{o}_{coul} is the variable electrostatic or "coulombic" term. An expression for the ΔG^{o}_{coul}, determined by taking into account the electrostatic work required to move ions through an interfacial potential gradient (Morel and Hering, 1993), is substituted for ΔG^{o}_{coul} such that

$$\Delta G^{o}_{tot} = \Delta G^{o}_{int} + \Delta ZF\psi_0 \qquad (B5.4f)$$

where ΔZ is the change in the charge of the species resulting from sorption. Since $\Delta G^{o}_{r} = -RT \ln K^{o}$, Equation (5.4f) can be rewritten as:

$$K^{int} = K^{cond} \exp(\Delta ZF\psi_0/RT) \qquad (B5.4g)$$

where K^{int} is referred to as an intrinsic equilibrium constant. The exponential term in Equation (5.4g) is referred to as an electrostatic or coulombic correction factor. This term allows one to consider the effects of surface charge variations on surface complexation reactions.

If the specific and nonspecific interactions for the surface acidity reactions (Equations [5.4a] and [5.4b]) are separated, one obtains

$$K^{int}_{a1} = \frac{[SOH](H^{+}_{s})}{[SOH_2]} = K^{cond}_{a1} \exp(-F\psi_0/RT) \qquad (B5.4h)$$

$$K^{int}_{a2} = \frac{[SOH^{-}](H^{+}_{s})}{[SOH]} = K^{cond}_{a2} \exp(-F\psi_0/RT) \qquad (B5.4i)$$

where H^{+}_{s} is a proton released at the surface but not moved to the bulk solution. This differentiation is made since K^{int} refers only to specific chemical interactions at the surface.

---

where brackets indicate concentrations in mol L^{-1}, parentheses indicate activity, ψ_i is the surface potential in V in the ith surface plane, F is the Faraday constant in C mol^{-1}, R is the gas constant in J $mol^{-1} K^{-1}$, and T is the absolute temperature in degrees Kelvin. The log of the intrinsic equilibrium constants can be obtained by plotting the log of the conditional equilibrium constants versus surface charge (σ) and extrapolating to zero surface charge (Stumm et al., 1980). The term $e_i^{-F\psi_i/RT}$ considers surface charge effects on surface complexation. Surface complexation models also contain several adjustable parameters, including K_i, equilibrium constants; C_i,

capacitance density for the ith surface plane; and $[SOH]_T$, the total number of reactive surface hydroxyl groups. Details on the determination of these parameters can be found in Goldberg (1992).

(a) Constant Capacitance Model (CCM)

The CCM model was formulated by Schindler, Stumm, and coworkers (Schindler and Gamsjager, 1972; Hohl and Stumm, 1976). It assumes that (Goldberg, 1992): (i) all surface complexes are inner-sphere and anions are adsorbed via a ligand exchange mechanism, (ii) a constant ionic strength reference state determines the activity coefficient of the aqueous species in the conditional equilibrium constant, and (iii) there is a linear relationship between surface charge and surface potential (ψ_0), which is expressed as

$$\sigma_p = (CAa/F)\psi_0 \qquad (5.30)$$

where C is capacitance density in F m^{-2}, A is specific surface area in m^2 g^{-1}, a is suspension density in g L^{-1}, and σ is expressed in units

of mol L^{-1}. The balance of surface charge equation is

$$\sigma_p = \sigma_0 \qquad (5.31)$$

An illustration of the CCM is shown in Figure 5.17, and the application of metal adsorption data to the CCM is shown in Figure 5.18.

(b) Triple-Layer Model (TLM)

The TLM was developed by Davis and Leckie (1978, 1980). It consists of two capacitance layers and a diffuse layer (Figure 5.19). The assumption of the TLM is that all metals and ligands are retained as outer-sphere complexes. Only H$^+$ and OH$^-$ are adsorbed as inner-sphere complexes. The balance of the surface charge equation for the TLM is

$$\sigma_p = \sigma_0 + \sigma_\beta + \sigma_{dh} \qquad (5.32)$$

where σ_β is the surface charge in the plane separating the inner Stern from the outer Stern layer, and σ_β and σ_{dh} are referred to as σ_1 and σ_2, respectively (Figure 5.19). The charge in the head end of the diffuse layer is σ_{dh}.

As illustrated in Figure 5.19a, the PDIs at the surface, H$^+$ and OH$^-$, are at the 0-plane next to the surface, and all other metals, ligands, major cations (C$^+$), and anions (A$^-$) are at the β-plane. The diffuse layer begins at the d-plane and

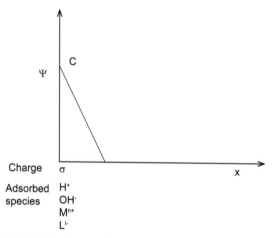

FIGURE 5.17 Placement of ions, potential, charge, and capacitance for the constant capacitance model. C refers to capacitance density. *From Westall (1986) with permission.*

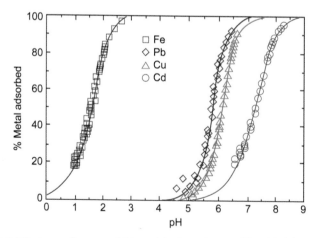

FIGURE 5.18 Fit of the constant capacitance model to metal adsorption on silica. Model results are represented by solid lines. *From Schindler et al. (1976), with permission.*

(A)

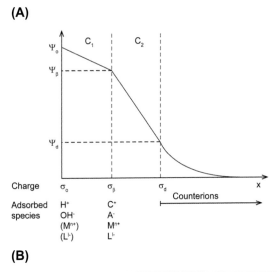

(B)

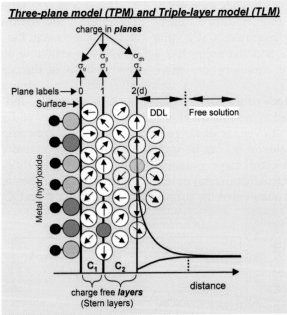

FIGURE 5.19 (a) Placement of ions, potentials, charges, and capacitances for the triple-layer model. Parentheses represent ion placement allowed only in the modified triple-layer model. *From Westall (1980), with permission.* (b) A description of the locations of charges in planes, the charge-free layers, and the diffuse layer in the three-plane model (TPM) and triple-layer model (TLM) *(adapted from Hiemstra and van Riemsdijk (2006), with permission).* A description of the structure of the electric double layer, including the planes and layers, on a metal (hydr)oxide surface is highlighted. Water molecules at the surface orientate on the charges and are also affected by thermal motion (Hiemstra and van Riemsdijk, 2006). The *arrows* indicate the orientation of the water dipoles; for example, the protons would be further from the *blue cations* while the oxygens would be closer to the cations. Hydrated cations sit at the 1-plane, and their charge is what causes the charge in the 1-plane. Capacitances (C_1 and C_2) have been determined experimentally to be similar; data suggest that the head end of the diffuse double layer (DDL) begins approximately 0.9 nm from the metal (hydr)oxide surface, see Hiemstra and van Riemsdijk (2006) and Venema et al. (1996) for additional details.

extends into the bulk solution. The TLM includes three equations, representing the three layers, that relate potential to charge (Davis et al., 1978):

$$\psi_0 - \psi_\beta = \sigma_0/C_1 \tag{5.33}$$

$$\psi_\beta - \psi_d = (\sigma_0 + \sigma_\beta)/C_2 \tag{5.34}$$

$$\text{where } \sigma_\beta = \sigma_p - \sigma_0 - \sigma_{dh} \tag{5.34b}$$

$$\sigma_{ddl} = -(8RTC\varepsilon_o\varepsilon_r)^{1/2}\sinh(F\psi_d/2RT) \tag{5.35}$$

where σ has the units $C\,m^{-2}$; C_1 and C_2 are the capacitance densities in the 0- and β-planes, respectively; ε_o is the permittivity of vacuum; ε_r is the dielectric constant of water; and C is the concentration of a 1:1 background electrolyte (e.g., $NaClO_4$; both the cation and anion have a charge of 1). For 1:2 background electrolytes, for example, Na_2SO_4 (the cation has a charge of 1 and the anion a charge of 2), and for 2:1 background electrolytes, for example, $CaCl_2$ (the cation has a charge of 2 and the anion has a charge of 1), the above equations are more complicated.

The TLM has two characteristics not included in other surface complexation models: the chemical constants are applicable over a wide range of ionic strength and the value of ψ_d can be used as an estimate of the electrokinetic potential (Goldberg, 1992). The electrokinetic (zeta) potential refers to the electrical potential at the surface

of shear between immobile liquid attached to a charged particle and mobile liquid further removed from the particle (SSSA, 2008).

The modified TLM was developed by Hayes and Leckie (1986, 1987) and allows metals and ligands to be adsorbed as inner-sphere complexes. Under these conditions, the surface balance equation is

$$\sigma_p = \sigma_0 + \sigma_\beta + \sigma_{dh} \tag{5.36}$$

An application of the modified TLM to selenate adsorption on goethite is shown in Figure 5.20.

(c) Stern Variable Surface Charge–Variable Surface Potential (VSC–VSP) Model

This model was formulated by Bowden et al. (1977, 1980) and Barrow et al. (1980, 1981). It is a hybrid between the CCM and TLM and is commonly known as the four-layer model. The assumptions of the VSC–VSP model are that H^+, OH^-, and strongly adsorbed oxyanions such as arsenate and metals develop inner-sphere complexes and major cations and anions form outer-sphere complexes. The balance of the surface charge equation for this model is the same as for the modified TLM (Equation [5.36]).

Figure 5.21 illustrates the VSC–VSP model. Hydrogen ions and OH^- are in the 0-plane,

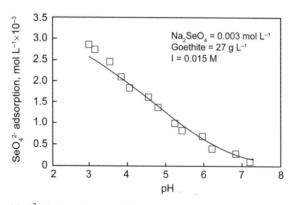

FIGURE 5.20 Adsorption of SeO_4^{2-} (selenate) on goethite *versus* pH. The symbols represent the experimental data and the solid line represents TLM conformity, assuming nonspecific adsorption with constant C_1 and $C_2 = 0.13$. *From Zhang and Sparks (1990b), with permission.*

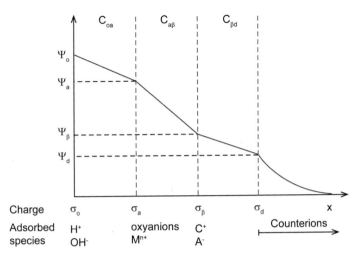

FIGURE 5.21 Placement of ions, potentials, charges, and capacitances for the Stern VSC−VSP model. The term σ_a is the charge in the a-plane, which represents strongly adsorbed ions a short distance away from the 0-plane. *From Bowden et al. (1980), with permission.*

strongly adsorbed oxyanions and metals are in the a-plane, major cations and anions are present as outer-sphere complexes and are in the β-plane, and the d-plane is the start of the diffuse double layer. The surface functional group is OH−S−OH$_2$, or one protonation or dissociation can occur for every two surface OH groups (Barrow et al., 1980; Goldberg, 1992). The VSC−VSP model defines no surface reactions, equilibrium constant expressions, and specific surface species. Consequently, one could consider it not to be a chemical model. There are four charge potential equations and four charge balance equations that are given in Goldberg (1992).

(d) Generalized Two-Layer Model

This model was promulgated by Dzombak and Morel (1990) and is based on the DDL description of Stumm and coworkers (Huang and Stumm, 1973). It assumes that all surface complexes are inner-sphere. The description of surface charge is identical to that employed in the CCM (Equation (5.30)). Similar to the TLM and the VSC−VSP models, the generalized two-layer model includes a diffuse layer and derives

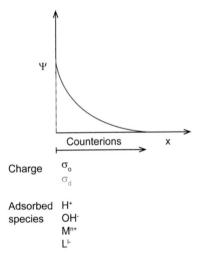

FIGURE 5.22 Placement of ions, potential, and charges for the generalized two-layer model. This model does not define a Stern layer. *From D.A. Dzombak and F.M.M. Morel (1990), with permission.*

its name from two surface planes of charge (Figure 5.22). Moreover, all surface complexes are found in the surface plane and no Stern layer is used. The diffuse layer starts at the d-plane and goes to the bulk solution. Metal ions are adsorbed on two types of sites, a small set of high-affinity sites and a large set of low-affinity sites.

(e) One-pK Stern Model

This model was proposed by Bolt and van Riemsdijk, (1982) and formulated by van Riemsdijk and co-workers (1986, 1987; Hiemstra et al., 1987; van Riemsdijk and Zee, 1991). The surface functional group is a singly coordinated oxygen atom that carries one or two H^+, resulting in the surface sites SOH and $(SOH)_2$, respectively. It is similar to the Stern model and assumes that H^+ and OH^- form inner-sphere complexes and are in the 0-plane and major cations and anions, metals, and ligands form outer-sphere complexes at the Stern plane, or d-plane. The Stern plane is at the start of the DDL (Figure 5.23). The surface charge balance equation for the one-pK model is (Adapted from Goldberg, 1992), where the net charge is equal to the charge at the surface 0-plane, σ_0, and the diffuse head plane, σ_{dh}. The 0-plane charge is impacted by

permanent charge in the mineral as well as H^+ and OH^- inner-sphere complexes.

$$\sigma_p = \sigma_0 + \sigma_{dh} \tag{5.37}$$

(f) CD-MUSIC Model

The CD-MUSIC model was introduced by Hiemstra and van Riemsdijk (1996) for ion adsorption to goethite and has since then been applied for many other sorbents such as ferrihydrite (Tiberg et al., 2013; Hiemstra and Zhao, 2016), calcite (Wolthers et al., 2008), and rutile (Ridley et al., 2009), as well as Al oxides (Hiemstra et al., 1999). Unlike the other surface complexation models discussed above, the CD-MUSIC model is not based on a certain fixed description of the interface using an assumed set of charge–potential relationships. Instead, the starting point of the model is a structural analysis of the solid–solution interface, the identification of reactive surface sites (based on the number of coordinating metal ions), and their

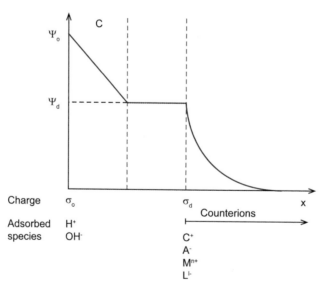

FIGURE 5.23 Placement of ions, potentials, charges, and capacitances for the one-pK Stern model. *From Westall (1986). with permission.*

relative contribution to the surface acid–base characteristics. Essential in the approach is the relation between electrostatics and structure.

Ligand exchange in the CD-MUSIC model has been rationalized with the introduction of the bond valence concept, and the charge of the adsorbed ions is distributed over the ligands, which in turn are distributed over the interfacial locations (i.e., ions are positioned across different electrostatic planes), as depicted in Figure 5.24.

Most commonly, the valence of an adsorbed ion is spread out over the 0- and β-planes, which results in a change in surface charge dependent on the exact coordination. To illustrate how this works, Figure 5.24 shows the charge distribution of four different oxyanion surface complexes for the special case in which the central atom distributes its valence symmetrically over its surrounding oxygen ligands (Rietra et al., 1999). For example, for a bidentate AsO_4 surface complex,

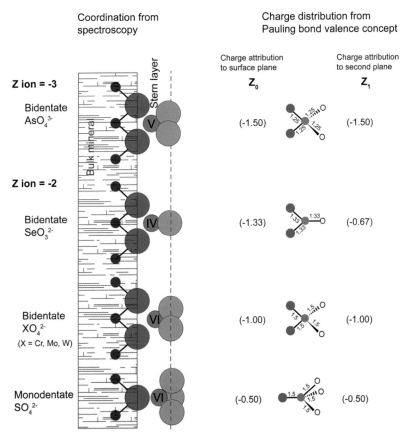

FIGURE 5.24 Examples of how the surface coordination of oxyanions may affect the surface charge when the charge of the central atom is distributed symmetrically over the surrounding oxygen ligands. The allocation of charge of the ion (Z ion) = Z_0 + Z_i over the two electrostatic planes. Charge in the planes can be calculated with the Pauling bond valence concept, where Z_i = $n_i(v - 2)$, where n_i is the number of ligands and v is the Pauling bond valence (i.e., the valence of the central ion divided by the coordination number). *From Rietra et al. (1999), with permission.*

BOX 5.5

Example of CD-MUSIC model

Investigate the use of Hiemstra and Zhao's CD-MUSIC model for ferrihydrite (Hiemstra and Zhao, 2016) for calculating the effect of carbonate on PO_4 adsorption to ferrihydrite in the pH range from 7 to 11. The example is from Mendez and Hiemstra (2019). The conditions are the following:

Ionic strength = 0.5 M $NaNO_3$

Concentration Fe (as ferrihydrite) = 4.88 mM

Total added PO_4: 0.251 mM

Specific surface area of ferrihydrite: 765 $m^2\,g^{-1}$

Concentration of CO_3 in system with carbonate: 30 mM

Temperature = 25°C

Stern layer capacitances: $C_1 = 1.19$ F m^{-2}, $C_2 = 0.92$ F m^{-2}.

To start with, let us review the surface reactions that can occur in this system. The CD-MUSIC model contains two types of sites, singly coordinated sites having a site density of 5.8 sites nm^{-2} (designated as $\equiv FeOH^{\frac{1}{2}-}$) and triply coordinated sites with 1.4 sites nm^{-2} ($\equiv Fe_3O^{\frac{1}{2}}$). Further, the $\equiv FeOH^{\frac{1}{2}-}$ sites are divided into two groups called $\equiv FecOH^{\frac{1}{2}-}$ and $\equiv FeeOH^{\frac{1}{2}-}$, which have 3.0 and 2.8 sites nm^{-2}, respectively. The $\equiv FecOH^{\frac{1}{2}-}$ groups can form corner-sharing bidentate complexes, while the $\equiv FeeOH^{\frac{1}{2}-}$ groups form edge-sharing bidentate complexes. Both groups are able to form monodentate complexes. A monodentate complex with an oxyanion can be formulated in a general way as follows:

$$\equiv FeOH^{\frac{1}{2}-} + nH^+$$
$$+ AO_y^{x-} \leftrightarrow \equiv FeOAO_{y-1}H_{(n+1)}^{-x-\frac{1}{2}+n} + H_2O$$
$$(B5.5a)$$

The reaction, for example, for a corner-sharing bidentate complex is instead written as:

$$2 \equiv FecOH^{\frac{1}{2}-} + nH^+$$
$$+ AO_y^{x-} \leftrightarrow (\equiv FecO)_2AO_{y-2}H_{(n+1)}^{-x-1+n} + 2H_2O$$
$$(B5.5b)$$

where AO_y^{x-} is an oxyanion having y oxygens and an overall charge of x (in its most deprotonated state). The surface complexation constants defining these reactions contain electrostatic interaction factors for the 0- and β-planes. Table Box 5.5 shows the full set of reactions needed to describe this system with Hiemstra and Zhao's CD-MUSIC model.

TABLE B5.5　CD-MUSIC model reactions and equilibrium constants for the example in Box 5.5. Reactions and constants are from Mendez and Hiemstra (2019).

Ion	Surface complex	No. H+ in reaction[a]	$(\Delta z_0, \Delta z_\beta)$[b]	log K[c]
H^+	$\equiv FeOH_2^{\frac{1}{2}+}$	1	(1,0)	8.1
	$\equiv Fe_3OH_2^{\frac{1}{2}+}$	1	(1,0)	8.1
Na^+	$\equiv FeOH^{\frac{1}{2}-}\cdots Na^+$	0	(0,1)	−0.6
	$\equiv Fe_3OH^{\frac{1}{2}-}\cdots Na^+$	0	(0,1)	−0.6
NO_3^-	$\equiv FeOH_2^{\frac{1}{2}+}\cdots NO_3^-$	1	(1,−1)	7.42
	$\equiv Fe_3OH_2^{\frac{1}{2}+}\cdots NO_3^-$	1	(1,−1)	7.42
CO_3^{2-}	$(\equiv FeOc)_2CO^-$	2	(0.66, −0.66)	21.73
	$(\equiv FeOc)_2CONa$	2	(0.65, 0.35)	22.38
	$\equiv FeOCO_2^{2-}$	1	(0.34, −1.34)	11.60
PO_4^{3-}	$\equiv FeOPO(OH)_2^{\frac{1}{2}-}$	3	(0.33, −0.33)	29.84
	$\equiv FeOPO_2OH^{1\frac{1}{2}-}$	2	(0.28, −1.28)	26.36
	$(\equiv FeOc)_2POOH^-$	3	(0.65, −0.65)	33.52
	$(\equiv FeOc)_2PO_2^{2-}$	2	(0.46, −1.46)	28.31

[a] This is n in Equations (B5.5a) and (B5.5b).
[b] Values of the electrostatic interaction factors $exp(-F\Psi_0/RT)$ and $exp(-F\Psi_\beta/RT)$, respectively (see Equations in Box 5.4).

This problem can be solved using Visual MINTEQ ver. 4.0 (https://vminteq.lwr.kth.se/) or by using another software code that incorporates the CD-MUSIC model (e.g., PHREEQC, Orchestra). In theory, it is possible to enter all the above

BOX 5.5 *(cont'd)*

Example of CD-MUSIC model

reactions and parameters into the code one by one and then run the program (in the case of Visual MINTEQ, step-by-step instructions are available in the help file). However, usually when performing calculations that include many species, the use of predefined databases is preferred, both because this is much more time efficient and because it minimizes the number of simple user errors. In Visual MINTEQ 4.0 there is a predefined database for Hiemstra and Zhao's CD-MUSIC model version, and below we show how this database can be used to solve this problem. We will start with the scenario when there is no carbonate in the system.

1 Fix the pH to the start pH of the simulation (i.e., 7).
2 Add all the ions as components to the problem on the main menu. Total concentrations of Na^+, NO_3^-, and PO_4^{3-} should be added.
3 Choose "Adsorption − Surface complexation reactions" from the main menu. Add one surface and choose "Fh−Hiem (Hiemstra and Zhao 2016)" to choose Hiemstra and Zhao's version of the CD−MUSIC model for ferrihydrite and the associated database of reactions.
4 Next add the Fe concentration (4.88 mM) and the specific surface area (765 m^2 g^{-1}). The interface will automatically calculate the correct Stern layer capacitances, based on Hiemstra's assumptions on ferrihydrite nanoparticle dimensions, as well as the resulting concentrations of the four different surface sites included in the model. The page should look like that in Figure B5.5a.
5 Go back to the main menu. The next step is to set up a sweep in which the pH is systematically varied between 7 and 11. To do that, choose "Multi−Problem/Sweep" from the main menu and make sure that the option

"Sweep: one parameter is varied" is selected. Specify the start pH value (7) and the increment (i.e., = step size). Here we choose 0.1 as the increment. This results in 41 data points over the pH range of interest (7−11), which we write into the box "State the number of problems." Finally, in the lower part of the menu we define the output table. To be able to plot the results in terms of % adsorbed vs. pH, we need to add columns for "Total dissolved" and "Total sorbed" PO_4. See Figure B5.5b.
6 Go back to the main menu. Click "Run" to execute the problem. The main output page will be automatically opened after the completion of the run. However, this output page will only display detailed results for the first data point. To see the table we designed in point 5, click "Selected Sweep Results."
7 The Selected Sweep Results page contains a button for exporting the results to Excel. Do this and save the Excel file.
8 Go back to the main menu again. Now add 30 mM CO_3 to the problem, leaving all the other settings unchanged. Click "Run" and export the new results to a new Excel file.
9 The results can now be treated in Excel to produce graphs on % adsorbed vs. pH. These can be compared to actual measurements performed by Mendez and Hiemstra (2019) for these same systems. The result is shown in Figure B5.5c. As is seen, the model gives an excellent description of pH-dependent PO_4 adsorption in the absence of carbonate. In the presence of carbonate the model correctly predicts a competitive effect, although in this case the fit to the data is not perfect.

Additional instructional videos are available at https://www.youtube.com/channel/UCk0 OcWbAM4eqwiiT_3ELNTg.

BOX 5.5 *(cont'd)*

Example of CD-MUSIC model

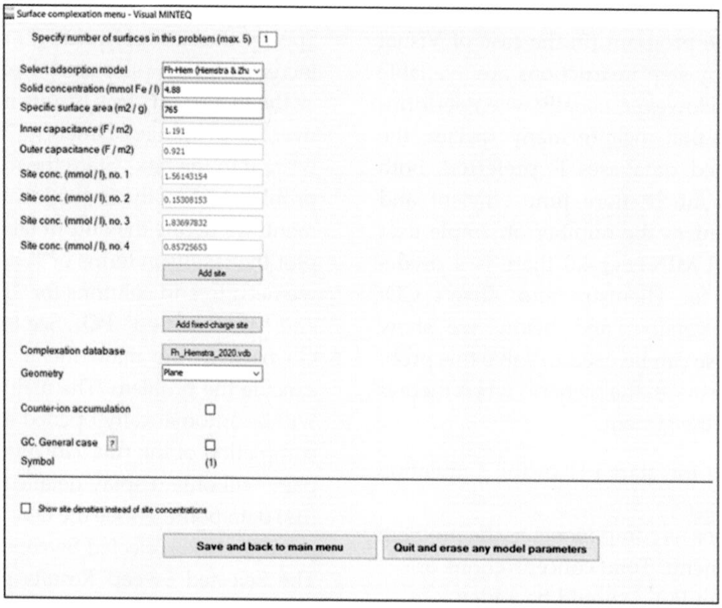

FIGURE B5.5A   A screenshot of the Visual MINTEQ surface complexation menu.

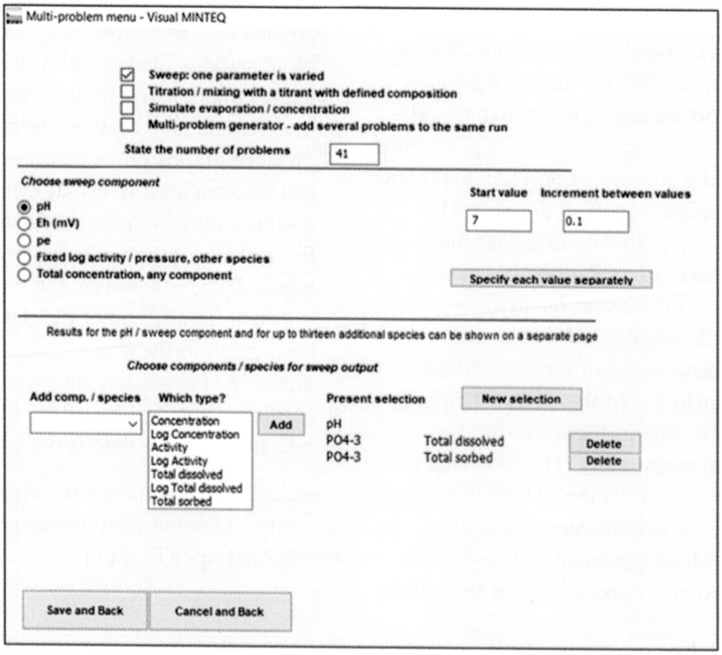

FIGURE B5.5B   A screenshot of the Visual MINTEQ Multi-Problem/Sweep menu.

BOX 5.5   *(cont'd)*

Example of CD-MUSIC model

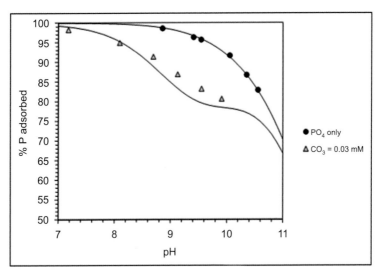

FIGURE B5.5C   Results of the surface complexation model for phosphate adsorption with and without the presence of carbonate.

the pentavalent As atom distributes its valence equally, resulting in 1.25 valence units per oxygen. However, because each oxygen has a nominal valence of -2, the two oxygen ligands facing the surface get a net charge of -1.50, whereas the two outer ligands also get a net charge of -1.50. To get the total change in surface charge, one also needs to consider the proton consumption at the surface due to ligand exchange and any protonation of the outer surface ligands.

To find out how adsorbed ions should be positioned between the planes, and how the charge of the central atom is distributed (so-called CD factors), any method or a combination

of the following three methods is most commonly used:

1. Characterization of the surface coordination using information from, for example, EXAFS spectroscopy and FTIR spectroscopy are often used to constrain the type of complexes that dominate.
2. Molecular orbital/density functional theory (MO/DFT) calculations may permit estimates of the most likely coordination modes and bond distances/CD factors.
3. By fitting to experimental data. This, however, requires the adsorption reaction to be well

characterized (point 1) to reduce the degrees of freedom.

The most common interface model used with the CD-MUSIC model historically has been the three-plane model (TPM). The charge–potential relationships of TPM are mathematically identical to those of the triple-layer model (TLM); see Table 5.4 and Figure 5.19. The main difference between TPM and TLM is that the former uses a one-pK description of the acid–base chemistry similar to the 1-pK Stern model. In addition, while the outer Stern layer capacitance in the TLM, C_2, is usually fixed at 0.2 F m^{-2}, no such assumption is made in the TPM; instead, this value can in principle be fitted from acid–base titration data. In the most recent models for ferrihydrite (which is exemplified in Box 5.5) adsorbed H$^+$ and OH$^-$ are placed in the 0-plane, while the inner-sphere surface complexes are placed across the 0- and β-planes, as described in Figure 5.19b. The outer-sphere complexes are usually assumed to affect the charge of the β-plane, although they often are thought to affect also σ_0 by means of hydrogen

bonding. Finally, the model also assumes that electrolyte ions, to a certain extent, affect the potential gradient by modifying the β- or d-plane charge.

5.6 Sorption of Metal Cations

Key points: Sorption of trace metal cations is pH dependent and related to the first hydrolysis constant of metals. The selectivity of monovalent alkali metals on metal oxides occurs in the reverse order to the one observed on 2:1 phyllosilicates.

The sorption of many metals and trace metals cations is pH dependent and is characterized by a narrow pH range where sorption increases to nearly 100%, traditionally known as an adsorption edge (Figure 5.25). The pH position of the adsorption edge for a particular metal cation is related to its hydrolysis or acid–base characteristics. In addition to pH, the sorption of metals is dependent on sorptive concentration, surface coverage, and the type of the sorbent.

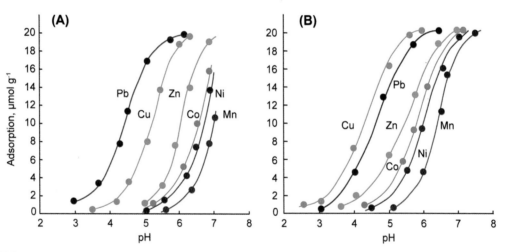

FIGURE 5.25   Sorption of a range of metals on (a) hematite and (b) goethite when the metals were added at a rate of 20 μmol g^{-1} of adsorbate. The values for the pK$_1$ for the dissociation of the metals to give the monovalent MOH$^+$ ions are Pb, 7.71; Cu, 8; Zn, 8.96; Co, 9.65; Ni, 9.86; and Mn, 10.59. *From McKenzie (1980), with permission.*

One can measure the relative affinity of a cation for a sorbent or selectivity. The properties of the cation, sorbent, and solvent all affect selectivity. With monovalent alkali metal cations, electrostatic interactions predominate, and the general order of selectivity is $Li^+ < Na^+ < K^+ < Rb^+ < Cs^+$ (Kinniburgh and Jackson, 1981). This order is related to the size of the hydrated radius. The ion in the above group with the smallest hydrated radius, Cs^+, can approach the surface the closest and be held the most tightly. However, on some hydrous oxides, the reverse order is often observed (Table 5.5). This has been particularly noted for some hydrous metal oxides. The reason for this selectivity is not well understood but may be related to the effect of the solid on water that is present on the oxide surface (Kinniburgh and Jackson, 1981) or to variation in the solution matrix.

With divalent ions, there is little consistency in the selectivity order. Table 5.6 shows the selectivity of alkaline earth cations and divalent transition and heavy metal cations on metal oxide surfaces. The type of surface and the pH both appear to have major effects on the selectivity sequence. Differences in the H^+/Mn^+ could cause reversals in selectivity since the ion with the higher H^+/Mn^+ stoichiometry would be favored at higher pH (Kinniburgh and Jackson, 1981). Divalent transition and heavy metal cations, both of which are often sorbed as inner-sphere complexes (Table 5.1), are more strongly sorbed than alkaline earth cations.

Studies on the sorption of metals on bacterial species have appeared. Figure 5.26 shows data for Cd sorption on various gram-positive and gram-negative bacterial species. The pH-dependent sorption behavior is similar for the different bacterial species, with the sorption edge strongly resembling metal cation sorption behavior.

5.7 Sorption of Anions

Key points: The adsorption of anions is closely related to solution pH, termed the adsorption envelope, which varies with the dissociation constant (pK_a) of the conjugate acid of the anion. Negligible adsorption occurs at pH values greater than the point of zero charge (PZC).

Anion sorption varies with pH, usually increasing with pH and reaching a maximum close to the pK_a for anions of monoprotic conjugate acids (a compound that can donate one proton), and slope breaks have been observed at pK_a values for anions of polyprotic (a compound that can donate more than one proton) conjugate acids (Hingston, 1981). This relationship is traditionally referred to as an adsorption envelope (Figure 5.27). In Figure 5.27 one sees that for silicate and fluoride, sorption increased with pH and reached a maximum near the pK_a (pK_1 in Figure 5.27) of the acid and then decreased with pH. For fluoride, the dominant species

TABLE 5.5 Selectivity sequences for the alkali metal cations on various hydrous metal oxides.

Sequence	Oxide	Reference
Cs > Rb > K > Na > Li	Si gel	Helfferich (1962)
K > Na > Li	Si gel	Tien (1965)
Cs > K > Li	SiO_2	Altug and Hair (1967)
Cs > K > Na > Li	SiO_2	Abendroth (1970)
Li > Na > K	SiO_2	Bartell and Fu (1929)
K > Na > Li	Al_2O_3	Churms (1966)
Li > K ~ Cs	Fe_2O_3	Breeuwsma and Lyklema (1971)
Li > Na > K ~ Cs	Fe_2O_3	Dumont and Watillon (1971)
Cs ~ K > Na > Li	Fe_3O_4	Venkataramani et al. (1978)
Li > Na > Cs	TiO_2	Bérubé and de Bruyn (1968)
Li > Na > K	$Zr(OH)_4$	Britz and Nancollas (1969)

All cations are monovalent.
From Kinniburgh and Jackson (1981), with permission.

TABLE 5.6 Selectivity sequences for the adsorption or coprecipitation of alkaline earth cations and of divalent transition and heavy metal cations on various hydrous metal oxides.[a]

Sequence	Oxide	Reference
Alkaline earth cations		
Ba > Ca	Fe gel	Kurbatov et al. (1945)
Ba > Ca > Sr > Mg	Fe gel	Kinniburgh et al. (1976)
Mg > Ca > Sr > Ba	α-Fe$_2$O$_3$	Breeuwsma and Lyklema (1971)
Mg > Ca > Sr > Ba	Al gel	Kinniburgh et al. (1976)
Mg > Ca > Sr > Ba	Al$_2$O$_3$	Huang and Stumm (1973)
Ba > Sr > Ca	α-Al$_2$O$_3$	Belot et al. (1966)
Ba > Sr > Ca > Mg	MnO$_2$	Gabano et al. (1965)
Ba > Sr > Ca > Mg	α-MnO$_2$	Murray (1975)
Ba > Ca > Sr > Mg	SiO$_2$	Tadros and Lyklema (1969)
Ba > Sr > Ca	SiO$_2$	Malati and Estefan (1966)
Divalent transition and heavy metal cations		
Pb > Zn > Cd	Fe gel	Gadde and Laitinen (1974)
Zn > Cd > Hg	Fe gel	Bruninx (1975)
Pb > Cu > Zn > Ni > Cd > Co	Fe gel	Kinniburgh et al. (1976)
Cu > Zn > Co > Mn	α-FeOOH	Grimme (1968)
Cu > Pb > Zn > Co > Cd	α-FeOOH	Forbes et al. (1976)
Cu > Zn > Ni > Mn	Fe$_3$O$_4$	Venkataramani et al. (1978)
Cu > Pb > Zn > Ni > Co > Cd	Al gel	Kinniburgh et al. (1976)
Cu > Co > Zn > Ni	MnO$_2$	Kozawa (1959)
Co > Cu > Ni	MnO$_2$	Murray et al. (1968)
Pb > Zn > Cd	MnO$_2$	Gadde and Laitinen (1974)
Co $\approx$ Mn > Zn > Ni	MnO$_2$	Murray (1975)
Cu > Zn > Co > Ni	δ-MnOOH	McKenzie (1972)
Co > Cu > Zn > Ni	α-MnO$_3$	McKenzie (1972)
Co > Zn	δ-MnO$_2$	Loganathan and Burau (1973)
Zn > Cu > Co > Mn > Ni	Si gel	Vydra and Galba (1969)
Zn > Cu > Ni $\approx$ Co > Mn	Si gel	Taniguechi et al. (1970)
Cu > Zn > Co > Fe > Ni > Mn	SnO$_2$	Donaldson and Fuller (1968)

All cations are divalent.
From Kinniburgh and Jackson (1981), with permission.

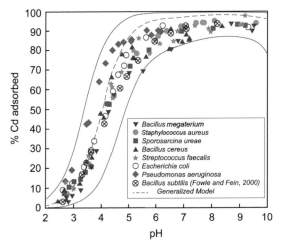

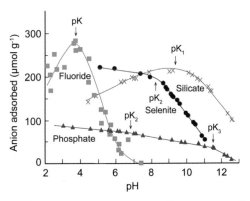

FIGURE 5.26 Cadmium sorption onto pure cultures of various gram-positive and gram-negative bacterial species. Each data point represents individual batch experiments with $10^{-4.1}$ M Cd and 1.0 g L^{-1} (dry wt.) bacteria. The dotted curve represents the sorption behavior calculated using the average parameters given in Table 1 of Yee and Fein (2001). The solid lines represent the upper and lower uncertainty limits calculated from the errors in pK$_a$, site concentration, and log K$_{ads}$ measurements. *From Yee and Fein (2001), with permission.*

FIGURE 5.27 Sorption of a range of anions on goethite. Two samples of goethite were used, and the level of addition of sorbing ion differed between the different ions. *From Hingston et al. (1972), with permission.*

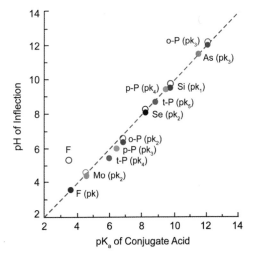

FIGURE 5.28 Relationship between pKa and pH at the change in slope of sorption envelopes. Sorbents: ●, goethite; ○, gibbsite. Sorbates: F, fluoride; Mo, molybdate; t-P, tripolyphosphate; p-P, pyrophosphate; o-P, orthophosphate; Se, selenite; Si, silicate; As, arsenate. *From Hingston et al. (1972), with permission.*

were HF and F$^-$, and for silicate, the species were H$_4$SiO$_4$ and H$_3$SiO$_4^-$. With selenite and phosphate, sorption decreased with increased pH, with the pH decrease more pronounced above pK$_2$. The ion species for selenite and phosphate were HSeO$_3^-$ and SeO$_3^{2-}$, and H$_2$PO$_4^-$ and HPO$_4^{2-}$, respectively (Barrow, 1985). An example of the correlation between adsorption maxima (pH of inflection) and the pK$_a$ values for conjugate acids is shown in Figure 5.28.

Generally speaking, the anions NO$_3^-$, Cl$^-$, and ClO$_4^-$ are sorbed as outer-sphere complexes and sorbed on surfaces that exhibit a positive charge. Sorption is sensitive to ionic strength. Some researchers have also concluded that SO$_4^{2-}$ (Zhang and Sparks, 1990a) can be sorbed as an outer-sphere complex; however, there is other evidence that SO$_4^{2-}$ can also be sorbed as an inner-sphere complex (Table 5.1). There is direct

spectroscopic evidence to show that selenate is sorbed as an outer-sphere complex and selenite is sorbed as an inner-sphere complex (Hayes et al., 1987). XAFS spectroscopy analyses showed that the sorbed selenite was bonded to the surface as a bidentate species with two Fe atoms 3.38 Å from the selenium atom. Selenate had no Fe atom in the second coordination shell of Se, which indicated that its hydration sphere was retained on sorption.

Most other anions such as molybdate, arsenate, arsenite, phosphate, and silicate appear to be strongly sorbed as inner-sphere complexes, and sorption occurs through a ligand exchange mechanism (Table 5.1). The sorption maximum is often insensitive to ionic strength changes. Sorption of anions via ligand exchange results in a shift in the PZC of the oxide to a more acid value.

5.8 Surface Precipitation

Key points: Incorporation of cations or anions into newly formed minerals is the general process of surface precipitation. The newly formed minerals are composed of ions present in the soil porewater, typically adjacent to a preexisting mineral surface. Surface precipitation can impede the transport/mobility of ions, such as potentially toxic metals, in soils.

As the amount of metal cation or anion sorbed on a surface (surface coverage or loading, which is affected by the pH at which sorption occurs) increases, sorption can proceed from mononuclear adsorption to surface precipitation (a three-dimensional phase). There are several thermodynamic reasons for surface precipitate formation: (i) the solid surface may lower the energy of nucleation by providing sterically similar sites (McBride, 1991); (ii) the activity of the surface precipitate is <1 (Sposito, 1986); and (iii) the solubility of the surface precipitate is lowered because the dielectric constant of the solution near the surface is less than that of the bulk solution (O'Day et al., 1994a). There are several types of surface precipitates. They can arise via polymeric metal complexes (dimers, trimers, etc.) that form on mineral surfaces and via the sorption of aqueous polymers (Chisholm-Brause

et al., 1990a). Homogeneous precipitates can form on a surface when the solution becomes saturated and the surface acts as a nucleation site. When adsorption attains monolayer coverage, sorption continues on the newly created sites, causing a precipitate on the surface (Farley et al., 1985; McBride, 1991). When the precipitate consists of chemical species derived from both the aqueous solution and the dissolution of the mineral, it is referred to as a coprecipitate. The composition of the coprecipitate varies between that of the original solid and a pure precipitate of the sorbing metal. The ionic radius of the sorbing metal and sorbent ions must be similar for coprecipitates to form. Thus Co(II), Mn(II), Ni(II), and Zn(II) form coprecipitates on sorbents containing Al(III) and Si(IV) but not Pb(II), which is considerably larger (0.12 nm). Coprecipitate formation is most limited by the rate of mineral dissolution rather than the lack of thermodynamic favorability (McBride, 1994; Scheidegger et al., 1998). If the formation of a precipitate occurs under solution conditions that would, in the absence of a sorbent, be undersaturated with respect to any known solid phase, this is referred to as surface-induced precipitation (Towle et al., 1997).

Thus there is often a continuum between surface complexation (adsorption) and surface precipitation. This continuum depends on several factors: (i) ratio of the number of surface sites versus the number of metal ions in solution; (ii) the strength of the metal oxide bond; and (iii) the degree to which the bulk solution is undersaturated with respect to the metal hydroxide precipitate (McBride, 1991). At low surface coverages, surface complexation (e.g., outer- and inner-sphere adsorption) tends to dominate. As surface coverage increases, nucleation occurs and results in the formation of distinct entities or aggregates on the surface. As surface loadings increase further, surface precipitation becomes the dominant mechanism (Figure 5.29). For

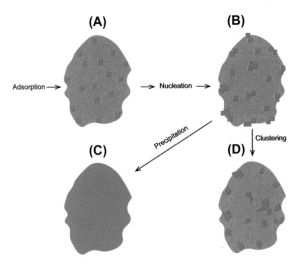

FIGURE 5.29 An illustration of metal ion sorption reactions on (hydr)oxide. (*A*) At low surface coverage, isolated site binding (adsorption) is the dominant sorption mechanism; (*B*) with increased metal loading, metal hydroxide nucleation begins. Further increases in metal loadings result in (*C*) surface precipitation or (*D*) surface clusters. *From Fendorf (1992), with permission.*

example, Fendorf et al. (1994) and Fendorf and Sparks (1994) used XAFS, FTIR, and HRTEM to study Cr(III) sorption on Si oxide. At low Cr(III) surface coverage (<20%), adsorption was the dominant process, with an inner-sphere monodentate complex forming. As Cr(III) surface coverage increased (>20%), surface precipitation occurred and was the dominant process. Table 5.1 shows that for a number of ions, as surface coverage increases, surface precipitates form.

Using *in situ* XAFS, it has been shown by a number of scientists that multinuclear metal hydroxide complexes and surface precipitates Co(II), Cr(III), Cu(II), Ni(II), and Pb(II) can form on metal oxides, phyllosilicates, soil clays, and soils (Chisholm-Brause et al., 1990a, b; Roe et al., 1991; Charlet and Manceau, 1992; Fendorf

et al., 1994a; O'Day et al., 1994a,b; Papelis and Hayes, 1996; Scheidegger et al., 1996, 1997, 1998; Towle et al., 1997; Elzinga and Sparks, 1999; Roberts et al., 1999; Thompson et al., 1999a,b; Ford and Sparks, 2000; Scheckel and Sparks, 2001). These metal hydroxide phases occur at metal loadings below a theoretical monolayer coverage and in a pH range well below the pH where the formation of metal hydroxide precipitates would be expected according to the thermodynamic solubility product (Scheidegger and Sparks, 1996).

5.8.1 Formation of Layered Double Hydroxides (LDHs)

Scheidegger et al. (1997) were the first to show that sorption of metals, such as Ni, on an array of phyllosilicates and Al oxide could result in the formation of mixed metal–Al hydroxide surface precipitates that appear to be coprecipitates. The precipitate phase shares structural features common to the hydrotalcite group of minerals and the layered double hydroxides (LDHs) observed in catalyst synthesis. The LDH structure is built of stacked sheets of edge-sharing metal octahedra containing divalent and trivalent metal ions separated by anions between the interlayer spaces. The general structural formula can be expressed as $[Me^{2+}_{1-x} Me^{3+}_x (OH)_2]^{x+} \cdot (x/n)$ $A^{-n} \cdot -m$ H_2O, where, for example, Me^{2+} could be Mg(II), Ni(II), Co(II), Zn(II), Mn(II), and Fe(II) and Me^{3+} is Al(III), Fe(III), and Cr(III).

The LDH structure exhibits a net positive charge x per formula unit, which is balanced by an equal negative charge from interlayer anions A^{n-} such as Cl^-, Br^-, I^-, NO_3^-, OH^-, ClO_4^-, and CO_3^{2-}; water molecules occupy the remaining interlayer space (Allman, 1970; Taylor, 1984). The minerals takovite, $Ni_6 Al_2(OH)_{16}$ $CO_3 \cdot H_2O$, and hydrotalcite, $Mg_6 Al_2(OH)_{16}$

$CO_3 \cdot H_2O$, are among the most common natural mixed-cation hydroxide compounds containing Al (Taylor, 1984). Figure 5.30 shows an Ni—Al LDH phase.

XAFS data showing the formation of Ni—Al LDH phases on soil components are shown in Figure 5.31 and Table 5.7 (Scheidegger et al., 1997). Radial structure functions (RSFs), collected from XAFS analyses, for Ni sorption on pyrophyllite, kaolinite, gibbsite, and montmorillonite were compared to the spectra of crystalline $Ni(OH)_2$ and takovite. All spectra showed a peak at $R \approx$ 1.8 Å, which represents the first coordination shell of Ni. A second peak representing the second Ni shell was observed in the spectra of the Ni sorption samples and takovite (Figure 5.31). The structural parameters, derived from XAFS analyses, for the various sorption samples and takovite and $Ni(OH)_2$ are shown in Table 5.7. In the first coordination shell Ni is surrounded by six O atoms, indicating that Ni(II) is in an octahedral environment. The Ni—O distances for the Ni

sorption samples are 2.02—2.03 Å and similar to those in takovite (2.03 Å). The Ni—O distances in crystalline $Ni(OH)_{2(s)}$ are distinctly longer (2.06 Å in this study).

For the second shell, best fits were obtained by including both Ni and Si or Al as second-neighbor backscatter atoms. Since Si and Al differ in atomic number by 1 (atomic number = 14 and 13, respectively), backscattering is similar. They cannot be easily distinguished from each other as second-neighbor backscatterers. There are 2.8 (for montmorillonite) to 5.8 (gibbsite) Ni second-neighbor (N) atoms, indicative of Ni surface precipitates. The Ni—Ni distances for the sorption samples were 3.00—3.03 Å, which are similar to those for takovite (3.01 Å), the mixed Ni—Al LDH phase, but they are much shorter than those in crystalline $Ni(OH)_2$ (3.09 Å). There are also 1.8—2.7 Si/Al second-neighbor atoms at 3.02—3.07 Å. The bond distances are in good agreement with the Ni—Al distances observed in takovite (3.03 Å).

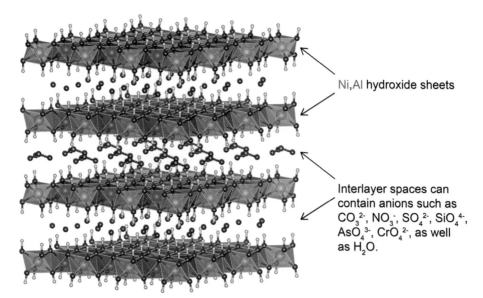

Ni,Al hydroxide sheets

Interlayer spaces can contain anions such as CO_3^{2-}, NO_3^-, SO_4^{2-}, SiO_4^{4-}, AsO_4^{3-}, CrO_4^{2-}, as well as H_2O.

FIGURE 5.30 The structure of Ni—Al LDH illustrating brucite-like octahedral layers in which Al^{3+} substitutes for Ni^{2+}. This replacement of a divalent ion with a trivalent one creates a net positive charge in the sheet, which is balanced by hydrated anions in the interlayer space. *From Siebecker et al. (2018), with permission.*

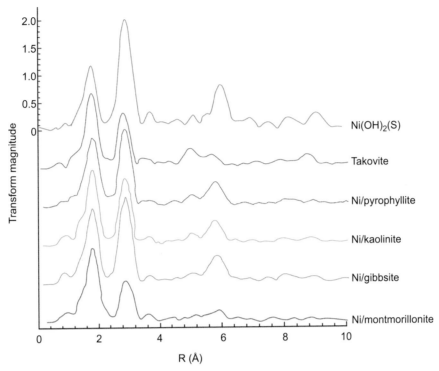

FIGURE 5.31   Radial structure functions (RSFs) produced by forward Fourier transforms of Ni sorbed on pyrophyllite, kaolinite, gibbsite, and montmorillonite compared to the spectrum of crystalline $Ni(OH)_{2(s)}$ and takovite. The spectra are uncorrected for phase shift. *From Scheidegger et al. (1997), with permission.*

TABLE 5.7   Structural information derived from EXAFS analysis for Ni sorption on various sorbents and for Ni hydroxide.[a]

	Γ (μmol/m^2)	Ni–O			Ni–Ni			Ni–Si,Al			CN(Ni)/CN(Si,Al)
		R (Å)	CN	$2\sigma^2$	R (Å)	CN	$2\sigma^2$	R (Å)	CN	$2\sigma^2$	
$Ni(OH)_2$	–	2.06	6.0	0.01	3.09	6.0	0.010	–	–	–	–
Takovite	–	2.03	6.0	0.01	3.01	3.1	0.009	3.03	1.1	0.009	2.8
Pyrophyllite	3.1	2.02	6.1	0.01	3.00	4.8	0.009	3.02	2.7	0.009	1.8
Kaolinite	19.9	2.03	6.1	0.01	3.01	3.8	0.009	3.02	1.8	0.009	2.2
Gibbsite	5.0	2.03	6.5	0.01	3.02	5.0	0.009	3.05	1.8	0.09	2.7
Montmorillonite	0.35	2.03	6.3	0.01	3.03	2.8	0.011	3.07	2.0	0.015	1.4

[a] *Interatomic distances (R, Å), coordination numbers (CN), and Debye–Waller factors ($2\sigma^2$, nm^2). The reported values are accurate to within $R \pm 0.02$ Å, CN(Ni–O) $\pm$ 20%, CN(Ni–Ni) $\pm$ 40%, and CN(Ni–Si,Al) $\pm$ 40%.*
From Scheidegger et al. (1997), with permission.

Mixed Co–Al and Zn–Al hydroxide surface precipitates also can form on aluminum-bearing metal oxides and phyllosilicates (Towle et al., 1997; Thompson et al., 1999a; Ford and Sparks, 2000). This is not surprising as Co^{2+}, Zn^{2+}, and Ni^{2+} all have similar radii (74.5, 74, and 69 pm, respectively), enhancing substitution for Al^{3+} (53.5 pm) in the mineral structure and allowing the formation of a coprecipitate. However, surface precipitates have not been observed with Pb^{2+}, as Pb^{2+} is too large (119 pm) to substitute for Al^{3+} in mineral structures.

Metal hydroxide precipitate phases can also form in the presence of non-Al-bearing minerals (Scheinost et al., 1999). Using diffuse reflectance spectroscopy (DRS), which is quite sensitive for discriminating between Ni–O bond distances,

it was shown that α-Ni(OH)$_2$ formed upon Ni^{2+} sorption to talc and silica (Figure 5.32a).

The mechanism for the formation of metal hydroxide surface precipitates is not clearly understood. It is clear that the type of metal ion determines whether metal hydroxide surface precipitates form, and the type of surface precipitate formed, that is, metal hydroxide or mixed metal hydroxide, is dependent on the sorbent type. The precipitation could be explained by the combination of several processes (Yamaguchi et al., 2001). First, the electric field of the mineral surface attracts metal ions, for example, Ni, through adsorption, leading to a local supersaturation at the mineral–water interface. Second, the solid phase may act as a nucleation center for polyhydroxy species and catalyze the

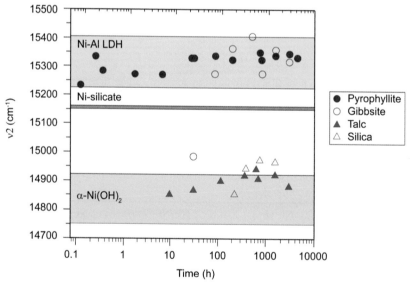

FIGURE 5.32A   Fitted v2 band positions of the Ni-reacted minerals (dots and triangles) over time using diffuse reflectance spectroscopy (DRS). The v2 band is attributed to crystal field splitting induced within the incompletely filled 3d electronic shell of Ni^{2+} through interaction with the negative charge of nearest-neighbor oxygen ions. For talc and silica, the v2 band was at $\sim$14,900 cm^{-1}, indicating the formation of an α-Ni(OH)$_2$ phase, while for Al-containing pyrophyllite and gibbsite, it appeared at $\sim$15,300 cm^{-1}, indicating an Ni-Al LDH phase. *From Scheinost et al. (1999), with permission.*

precipitation process (McBride, 1994). Third, the physical properties of water molecules adsorbed at the mineral surface are different from those of free water (Sposito, 1989), causing a lower solubility of metal hydroxides at the mineral–water interface. With time, Al, which is released by weathering of the mineral surface, slowly diffuses into the octahedral layer of the mineral and partially replaces the metal (e.g., Ni) in the octahedral sites. An Ni–Al LDH, which is thermodynamically favored over α-Ni hydroxide, is formed.

The formation of metal hydroxide surface precipitates appears to be an important way to sequester metals. As the surface precipitates age, metal release is greatly reduced.

Thus the metals are less prone to leaching and being taken up by plants. This is due to the silication of the interlayer of the LDH phases creating precursor phyllosilicate surface precipitates. Li et al. (2012) examined the formation of Zn LDH precipitates on a variety of aluminum oxides. It was clearly demonstrated that minerals that are more soluble would form LDH surface precipitates much more quickly than less

soluble minerals. This highlights the important role of mineral surface dissolution in the formation of surface precipitates. Surface precipitates were directly imaged using high-resolution TEM (Figure 5.32b). More details on the metal release rates from surface precipitates and the mechanisms for the metal sequestration are given in Chapter 7.

5.8.2 Heteroepitaxial Growth

The formation of LDH during *in situ* reactions indicates that dissolved Al^{3+} adsorbs onto surface complexed Ni, that is, Ni that is already adsorbed to an Al-rich mineral's surface (Siebecker et al., 2014, 2018). The LDH then continually grows outward from the mineral surface as dissolved Ni^{2+} and Al^{3+} simultaneously adsorb to the continuingly forming LDH precipitate. This mechanism can be termed "heteroepitaxial growth." Isomorphous substitution is not the dominant mechanism for LDH formation whether it be Al^{3+} substitution into a precursor α-$Ni(OH)_2$ or Ni^{2+} substitution into an Al-rich sorbent (Siebecker et al., 2018).

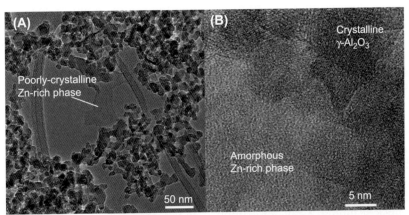

FIGURE 5.32B (A) Low magnification of a poorly crystalline Zn-rich phase distributed between gamma-alumina particles. (B) High-resolution TEM of a Zn-rich phase (amorphous) adjacent to crystalline gamma-alumina particles. The Zn-rich phase shows no lattice fringes while the gamma-alumina lattice can be viewed as dark stripes. *From Li et al. (2012), with permission.*

Epitaxial growth is when one crystal develops on or is deposited on top of another crystal. There are several types of epitaxial growth. During the formation of surface precipitated LDH, the developing crystals are LDH and the base crystals are the Al-rich clay mineral or a metal oxide sorbent. The LDH develops on top of the Al-rich sorbent (Siebecker et al., 2018). The epitaxial growth of LDH can be considered heteroepitaxial because the two materials (i.e., the LDH and the sorbent) are of different chemical compositions. Homoepitaxial growth is when the developing crystal is of the same material as the sorbent (Atkins et al., 2006). Thus in this process of LDH formation LDH is continuously built outward via the constant adsorption of the metal cations from solution, and the aqueous cations (e.g., Ni^{2+} and Al^{3+}) undergo hydrolysis during the precipitation reaction to form the LDH mineral.

During the initial stages of adsorption, bidentate Ni adsorption onto octahedral Al-edge sites would provide new, unobstructed adsorption sites. Then Ni-hydroxide polymers (e.g., dimers or trimers of octahedrally coordinated Ni atoms) continue attaching epitaxially from the Al-edge sites of the sorbent. The LDH does not necessarily have to grow linearly outward from the surface of the mineral (Figure 33b). Sorptive Ni and/or Al atoms can bind to any of the multiple exposed octahedral edges of previously bound Ni and/or Al atoms (Siebecker et al., 2018). Subsequent polymerization of aqueous metal ions (sorptive ions) likely proceeds via Me—OH—Me, Al—OH—Al, and Me—OH—Al olation reactions (Scheinost et al., 1999), where the ions form polymeric hydroxides on the surface of the sorbent. Upon the hydrolysis of metal ions, bridges between partially hydrolyzed metals can begin to form via —OH— or —O— bonding (Siebecker et al., 2018). Quick-scanning X-ray absorption spectroscopy (QXAS) indicates that the readsorption and hydrolysis of dissolved Al must be fast for Ni—Al LDH to form in real time in a flow cell where products are continuously removed from the system (Siebecker et al., 2014).

The nonlinear growth and bonding (as seen in Figure 5.33a) would be expected because of the amorphous nature that LDH possess when they form on phyllosilicates clays (Siebecker et al., 2018). Neither HRTEM nor powder X-ray diffraction identified any crystalline peaks associated with Ni—Al LDH even after 5 years of Ni reacting with pyrophyllite (Livi et al., 2009).

5.8.3 Stability of Layered Double Hydroxides

Metal—Al LDHs have been shown to be more thermodynamically stable versus their single-metal, isostructural-layered hydroxide phases (e.g., $Ni(OH)_2$) (Peltier et al., 2006). The increase in stability is related to the elemental composition of the metal hydroxide sheet and the type of interlayer anion. Changes to the interlayer anion and interlayer composition can increase LDH stability, which in turn decreases LDH solubility; this is important to inhibit contaminant transport. Because interlayer anions affect the solubility of an LDH, it will also affect heavy metal release into the soil or sediment solution. For example, substituting NO_3^- with CO_3^{2-} could decrease the solubility of an LDH by two orders of magnitude (Peltier et al., 2006; Siebecker et al., 2018).

In the LDH structure the interlayer can be occupied by water molecules or anions, for example, NO_3^-, SO_4^{2-}, CO_3^{2-}, SiO_4^{4-}, PO_4^{3-}, AsO_4^{3-}, or CrO_4^{2-} (Figure 5.30). The type of interlayer anion between the hydroxide sheets greatly affects the stability of the LDH (Allada et al., 2002; Peltier et al., 2006). For example, the stability of Ni—Al LDH solids were found to increase in the order of nitrate < sulfate < carbonate < and silicate as the interlayer anions. The increase in stability is related to decreasing enthalpies of formation determined by calorimetry. Additionally, the carbonate, sulfate, and nitrate phases are all less

(A)

(B)

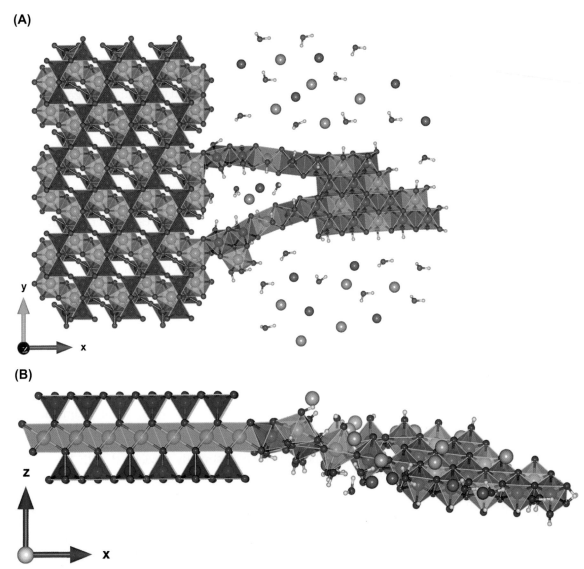

FIGURE 5.33 Top (Panel A) and side (Panel B) views of proposed heteroepitaxial LDH growth on a pyrophyllite edge site. Pyrophyllite is on the left side and the growing LDH is on the right side of each figure. During the formation of Ni–Al LDH, Ni adsorbs to the Al edge site. Additionally, Al dissolves from the pyrophyllite, and subsequently, Al or Al polymers immediately adsorb onto the surface adjacent to surface-bound Ni. In this fashion the double-metal octahedral layer grows outward from the pyrophyllite surface. Ni and Al simultaneously adsorb to the newly forming LDH precipitate in a way that can be termed heteroepitaxial growth. *From Siebecker et al. (2018), with permission.*

soluble than the single-metal hydroxide $Ni(OH)_2$ by several orders of magnitude. Silica substitution into the interlayer can additionally decrease the enthalpy of formation of the LDH versus carbonate interlayer. Silica substitution is the most stable and is thermodynamically favorable (Peltier et al., 2006; Siebecker et al., 2018).

5.9 Speciation of Metal-Contaminated Soils

Key points: Speciation means the chemical form of an element. There are many aspects to speciation, including surface complexes, surface precipitation, and organic matter complexation, all of which impact the ability of an element to remain soluble.

The ability to speciate metal-contaminated soils is critical in developing viable and cost-effective remediation strategies and in predicting the mobility and bioavailability of the metals. Metals in contaminated soils can be found in the mineral and sorbed phases. Total metal contents of soils, while indicating the degree of contamination, reveal no information on speciation. Chemical extraction techniques (e.g., sequential extractions) provide useful information on the quantities of metals that could be associated with different phases, but transformations in the phases may occur.

X-ray diffraction, thermal gravimetric analysis (TGA), and X-ray photoelectron spectroscopy (XPS) have been used to characterize metals in contaminated soils. Even though they provide a more accurate characterization of metal speciation than extraction methods, they may introduce artifacts due to sample alterations. Additionally, their detection limits are often far above the background concentrations of the metals contained in the soils.

Electron microscopy, coupled with energy-dispersive X-ray fluorescence spectroscopy (EDXS), can enable one to obtain both quantitative (elemental composition) and qualitative

information (metal distribution) with good spatial resolution (<1 μm). However, only elemental concentrations are obtained. Thus one could not, for example, distinguish between S and SO_4^{2-}. The elements also must be in nearly crystalline phases in concentrations greater than 1%. This limits its utility in many contaminated systems. Bulk XAFS has been useful in characterizing metal-bearing soil minerals and soils. However, one can probe an area of only several millimeters, providing only an average speciation of metals in a sample. This can be a problem in heterogeneous soil samples where several metal species can be present over a small area. Moreover, in samples where the metal may be present in numerous phases, the detection limit for minor species is problematic.

To overcome the above difficulties, major advances in the use of spatially resolved micro-XAFS (μ-XAFS; both μ-XANES and μ-EXAFS) combined with μ-synchrotron- fluorescence (μ-SXRF) spectroscopy have occurred. With these techniques, one can study metal speciation in soils on a micron scale (Manceau et al., 2000; Roberts et al., 2002).

An example of using these techniques to speciate Zn in a subsoil is shown in Figure 5.34a. From the μ-SXRF maps, one sees that Zn is strongly associated with Mn in the center of the sample, with Fe in other portions of the sample, and with neither Mn nor Fe in some portions of the sample (Figure 5.34a — Panel A). This microscale heterogeneity in elemental associations suggests Zn could be present in different phases over a small sample area. Bulk XAFS analyses suggested that Zn was bound as an inner-sphere complex to Al, Fe, and Mn oxides but the data interpretation was difficult (Figure 5.34a — Panel B).

μ-XAFS data were collected on the three regions of the sample (labeled 1, 2, and 3 on the Zn map, Figure 5.34a — Panel A). Spot 1 on the map showed that Zn was octahedrally coordinated and sorbed to an Al oxide phase (Figure 5.34a — Panel C). Spots 2 and 3, with

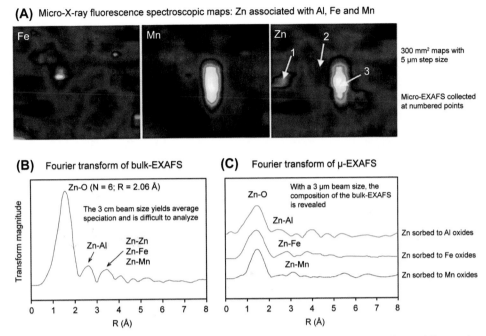

FIGURE 5.34A   Synchrotron-based speciation of Zn in a smelter-contaminated subsurface soil using (*A*) μ-synchrotron X-ray fluorescence (μ-SXRF) spectroscopy, (*B*) bulk XAFS, and (*C*) μ-XAFS techniques. *From Roberts (2001), with permission.*

Zn in a tetrahedral coordination, revealed that Zn was primarily sorbed to Fe oxide (Spot 2) and to Mn oxide (Spot 3) (Figure 5.34 – Panel C). These results suggest that Zn^{2+} could have been released from Zn-bearing mineral phases in the topsoil (primarily sphalerite, ZnS, and franklinite, $ZnFe_2O_4$) and transported to the subsoil where the Zn^{2+} was partially readsorbed to Al, Fe, and Mn oxides.

The formation of secondary mineral precipitates can have a profound effect on the chemical forms of trace metals in soils. For example, in a study by Nachtegaal et al. (2005) it was shown that in soils contaminated with Zn by smelter operations in Belgium that newly formed Zn precipitates accounted for 50%–70% of the Zn species in the soil. The chemical species of the precipitates included both Zn−Al LDH and Zn-phyllosilicate (Figure 5.34b). The incorporation of Zn into newly formed precipitates leads to significant natural attenuation of the exchangeable (bioavailable) Zn fraction. At lower soil pH conditions, however, the conditions are less favorable to the formation and stability of Zn-enriched precipitates.

5.10 Techniques to Determine Molecular-Scale Sorption Mechanisms

Key points: It has become increasingly recognized that if we are going to predict and model fate/transport, toxicity, speciation (form of), bioavailability, and risk assessment of plant nutrients, toxic metals, oxyanions, radionuclides, and organic chemicals at the landscape scale, we must have fundamental information at multiple scales, and our research efforts must be multi- and interdisciplinary.

(A)

(B)

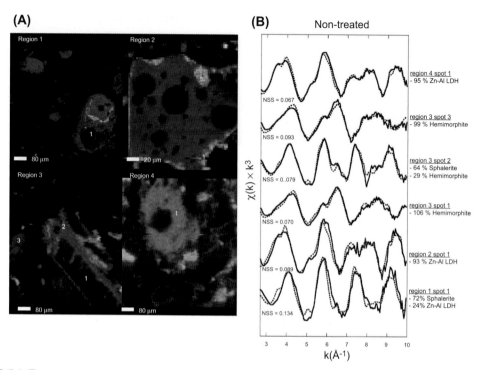

FIGURE 5.34B   (A) Microfocused XRF maps for Belgian soils affected by Zn smelting operations. The numbers in the tricolor maps indicate the locations where microfocused-EXAFS spectra were collected. In the tricolor maps red indicates the presence of Fe, green is Cu, and blue is Zn. (B) Microfocused EXAFS spectra from the numbered locations. The solid lines are the EXAFS spectra, and the dotted lines are the best fits obtained through linear combination fitting. *From Nachtegaal et al. (2005), with permission.*

The sorption phenomenon in soils has been discussed extensively in this chapter. This section aims to provide an introduction to understanding how collected empirical data can be used to understand specific adsorption complexes. These complexes can be incorporated into adsorption/desorption models as well as newly developed surface complexation models. Emphasis is given to synchrotron-based X-ray absorption spectroscopy, which helps identify sorption processes at the atomic and molecular scales (Figure 5.35).

With the advent of state-of-the-art analytical techniques, some of which are synchrotron based (see discussions that follow), one can elucidate reaction mechanisms on a small scale. This has been one of the major advances in the environmental sciences since the early 1990s. The use of small-scale techniques in environmental research has resulted in a new multidisciplinary field of

study that environmental soil chemists are actively involved in: molecular environmental science. Molecular environmental science can be defined as the study of the chemical and physical forms and distribution of contaminants in soils, sediments, waste materials, natural waters, and the atmosphere at the molecular level.

There are a number of areas in environmental soil chemistry where the application of molecular environmental science has resulted in major frontiers. These include the speciation of contaminants, which is essential for understanding release mechanisms, spatial resolution, chemical transformations, toxicity, bioavailability, and ultimate impacts on human health; mechanisms of microbial transformations, for example, bioremediation; phytoremediation; development of predictive models; effective remediation and waste management strategies; and risk assessment.

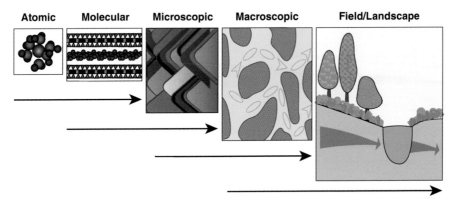

FIGURE 5.35 Illustration of the various spatial scales that environmental scientists are interested in. *From Bertsch and Hunter (1998), with permission.*

5.10.1 Electromagnetic Spectrum of Light

The use of intense light to understand the mechanisms of soil chemical reactions and processes has revolutionized the field of environmental soil chemistry or, more appropriately, molecular environmental soil chemistry.

The electromagnetic spectrum of light is shown in Figure 5.36. Electromagnetic radiation has both particle and wave properties such that light at a particular wavelength corresponds to a particular scale of detection (O'Day, 1999). For example, longer wave radiation detects larger objects while shorter wave radiation detects smaller objects. Light employed to see an object must have a wavelength similar to the object's size. Light has wavelengths longer or shorter than visible light. On the longer side are radio waves, microwaves, and infrared radiation. Shorter wavelength light includes ultraviolet, X-rays, and gamma rays. The shorter the wavelength, the greater the frequency and the more energetic or intense the light is. Light generated at shorter wavelengths such as X-rays is not visible to the human eye and must be detected via special means.

Each region of the spectrum is characterized by a range of wavelengths and photon energies that will determine the degree to which light will penetrate and interact with matter. At wavelengths from 10^{-7} to 10^{-10} m, one can explore the atomic structure of solids, molecules, and biological structures. Atoms, molecules, proteins, chemical bond lengths, and minimum distances between atomic planes in crystals fall within this wavelength range and can be detected. The binding energies of many electrons in atoms, molecules, and biological systems fall in the range of photon energies between 10 and 10,000 eV. When absorbed by an atom, a photon causes an electron to separate from the atom or can cause the release or emission of other photons. By detecting and analyzing such e^- or photon emissions, scientists can better understand the properties of a sample.

5.10.2 Synchrotron Radiation

Intense light can be produced at a synchrotron facility. Synchrotron radiation is produced over a wide range of energies from the infrared region with energies <1 eV to the hard X-ray region with energies of 100 keV or more. In the United States major facilities are found at National Laboratories. These include the National Synchrotron Light Source II (NSLS-II) at Brookhaven National Laboratory (Figure 5.37), the Advanced Photon Source (APS) at Argonne National

The Electromagnetic Spectrum

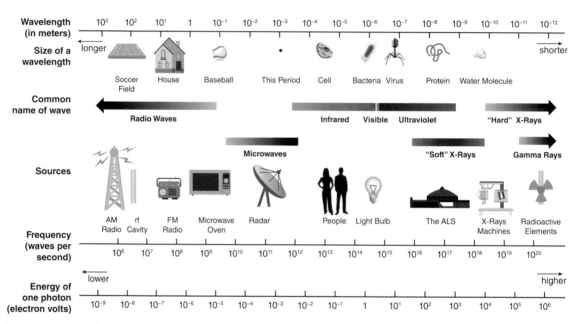

FIGURE 5.36   Electromagnetic spectrum of light covering a wide range of wavelengths and photon energies. *From Advanced Light Source, Lawrence Berkeley National Laboratory with permission.*

Laboratory, the Advanced Light Source (ALS) at Lawrence Berkeley National Laboratory, and the Stanford Synchrotron Radiation Laboratory (SSRL) at Stanford University.

Synchrotrons are large machines (Figure 5.38). The APS has a storage ring that is 1104 m in circumference (Figure 5.37), while the NSLS storage ring is 170 m in circumference. In the synchrotron charged particles, either e^- or positrons, are injected into a ring-shaped vacuum chamber maintained at an ultrahigh vacuum ($\sim 10^{-9}$ Torr). The particles enter the ring by way of an injection magnet and then travel around the ring at or near the speed of light, steered by bending

FIGURE 5.37   The National Synchrotron Light Source II, Brookhaven National Laboratory. *From Brookhaven National Laboratory (www.bnl.gov) with permission.*

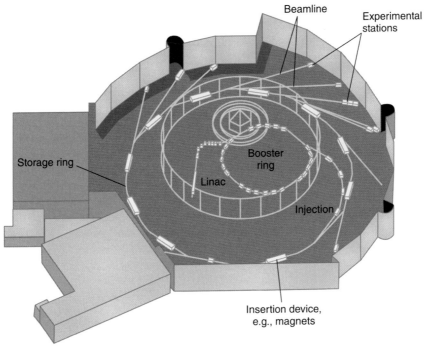

FIGURE 5.38 Schematic diagram of a synchrotron X-ray source. Please see the main text for component descriptions. *From Advanced Light Source, Lawrence Berkeley National Laboratory With permission.*

magnets. Additional magnets focus and shape the particle beam as it travels around the ring. Synchrotron radiation or light is emitted when the charged particles go through the bending magnets, or through insertion devices, which are additional magnetic devices called wigglers or undulators inserted into straight sections of the ring. Beamlines allow the X-rays to enter experimental stations, which are shielded rooms that contain instrumentation for conducting experiments (Schulze and Bertsch, 1999).

Synchrotron radiation has enabled soil and environmental scientists to employ a number of spectroscopic and microscopic analytical techniques to understand chemical reactions and processes at molecular and smaller scales. Spectroscopies (Table 5.8) reveal chemical information and deal with the interaction of electromagnetic radiation with matter. A large number of spectroscopic techniques are a

function of both large frequency or energy ranges of electromagnetic radiation involved and the approach used for probing the interaction over a given frequency range (Bertsch and Hunter, 1998). Microscopic techniques (Table 5.8) provide spatial information and arise from the interaction of energy with matter that either focuses or rasters radiation in some way to produce an image (O'Day, 1999).

5.10.3 X-Ray Absorption Spectroscopy

One of the most widely used synchrotron-based spectroscopic techniques is X-ray absorption spectroscopy (XAS). XAS can be used to study most elements in crystalline or noncrystalline solid, liquid, or gaseous states over a concentration range of a few milligrams per liter to the pure element. It is also an in situ technique, which means that one can study reactions in

TABLE 5.8 Summary of selected analytical methods for molecular environmental soil chemistry.

Analytical method	Source	Signal
Absorption, emission, and relaxation spectroscopies		
IR[a] and FTIR	Infrared radiation	Transmitted infrared radiation
Synchrotron XAS (XANES and EXAFS)	Synchrotron X-rays	Transmitted or fluorescent X-rays; electron yield
Synchrotron microanalysis (XRF, XANES)	Synchrotron X-rays	Fluorescent X-rays
EELS (also called PEELS)	Electrons	Electrons
XPS and Auger spectroscopy	X-rays	Electrons
Resonance spectroscopies		
NMR	Radio waves (+ magnetic field)	Radio waves
ESR (also called EPR)	Microwaves (+ magnetic field)	Microwaves
Scattering and ablation		
X-ray scattering (small angle, SAXS; wide angle, WAXS)	X-rays (synchrotron or laboratory)	Scattered X-rays
SIMS	Charged ion beam	Atomic mass
LA-ICP-MS	Laser	Atomic mass
Microscopies		
STM	Tunneling electrons	Electronic perturbations
AFM (also called SFM)	Electronic force	Force perturbation
HR-TEM and STEM	Electrons	Transmitted or secondary electrons
SEM with EDS or WDS chemical analysis	Electrons	Secondary or backscattered electrons; fluorescent X-rays

[a] Abbreviations are AFM, atomic force microscopy (also known as scanning force microscopy, SFM); EDS, energy dispersive spectrometry; EELS, electron energy loss spectroscopy; EM, electron microscopy; EPR, electron paramagnetic resonance (also known as ESR); ESR, electron spin resonance (also known as EPR); EXAFS, extended X-ray absorption fine structure; FTIR, Fourier transform infrared; HR-TEM, high-resolution transmission electron microscopy; IR, infrared; LA-ICP-MS, laser ablation inductively coupled plasma mass spectrometry; NMR, nuclear magnetic resonance; PEELS, parallel electron energy loss spectroscopy; SAXS, small-angle X-ray scattering; SEM, scanning electron microscopy; SIMS, secondary ion mass spectrometry; STEM, scanning transmission electron microscopy; STM, scanning tunneling microscopy; WAXS, wide-angle X-ray scattering; WDS, wavelength dispersive spectrometry; XANES, X-ray absorption near-edge structure; XAS, X-ray absorption spectroscopy; XPS, X-ray photoelectron spectroscopy; XRF, X-ray fluorescence. From O'Day (1999), with permission.

the presence of water. This is a major advantage over many molecular scale techniques, which are ex situ, often requiring drying of the sample material, placing it in an ultrahigh vacuum (UHV), heating the sample, or employing particle bombardment. Such conditions can alter the sample, creating artifacts, and do not simulate most natural soil conditions. XAS is an element-specific, bulk method that yields information about the local structural and compositional environment of an absorbing atom. It "sees" only the two or three closest shells of neighbors around an absorbing atom (<6 Å; note that Å will be used rather than nm to describe XAS analyses since Å is the standard unit used in the XAS scientific literature) due to the short electron mean free path in most substances. Using XAS, one can ascertain important soil chemical information such as the oxidation state, information on the next nearest neighbors, bond distances (accurate to ± 0.02 Å), and coordination numbers (accurate to $\pm 15\%-20\%$) (Brown et al., 1995).

An XAS experiment, which results in a spectrum (Figure 5.39), consists of exposing a sample to an incident monochromatic beam of synchrotron X-rays, which is scanned over a range of energies below and above the absorption edge (K, L, M) of the element of interest. When X-rays interact with matter, a number of processes can occur: X-ray scattering production of optical photons, production of photoelectrons and Auger electrons, production of fluorescence X-ray photons, and positron–electron pair production can all take place.

In the X-ray energy range of 0.5 to 100 keV photoelectron production dominates and causes X-ray attenuation by matter. When the energy of the incident X-ray beam (hv) is less than the binding energy (E_b) of a core electron on the element of interest, absorption is minimal. However, when hv $\approx E_b$, electron transitions to unoccupied bound energy levels arise, contributing to the main absorption edge and causing features below the main edge, referred to as the preedge portion of the spectrum (Figure 5.39). As hv increases beyond E_b, electrons can be ejected to unbound levels and stay in the vicinity of the absorber for a short time with excess kinetic energy. In the energy region extending from just above to about 50 eV above E_b, the absorption edge electrons are multiply scattered among neighboring atoms (Figure 5.40a), which produces the XANES (X-ray absorption near-edge structure) portion of the spectrum (Figure 5.40a). Fingerprint

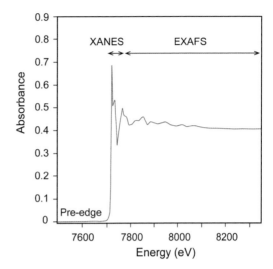

FIGURE 5.39 Co K-edge X-ray absorption spectrum of $CoCO_3$ recorded in the transmission mode showing the XANES and EXAFS regions. The preedge region is from a few eV to ≈ 10 eV below the main absorption edge and shows a small preedge feature due to a 1s → 3d bound-state electron transition. *From Xu (1993), with permission.*

information, such as oxidation states, can be gleaned from this portion of the XAS spectrum. When hv is about 50–1000 eV above E_b and the absorption edge, electrons are ejected from the absorber, singly or multiply scattered from first- or second-neighbor atoms back to the absorber, and then leave the vicinity of the absorber (Figure 5.40b), creating the EXAFS (extended X-ray absorption fine structure) portion (Figure 5.39) of the spectrum (Brown et al., 1995). The EXAFS spectrum is caused by interference between outgoing and backscattered photoelectrons, which modulates the atomic absorption coefficient (Figure 5.39). Analyses of the EXAFS spectrum provide information on bond distances, coordination number, and next nearest neighbors (Brown et al., 1995).

XAS experiments can be conducted in several modes that differ in the type of detected particle: transmission (X-rays transmitted through the sample), fluorescence (fluorescent X-rays

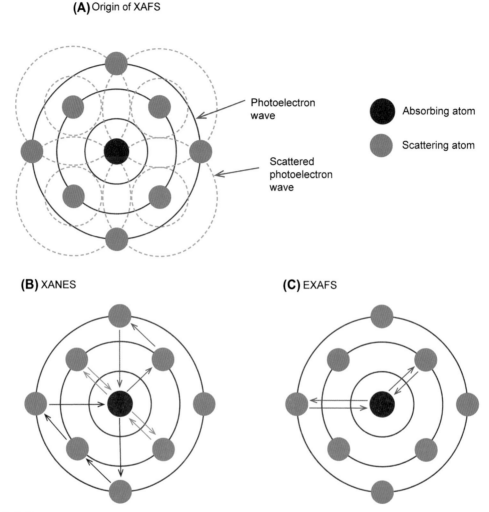

FIGURE 5.40 Schematic representation of the origin of X-ray absorption fine structure (XAFS) created by outgoing (*solid red lines*) and reflected (*dotted green lines*) photoelectron waves (*A*); multiple scattering of electrons between the absorbing atom and surrounding atoms leads to the X-ray absorption near-edge structure (XANES) (*B*), (*C*) and mostly single scattering of electrons leads to the extended X-ray absorption fine structure (EXAFS) part of the XAFS.

emitted due to the absorption of the incident X-ray beam), or electron-yield (emitted photons). In a transmission experiment the incident (I_0) and transmitted (I_1) X-ray intensities are recorded as a function of increasing incident X-ray energy (E) to yield an absorption spectrum, which is plotted as ln (I_0/I_1) versus E (in eV)

(Figure 5.39). The relationship between these intensities and the linear absorption coefficient μ (in cm^{-1}) of a sample of thickness x (in cm) is ln (I_0/I_1) = μx. In a fluorescence experiment the X-ray fluorescence from the sample, I_f, can be measured and ratioed with I_0 as I_f/I_0, which is proportional to μ for dilute samples.

Fluorescence methods are preferred for elements that may be contained in low concentrations on mineral surfaces (Brown et al., 1995).

Samples are loaded into a mylar-windowed sample holder made of low-Z materials (aluminum or Teflon) (Figure 5.41). The Lytle detector, which is a solid-angle, gas-filled ion chamber detector, is frequently used in fluorescence experiments. Figure 5.42 shows the experimental apparatus for fluorescence XAS measurements using a Lytle detector.

The XANES region of the spectrum, while not providing as much quantitative information as the EXAFS region, is often more intense and can provide qualitative or semiquantitative information on the oxidation state of the measured element (Brown et al., 1995). Such information can be obtained by comparing the features of the XANES spectrum of the sample

with features of XANES spectra for well-characterized reference compounds (Figure 5.43). Some species, such as Cr, yield remarkably different, easily recognizable XANES spectra (Figure 5.44). In Figure 5.43 it is easy to differentiate Cr(III) from Cr(VI) as there is a prominent preedge feature for Cr(VI) that is absent for Cr(III).

Analysis of an EXAFS spectrum is illustrated in Box 5.6 and involves extracting structural parameters, including interatomic distances (R); coordination numbers (CN); and identity of first, second, and more distant shells of neighbors around an absorber (see Brown et al., 1995 and O'Day et al., 1992 for additional details).

To derive accurate structural parameters, it is also necessary to obtain experimental EXAFS data for model or reference compounds that have known structures and contain the absorber

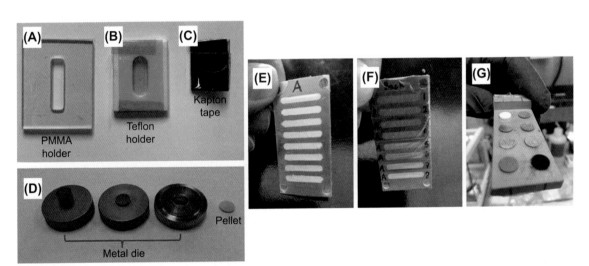

FIGURE 5.41 Common sample preparation methods for collecting X-ray absorption spectroscopic data. The plastic sample holders (panels A and B) can hold wet or dry samples and are made of polymethyl methacrylate (PMMA) or Teflon. Kapton tape (C) is useful for spreading thin powders over multiple layers of tape by folding the tape. Pellet presses and dies (D) are useful for making pellets from dry samples. The pellets can then be mounted to sample holders and typically are 7 mm in diameter. Panels E and F show an aluminum sample holder used at SSRL. Dry powders were mounted from behind with XRF film on the front, attached with double-sided tape. Samples are mounted individually and sealed from behind to avoid cross-contamination. Panel G shows pellets mounted on a sample holder for analysis at the APS. Proper sample preparation is of utmost importance for XAS data collection. *From Mo et al. (2021), with permission.*

(A)

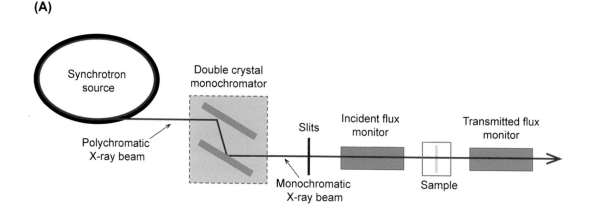

(B)

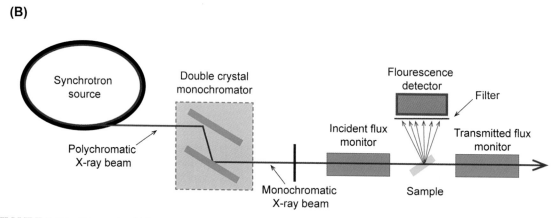

FIGURE 5.42   Schematic of (A) transmission and (B) fluorescence experimental setups to collect XAS data. In (A) the incident X-rays pass through the sample, and the flux is detected both before (I_0, incident flux) and after (I_t, transmitted flux) the sample. For transmission XAS, the raw absorption signal is plotted as the natural logarithm of the incident flux divided by the transmitted flux [$\mu x = \ln(I_0/I_t)$]. In (B) the sample is rotated 45° to the fluorescence detector. Fluorescence X-rays are emitted in all directions, and those that are emitted toward the detector are recorded. In general, a Z-1 filter is placed in front of the fluorescence detector to decrease the number of scattered X-rays, thus improving the signal-to-noise ratio. Fluorescence XAS data are plotted as $\mu x = I_f/I_0$. Often a reference metal foil is placed behind the transmitted flux monitor, and an additional flux monitor will record transmitted flux that passes through the metal foil. The data from the reference foil can be used as a calibration of X-ray energy. *From Fendorf et al. (1994), with permission.*

and nearest-neighbor backscatters of interest. More detail on XAS methodology, sample preparation, and data analyses can be found in a number of excellent sources (Brown et al., 1988, 1995; Brown, 1990; Fendorf and Sparks, 1996; Bertsch and Hunter, 1998; Fendorf, 1999; O'Day, 1999; Schulze and Bertsch, 1999; Kelly et al., 2008; Calvin, 2013).

5.10.4 Other Molecular-Scale Spectroscopic and Microscopic Techniques

As shown in Table 5.8, there are a number of in situ and ex situ analytical methods that are used in molecular environmental science. The principal invasive ex situ techniques used for

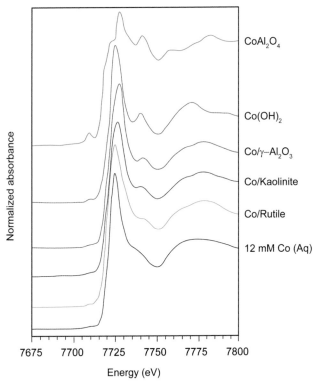

FIGURE 5.43 Co K-edge XANES spectra of CoAl$_2$O$_4$ spinel, crystalline Co(OH)$_2$, three samples with aqueous Co(II) sorbed on γ-Al$_2$O$_3$, kaolinite [Si$_2$Al$_2$O$_5$(OH)$_4$], and rutile (TiO$_2$), and a 12 mM aqueous Co(NO$_3$)$_2$ solution. *From Chisholm-Brause et al. (1990a), with permission.*

soil and aquatic systems are XPS, Auger electron spectroscopy (AES), and SIMS. Each of these techniques yields detailed information about the structure and bonding of minerals and the chemical species present on the mineral surfaces. XPS is the most widely used non–in situ surface-sensitive technique. It has been used to study the sorption mechanisms of inorganic cations and anions in soil and aquatic systems. Examples of in situ techniques are ESR, FTIR, NMR, XAS, and Mössbauer spectroscopies.

ESR spectroscopy is a technique for detecting paramagnetism. Electron paramagnetism occurs in all atoms, ions, organic free radicals, and molecules with an odd number of electrons. ESR is based upon the resonant absorption of microwaves by paramagnetic substances and describes the interaction between an electronic spin subjected to the influence of a crystal field and an external magnetic field (Calas, 1988). The method is applicable to transition metals of Fe^{3+}, Cu^{2+}, Mn^{2+}, V^{4+}, and Mo and has been used widely to study metal ion sorption on soil mineral components (McBride, 1982; McBride et al., 1984; Bleam and McBride, 1986) and soil organic matter (Senesi and Sposito, 1984; Senesi et al., 1985).

The application of IR spectroscopy to the study of soil chemical processes and reactions has a long history. The introduction of Fourier transform techniques has made a significant contribution to the development of new

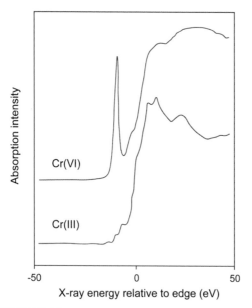

FIGURE 5.44   Comparison of XANES spectra for Cr(III) and Cr(VI). *From Fendorf et al. (1994), with permission.*

investigation techniques such as diffuse reflectance infrared Fourier transform (DRIFT) and attenuated total reflectance (ATR) spectroscopy. IR spectroscopy now extends far beyond classical chemical analysis and is applied successfully to study the sorption processes of inorganic and organic soil components. These techniques, and other vibrational spectroscopies, such as Raman, are the subject of numerous reviews (Hair, 1967; Bell, 1980; McMillan and Hofmeister, 1988; Johnston et al., 1993; Piccolo, 1994).

The use of NMR spectroscopy to study surfaces has a shorter history and fewer applications than vibrational spectroscopies. The primary reason is that the sensitivity of NMR is much lower than that of IR. Properties that might be exploited are the chemical shift, NMR relaxation times, and magnetic couplings to nuclei that are characteristic of a surface (Wilson, 1987). Most NMR studies in the field of soil science concentrate on the characterization of

soil organic matter and soil humification processes and therefore involve ^{1}H, ^{13}C, and ^{15}N NMR. Reviews on these and related topics are available (Wershaw and Mikita, 1987; Wilson, 1987; Johnston et al., 1993; Hatcher et al., 1994). Because it is virtually impossible to obtain any useful molecular information by observing the nucleus of a paramagnetic metal directly, studies of cation exchange have focused on diamagnetic metals, such as Cd^{2+}, which have a spin of $1/2$ and an acceptable natural abundance (e.g., 12% and 13%, respectively, for the two NMR-active isotopes, 113Cd and 111Cd). NMR is essentially a bulk spectroscopic technique. The advent of high-resolution, solid-state NMR techniques, such as magic angle sample spinning (MAS) and cross-polarization (CP), along with more sensitive, high-magnetic-field, user-friendly, pulsed NMR spectrometers has brought increased applications to heterogeneous aqueous systems (Johnston et al., 1993). In particular, 27Al and 29Si NMR in zeolites and other minerals have proven valuable for the structural elucidation of samples whose disorder has prevented diffraction techniques from being very useful (Altaner et al., 1988; Herrero et al., 1989; Woessner, 1989).

Spatial information on soil chemical processes can be gleaned from microscopic analyses. Scanning electron microscopy, TEM, and HRTEM are well-established methods for acquiring both chemical and micromorphological data on soils and soil materials. TEM can provide the spatial resolution of surface alterations and the amorphous nature or degree of crystallinity of sorbed species (ordering). It can also be combined with electron spectroscopies to determine elemental analysis.

From the very inception of the STM in 1981, it was apparent that this technique would revolutionize the study of mineral surfaces and surface-related phenomena. Indeed, by the end of the 1980s, applications of STM were

BOX 5.6

Analysis of EXAFS data

This box serves as a practical example for determining interatomic distances and coordination numbers from EXAFS data. Initially (Figure B5.6, Panel A), the raw data of β-nickel hydroxide (β-Ni(OH)$_2$) are plotted. The data were collected in fluorescence mode and are plotted as I$_f$/I$_0$, or intensity fluorescence X-rays divided by the intensity of incident X-rays. In the raw data the height of the edge step is proportional to the concentration of the element in the sample, until the sample is too concentrated and self-absorption (overabsorption) of X-rays occur. XAS data collected in florescence mode should not exceed approximately 0.5%–1% of the concentration of the element of interest to avoid self-absorption. Prior to collecting XAS data, careful planning is needed to consider whether fluorescence or transmission data will be acquired.

The next step (Panel B) is to normalize the data. Normalization sets the edge-jump (i.e., the position of the line before the "step" feature and the position of the line after the "step" feature) to be equal to one. That is, the preedge line is set to equal zero and the postedge line is set to be one. This effectively normalizes the data for differences in concentration between samples. Pre- and postedge normalization lines are one part of background subtraction. Often the sharp peak in the XANES region is termed the "white line."

In order to convert the normalized data to EXAFS data, a spline function is also used fit to the data. The spline should follow only the low-frequency oscillations in the EXAFS data. The low-frequency oscillations are not attributed to photoelectron scattering. Conversion of the normalized data in energy (electronvolts, eV) to EXAFS data in k-space (Å^{-1}) (Panel C) is k = [0.262(E-E$_0$)$^{0.5}$] or (E − E$_0$) ≈ 3.81k^2, where (E − E$_0$) is in units of electronvolts, k is in units of Å^{-1}, and χ(E) = (μ(E)−μ$_0$(E)/μ$_0$(E). For further detailed information about EXAFS data reduction, see Calvin (2013) and Kelly et al. (2008).

Next, the EXAFS region is amplified by weighting with either a power of 1, 2, or 3 (Panel C). Generally for transition metals or heavy elements, it is common to amplify by a power (k-weight) of 3, while for light elements, such as P, S, or K, a lower k-weight is often used. In Panel C a k-weight of 3 is used, and the data from 3 to 13 Å^{-1} are potted. Once the EXAFS data are plotted, a region (also termed a "window") over which to apply the Fourier transformation (FT) is chosen, generally 2.5 or 3 Å^{-1} to 12 Å^{-1}. Including noisy data in the FT will decrease the quality of the FT plot and the resulting fit of the atomic shells.

In Panel D the fits (*dashed red lines*) of the magnitude (*black line*) and real portion (*blue line*) of the FT are plotted. The FT data are fit using a structural model. The structural model should be similar to the hypothesized structure of the sample. In this case theophrastite was used. Many structural models can be found in databases (e.g., the American Mineralogist Crystal Structure Database). In general, the first shell (oxygen) is fit before the second shell. After reasonable fitting parameters are determined, then the second shell is fit. Reasonable fitting parameters need to be evaluated by the scientist performing the analysis and should be reported along with error values. See Calvin (2013) and Kelly et al. (2008) for a discussion of reasonable fitting values. In general, there are five values that need to be reported when fitting EXAFS data: coordination number (CN), interatomic distance (R), sigma square values (the disorder term), any changes in E$_0$, and the amplitude reduction factor (S$_0^2$). As can be seen in Panel D, there is another scattering shell located at about 5.8 Å that is not included in the fit. Typically, EXAFS is only sensitive up to 5 or 6 Å at most. Here, the multiple forward scattering paths are located at a similar distance in the hydroxide structure, leading to the formation of a significant peak at 5.8 Å. The contributions to the EXAFS signal for the first and second shells are shown in Panel E.

Continued

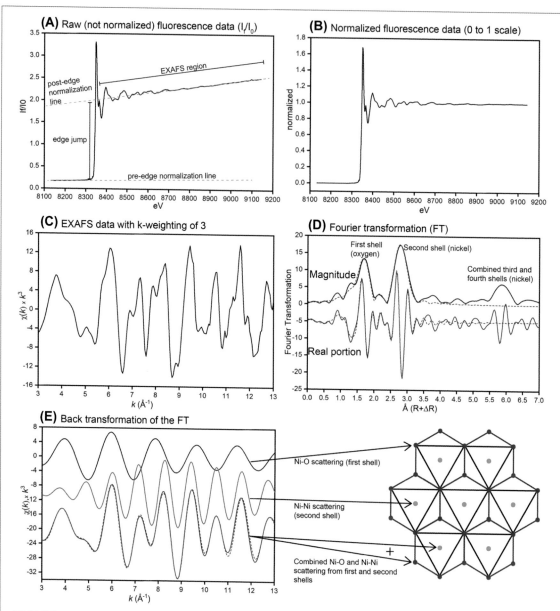

FIGURE B5.6   Steps involved in analysis of an EXAFS spectrum, using the Ni K-edge EXAFS data of β-Ni(OH)$_2$. Subtraction of the background (both pre- and postabsorption edge lines in blue) is necessary to convert into normalized data. The normalized data (B) are converted into (C) the EXAFS spectrum [χ(k) x k^3) vs. k (Å^{-1})]. In (D) the Fourier transformation (FT) of the EXAFS spectrum is shown (both magnitude and real portions). The FT is a radial structure function containing the peaks (i.e., shells) associated with different atoms at different distances from the central absorption Ni atoms. The back-FT (E) is a filtering technique to identify scattering paths that contribute to the EXAFS spectra. The Ni–O scattering can be noted with a more intense amplitude at a lower energy (k-space), as well as lower frequency. The Ni–Ni scattering has a higher amplitude at a higher energy as well as higher frequency. This is one reason why the EXAFS data are k-weighted: to amplify the scattering of elements at a higher energy, which correspond to the nearest neighboring atoms. Based on the fitting results of the FT (dashed red lines), the first peak in the FT is caused by the backscattering of the photoelectron from the six nearest neighboring oxygen atoms at 2.06 Å from the Ni atoms (Table B5.6). The second shell peak is caused by backscattering from the six second neighboring Ni atoms at 3.12 Å^{-1} (see O'Day (1992) for a discussion based on cobalt).

BOX 5.6   (cont'd)

Analysis of EXAFS data

The EXAFS signal is a summation of sine waves. When the sine wave is converted via the Fourier transformation (FT), a peak is the result. The reverse FT of the peak, e.g., the first shell in Panel D, can be viewed independently in Panel E. In general, oxygen in the first shell will contribute more significantly at low k regions. The scattering due to the heavier elements, e.g., Ni in the second shell, contributes more significantly at high k regions and with higher frequency (the *gray line*). When the *gray* and *black lines* are summed together, the *blue line* results, and the *blue line* is the EXAFS spectra shown in Panel C of just the first two shells.

The summative nature of the EXAFS spectra is evident in the *blue line*: when the *black line* moves down and the *gray line* moves up, a small shoulder feature is present in the *blue line* (e.g., 5 Å^{-1}). When both the *black* and *gray* lines have peaks at the same energy, the *blue line* displays a more intense peak (e.g., at 6 Å^{-1}). The oxygen peak at 7.8 Å^{-1} is coupled with the downward Ni peak at 7.7 Å^{-1}, leading to the summed spectra (*blue line*) having a valley at 7.7 Å^{-1}. The valley at 8.9 Å^{-1} in the *blue line* is especially low because both the O and Ni scattering interact via "constructive" interference, making the peak

lower. The EXAFS signal is a combination of both constructive and destructive interference due to the variety of photoelectron scattering waves all occurring simultaneously. Thus the EXAFS spectrum is the molecular "signature" of an element in a particular coordination environment.

TABLE B5.6   EXAFS shell-by-shell fitting results of β-Ni(OH)$_2$. The fitting range was carried out over 3-13 Å^{-1} in k-space and 1-3.4 Å in the Fourier transformation (R-space). The sigma square values and ±R (Å) are given in 10^{-3} units. Fitting was done with a chi of k weighting of 3. The amplitude reduction factor (S_0^2) was fixed at 0.9. The R-factor was 0.0078.

Path	CN	R (Å)	σ^2 (Å^2)	ΔE (eV)	±CN	±R (Å)	±σ^2 (Å^2)	±ΔE (eV)
Ni—O	5.94	2.063	3.5	−1.4	0.38	5.3	0.87	0.6
Ni—Ni	5.97	3.118	3.8	−1.4	0.65	4.5	7.5	0.6

Other detailed instructions for shell-by-shell fitting are also available here (https://www.youtube.com/watch?v=0Kex6KXeGG8). To follow along in this process, data in this Box can be found here (http://www1.udel.edu/soilchem/data.zip).

---

beginning to appear in the earth sciences literature (Hochella et al., 1989; Eggleston and Hochella, 1990).

However, the major event for the environmental science community came with the development of SFM, also known as AFM. SFM allows the imaging of mineral surfaces in air or immersed in solution, at the subnanometer scale

resolution (Maurice, 1996). SFM employs an atomic-sized tip positioned by a microcantilever and rastered over a surface (O'Day, 1999). Applications include determining the molecular-to-atomic scale structure of mineral surfaces (Johnsson et al., 1991); probing forces at the mineral—water interface (Ducker et al., 1992); visualizing sorption of hemimicelles and macromolecular organic substances such as humic and fulvic acid (Manne et al., 1994; Maurice, 1996); determining clay particle thicknesses and morphology of clay-sized particles (Hartman et al., 1990; Maurice et al., 1995); imaging soil bacteria (Grantham and Dove, 1996); and measuring directly the kinetics of growth, dissolution, heterogeneous nucleation, and redox processes (Dove and Hochella, 1993; Junta and Hochella, 1994; Fendorf et al., 1996). Excellent references on SFM methodology and applications can be found in a number of reviews (Hochella et al., 1998; Maurice, 1999).

Suggested Reading

Anderson, M.A., Rubin, A.J. (Eds.), 1981. Adsorption of Inorganics at Solid–Liquid Interfaces. Ann Arbor Science Publishers, Ann Arbor, MI.

Barrow, N.J., 1985. Reactions of anions and cations with variable-charged soils. Adv. Agron. 38, 183–230.

Bolt, G.H., De Boodt, M.F., Hayes, M.H.B., McBride, M.B. (Eds.), 1991. Interactions at the Soil Colloid–Soil Solution Interface. NATO ASI Ser. E, vol. 190. Kluwer Academic, Dordrecht, The Netherlands.

Goldberg, S., 1992. Use of surface complexation models in soil chemical systems. Adv. Agron. 47, 233–329.

Hingston, F.J., 1981. A review of anion adsorption. In: Anderson, M.A., Rubin, A.J. (Eds.), Adsorption of Inorganics at Solid–Liquid Interfaces", 51–90. Ann Arbor Science Publishers, Ann Arbor, MI.

Kinniburgh, D.G., Jackson, M.L., 1981. Cation adsorption by hydrous metal oxides and clay. In: Anderson, M.A., Rubin, A.J. (Eds.), Adsorption of Inorganics at Solid–Liquid Interfaces". Ann Arbor Science Publishers, Ann Arbor, MI, pp. 91–160.

Lewis-Russ, A., 1991. Measurement of surface charge of inorganic geologic materials: techniques and their consequences. Adv. Agron. 46, 199–243.

Lyklema, J., 1991. Fundamentals of Interface and Colloid Science, vol. 1. Academic Press, London.

Parfitt, R.L., 1978. Anion adsorption by soils and soil materials. Adv. Agron. 30, 1–50.

Scheidegger, A.M., Sparks, D.L., 1996. A critical assessment of sorption–desorption mechanisms at the soil mineral/water interface. Soil Sci. 161, 813–831.

Schindler, P.W., 1981. Surface complexes at oxide–water interfaces. In: Anderson, M.A., Rubin, A.J. (Eds.), Adsorption of Inorganics at Solid–Liquid Interfaces", 1–49. Ann Arbor Science Publishers, Ann Arbor, MI.

Schulthess, C.P., Sparks, D.L., 1991. Equilibrium-based modeling of chemical sorption on soils and soil constituents. Adv. Soil Sci. 16, 121–163.

Singh, U., Uehara, G., 1999. Electrochemistry of the double-layer: principles and applications to soils. In: Sparks, D.L. (Ed.), Soil Physical Chemistry", second ed. CRC Press, Boca Raton, FL, pp. 1–46.

Sparks, D.L. (Ed.), 1999. Soil Physical Chemistry, second ed. CRC Press, Boca Raton, FL.

ACS Symp. Ser. 715. In: Sparks, D.L., Grundl, T.J. (Eds.), 1998. Mineral–Water Interfacial Reactions. American Chemical Society, Washington, DC.

Sposito, G., 1984. The Surface Chemistry of Soils. Oxford University Press, New York.

Stumm, W. (Ed.), 1987. Aquatic Surface Chemistry. Wiley, New York.

Stumm, W., 1992. Chemistry of the Solid–Water Interface. Wiley, New York.

Stumm, W., Morgan, J.J., 1996. Aquatic Chemistry, third ed. Wiley, New York.

Theng, B.K.G. (Ed.), 1980. Soils with Variable Charge. New Zealand Society of Soil Science, Palmerston.

Uehara, G., Gillman, G., 1981. The Mineralogy, Chemistry, and Physics of Tropical Soils with Variable Charge Clays. Westview Press, Boulder, CO.

van Olphen, H., 1977. An Introduction to Clay Colloid Chemistry, Second edition. Wiley, New York.

Problem Set

Q1. Define: (i) adsorption, (ii) sorption, (iii) adsorbate, (iv) adsorbent, and (v) adsorptive.

Q2. List the physical and chemical forces involved in adsorption.

Q3. What is a surface functional group? What are major inorganic surface functional groups in soils, and where are they located?

Q4. What is the difference between a Lewis acid and a Lewis base? A water molecule attaches to an Fe atom at the goethite surface. Which is the Lewis acid and which is the Lewis base?

Q5. What is an amphoteric functional group?

Q6. Draw an example of an outer-sphere and inner-sphere surface complex. Draw examples of monodentate and bidentate inner-sphere surface complexes between metal(loid)s and mineral surfaces. Label atoms. What is the difference between mononuclear and binuclear surface complexes?

Q7. List five environmental factors that can affect the type of sorption complex formed during a sorption reaction.

Q8. What is an adsorption isotherm?

Q9. Your lab mate completes a sorption isotherm and finds their data to be similar to the L-type isotherm curve. They note the steep slope of the curve at low adsorptive concentrations and attribute this to mineral surface precipitation. Do you agree with their

which sorption complex formed and the mechanism by which it happened. How might the formation of the sorption complex influence metal mobility in natural systems?

Q11. Create a sorption isotherm. Use the following data to construct an isotherm, and determine the basic parameters based on the Freundlich and Langmuir equations. (i) Is there any justification for using one model over the other? (ii) Name several possible uses for the data from these isotherms. (iii) What are some of the experimental factors that one should consider when conducting an adsorption isotherm experiment?

The mass of soil in each reaction tube is 0.60 g, and the soil/solution ratio = 0.1 g mL^{-1}.

Initial Concentration of Pb (mg L^{-1})	75.4	100	148	245	347	498	707	1204	1306	1748
Final Concentration (mg L^{-1})		0.223	0.383	0.82	2.11	6.36	15.8	45	278	370	570

conclusion? Explain why by giving reasons for or against.

Q10. You performed a sorption experiment involving Ni and pyrophyllite. The initial experimental conditions are pH 7.5, [Ni] = 3 mM, ionic strength I = 0.1M NaNO$_3$. To characterize the molecular environment of the sorption complexes, you take the solid to a synchrotron facility and perform EXAFS spectroscopy. From the fitting of the data, you find that Ni, Si, and Al contributed to the second shell. Knowing that the amount of Ni added to the system was below the thermodynamic solubility product of Ni(OH)$_2$ (i.e., Ni won't precipitate out as Ni(OH)$_2$), speculate

Q12. What is the electrical double layer?

Q13. What is the difference between surface charge and surface potential?

Q14. At constant surface potential, would the double layer be thicker with monovalent or divalent ions? Why? At constant surface potential, would the double layer be thicker with increasing ion concentration? Why?

Q15. What is the thickness (in nanometers) of the diffuse double layer of 0.001 M NaCl?

Q16. What is the thickness (in nanometers) of the diffuse double layer of 0.001 M Al$_2$(SO$_4$)$_3$?

Q17. What is the thickness (in nanometers) of the diffuse double layer of 0.1 M NaCl?

References

Abendroth, R.P., 1970. Behavior of a pyrogenic silica in simple electrolytes. J. Colloid Interface Sci. 34 (4), 591–596.

Allada, R.K., Navrotsky, A., Berbeco, H.T., Casey, W.H., 2002. Thermochemistry and aqueous solubilities of hydrotalcite-like solids. Science 296 (5568), 721–723.

Allmann, R., 1970. Double layer structures with layer ions [Me(II)1-X Me(III)X (OH)$_2$]X+ of brucite type. Chimia 24 (3), 99–108.

Altaner, S.P., Weiss, C.A., Kirkpatrick, R.J., 1988. Evidence from ^{29}Si NMR for the structure of mixed-layer illite/smectite clay minerals. Nature 331 (6158), 699–702.

Altug, I., Hair, M.L., 1967. Cation exchange in porous glass. J. Phys. Chem. 71 (13), 4260–4263.

Atkins, P., Overton, T., Rourke, J., Weller, M., Armstrong, F., 2006. Shriver and Atkins' Inorganic Chemistry, Fourth edition. Oxford University Press, Oxford.

Axe, L., Bunker, G.B., Anderson, P.R., Tyson, T.A., 1998. An XAFS analysis of strontium at the hydrous ferric oxide surface. J. Colloid Interface Sci. 199 (1), 44–52.

Bargar, J.R., Brown, G.E., Parks, G.A., 1997. Surface complexation of Pb(II) at oxide-water interfaces. I. XAFS and bond-valence determination of mononuclear and polynuclear Pb(II) sorption products on aluminum oxides. Geochim. Cosmochim. Acta. 61 (13), 2617–2637.

Barrow, N.J., 1985. Reactions of anions and cations with variable-charged soils. In: Sparks, D.L. (Ed.), Advances in Agronomy, Vol 38. Academic Press, Cambridge, MA, pp. 183–230.

Barrow, N.J., Bowden, J.W., Posner, A.M., Quirk, J.P., 1980. An objective method for fitting models of ion adsorption on variable charge surfaces. Aust. J. Soil Res. 18 (1), 37–47.

Barrow, N.J., Bowden, J.W., Posner, A.M., Quirk, J.P., 1981. Describing the adsorption of copper, zinc and lead on a variable charge mineral surface. Aust. J. Soil Res. 19 (4), 309–321.

Bartell, F.E., Fu, Y., 1929. Adsorption from aqueous solutions by silica. J. Phys. Chem. 33, 676–687.

Bell, A.T., 1980. Applications of Fourier transform infrared spectroscopy to studies of adsorbed species. In: Bell, A.T., Hair, M.L. (Eds.), ASC Symposium Series No. 137. ASC Symposium Series, Washington, DC, pp. 13–35.

Belot, Y., Gailledreau, C., Rzekiecki, R., 1966. Retention of strontium-90, calcium-45, and barium-140 by aluminum oxide of large area. Health Phys. 12, 811–823.

Bertsch, P.M., Hunter, D.B., 1998. Elucidating fundamental mechanisms in soil and environmental chemistry: the role of advanced analytical, spectroscopic, and microscopic methods. In: Huang, P.M. (Ed.), Future Prospects for Soil Chemistry. SSSA Spec. Publ. No. 55. Soil Science Society of America, Madison, WI, pp. 103–122.

Bérubé, Y.G., de Bruyn, P.L., 1968. Adsorption at the rutile solution interface. I. Thermodynamic and experimental study. J. Colloid Interface Sci. 27, 305–318.

Bleam, W.F., Mcbride, M.B., 1986. The chemistry of adsorbed Cu(II) and Mn(II) in aqueous titanium dioxide suspensions. J. Colloid Interface Sci. 110 (2), 335–346.

Bochatay, L., Persson, P., Sjoberg, S., 2000. Metal ion coordination at the water-manganite (σ-MnOOH) interface I. An EXAFS study of cadmium(II). J. Colloid Interface Sci. 229 (2), 584–592.

Bolt, G.H. (Ed.), 1979. Soil Chemistry. B: Physico-Chemical Models. Elsevier, Amsterdam.

Bolt, G.H., van Riemsdijk, W.H., 1982. Ion adsorption on inorganic variable charge constituents. In: Bolt, G.H. (Ed.), Soil Chemistry. Part B. Physico-Chemical Methods. Elsevier, Amsterdam, pp. 459–503.

Bowden, J.W., Nagarajah, S., Barrow, N.J., Posner, A.M., Quirk, J.P., 1980. Describing the adsorption of phosphate, citrate and selenite on a variable-charge mineral surface. Aust. J. Soil Res. 18 (1), 49–60.

Bowden, J.W., Posner, A.M., Quirk, J.P., 1977. Ionic adsorption on variable charge mineral surfaces – theoretical-charge development and titration curves. Aust. J. Soil Res. 15 (2), 121–136.

Breeuwsma, A., Lyklema, J., 1971. Interfacial electrochemistry of hematite (α-Fe$_2$O$_3$). Disc. Faraday Soc. 522, 3224–3233.

Britz, D., Nancollas, G.H., 1969. Thermodynamics of cation exchange of hydrous zirconia. J. Inorg. Nucl. Chem. 31, 3861–3868.

Brown Jr., G.E., 1990. Spectroscopic studies of chemisorption reaction mechanisms at oxide-water interfaces. In: Hochella, M.F., White, A.F. (Eds.), Mineral-Water Interface Geochemistry. Mineralogical Society of America, Washington, DC, pp. 309–353.

Brown, G.E., Calas, G., Waychunas, G.A., Petiau, J., 1988. X-ray absorption spectroscopy; applications in mineralogy and geochemistry. Rev. Miner. Geochem. 18 (1), 431–512.

Brown Jr., G.E., Foster, A.L., Ostergren, J.D., 1999. Mineral surfaces and bioavailability of heavy metals: A molecular-scale perspective. Proc. Natl. Acad. Sci. U.S.A. 96 (7), 3388–3395.

Brown Jr., G.E., Parks, G.A., O'Day, P.A., 1995. Sorption at mineral-water interfaces: Macroscopic and microscopic perspectives. In: Vaughan, D.J., Pattrick, R.A.D. (Eds.), Mineral Surfaces. Chapman & Hall, London, pp. 129–183.

Bruninx, E., 1975. The coprecipitation of Zn, Cd and Hg with ferric hydroxide. Phillips Res. Rep. 30, 177–191.

Calas, G., 1988. Electron paramagnetic resonance. In: Hawthorne, F.C. (Ed.), Reviews in Mineralogy, Vol 18. Mineralogical Society of America, Washington, DC, pp. 513–563.

Calvin, S., 2013. XAFS for Everyone. CRC Press, New York, NY.

Chapman, D.L., 1913. A contribution to the theory of electrocapillarity. Phil. Mag. 25 (6), 475–481.

Charlet, L., Manceau, A., 1992. X-ray absorption spectroscopic study of the sorption of Cr(III) at the oxide-water interface. II: Adsorption, coprecipitation and surface precipitation on ferric hydrous oxides. J. Colloid Interface Sci. 148, 443–458.

Chiou, C.T., Freed, V.H., Schmedding, D.W., Kohnert, R.L., 1977. Partition coefficient and bioaccumulation of selected organic compounds. Environ. Sci. Technol. 11 (5), 475–478.

Chiou, C.T., Peters, L.J., Freed, V.H., 1979. A physical concept of soil-water equilibria for nonionic organic compounds. Science 206, 831–832.

Chiou, C.T., Porter, P.E., Schmedding, D.W., 1983. Partition equilibria of nonionic organic compounds between soil organic matter and water. Environ. Sci. Technol. 17 (4), 227–231.

Chisholm-Brause, C.J., O'day, P.A., Brown, G.E., Parks, G.A., 1990a. Evidence for multinuclear metal-ion complexes at solid water interfaces from X-ray absorption spectroscopy. Nature 348 (6301), 528–531.

Chisholm-Brause, C.J., Roe, A.L., Hayes, K.F., Brown, J.G.E., Parks, G.A., Leckie, J.O., 1990b. Spectroscopic investigation of Pb(II) complexes at the γ-Al$_2$O$_3$/water interface. Geochimica Et Cosmochimica Acta 54, 1897–1909.

Churms, S.C., 1966. The effect of pH on the ion-exchange properties of hydrated alumina. Part 1. Capacity and selectivity. J. S. Afr. Chem. Inst. 19, 98–107.

d'Espinose de la Caillerie, J.B., Kermarec, M., Clause, O., 1995. Impregnation of gamma-alumina with Ni(II) or Co(II) ions at neutral pH — hydrotalcite-type coprecipitate formation and characterization. J. Am. Chem. Soc. 117 (46), 11471–11481.

Davis, J.A., James, R.O., Leckie, J.O., 1978. Surface ionization and complexation at the oxide/water interface. I. Computation of electrical double layer properties in simple electrolytes. J. Colloid Interface Sci. 63, 480–499.

Davis, J.A., Leckie, J.O., 1978. Surface ionization and complexation at the oxide/water interface. II. Surface properties of amorphous iron oxyhydroxide and adsorption of metal ions. J. Colloid Interface Sci. 67, 90–107.

Davis, J.A., Leckie, J.O., 1980. Surface ionization and complexation at the oxide/water interface. III. Adsorption of anions. J. Colloid Interface Sci. 74, 32–43.

Donaldson, J.D., Fuller, M.J., 1968. Ion exchange properties of tin (IV) materials. I. Hydrous tin (IV) oxide and its cation exchange properties. J. Inorg. Nucl. Chem. 30, 1083–1092.

Dove, P.M., Hochella Jr., M.F., 1993. Calcite precipitation mechanisms and inhibition by orthophosphate — In situ observations by scanning force microscopy. Geochim. Cosmochim. Acta. 57 (3), 705–714.

Ducker, W.A., Senden, T.J., Pashley, R.M., 1992. Measurement of forces in liquids using a force microscope. Langmuir. 8, 1831–1836.

Dumont, F., Watillon, A., 1971. Stability of ferric oxide hydrosols. Discuss. Faraday Soc. 52, 352–360.

Dzombak, D.A., Morel, F.M.M., 1990. Surface Complexation Modeling. Hydrous Ferric Oxide. Wiley, New York, NY.

Eggleston, C.M., Hochella, M.F., 1990. Scanning tunneling microscopy of sulfide surfaces. Geochim. Cosmochim. Acta. 54, 1511–1517.

Elrashidi, M.A., O' Connor, G.A., 1982. Influence of solution composition on sorption of zinc. Soil Sci. Soc. Am. J. 46 (6), 1153–1158.

Elzinga, E.J., Sparks, D.L., 1999. Nickel sorption mechanisms in a pyrophyllite-montmorillonite mixture. J. Colloid Interface Sci. 213 (2), 506–512.

Farley, K.J., Dzombak, D.A., Morel, F.M.M., 1985. A surface precipitation model for the sorption of cations on metal oxides. J. Colloid Interface Sci. 106 (1), 226–242.

Fendorf, S., Eick, M.J., Grossl, P., Sparks, D.L., 1997. Arsenate and chromate retention mechanisms on goethite.1. Surface structure. Environ. Sci. Technol. 31 (2), 315–320.

Fendorf, S.E., 1992. Oxidation and Sorption Mechanisms of Hydrolyzable Metal Ions on Oxide Surfaces. Ph.D. Dissertation, University of Delaware, Newark, DE.

Fendorf, S.E., Lamble, G.M., Stapleton, M.G., Kelley, M.J., Sparks, D.L., 1994. Mechanisms of chromium (III) sorption on silica: 1. Cr(III) surface structure derived by extended x-ray absorption fine structure spectroscopy. Environ. Sci. Technol. 28 (2), 284–289.

Fendorf, S.E., Li, G.C., Gunter, M.E., 1996. Micromorphologies and stabilities of chromium(III) surface precipitates elucidated by scanning force microscopy. Soil Sci. Soc. Am. J. 60 (1), 99–106.

Fendorf, S.E., Sparks, D.L., 1994. Mechanisms of chromium (III) sorption on silica: 2. Effect of reaction conditions. Environ. Sci. Technol. 28 (2), 290–297.

Fendorf, S.E., Sparks, D.L., 1996. XAFS spectroscopy. In: Sparks, D.L. (Ed.), Methods of Soil Analysis: Part 3. Chemical Methods and Processes. SSSA, Madison, WI, pp. 377–416.

Fendorf, S.E., 1999. Fundamental aspects and applications of x-ray absorption spectroscopy in clay and soil science. In: Schulze, D.G., Stucki, J.W., Bertsch, P.M. (Eds.), Synchrotron X-ray Methods in Clay Science. The Clay Minerals Society, Boulder, CO, pp. 19–68.

Fendorf, S.E., Sparks, D.L., Lamble, G.M., Kelley, M.J., 1994. Applications of X-ray absorption fine structure spectroscopy to soils. Soil Sci. Soc. Am. J. 58 (6), 1583–1595.

Forbs, E.A., Posner, A.M., Quirk, J.P., 1976. The specific adsorption of divalent Cd, Co, Cu, Pb, and Zn on Goethite. J. Soil Sci. 27, 154–166.

Ford, R.G., Sparks, D.L., 2000. The nature of Zn precipitates formed in the presence of pyrophyllite. Environ. Sci. Technol. 34 (12), 2479–2483.

Fried, M., Shapiro, G., 1956. Phosphate supply pattern of various soils. Soil Sci. Soc. Am. Proc. 20, 471–475.

Gabano, J.P., Étienne, P., Laurent, J.F., 1965. Étude des proprietes de surface du bioxyde de manganese. Electrochim. Acta. 10, 947–963.

Gadde, R.R., Laitinen, H.A., 1974. Studies of heavy metal adsorption by hydrous iron and manganese oxides. Anal. Chem. 46, 2022–2026.

Goldberg, S., 1992. Use of surface complexation models in soil chemical systems. Adv. Agron. 47, 233–329.

Goldberg, S., 2013. Surface complexation modeling. In: Hillel, D. (Ed.), Reference Module in Earth Systems and Environmental Sciences. Elsevier, Amsterdam, pp. 1–14.

Goldberg, S., Criscenti, L.J., Turner, D.R., Davis, J.A., Cantrell, K.J., 2007. Adsorption–desorption processes in subsurface reactive transport modeling. Vadose Zone J. 6 (3), 407–435.

Gouy, G., 1910. Sur la constitution de la charge électrique à la surface d'un électrolyte. Ann. Phys (Paris), [IV] 9, 457–468.

Grahame, D.C., 1947. The electrical double layer and the theory of electrocapillarity. Chem. Rev. 41, 441–501.

Grantham, H.C., Dove, P.M., 1996. Investigation of bacterial-mineral interactions using fluid tapping mode atomic force microscopy. Geochim. Cosmochim. Acta. 60, 2473–2480.

Grimme, H., 1968. Die adsorption von Mn, Co, Cu and Zn durch goethit aus verdünnten lösungen. Z. Pflanzenernähr. Düng. Bodenkunde. 121, 58–65.

Hair, M.I., 1967. Infrared Spectroscopy in Surface Chemistry. Marcel Dekker, New York, NY.

Harris, D.C., 1987. Quantitative Chemical Analysis, Second edition. Freeman, New York, NY.

Harter, R.D., Smith, G., 1981. Langmuir equation and alternate methods of studying "adsorption" reactions in soils. In: Dowdy, R.H., Ryan, J.A., Volk, V.V., Baker, D.E. (Eds.), Chemistry in the Soil Environment. American Society of Agronomy/Soil Science Society of America, Madison, WI, pp. 167–182.

Hartman, H., Sposito, G., Yang, A., Manne, S., Gould, S.A.C., Hansma, P.K., 1990. Molecular-scale imaging of clay mineral surfaces with atomic force microscope. Clays Clay Miner. 38, 337–342.

Hatcher, P.G., Bortiatynski, J.M., Knicker, H., 1994. NMR techniques (C, H, and N) in soil chemistry, pp. 23–44 in 15th World Congress of Soil Science, Transactions. Vol 3a. Commission II, Acapulco, Mexico.

Hayes, K.F., Katz, L.E., 1996. Application of x-ray absorption spectroscopy for surface complexation modeling of metal ion sorption. In: Brady, P.V. (Ed.), Physics and Chemistry of Mineral Surfaces. CRC Press, Inc., Boca Raton, FL, pp. 147–223.

Hayes, K.F., Leckie, J.O., 1986. Mechanism of lead ion adsorption at the goethite-water interface. ACS Symp. Ser. 323, 114–141.

Hayes, K.F., Leckie, J.O., 1987. Modeling ionic strength effects on cation adsorption at hydrous oxide/solution interfaces. J. Colloid Interface Sci. 115, 564–572.

Hayes, K.F., Roe, A.L., Brown Jr., G.E., Hodgson, K.O., Leckie, J.O., Parks, G.A., 1987. In situ x-ray absorption study of surface complexes: selenium oxyanions on α-FeOOH. Science. 238 (4828), 783–786.

Helfferich, F., 1962. Ion Exchange. McGraw-Hill, New York, NY.

Herrero, C.P., Sanz, J., Serratosa, J.M., 1989. Dispersion of charge deficits in the tetrahedral sheet of phyllosilicates. Analysis from 29Si NMR spectra. J. Phys. Chem. 93, 4311–4315.

Hiemstra, T., van Riemsdijk, W.H., 1996. A surface structural approach to ion adsorption: the charge distribution (CD) model. J. Colloid Interface Sci. 179 (2), 488–508.

Hiemstra, T., van Riemsdijk, W.H., 2006. On the relationship between charge distribution, surface hydration, and the structure of the interface of metal hydroxides. J. Colloid Interface Sci. 301 (1), 1–18.

Hiemstra, T., van Riemsdijk, W.H., Bolt, G.H., 1989. Multisite proton adsorption modeling at the solid/solution interface of (hydr)oxides: a new approach: I. Model description and evaluation of intrinsic reaction constants. J. Colloid Interface Sci. 133 (1), 91–104.

Hiemstra, T., van Riemsdijk, W.H., Bruggenwert, M.G.M., 1987. Proton adsorption mechanism at the gibbsite and aluminum oxide solid/solution interface. Neth. J. Agric. Sci. 35, 281–293.

Hiemstra, T., Zhao, W., 2016. Reactivity of ferrihydrite and ferritin in relation to surface structure, size, and nanoparticle formation studied for phosphate and arsenate. Environ. Sci. Nano. 3 (6), 1265–1279.

Hingston, F.J., 1981. A review of anion adsorption. In: Anderson, M.A., Rubin, A.J. (Eds.), Adsorption of Inorganics at Solid-Liquid Interfaces. Ann Arbor Science, Ann Arbor, MI, pp. 51–90.

Hingston, F.J., Posner, A.M., Quirk, J.P., 1972. Anion adsorption by goethite and gibbsite. I. The role of the proton in determining adsorption envelopes. J. Soil Sci. 23, 177–192.

Hochella, M.F., Eggleston, C.M., Elings, V.B., Parks, G.A., Brown, G.E., Kjoller, K.K., Wu, C.M., 1989. Mineralogy in two dimensions: scanning tunneling microscopy of semiconducting minerals with implications for geochemical reactivity. Am. Mineral. 74, 1233–1246.

Hochella, M.F., Jr, Rakovan, J.F., Rosso, K.M., Bickmore, B.R. and Rufe, E. (1998). New directions in mineral surface geochemical research using scanning probe microscopies. In: Sparks, D.L., Grundl, T.J. (Eds.), Mineral-Water Interfacial Reactions. ACS Symp. Ser., 715, pp. 37–56. Am. Chem Soc., Washington, D.C.

Hohl, H., Sigg, L., and Stumm, W. (1980). Characterization of Surface Chemical Properties of Oxides in Natural Waters: The Role of Specific Adsorption in Determining the Surface Charge. In: Particulates in Water, Vol. 189, pp. 1–31. American Chemical Society.

Hohl, H., Stumm, W., 1976. Interaction of Pb^{2+} with hydrous γ-Al_2O_3. J. Colloid Interface Sci. 55, 281–288.

Huang, C.P., Stumm, W., 1973. Specific adsorption of cations on hydrous γ-Al_2O_3. J. Colloid Interface Sci. 43, 409–420.

Hug, S.J., 1997. In situ Fourier transform infrared measurements of sulfate adsorption on hematite in aqueous solutions. J. Colloid Interface Sci. 188 (2), 415–422.

Hiemstra, T., Yong, H., Van Riemsdijk, W.H., 1999. Interfacial charging phenomena of aluminum (hydr)oxides. Langmuir. 15 (18), 5942–5955.

Johnsson, P.A., Eggleston, C.M., Hochella, M.F., 1991. Imaging molecular-scale structure and microtopography of hematite with atomic force microscope. Am. Mineral. 76, 1442–1445.

Johnston, C.T., Sposito, G., Earl, W.L., 1993. Surface spectroscopy of environmental particles by Fourier-transform infrared and nuclear magnetic resonance spectroscopy. In: Buffle, J., Leeuwen, H.P.v. (Eds.), Environmental Particles. Lewis Publ., Boca Raton, FL, pp. 1–36.

Junta, J.L., Hochella, M.F., 1994. Manganese (II) oxidation at mineral surfaces: a microscopic and spectroscopic study. Geochim. Cosmochim. Acta. 58, 4985–4999.

Katz, L.E., Boyle-Wight, E.J., 2001. Application of spectroscopic methods to sorption model parameter estimation. In: Selim, H.M., Sparks, D.L. (Eds.), Physical and Chemical Processes of Water and Solute Transport/Retention in Soils. SSSA Spec. Publ. 56. Soil Science Society of America, Madison, WI, pp. 213–256.

Kelly, S.D., Hesterberg, D., Ravel, B., 2008. Analysis of soils and minerals using X-ray absorption spectroscopy. In: Ulery, A.L., Drees, L.R. (Eds.), Methods of Soil Analysis, Part 5, Mineralogical Methods. Soil Science Society of America, Madison, WI, pp. 387–463.

Kinniburgh, D.G., Jackson, M.L., 1981. Cation adsorption by hydrous metal oxides and clay. In: Anderson, M.A., Rubin, A.J. (Eds.), Adsorption of Inorganics at Solid-Liquid Interfaces. Ann Arbor Science Publishers, Ann Arbor, MI, pp. 91–160.

Kinniburgh, D.G., Jackson, M.L., Syers, J.K., 1976. Adsorption of alkaline earth, transition, and heavy metal cations by hydrous oxide gels of iron and aluminum. Soil Sci. Soc. Am. J. 40, 796–799.

Kozawa, A., 1959. On an ion exchange property. J. Electrochem. Soc. 106, 552–556.

Kurbatov, J.D., Kulp, J.L., Mack, E., 1945. Adsorption of strontium and barium ions and their exchange on hydrous ferric oxide. J. Am. Chem. Soc. 67, 1923–1929.

Langmuir, I., 1918. The adsorption of gases on plane surfaces of glass, mica, and platinum. J. Am. Chem. Soc. 40, 1361–1382.

Li, W., Livi, K.J.T., Xu, W.Q., Siebecker, M.G., Wang, Y.J., Phillips, B.L., Sparks, D.L., 2012. Formation of crystalline Zn-Al layered double hydroxide precipitates on γ-alumina: the role of mineral dissolution. Environ. Sci. Technol. 46 (21), 11670–11677.

Livi, K.J.T., Senesi, G.S., Scheinost, A.C., Sparks, D.L., 2009. Microscopic examination of nanosized mixed Ni-Al hydroxide surface precipitates on pyrophyllite. Environ. Sci. Technol. 43 (5), 1299–1304.

Loganthan, P., Burau, R.G., 1973. Sorption of heavy metal ions by a hydrous manganese oxide. Geochim. Cosmochim. Acta. 37, 1277–1293.

Malati, M.A., Estefan, S.F., 1966. The role of hydration in the adsorption of alkaline earth ions onto quartz. J. Colloid Interface Sci. 24, 306–307.

Manceau, A., Lanson, B., Schlegel, M.L., Harge, J.C., Musso, M., Eybert-Berard, L., Hazemann, J.-L., Chateigner, D., Lamble, G.M., 2000. Quantitative Zn speciation in smelter-contaminated soils by EXAFS spectroscopy. Am. J. Sci. 300 (4), 289–343.

Manceau, A., Marcus, M.A., Tamura, N., 2002. Quantitative speciation of heavy metals in soils and sediments by synchrotron X-ray techniques. In: Fenter, P., Sturchio, N.C. (Eds.), Applications of Synchrotron Radiation in Low-temperature Geochemistry and Environmental Science, Vol 49. Mineralogical Society of America, Washington, DC, pp. 341–428.

Manne, S., Cleveland, J.P., Gaub, H.E., Stucky, G.D., Hansma, P.K., 1994. Direct visualization of surfactant hemimicelles by force microscopy of the electrical double layer. Langmuir. 10, 4409–4413.

Maurice, P.A., 1996. Scanning probe microscopy of mineral surfaces. In: Huang, P.M., Senesi, N., Buffle, J. (Eds.), Environmental Particles, Structure and Surface Reactions of Soil Particles, Vol 4. Wiley, New York, NY, pp. 109–153.

Maurice, P.A., 1999. Atomic force microscopy as a tool for studying the reactivities of environmental particles. In: Sparks, D.L., Grundl, T.J. (Eds.), Mineral-Water Interfacial Reactions. ACS Symposium Series Vol. 715. American Chemical Society, Washington, DC, pp. 57–66.

Maurice Jr., P.A., Hochella, M.F., Parks, G.A., Sposito, G., Schwertmann, U., 1995. Evolution of hematite surface microtopography upon dissolution by simple organic acids. Clays Clay Miner 43 (1), 29—38.

McBride, M.B., 1982. Cu^{2+} adsorption characteristics of aluminum hydroxide and oxyhydroxides. Clays Clay Miner. 30, 21—28.

McBride, M.B., 1991. Processes of heavy and transition metal sorption by soil minerals. In: Bolt, G.H., Boodt, M.F.D., Hayes, M.H.B., McBride, M.B. (Eds.), Interactions at the Soil Colloid-Soil Solution Interface, Vol 190. Kluwer Academic Publishers, Dordrecht, pp. 149—176.

McBride, M.B., 1994. Environmental Chemistry of Soils. Oxford University Press, New York, NY.

McBride, M.B., Fraser, A.R., McHardy, W.J., 1984. Cu^{2+} interaction with microcrystalline gibbsite. Evidence for oriented chemisorbed copper ions. Clays Clay Miner. 32, 12—18.

McKenzie, R.M., 1972. The sorption of some heavy metals by the lower oxides of manganese. Geoderma. 8, 29—35.

McKenzie, R.M., 1980. The adsorption of lead and other heavy metals on oxides of manganese and iron. Aust. J. Soil Res. 18, 61—73.

McMillan, P.F., Hofmeister, A.M., 1988. Infrared and Raman spectroscopy. In: Hawthorne, F.C. (Ed.), Spectroscopic Methods in Mineralogy and Geology. Reviews in Mineralogy Vol. 18. Mineralogical Society of America, Washington DC, pp. 573—630.

Mendez, J.C., Hiemstra, T., 2019. Carbonate adsorption to ferrihydrite: competitive interaction with phosphate for use in soil systems. ACS Earth Space Chem. 3 (1), 129—141.

Mo, X., Siebecker, M.G., Gou, W., Li, L., Li, W., 2021. A review of cadmium sorption mechanisms on soil mineral surfaces revealed from synchrotron-based X-ray absorption fine structure spectroscopy: implications for soil remediation. Pedosphere. 31 (1), 11—27.

Morel, F.M.M., Hering, J.G., 1993. Principles and Applications of Aquatic Chemistry. John Wiley and Sons Inc., New York, NY.

Murray, J.W., 1975. The interaction of cobalt with hydrous manganese dioxide. Geochim. Cosmochim. Acta. 39, 635—647.

Murray, D.J., Healy, T.W., Fuerstenau, D.W., 1968. The adsorption of aqueous metals on colloidal hydrous manganese oxide. In: Walter, J., Weber, Jr., Egon, Matijevic (Eds.), Adsorption from Aqueous Solution. Adv. in Chem. Series 79. American Chemical Society, Washington, DC, pp. 74—81.

Nachtegaal, M., Marcus, M.A., Sonke, J.E., Vangronsveld, J., Livi, K.J.T., Van der Lelie, D., Sparks, D.L., 2005. Effects of in situ remediation on the speciation and bioavailability of zinc in a smelter contaminated soil. Geochim. Cosmochim. Acta. 69 (19), 4649—4664.

O'Day, P.A., 1992. Structure, Bonding, and Site Preference of Cobalt(II) Sorption Complexes on Kaolinite and Quartz From Solution and Spectroscopic Studies. Ph.D. Dissertation, Stanford University, Stanford, CA, p. 208.

O'Day, P.A., 1999. Molecular environmental geochemistry. Rev. Geophys. 37 (2), 249—274.

O'Day, P.A., Brown, G.E., Parks, G.A., 1994c. X-ray absorption spectroscopy of cobalt(II) multinuclear surface complexes and surface precipitates on kaolinite. J. Colloid Interface Sci. 165 (2), 269—289.

O'Day, P.A., Chisholm-Brause, C.J., Towle, S.N., Parks, G.A., Brown, G.E., 1996. X-ray absorption spectroscopy of Co(II) sorption complexes on quartz (α-SiO_2) and rutile (TiO_2). Geochim. Cosmochim. Acta. 60, 2515—2532.

O'Day, P.A., Parks, G.A., Brown, J.G.E., 1994a. Molecular structure and binding sites of cobalt(II) surface complexes on kaolinite from X-ray absorption spectroscopy. Clays Clay Miner 42, 337—355.

O'Day, P.A., Rehr, J.J., Zabinsky, S.I., Brown, G.E., 1994b. Extended X-ray absorption fine structure (EXAFS) analysis of disorder and multiple-scattering in complex crystalline solids. J. Am. Chem. Soc. 116, 2938—2949.

Olsen, S.R., Watanabe, F.S., 1957. A method to determine a phosphorus adsorption maximum of soils as measured by the Langmuir isotherm. Soil Sci. Soc. Am. Proc. 21, 144—149.

Overbeek, J.T., 1952. Electrochemistry of the double-layer. Colloid Sci. 1, 115—193.

Papelis, C., Hayes, K.F., 1996. Distinguishing between interlayer and external sorption sites of clay minerals using X-ray absorption spectroscopy. Colloids Surf. A Physicochem. Eng. Asp. 107, 89—96.

Peltier, E., Allada, R., Navrotsky, A., Sparks, D.L., 2006. Nickel solubility and precipitation in soils: A thermodynamic study. Clays Clay Miner. 54 (2), 153—164.

Piccolo, A., 1994. Advanced infrared techniques (FT-IR, DRIFT, and ATR) applied to organic and inorganic soil materials, Vol 3a. Paper presented at the 15th World Congress of Soil Science, Acapulco, Mexico, pp. 3—22. Transactions. Commission II.

Ridley, M.K., Hiemstra, T., van Riemsdijk, W.H., Machesky, M.L., 2009. Inner-sphere complexation of cations at the rutile—water interface: a concise surface structural interpretation with the CD and MUSIC model. Geochim. Cosmochim. Acta. 73 (7), 1841—1856.

Rietra, R.P.J.J., Hiemstra, T., van Riemsdijk, W.H., 1999. The relationship between molecular structure and ion adsorption on variable charge minerals. Geochim. Cosmochim. Acta. 63 (19), 3009—3015.

Roberts, D.R., 2001. Speciation and Sorption Mechanisms of Metals in Soils Using Bulk and Micro-Focused Spectroscopic and Microscopic Techniques. Ph.D. Dissertation, University of Delaware, Newark, DE, p. 171.

Roberts, D.R., Scheidegger, A.M., Sparks, D.L., 1999. Kinetics of mixed Ni-Al precipitate formation on a soil clay fraction. Environ. Sci. Technol. 33 (21), 3749–3754.

Roberts, D.R., Scheinost, A.C., Sparks, D.L., 2002. Zinc speciation in a smelter-contaminated soil profile using bulk and microspectroscopic techniques. Environ. Sci. Technol. 36 (8), 1742–1750.

Roe, A.L., Hayes, K.F., Chisholm-Brause, C., Brown, G.E., Parks, G.A., Hodgson, K.O., Leckie, J.O., 1991. In situ X-ray absorption study of lead ion surface complexes at the goethite-water interface. Langmuir. 7 (2), 367–373.

Scheckel, K.G., Sparks, D.L., 2001. Dissolution kinetics of nickel surface precipitates on clay mineral and oxide surfaces. Soil Sci. Soc. Am. J. 65 (3), 685–694.

Scheidegger, A.M., Lamble, G.M., Sparks, D.L., 1996. Investigation of Ni sorption on pyrophyllite: An XAFS study. Environ. Sci. Technol. 30 (2), 548–554.

Scheidegger, A.M., Lamble, G.M., Sparks, D.L., 1997. Spectroscopic evidence for the formation of mixed-cation hydroxide phases upon metal sorption on clays and aluminum oxides. J. Colloid Interface Sci. 186 (1), 118–128.

Scheidegger, A.M., Sparks, D.L., 1996. Kinetics of the formation and the dissolution of nickel surface precipitates on pyrophyllite. Chem. Geol. 132 (1–4), 157–164.

Scheidegger, A.M., Strawn, D.G., Lamble, G.M., Sparks, D.L., 1998. The kinetics of mixed Ni-Al hydroxide formation on clay and aluminum oxide minerals: a time-resolved XAFS study. Geochim. Cosmochim. Acta. 62 (13), 2233–2245.

Scheinost, A.C., Ford, R.G., Sparks, D.L., 1999. The role of Al in the formation of secondary Ni precipitates on pyrophyllite, gibbsite, talc, and amorphous silica: a DRS study. Geochim. Cosmochim. Acta. 63 (19–20), 3193–3203.

Schindler, P.W., 1981. Surface complexes at oxide-water interfaces. In: Anderson, M.A., Rubin, A.J. (Eds.), Adsorption of Inorganics at Solid-Liquid Interfaces. Ann Arbor Science Publishers, Ann Arbor, MI, pp. 1–49.

Schindler, P.W., Fürst, B., Dick, B., Wolf, P.U., 1976. Ligand properties of surface silanol groups. I. Surface complex formation with Fe^{3+}, Cu^{2+}, Cd^{2+}, and Pb^{2+}. J. Colloid Interface Sci. 55, 469–475.

Schindler, P.W., Gamsjager, H., 1972. Acid-base reactions of the TiO_2 (anatase)-water interface and the point of zero charge of TiO_2 suspensions. Kolloid Z. Z. Polym. 250, 759–763.

Schofield, R.K., 1947. Calculation of surface areas from measurements of negative adsorption. Nature. 160, 408–412.

Schulze, D.G., Bertsch, P.M., 1999. Overview of synchrotron X-ray sources and synchrotron X-rays. In: Schulze, D.G., Stucki, J.W., Bertsch, P.M. (Eds.), Synchrotron X-Ray Methods in Clay Science, Vol 9. Clay Minerals Society, Boulder, CO, pp. 1–18.

Senesi, N., Bocian, D.F., Sposito, G., 1985. Electron spin resonance investigation of copper(II) complexation by soil fulvic acid. Soil Sci. Soc. Am. J. 49, 114–119.

Senesi, N., Sposito, G., 1984. Residual copper(II) complexes in purified soil and sewage sludge fulvic acids: An electron spin resonance study. Soil Sci. Soc. Am. J. 48, 1247–1253.

Shuman, L.M., 1975. The effect of soil properties on zinc adsorption by soils. Soil Sci. Soc. Am. Proc. 39, 455–458.

Siebecker, M., Li, W., Khalid, S., Sparks, D., 2014. Real-time QEXAFS spectroscopy measures rapid precipitate formation at the mineral–water interface. Nat. Commun. 5, 6003.

Siebecker, M.G., Li, W., Sparks, D.L., 2018. Chapter One – The important role of layered double hydroxides in soil chemical processes and remediation: What we have learned over the past 20 years. In: Sparks, D.L. (Ed.), Advances in Agronomy, Vol 147. Academic Press, Cambridge, MA, pp. 1–59.

Singh, U., Uehara, G., 1986. Electrochemistry of the double-layer: principles and applications to soils. In: Sparks, D.L. (Ed.), Soil Physical Chemistry. CRC Press, Boca Raton, FL, pp. 1–38.

Sparks, D.L., 1989. Kinetics of Soil Chemical Processes. Academic Press, San Diego, CA.

Sposito, G., 1984. The Surface Chemistry of Soils. Oxford University Press, New York, NY.

Sposito, G., 1986. Distinguishing adsorption from surface precipitation. In: Geochemical Processes at Mineral Surfaces. ACS Symposium Series 323. In: Davis J.A., Hayes, K.F., (Eds.), American Chemical Society, Washington, DC, pp. 217–229.

Sposito, G., 1989. The Chemistry of Soils. Oxford University Press, New York, NY.

SSSA, 2008. Glossary of Soil Science Terms. Soil Science Society of America, Madison, WI.

Stern, O., 1924. Zur Theorie der Elektrolytischen Doppelschicht. Z. Elektrochem. 30, 508–516.

Strawn, D.G., Sparks, D.L., 1999. The use of XAFS to distinguish between inner- and outer-sphere lead adsorption complexes on montmorillonite. J. Colloid Interface Sci. 216 (2), 257–269.

Stumm, W. (Ed.), 1987. Aquatic Surface Chemistry. Wiley (Interscience), New York, NY.

Stumm, W., 1992. Chemistry of the Solid-Water Interface. Wiley, New York, NY.

Stumm, W., Kummert, K., Sigg, L., 1980. A ligand exchange model for the adsorption of inorganic and organic ligands at hydrous oxide interfaces. Croat. Chem. Acta. 53, 291–312.

Stumm, W., Morgan, J.J., 1981. Aquatic Chemistry, Second edition. Wiley, New York, NY.

Su, C.M., Suarez, D.L., 1995. Coordination of adsorbed boron — A FTIR spectroscopic study. Environ. Sci. Technol. 29 (2), 302—311.

Su, C., Suarez, D.L., 1997. In situ infrared speciation of adsorbed carbonate on aluminum and iron oxides. Clays Clay Miner. 45 (6), 814—825.

Sun, X.H., Doner, H.E., 1996. An investigation of arsenate and arsenite bonding structures on goethite by FTIR. Soil Sci. 161 (12), 865—872.

Tadros, T.F., Lyklema, J., 1969. The electrical double layer on silica in the presence of bivalent counter-ions. J. Electroanal. Chem. 22, 1.

Taniguechi, K., Nakajima, M., Yoshida, S., Tarama, K., 1970. The exchange of the surface protons in silica gel with some kinds of metal ions. Nippon Kagaku Zasshi 91, 525—529.

Taylor, R.M., 1984. The rapid formation of crystalline double hydroxy salts and other compounds by controlled hydrolysis. Clay Miner. 19 (4), 591—603.

Tejedor-Tejedor, M.I., Anderson, M.A., 1990. Protonation of phosphate on the surface of goethite as studied by CIR-FTIR and electrophoretic mobility. Langmuir. 6 (3), 602—611.

Thompson, H.A., Parks, G.A., Brown, G.E., 1999a. Ambient-temperature synthesis, evolution, and characterization of cobalt-aluminum hydrotalcite-like solids. Clays Clay Miner. 47 (4), 425—438.

Thompson, H.A., Parks, G.A., Brown, G.E., 1999b. Dynamic interactions of dissolution, surface adsorption, and precipitation in an aging cobalt(II)-clay-water system. Geochimica Et Cosmochimica Acta 63 (11—12), 1767—1779.

Tiberg, C., Sjöstedt, C., Persson, I., Gustafsson, J.P., 2013. Phosphate effects on copper(II) and lead(II) sorption to ferrihydrite. Geochim. Cosmochim. Acta 120, 140—157.

Tien, H.T., 1965. Interaction of alkali metal cations with silica gels. J. Phys. Chem. 69, 350—352.

Towle, S.N., Bargar, J.R., Brown, G.E., Parks, G.A., 1997. Surface precipitation of Co(II)(aq) on Al_2O_3. J. Colloid Interface Sci. 187 (1), 62—82.

Trainor, T.P., Brown, G.E., Parks, G.A., 2000. Adsorption and precipitation of aqueous Zn(II) on alumina powders. J. Colloid Interface Sci. 231 (2), 359—372.

Uehara, G., Gillman, G., 1981. The Mineralogy, Chemistry, and Physics of Tropical Soils With Variable Charge Clays. Westview Press, Boulder, CO.

van Bemmelen, J.M., 1888. Die Absorptionsverbindungen und das Absorptionsvermögen der Ackererde. Landw. Vers. Stat. 35, 69—136.

van Olphen, H., 1977. An Introduction to Clay Colloid Chemistry, Second edition. Wiley, New York, NY.

van Raji, B., Peech, M., 1972. Electrochemical properties of some Oxisols and Alfisols of the tropics. Soil Sci. Soc. Am. Proc. 36, 587—593.

van Riemsdijk, W.H., Bolt, G.H., Koopal, L.K., Blaakmeer, J., 1986. Electrolyte adsorption on heterogeneous surfaces: adsorption models. J. Colloid Interface Sci. 109, 219—228.

van Riemsdijk, W.H., Wit, J.C.M.d., Koopal, L.K., Bolt, G.H., 1987. Metal ion adsorption on heterogeneous surfaces: adsorption models. J. Colloid Interface Sci. 116, 511—522.

van Riemsdijk, W.H., Zee, S.E.A.T.M.v.d., 1991. Comparison of models for adsorption, solid solution and surface precipitation. In: Bolt, G.H., Boodt, M.F.D., Hayes, M.H.B., Bride, M.B. (Eds.), Interactions at the Soil Colloid-Soil Solution Interface. Kluwer Academic Publishers, Dordrecht, pp. 241—256.

Veith, J.A., Sposito, G., 1977. On the use of the Langmuir equation in the interpretation of "adsorption" phenomena. Soil Sci. Soc. Am. J. 41, 497—502.

Venema, P., Hiemstra, T., van Riemsduk, W.H., 1996. Comparison of different site binding models for cation sorption: description of pH dependency, salt dependency, and cation—proton exchange. J. Colloid Interface Sci. 181 (1), 45—59.

Venkataramani, B., Venkateswarlu, K.S., Shankar, J., 1978. Sorption properties of oxides. III. Iron oxides. J. Colloid Interface Sci. 67, 187—194.

Verwey, E.J.W., 1935. The electric double layer and the stability of lyophobic colloids. Chem. Rev. 16, 363—415.

Verwey, E.J.W., Overbeek, J.T.G., 1948. Theory of the Stability of Lyophobic Colloids. Elsevier, New York, NY.

Vydra, F., Galba, J., 1969. Sorption von metallkomplexen an silicagel. V. Sorption von hydrolysenprodukten des Co^{2+}, Mn^{2+}, Ni^{2+}, Zn^{2+} an silicagel. Colln. Czech. Chem. Comm. 34, 3471—3478.

Wann, S.S., Uehara, G., 1978. Surface charge manipulation in constant surface potential soil colloids: I. Relation to sorbed phosphorus. Soil Sci. Soc. Am. J. 42, 565—570.

Waychunas, G.A., Rea, B.A., Fuller, C.C., Davis, J.A., 1993. Surface chemistry of ferrihydrite: Part 1. EXAFS studies of the geometry of coprecipitated and adsorbed arsenate. Geochim. Cosmochim. Acta. 57, 2251—2269.

Wershaw, R.L., Mikita, M.A., 1987. NMR of Humic Substances and Coal. Lewis Publishers, Chelsea, MI.

Westall, J.C., 1980. Chemical equilibrium including adsorption on charged surfaces. Adv. Chem. Ser. 189, 33—44.

Westall, J.C., 1986. Reactions at the oxide-solution interface: chemical and electrostatic models. ACS Symp. Ser. 323, 54—78.

Westall, J.C., Hohl, H., 1980. A comparison of electrostatic models for the oxide/solution interface. Adv. Colloid Interface Sci. 12, 265—294.

Wilson, M.A., 1987. NMR Techniques and Applications in Geochemistry and Soil Chemistry. Pergamon Press, Oxford.

Wolthers, M., Charlet, L., Van Cappellen, P., 2008. The surface chemistry of divalent metal carbonate minerals; a critical assessment of surface charge and potential data using the charge distribution multi-site ion complexation model. Am. J. Sci. 308 (8), 905—941.

Woessner, D.E., 1989. Characterization of clay minerals by 27Al nuclear magnetic resonance spectroscopy. Am. Mineral. 74, 203—215.

Xia, K., Mehadi, A., Taylor, R.W., Bleam, W.F., 1997. X-ray absorption and electron paramagnetic resonance studies of Cu(II) sorbed to silica: surface-induced precipitation at low surface coverages. J. Colloid Interface Sci. 185 (1), 252—257.

Xu, N., 1993. Spectroscopic and Solution Chemistry Studies of Cobalt(II) Sorption Mechanisms at the Calcite—Water Interface. Ph.D. Dissertation, Stanford University, Stanford, CA, p. 143.

Yamaguchi, N.U., Scheinost, A.C., Sparks, D.L., 2001. Surface-induced nickel hydroxide precipitation in the presence of citrate and salicylate. Soil Sci. Soc. Am. J. 65 (3), 729—736.

Yee, N., Fein, J., 2001. Cd adsorption onto bacterial surfaces: a universal adsorption edge? Geochim. Cosmochim. Acta. 65 (13), 2037—2042.

Zhang, P.C., Sparks, D.L., 1990a. Kinetics and mechanisms of sulfate adsorption desorption on goethite using pressure-jump relaxation. Soil Sci. Soc. Am. J. 54 (5), 1266—1273.

Zhang, P.C., Sparks, D.L., 1990b. Kinetics of selenate and selenite adsorption/desorption at the goethite/ water interface. Environ. Sci. Technol. 24 (12), 1848—1856.

6

Ion Exchange Processes

6.1 Introduction

Key points: Ion exchange is a fundamental soil chemical process that impacts both cations and anions. This molecular-scale reaction influences field-scale processes, such as the plant availability and mobility of nutrients and contaminants in the soil. Ion exchange has both thermodynamic and kinetic considerations.

Ion exchange, the interchange between an ion in solution and another ion in the boundary layer between the solution and a charged surface (Glossary of Soil Science Terms, 2008), truly has been one of the hallmarks of soil chemistry. Since the pioneering studies of J. Thomas Way in the middle of the 19th century (Way, 1850), many important studies have occurred on various aspects of both cation and anion exchange in soils. The sources of cation exchange in soils are clay minerals, organic matter, and amorphous materials. The sources of anion exchange in soils are clay minerals, primarily 1:1 clay minerals such as kaolinite, and metal oxides and amorphous materials.

The ion exchange capacity is the maximum adsorption of readily exchangeable ions (diffuse ion swarm and outer-sphere complexes) on soil particle surfaces (Sposito, 2000). From a practical point of view, the ion exchange capacity (the sum of the CEC [defined earlier; see Box 6.1 for a description of CEC measurement] and the AEC [anion exchange capacity, which is the sum of total exchangeable anions that a soil can adsorb, expressed as $cmol_c$ kg^{-1} or $mmol_c$ kg^{-1}, where the subscript c indicates charge]; (Glossary of Soil Science Terms, 2008) of a soil is important since it determines the capacity of a soil to retain ions in a form such that they are available for plant uptake and not susceptible to leaching in the soil profile. This feature has important environmental and plant nutrient implications. As an example, NO_3^- is important for plant growth, but if it leaches, as it often does, it can move below the plant root zone and leach into groundwater, where it is deleterious to human health (see Chapter 1). If a soil has a significant AEC, nitrate can be held, albeit weakly. Sulfate can be significantly held in soils that have AEC and be available for plant uptake (sulfate accumulations are sometimes observed in subsoils where oxides as discrete particles or as coatings on clays impart positive charge or an AEC to the soil). However, in soils lacking the ability to retain anions sulfate can leach readily and is no longer available to support plant growth.

6.2 Characteristics of Ion Exchange

Key points: Ion exchange involves electrostatic (coulombic) interactions between the counterions at the planes between the bulk solution and a charged particle surface (i.e., the 0, 1, and

Environmental Soil Chemistry, Third Edition
https://doi.org/10.1016/B978-0-12-815880-7.00006-7

BOX 6.1

Measurement of cation exchange capacity (CEC)

The CEC of a soil is usually measured by saturating a soil or soil component with an index cation such as Ca^{2+} or NH_4^+, removing excess salts of the index cation with a dilute electrolyte solution, and then displacing the Ca^{2+} with another cation, such as Mg^{2+}. The amount of displaced Ca^{2+} is then measured, and the CEC is calculated. For example, let us assume that 5 g soil was used for CEC analysis, and displaced Ca was collected in an $Mg(NO_3)_2$ solution with a total volume of 100 mL. The concentration of Ca in the solution obtained using ICP-AES analysis was found to be 200 mg L^{-1}. The CEC would then be calculated as:

$$\frac{200 \text{ mg } Ca^{2+}}{L^{-1}} \times 0.1 \text{ L} = 20 \text{ mg } Ca^{2+} \quad \text{(B6.1a)}$$

$$20 \text{ mg } Ca^{2+} \times \frac{1 \text{ mmol } Ca^{2+}}{40.08 \text{ mg } Ca^{2+}} = 0.499 \text{ mmol } Ca^{2+}$$
$$\text{(B6.1b)}$$

$$0.499 \text{ mmol } Ca^{2+} \times \frac{2 \text{ mmol}_c}{1 \text{ mmol } Ca^{2+}} = 0.998 \text{ mmol}_c$$
$$\text{(B6.1c)}$$

$$0.998 \text{ mmol}_c / 0.005 \text{ kg soil} = 199.6 \text{ mmol}_c \text{ kg}^{-1}$$
$$= 19.96 \text{ cmol}_c \text{ kg}^{-1} = CEC$$
$$\text{(B6.1d)}$$

Both CEC units, i.e., $mmol_c$ kg^{-1} and $cmol_c$ kg^{-1}, are used in the literature; however, the $cmol_c$ kg^{-1} unit is preferred, and it is equivalent to the older CEC unit of meq/100 g that is no longer used but can still be found in the literature.

The CEC values of various soil minerals were provided in Chapter 2. The CEC of a soil generally increases with soil pH due to the greater negative charge that develops on organic matter and clay minerals such as kaolinite due to the deprotonation of functional groups as pH increases. Thus in measuring the CEC of variable charge soils and minerals if the index cation saturating solution is at a pH greater than the pH of the soil or mineral, the CEC can be overestimated (Sumner and Miller, 1996). The anion exchange capacity increases with decreasing pH as the variable charge surfaces become more positively charged due to the protonation of functional groups.

The magnitude of the CEC in soils is usually greater than the AEC. However, in soils that are highly weathered and acidic, e.g., some tropical soils, copious quantities of variable charge surfaces such as Fe/Al oxides and kaolinite may be present and the positive charge on the soil surface may be significant. These soils can exhibit a substantial AEC

2 planes) as well as counterions in the diffuse double layer surrounding the charged particle.

Ion exchange is usually rapid, diffusion-controlled, reversible, and stoichiometric, and in most cases, there is some selectivity by the exchanging surface for one ion over another. Exchange reversibility is indicated when the exchange isotherms for the forward and backward exchange reactions coincide (see the later section, Experimental Interpretations, for a discussion on exchange isotherms). Exchange irreversibility or hysteresis is also observed and can be attributed to colloidal aggregation and the formation of surface precipitates. Surface precipitates have been shown to form rapidly and sequester heavy metals in newly formed solid phases (Chapter 5). The formation of surface precipitates will make ions bound to exchange sites inaccessible (Van Bladel and Lau-delout, 1967; Verburg and Baveye, 1994; Sie-becker et al., 2014, 2018).

Stoichiometry means that any ions that leave the colloidal surface are replaced by an equiva-lent (in terms of ion charge on a monovalent charge basis) amount of other ions. This is due to the electroneutrality requirement. When an ion is displaced from the surface, the exchanger has a deficit in counterion charge that must be balanced by counterions in the diffuse ion cloud around the exchanger. The total coun-terion content in $cmol_c \ kg^{-1}$ (or $mmol_c \ kg^{-1}$) re-mains constant. For example, to maintain stoichiometry, two K^+ ions are necessary to replace one Ca^{2+} ion.

Since electrostatic forces are involved in ion exchange, Coulomb's law can be invoked to explain the selectivity or preference of the ion exchanger for one ion over another. This was dis-cussed in Chapter 5. However, in review, one can say that for a given group of elements from the periodic table with the same valence, ions with the smallest hydrated radius will be preferred, since ions are hydrated in the soil environment. Thus for the group 1 elements the general order of selectivity would be $Cs^+ >$ $Rb^+ > K^+ > Na^+ > Li^+ > H^+$. If one is dealing with ions of different valence, generally, the higher-charged ion will be preferred. For example, $Al^{3+} > Ca^{2+} > Mg^{2+} > K^+ = NH_4^+ >$ Na^+. In examining the effect of valence on selec-tivity polarization must be considered. Polariza-tion is the distortion of the electron cloud about an anion by a cation. The smaller the hydrated radius of the cation, the greater the polarization, and the greater its valence, the greater its polar-izing power. With anions, the larger they are, the more easily they can be polarized. The coun-terion with the greater polarization is usually preferred, and it is also least apt to form a com-plex with its coion. Helfferich (1962a) has given the following selectivity sequence, or lyotropic series, for some of the common cations: $Ba^{2+} >$

$Pb^{2+} > Sr^{2+} > Ca^{2+} > Ni^{2+} > Cd^{2+} > Cu^{2+} >$ $Co^{2+} > Zn^{2+} > Mg^{2+} > Ag^+ > Cs^+ > Rb^+ >$ $K^+ > NH_4^+ > Na^+ > Li^+$.

The rate of ion exchange in soils is dependent on the type and quantity of inorganic and organic components and the charge and radius of the ion being considered (Sparks, 1989). With clay minerals like kaolinite, where only external exchange sites are present, the rate of ion exchange is rapid. With 2:1 clay minerals that contain both external and internal exchange sites, particularly with vermiculite and micas, where partially collapsed interlayer space sites exist, the kinetics are slower. In these types of clay minerals, ions such as K^+ slowly diffuse into the partially collapsed interlayer spaces and the exchange can be slow and tortuous. The charge of the ion also affects the kinetics of ion exchange. Generally, the rate of exchange de-creases as the charge of the exchanging species increases (Helfferich, 1962b). More details on the kinetics of ion exchange reactions can be found in Chapter 7.

6.3 Cation Exchange Equilibrium Constants and Selectivity Coefficients

Key points: Many attempts to define an equi-librium exchange constant have been made since such a parameter would be useful for deter-mining the state of ionic equilibrium at different ion concentrations. Some of the better-known equations attempting to do this are the Kerr (1928), Vanselow (1932), and Gapon (1933) expressions.

In many studies it has been shown that the equilibrium exchange constants derived from these equations are not constant as the composi-tion of the exchanger phase (solid surface) changes. Thus it is often better to refer to them as selectivity coefficients rather than exchange constants.

6.3.1 Kerr Equation

In 1928 Kerr proposed an "equilibrium constant," given below, and correctly pointed out that the soil was a solid solution (a macroscopically homogeneous mixture with a variable composition; Lewis and Randall, 1961). For a binary reaction (a reaction involving two ions),

$$vACl_u(aq) + uBX_v(s) \rightleftharpoons uBCl_v(aq) + vAX_u(s)$$
$$(6.1)$$

where A^{u+} and B^{v+} are exchanging cations and X represents the exchanger, (aq) represents the solution or aqueous phase, and (s) represents the solid or exchanger phase.

Kerr (1928) expressed the "equilibrium constant" or, more correctly, a selectivity coefficient for the reaction in Equation (6.1) as

$$K_K = \frac{[BCl_v]^u \{AX_u\}^v}{[ACl_u]^v \{BX_v\}^u}$$
$$(6.2)$$

where brackets "[ ]" indicate the concentration in the aqueous phase in mol liter^{-1} and braces "{ }" indicate the concentration in the solid or exchanger phase in mol kg^{-1}.

Kerr (1928) studied Ca–Mg exchange and found that the K_K value remained relatively constant as the exchanger composition changed. This indicated that the system behaved ideally; that is, the exchanger phase activity coefficients for the two cations were each equal to 1 (Lewis and Randall, 1961). These results were fortuitous since Ca–Mg exchange is one of the few binary exchange systems in which ideality is observed.

6.3.2 Vanselow Equation

Albert Vanselow was a student of Lewis and was the first person to give ion exchange a truly thermodynamical context. Considering the binary cation exchange reaction in Equation (6.1),

Vanselow (1932) described the thermodynamic equilibrium constant as

$$K_{eq} = \frac{(BCl_v)^u (AX_u)^v}{(ACl_u)^v (BX_v)^u}$$
$$(6.3)$$

where the parentheses indicate the thermodynamic activity. It is not difficult to determine the activity of solution components, since the activity would equal the product of the equilibrium molar concentration of the cation multiplied by the solution activity coefficients of the cation, that is, $(ACl_u) = (C_A)(\gamma_A)$ and $(BCl_v) = (C_B)(\gamma_B)$. C_A and C_B are the equilibrium concentrations of cations A and B, respectively, and γ_A and γ_B are the solution activity coefficients of the two cations, respectively.

The activity coefficients of the electrolytes can be determined using Equation (4.15).

However, calculating the activity of the exchanger phase is not as simple. Vanselow defined the exchanger phase activity in terms of mole fractions, $\overline{N}_A$ and $\overline{N}_B$ for ions A and B, respectively. Thus according to Vanselow (1932), Equation (6.3) could be rewritten as

$$K_V = \frac{\gamma_B^u C_B^u \overline{N}_A^v}{\gamma_A^v C_A^v \overline{N}_B^u}$$
$$(6.4)$$

where

$$\overline{N}_A = \frac{\{AX_u\}}{\{AX_u\} + \{BX_v\}}$$

and

$$\overline{N}_B = \frac{\{BX_v\}}{\{AX_u\} + \{BX_v\}}$$
$$(6.5)$$

Vanselow (1932) assumed that K_V was equal to K_{eq}. However, he failed to realize one very important point. The activity of a "component of a homogeneous mixture is equal to its mole fraction only if the mixture is ideal" (Guggenheim, 1967), that is, $f_A = f_B = 1$, where f_A and f_B are the exchanger phase activity coefficients for cations A and B, respectively. If the mixture is not ideal, then the activity is a product of $\overline{N}$ and f. Thus K_{eq} is correctly written as

$$K_{eq} = \left(\frac{\gamma_B^u C_B^u \overline{N}_A^v f_A^v}{\gamma_A^v C_A^v \overline{N}_B^u f_B^u} \right) = K_V \left(\frac{f_A^v}{f_B^u} \right) \qquad (6.6)$$

where

$$f_A \equiv (AX_u)/\overline{N}_A \text{ and } f_B \equiv (BX_v)/\overline{N}_B \qquad (6.7)$$

Thus

$$K_V = K_{eq}\left(f_B^u / f_A^v \right) \qquad (6.8)$$

and K_V is an apparent equilibrium exchange constant or a cation exchange selectivity coefficient.

6.3.2 Other Empirical Exchange Equations

A number of other cation exchange selectivity coefficients have also been employed in environmental soil chemistry. Krishnamoorthy and Overstreet (1949) used a statistical mechanics approach and included a factor for the valence of the ions, 1 for monovalent ions, 1.5 for divalent ions, and 2 for trivalent ions, to obtain a selectivity coefficient K_{KO}. Gaines and Thomas (1953) and Gapon (1933) also introduced exchange equations that yielded selectivity coefficients (K_{GT} and K_G, respectively). For K–Ca exchange on a soil, the Gapon convention expresses exchange per mol of exchange sites rather than per mol of exchanging cation:

$$Ca_{1/2}\text{-soil} + K^+ \rightleftharpoons K\text{-soil} + \tfrac{1}{2}Ca^{2+} \qquad (6.9)$$

where there are chemically equivalent quantities of the exchanger phases and the exchangeable cations. The Gapon selectivity coefficient for K–Ca exchange would be expressed as

$$K_G = \frac{\{K\text{-soil}\}\left[Ca^{2+}\right]^{1/2}}{\{Ca_{1/2}\text{-soil}\}\left[K^+\right]} \qquad (6.10)$$

where the square brackets represent the concentration in the aqueous phase, expressed as

mol liter^{-1}, and the braces represent the concentration in the exchanger phase, expressed as mol kg^{-1}. The structure of this formula is the same as other equilibrium constants and association constants that have been discussed in Chapters 3 and 4. Specifically, it is K = (products)/(reactants). The selectivity coefficient obtained from the Gapon equation has been the most widely used in soil chemistry and appears to vary the least as exchanger phase composition changes. The various cation exchange selectivity coefficients for homovalent and heterovalent exchange are given in Table 6.1. When writing exchange reactions, the final ions on the soil (i.e., on the exchanger phase) are K for K–Na exchange and K for K–Ca exchange.

6.4 Thermodynamics of Ion Exchange

Key points: The purpose of calculating the thermodynamics of ion exchange reactions is to determine if a reaction is favorable or not. In order for a reaction to proceed, the Gibbs free energy must be negative. The favorability of an ion exchange reaction is influenced by the ion itself and the minerals present in the system.

6.4.1 Theoretical Background

Thermodynamic equations that provide a relationship between exchanger phase activity coefficients and the exchanger phase composition were independently derived by Argersinger et al. (1950) and Hogfeldt (Ekedahl et al., 1950; Hogfeldt et al., 1950). These equations, as shown later, demonstrated that the calculation of an exchanger phase activity coefficient, f, and the thermodynamic equilibrium constant, K_{eq}, were reduced to the measurement of the Vanselow selectivity coefficient, K_V, as a function of the exchanger phase composition (Sposito, 1981; Bourg and Sposito, 2011). Argersinger et al. (1950) defined f as $f = a/\overline{N}$, where a is the activity of the exchanger phase.

TABLE 6.1   Cation exchange selectivity coefficients for homovalent (K–Na) and heterovalent (K–Ca) exchange.

Selectivity coefficient	Homovalent exchange[a]	Heterovalent exchange[b]
Kerr	$K_K = \dfrac{\{K\text{-soil}\}[Na^+]}{\{Na\text{-soil}\}[K^+]}$[c]	$K_K = \dfrac{\{K\text{-soil}\}^2[Ca^{2+}]}{\{Ca\text{-soil}\}[K^+]^2}$
Vanselow[d]	$K_V = \dfrac{\{K\text{-soil}\}[Na^+]}{\{Na\text{-soil}\}[K^+]}$ Or $K_V = K_K$	$K_V = \left[\dfrac{\{K\text{-soil}\}^2[Ca^{2+}]}{\{Ca\text{-soil}\}[K^+]^2}\right] \times \left[\dfrac{1}{\{K\text{-soil}\}\{Ca\text{-soil}\}}\right]$ Or $K_V = K_K\left[\dfrac{1}{\{K\text{-soil}\}\{Ca\text{-soil}\}}\right]$
Krishnamoorthy–Overstreet (focuses on valence effects)	$K_{KO} = \dfrac{\{K\text{-soil}\}[Na^+]}{\{Na\text{-soil}\}[K^+]}$ Or $K_{KO} = K_K$	$K_{KO} = \left[\dfrac{\{K\text{-soil}\}^2[Ca^{2+}]}{\{Ca\text{-soil}\}[K^+]^2}\right] \times \left[\dfrac{1}{\{K\text{-soil}\}+1.5\{Ca\text{-soil}\}}\right]$
Gaines–Thomas[d] (focuses on equivalent fractions)	$K_{GT} = \dfrac{\{K\text{-soil}\}[Na^+]}{\{Na\text{-soil}\}[K^+]}$ Or $K_{GT} = K_K$	$K_{GT} = \left[\dfrac{\{K\text{-soil}\}^2[Ca^{2+}]}{\{Ca\text{-soil}\}[K^+]^2}\right] \times \left[\dfrac{1}{2[2\{Ca\text{-soil}\}+\{K\text{-soil}\}]}\right]$
Gapon	$K_G = \dfrac{\{K\text{-soil}\}[Na^+]}{\{Na\text{-soil}\}[K^+]}$ Or $K_G = K_K$	$K_G = \left[\dfrac{\{K\text{-soil}\}[Ca^{2+}]^{1/2}}{\{Ca_{1/2}\text{soil}\}[K^+]}\right]$

[a] The homovalent exchange reaction (K–Na exchange) is $Na\text{-soil} + K^+ \rightleftharpoons K\text{-soil} + Na^+$.
[b] The heterovalent exchange reaction (K–Ca exchange) is $Ca\text{-soil} + 2K^+ \rightleftharpoons 2K\text{-soil} + Ca^{2+}$, except for the Gapon convention, where the exchange reaction would be $Ca_{0.5}\text{-soil} + K^+ \rightleftharpoons K\text{-soil} + 0.5\ Ca^{2+}$.
[c] Square brackets represent the concentration in the aqueous phase, which is expressed in mol liter^{-1}; braces represent the concentration in the exchanger phase, which is expressed in mol kg^{-1}.
[d] Vanselow (1932) and Gaines and Thomas (1953) originally expressed both aqueous and exchanger phases in terms of activity. For simplicity, they are expressed here as concentrations.

Before thermodynamic parameters for exchange equilibria can be calculated, standard states for each phase must be defined. The choice of standard state affects the value of the thermodynamic parameters and their physical interpretation (Goulding, 1983a). Table 6.2 shows the different standard states and the effects of using them. Normally, the standard state for the adsorbed phase is the homoionic exchanger in equilibrium with a solution of the saturating cation at constant ionic strength.

Argersinger et al. (1950), based on Equation (6.8), assumed that any change in K_V with regard to exchanger phase composition occurred because of a variation in exchanger phase activity coefficients. This is expressed as

$$v \ln f_A = u \ln f_B = \ln K_{eq} - \ln K_V \qquad (6.11)$$

Taking differentials of both sides, realizing that K_{eq} is a constant, results in

$$vd \ln f_A = ud \ln f_B = -d \ln K_V \qquad (6.12)$$

Any change in the activity of BX_v (s) must be accounted for by a change in the activity of AX_u (s) such that the mass in the exchanger is conserved. This necessity, an application of the Gibbs–Duhem equation (Guggenheim, 1967), results in

$$\overline{N}_A d \ln f_A + \overline{N}_B d \ln f_B = 0 \qquad (6.13)$$

Equations (6.12) and (6.13) can be solved, resulting in

$$vd \ln f_A = \left(\dfrac{-v\overline{N}_B}{u\overline{N}_A + v\overline{N}_B}\right) d \ln K_V \qquad (6.14)$$

TABLE 6.2   Some of the standard states used in calculating the thermodynamic parameters of cation-exchange equilibria.[a]

Standard state			
Adsorbed phase	Solution phase	Implications	Reference
Activity = mole fraction when the latter = 1	Activity = molarity as concentration → 0	Can calculate f, K_V, etc., but all depend on ionic strength	Argersinger et al. (1950)
Homoionic exchanger in equilibrium with an infinitely dilute solution of the ion	Activity = molarity as concentration → 0	ΔG_{ex}^o expresses relative affinity of exchanger for cations	Gaines and Thomas (1953)
Activity = mole fraction when the latter = 0.5. Components not in equilibrium	Activity = molarity as concentration → 0	ΔG_{ex}^o expresses relative affinity of exchanger for cations when mole fraction = 0.5	Babcock (1963)

[a] *From Goulding (1983b), with permission.*

$$u d \ln f_B = \left(\frac{-v\overline{N}_A}{u\overline{N}_A + v\overline{N}_B} \right) d \ln K_V \quad (6.15)$$

where $\left(u\overline{N}_A / (u\overline{N}_A + v\overline{N}_B) \right)$ is equal to $\overline{E}_A$ or the equivalent fraction of $AX_u(s)$ and $\overline{E}_B$ is $\left(v\overline{N}_B / (u\overline{N}_A + v\overline{N}_B) \right)$ or the equivalent fraction of $BX_v(s)$ and the identity $\overline{N}_A$ and $\overline{N}_B = 1$.

In terms of $\overline{E}_A$, Equations (6.14) and (6.15) become

$$v d \ln f_A = - \left(1 - \overline{E}_A \right) d \ln K_V \quad (6.16)$$

$$u d \ln f_B = \overline{E}_A d \ln K_V \quad (6.17)$$

Integrating Equations (6.16) and (6.17) by parts, noting that $\ln f_A = 0$ at $\overline{N}_A = 1$, or $E_a = 1$, and similarly $\ln f_B = 0$ at $\overline{N}_A = 0$, or $\overline{E}_A = 0$,

$$-v \ln f_A = \left(1 - \overline{E}_A \right) \ln K_V - \int_{\overline{E}_A}^1 \ln K_V d\overline{E}_A \quad (6.18)$$

$$-u \ln f_B = -\overline{E}_A \ln K_V + \int_0^{\overline{E}_A} \ln K_V d\overline{E}_A \quad (6.19)$$

Substituting these into Equation (6.11) leads to

$$\ln K_{eq} = \int_0^1 \ln K_V d\overline{E}_A \quad (6.20)$$

Equation 6.20 is critical because it provides for the calculation of the thermodynamic equilibrium

exchange constant. Thus by plotting $\ln K_V$ vs $\overline{E}_A$ and integrating under the curve, from $\overline{E}_A = 0$ to $\overline{E}_A = 1$, one can calculate K_{eq}, or in ion exchange studies, K_{ex}, the equilibrium exchange constant. Other thermodynamic parameters can then be determined as given below:

$$\Delta G_{ex}^0 = - RT \ln K_{ex} \quad (6.21)$$

where ΔG_{ex}^0 is the standard Gibbs free energy of exchange.

Examples of how exchanger phase activity coefficients and K_{ex} and ΔG_{ex}^0 values can be calculated for binary exchange processes are provided in Boxes 6.2 and 6.3, respectively.

Using the van 't Hoff equation, one can calculate the standard enthalpy of exchange, ΔH_{ex}^0, as

$$\ln \frac{K_{ex_{T_2}}}{K_{ex_{T_1}}} = \left(\frac{-\Delta H_{ex}^0}{R} \right) \left(\frac{1}{T_2} - \frac{1}{T_1} \right) \quad (6.22)$$

where subscripts 1 and 2 denote temperatures 1 and 2, respectively. From this relationship:

$$\Delta G_{ex}^0 = \Delta H_{ex}^0 - T\Delta S_{ex}^0 \quad (6.23)$$

The standard entropy of exchange, ΔS_{ex}^0, can be calculated using

$$\Delta S_{ex}^0 = \frac{\left(\Delta H_{ex}^0 - \Delta G_{ex}^0 \right)}{T} \quad (6.24)$$

BOX 6.2

─────────

Calculation of exchanger phase activity coefficients

It would be instructive at this point to illustrate how exchanger phase activity coefficients would be calculated for the homovalent and heterovalent exchange reactions in Table 6.1. For the homovalent reaction, K–Na exchange, the f_K and f_{Na} values would be calculated as (Argersinger et al., 1950)

$$-\ln f_K = \left(1 - \bar{E}_K\right) \ln K_V - \int_{\bar{E}_K}^{1} \ln K_V d\bar{E}_K$$

(B6.2a)

$$-\ln f_{Na} = -\bar{E}_K \ln K_V + \int_{0}^{\bar{E}_K} \ln K_V d\bar{E}_K$$ (B6.2b)

and

$$\ln K_{ex} = \int_{0}^{1} \ln K_V d\bar{E}_K$$ (B6.2c)

For the heterovalent exchange reaction, K–Ca exchange, the f_K and f_{Ca} values would be calculated as (Ogwada and Sparks, 1986a)

$$2 \ln f_K = -\left(1 - \bar{E}_k\right) \ln K_V + \int_{\bar{E}_K}^{1} \ln K_V d\bar{E}_K$$

(B6.2d)

$$\ln f_{Ca} = \bar{E}_K \ln K_V - \int_{0}^{\bar{E}_K} \ln K_V d\bar{E}_K$$ (B6.2e)

and

$$\ln K_{ex} = \int_{0}^{1} \ln K_V d\bar{E}_K$$ (B6.2f)

Gaines and Thomas (1953) also described the thermodynamics of cation exchange and made two contributions. They included a term in the Gibbs–Duhem equation for the activity of water that may be adsorbed on the exchanger. This activity may change as the exchanger phase composition changes. Changes in water activity with exchanger composition variations were later shown to have little effect on K_{ex} calculations (Laudelout and Thomas, 1965), but they can affect the calculation of f values for zeolites (Barrer and Klinowski, 1974). Gaines and Thomas (1953) also defined the reference state of a component of the exchanger as the homoionic exchanger made up of the component in equilibrium with an infinitely dilute aqueous solution containing the components. Gaines and

Thomas (1953) defined the exchange equilibrium constant of Equation (6.1) as

$$K_{ex} = (BCl_v)^u g_A^v \bar{E}_A^v / (ACl_u)^v g_B^u \bar{E}_B^u$$ (6.25)

where g_A and g_B are the exchanger phase activity coefficients and are defined as

$$g_A \equiv (AX_u)/\bar{E}_A \text{ and } g_B \equiv (BX_v)/\bar{E}_B$$ (6.26)

Thus the Gaines and Thomas selectivity coefficient, K_{GT}, would be defined as

$$K_{GT} = (BCl_v)^u \bar{E}_A^v / (ACl_u)^v \bar{E}_B^u$$ (6.27)

Hogfeldt et al. (1950) also defined the exchanger phase activity coefficients in terms of the equivalent fraction rather than the Vanselow (1932) convention of mole fraction. Of course, for

BOX 6.3

Calculation of thermodynamic parameters for K−Ca exchange on a soil

Consider the general binary exchange reaction in Equation (6.1).

$$vACl_u(aq) + uBX_u(s) \rightleftharpoons uBCl_v(aq) + vAX_u(s) \tag{B6.3a}$$

If one is studying K−Ca exchange where A is K^+, B is Ca^{2+}, v is 2, and u is 1, then Equation (6.3a) becomes

$$2KCl + Ca\text{-soil} \rightleftharpoons CaCl_2 + 2K\text{-soil} \tag{B6.3b}$$

Using the experimental methodology given in the text, one can calculate K_v, K_{ex}, and ΔG^0_{ex} parameters for the K−Ca exchange reaction in Equation (6.3b) as shown in the calculations below. Assume the ionic strength (I) is 0.01 and the temperature at which the experiment was conducted is 298 K.

Ex-changer test	Solution (aq) phase concentration (mol liter^{-1})		Exchanger phase concentration (mol kg^{-1})		Mole fractions[a]				
	K^+	Ca^{2+}	K^+	Ca^{2+}	$\overline{N}_K$	$\overline{N}_{Ca}$	K_v^b	$\ln K_v$	$\overline{E}_K^c$
1	0	3.32×10^{-3}	0	1.68×10^{-2}	0	1.000	—	5.11^d	0
2	1×10^{-3}	2.99×10^{-3}	2.95×10^{-3}	1.12×10^{-2}	0.2086	0.7914	134.20	4.90	0.116
3	2.5×10^{-3}	2.50×10^{-3}	7.88×10^{-3}	1.07×10^{-2}	0.4232	0.5768	101.36	4.62	0.268
4	4.0×10^{-3}	1.99×10^{-3}	8.06×10^{-3}	5.31×10^{-3}	0.6030	0.3970	92.95	4.53	0.432
5	7.0×10^{-3}	9.90×10^{-4}	8.63×10^{-3}	2.21×10^{-3}	0.7959	0.2041	51.16	3.93	0.661
6	8.5×10^{-3}	4.99×10^{-4}	1.17×10^{-2}	1.34×10^{-3}	0.8971	0.1029	44.07	3.79	0.813
7	9.0×10^{-3}	3.29×10^{-4}	1.43×10^{-2}	1.03×10^{-3}	0.9331	0.0669	43.13	3.76	0.875
8	1.0×10^{-2}	0	1.45×10^{-2}	0	1.000	0.0000	—	3.70^d	1

[a]
$$\overline{N}_K = \frac{\{K^+\}}{\{K^+\} + \{Ca^{2+}\}}; \overline{N}_{ca} = \frac{\{Ca^{2+}\}}{\{K^+\} + \{Ca^{2+}\}} \tag{B6.3c}$$

where the braces indicate the exchanger phase composition in mol kg^{-1}; e.g., for exchanger test 2,

$$\overline{N}_K = \frac{(2.95 \times 10^{-3})}{(2.95 \times 10^{-3}) + (1.12 \times 10^{-2})} = 0.2086 \tag{B6.3d}$$

[b]
$$K_v = \frac{(\gamma_{Ca^{2+}})(C_{Ca^{2+}})\left((\overline{N}_K)^2\right)}{(\gamma_{K^+})^2(C_{K^+})^2(\overline{N}_{Ca})} \tag{B6.3e}$$

where γ is the solution phase activity coefficient calculated according to Equation (4.15) and C is the solution concentration; e.g., for exchanger test 2,

$$K_V = \frac{(0.6653)\left(2.99 \times 10^{-3} \text{mol liter}^{-1}\right)(0.2086)^2}{(0.9030)^2 \left(1 \times 10^{-3} \text{mol liter}^{-1}\right)^2 (0.7914)} = 134.20 \tag{B6.3f}$$

Continued

BOX 6.3   (cont'd)

Calculation of thermodynamic parameters for K–Ca exchange on a soil

[c] $\bar{E}_K$ is the equivalent fraction of K^+ on the exchanger:

$$\bar{E}_K = \frac{u\bar{N}_K}{u\bar{N}_K + v\bar{N}_{Ca}} = \frac{\bar{N}_K}{\bar{N}_K + 2\bar{N}_{Ca}} \tag{B6.3g}$$

e.g., for exchanger test 2,

$$\frac{0.2086}{0.2086 + (0.7914)(2)} = \frac{0.2086}{1.7914} = 0.116 \tag{B6.3h}$$

[d] Extrapolated $\ln K_V$ values. Values are extrapolated using the slope from the adjacent two points.

Using Equation (6.20),

$$\ln K_{eq} = \int_0^1 \ln K_V d\bar{E}_K$$

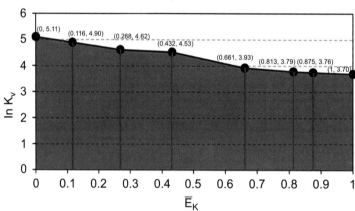

FIGURE B6.3   The shaded area represents $\ln K_{ex}$, which is then used to determine the Gibbs free energy of the exchange reaction. Importantly, the endpoints are extrapolated using the slope from the adjacent two points.

one can determine $\ln K_{ex}$ by plotting $\ln K_V$ vs $\bar{E}_K$ (Figure B6.3) and integrating under the curve by summing the areas of the trapezoids using the relationship

$$\frac{1}{2}\sum_{i=1}^{8}\left(\bar{E}_K^{i+1} - \bar{E}_K^i\right)\left(y^i + y^{i+1}\right) \tag{B6.3i}$$

where $\bar{E}_K^1 \dots \bar{E}_K^8$ are the experimental values of $\bar{E}_K$, $(\bar{E}_K^{i+1} - \bar{E}_K^i)$ gives the width of the ith trapezoid, and $y^1 \dots y^8$ represent the corresponding $\ln K_V$ values.

Accordingly, $\ln K_{ex}$ for the exchange reaction in Equation (6.3b) would be

BOX 6.3   (cont'd)

Calculation of thermodynamic parameters for K–Ca exchange on a soil

$$
\ln K_{ex} = \frac{1}{2}
\begin{bmatrix}
(0.116 - 0)(5.11 + 4.90) + (0.268) + (0.116) \\
\times (4.90 + 4.62) + (0.432 + 0.268)(4.62 + 4.53) \\
+(0.661 - 0.432)(4.53 + 3.93) + (0.813 - 0.661) \\
\times (3.93 + 3.79) + (0.875 - 0.813)(3.79 + 3.76) \\
+(1 - 0.875)(3.76 + 3.70)
\end{bmatrix}
$$

where $\ln K_{ex} = 4.31$ and $K_{ex} = 74.45$. From this value, one can then calculate ΔG_{ex}^0 using Equation (6.21):

$$\Delta G_{ex}^0 = RT \ln K_{ex}$$

Substituting $8.314 \; J \; mol^{-1} \; K^{-1}$ for R and assuming $T = 298 \; K$, $\Delta G_{ex}^0 = -(8.314 \; J \; mol^{-1} \; K^{-1})$ $(298 \; K)(4.31) = -10{,}678 \; J \; mol^{-1} = -10.68 \; kJ \; mol^{-1}$.

Since ΔG_{ex}^0 is negative, this would indicate that K^+ is preferred over Ca^{2+} on the soil.

homovalent exchange, mole and equivalent fractions are equal.

There has been some controversy as to whether the Argersinger et al. (1950) or the Gaines and Thomas (1953) approach should be used to calculate thermodynamic parameters, particularly exchanger phase activity coefficients. Sposito and Mattigod (1979) and Babcock and Doner (1981) have questioned the use of the Gaines and Thomas (1953) approach. They note that except for homovalent exchange, the g values are not true activity coefficients, since the activity coefficient is the ratio of the actual activity to the value of the activity under those limiting conditions when Raoult's law applies (Sposito and Mattigod, 1979). Thus for exchanger phases, an activity coefficient is the ratio of an actual activity to a mole fraction. Equivalents are formal quantities not associated with actual chemical species except for univalent ions.

Goulding (1983b) and Ogwada and Sparks (1986a) compared the two approaches for several exchange processes and concluded that while there were differences in the magnitude of the selectivity coefficients and adsorbed phase activity coefficients, the overall trends and conclusions concerning ion preferences were the same. Ogwada and Sparks (1986a) studied K–Ca exchange on soils at several temperatures and compared the Argersinger et al. (1950) and Gaines and Thomas (1953) approaches. The difference in the exchanger phase activity coefficients with the two approaches was small at low fractional K^+ saturation values but increased as fractional K^+ saturation increased (Figure 6.1). However, as seen in Figure 6.1, the minima, maxima, and inflections occurred at the same fractional K^+ saturations with both approaches.

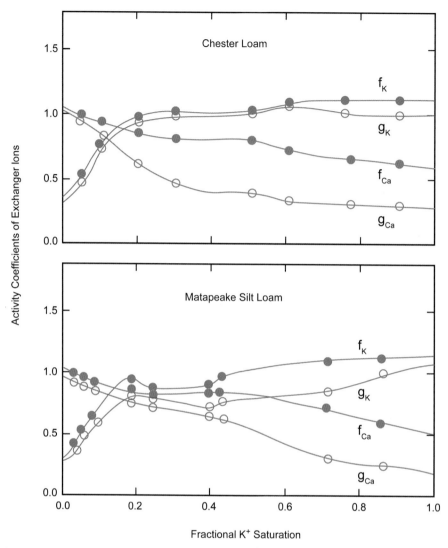

FIGURE 6.1　Exchanger phase activity coefficients for K^+ and Ca^{2+} calculated according to the Argersinger et al. (1950) approach (f_K and f_{Ca}, respectively) and according to the Gaines and Thomas (1953) approach (g_K and g_{Ca}, respectively) versus fractional K^+ saturation (percentage of exchanger phase saturated with K^+). *From Ogwada and Sparks (1986a), with permission.*

6.4.2 Experimental Interpretations

In conducting an exchange study to measure selectivity coefficients, exchanger phase activity coefficients, equilibrium exchange constants, and standard free energies of exchange the exchanger is first saturated to make it homoionic (i.e., one ion predominates on the exchanger). For example, if one wanted to study K–Ca exchange, that is, the conversion of the ion on the soil surface from Ca^{2+} to K^+, one would equilibrate the soil several times with 1 M $CaCl_2$ or $Ca(ClO_4)_2$ and then remove excess salts with water and perhaps an organic solvent such as methanol. After the soil is in a homoionic form, that is, Ca-soil, one would equilibrate the soil with a series of salt solutions containing a range of Ca^{2+} and K^+ concentrations (Box 6.3). For example, in the K–Ca exchange experiment described in Box 6.3, the Ca-soil would be reacted by shaking with or leaching with the varying solutions until equilibrium had been obtained, that is, the concentrations of K^+ and Ca^{2+} in the equilibrium (final) solutions are equal to the initial solution concentrations of K^+ and Ca^{2+} (for the leaching procedure only). To calculate the quantity of ions adsorbed on the exchanger (exchanger phase concentration in Box 6.3) at equilibrium, one would exchange the ions from the soil using a different electrolyte solution, for example, ammonium acetate, and measure the exchanged ions using inductively coupled plasma (ICP) spectrometry or some other analytical technique. Based on such an exchange experiment, one could then calculate the mole fractions of the adsorbed ions, the selectivity coefficients, and K_{ex} and ΔG_{ex}^0 as shown in Box 6.3.

From the data collected in an exchange experiment, exchange isotherms that show the relationship between the equivalent fraction of an ion on the exchanger phase $(\bar{E}_i)$ versus the equivalent fraction of that ion in solution (E_i) are often presented. In homovalent exchange, where the equivalent fraction in the exchanger phase is not affected by the ionic strength and exchange equilibria are also not affected by valence effects, a diagonal line through the exchange isotherm can be used as a nonpreference exchange isotherm $\Delta G_{ex}^0 = 0$; $\bar{E}_i = E_i$, where i refers to ion i (Jensen and Babcock, 1973). That is, if the experimental data lie on the diagonal, there is no preference for one ion over the other. If the experimental data lie above the nonpreference isotherm, the final ion or product is preferred, whereas if the experimental data lie below the diagonal, the reactant is preferred. In heterovalent exchange, however, ionic strength affects the course of the isotherm, and the diagonal cannot be used (Jensen and Babcock, 1973). By using Equation (6.28), which illustrates

divalent–univalent exchange, for example, Ca–K exchange, nonpreference exchange isotherms can be calculated (Sposito, 2000),

$$\bar{E}_{Ca} = 1$$
$$- \left\{ 1 + \frac{2}{\Gamma I} \left[\frac{1}{(1 - E_{Ca})^2} - \frac{1}{(1 - E_{Ca})} \right] \right\}^{-1/2}$$

$$(6.28)$$

where I = ionic strength of the solution, $\bar{E}_{Ca}$ = equivalent fraction of Ca^{2+} on the exchanger phase, E_{Ca} = equivalent fraction of Ca^{2+} in the solution phase, and $\Gamma = \gamma_K^2/\gamma_{Ca}$. If the experimental data lie above the curvilinear nonpreference isotherm calculated using Equation (6.28), then $K_{ex} > 1$ and the final ion or product is preferred (in this case, Ca^{2+}). If the data lie below the nonpreference isotherm, the initial ion or reactant is preferred (in this case, K^+). Thus from Figure 6.2 (Jensen and Babcock, 1973), one sees that K^+ is preferred over Na^+ and Mg^{2+} and Ca^{2+} is preferred over Mg^{2+}.

Table 6.3, from Jensen and Babcock (1973), shows the effect of ionic strength on thermodynamic parameters for several binary exchange systems of a Yolo soil from California. The K_{ex} and ΔG_{ex}^0 values are not affected by ionic strength. Although not shown in Table 6.3, the K_V was dependent on ionic strength with the K exchange systems (K–Na, K–Mg, K–Ca), and there was a selectivity of K^+ over Na^+, Mg^{2+}, and Ca^{2+} that decreased with increasing K saturation. For Mg–Ca exchange, the K_V values were independent of ionic strength and exchanger composition. This system behaved ideally.

It is often observed, particularly with K^+, that K_V values decrease as the equivalent fraction of cation on the exchanger phase or fractional cation saturation increases (Figure 6.3). Ogwada and Sparks (1986a) ascribed the decrease in the K_V with increasing equivalent fractions to the heterogeneous exchange sites and a decreasing specificity of the surface for K^+ ions. Jardine and Sparks (1984a, b) had shown earlier that there were different sites for K^+ exchange on soils.

One also observes with K^+ as well as other ions that the exchanger phase activity

6. Ion Exchange Processes

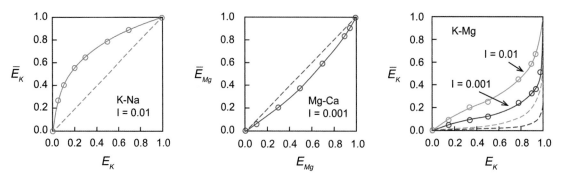

FIGURE 6.2 Cation exchange isotherms for several cation exchange systems. E = equivalent fraction in the solution phase, $\bar{E}$ = equivalent fraction on the exchanger phase. The broken lines represent nonpreference exchange isotherms. In the K–Na isotherm K is the final ion on the exchanger; for Mg–Ca, Mg is the final ion on the exchanger; and for K–Mg, K is the final ion on the exchanger. *From Jensen and Babcock (1973), with permission.*

TABLE 6.3 Effect of ionic strength on thermodynamic parameters for several cation-exchange systems.[a] The first element listed for the exchange process, for example, K in K–Na, Mg in Mg–Ca, K in K–Mg, and K in K–Ca exchange, represents the final ion on the exchanger phase (i.e., sorbed onto the soil).

Ionic strength (I)	Standards Gibbs free energy of exchange (ΔG_{ex}^0) (kJ mol^{-1})				Equilibrium exchange constant (K_{ex})			
	K–Na exchange	Mg–Ca exchange	K–Mg exchange	K–Ca exchange	K–Na exchange	Mg–Ca exchange	K–Mg exchange	K–Ca exchange
0.001	—	1.22	−7.78	—	—	0.61	22.93	—
0.010	−4.06	1.22	−7.77	−6.18	5.12	0.61	22.85	12.04
0.100	−4.03	—	—	—	5.08	—	—	—

[a] *From Jensen and Babcock (1973), with permission. The exchange studies were conducted on a Yolo loam soil.*

coefficients do not remain constant as exchanger phase composition changes (Figure 6.1). This indicates nonideality since if ideality existed, f_{Ca} and f_K would both be equal to 1 over the entire range of exchanger phase composition. A lack of ideality is probably related to the heterogeneous sites and the heterovalent exchange. Exchanger phase activity coefficients correct the equivalent or mole fraction terms for departures from ideality. They thus reflect the change in the status, or fugacity, of the ion held at exchange sites and the heterogeneity of the exchange process. Fugacity is the degree of freedom an ion has to leave the adsorbed state relative to a standard state of maximum freedom of unity. Plots of exchanger phase activity coefficients versus the equivalent fraction of an ion on the

exchanger phase show how this freedom changes during the exchange process, which tells something about the exchange heterogeneity. Selectivity changes during the exchange process can also be gleaned (Ogwada and Sparks, 1986a).

The ΔG_{ex}^0 values indicate the overall selectivity of an exchanger at constant temperature and pressure and independently of ionic strength. For K–Ca exchange, a negative ΔG_{ex}^0 would indicate that the product or K^+ is preferred. A positive ΔG_{ex}^0 would indicate that the reactant, that is, Ca^{2+}, is preferred. Some ΔG_{ex}^0 values as well as ΔH_{ex}^0 parameters for exchange on soils and soil components are shown in Table 6.4. It is important to ensure that the exchange reaction is reversible when interpreting exchange

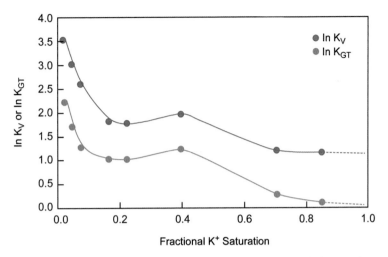

FIGURE 6.3 Natural logarithm of Vanselow selectivity coefficients (K_V, •) and Gaines and Thomas selectivity coefficients (K_{GT}, •) as a function of fractional K^+ saturation (percentage of exchanger phase saturated with K^+) on Chester loam soil at 298 K. At a low K saturation, there are some very specific binding sites for K. Thus a higher K_V value is seen with lower K in solution. If K_V changes, it shows that the system is not behaving in an ideal fashion. *From Ogwada and Sparks (1986a), with permission.*

constants. Being reversible is the basis for the theory. For example, in some minerals, such as 2:1 clay minerals, the interlayer can collapse around K^+, and thus K^+ would not be further exchangeable.

The binding strengths of an ion on a soil or soil component exchanger can be determined from ΔH^0_{ex} values. Enthalpy expresses the gain or loss of heat during the reaction. If the reaction is exothermic, the enthalpy is negative, and heat is lost to the surroundings. If it is endothermic, the enthalpy change is positive, and heat is gained from the surroundings. A negative enthalpy change implies stronger bonds in the reactants. Enthalpies can be measured using the van 't Hoff equation (Equation 6.22) or one can use calorimetry.

6.5 Relationship Between Thermodynamics and Kinetics of Ion Exchange

Key points: While the Gibbs free energy of ion exchange reactions dictates whether a reaction is favorable or not, the kinetics of the reaction will tell us if a reaction is environmentally significant. Additionally, thermodynamic exchange parameters can be obtained through kinetic experiments.

If a reaction is reversible, then $k_1/k_{-1} = K_{ex}$, where k_1 is the forward reaction rate constant and k_{-1} is the backward reaction rate constant (Sparks, 1989). However, this relationship is valid only if mass transfer or diffusion processes are not rate limiting; that is, one must measure the actual chemical exchange reaction (see Chapter 7 for a discussion on mass transfer and chemical exchange reaction processes). Ogwada and Sparks (1986b) found that this assumption is not valid for most kinetic techniques. Only if mixing is very rapid does diffusion become insignificant, and at such mixing rates, one must be careful not to alter the surface area of the adsorbent. Calculations of energies of activation for the forward and backward reactions, E_1 and E_{-1}, respectively, using a kinetics approach, are given below:

$$d \ln k_1 / dT = E_1/RT^2 \qquad (6.29)$$

$$d \ln k_{-1} / dT = E_{-1}/RT^2 \qquad (6.30)$$

Substituting,

$$d \ln k_1/dT - d \ln k_{-1}/dT = d \ln k_{ex}/dT \qquad (6.31)$$

TABLE 6.4 Standard Gibbs free energy of exchange (ΔG^0_{ex}) and standard enthalpy of exchange (ΔH^0_{ex}) values for binary exchange processes on soils and soil components.[a] The first element listed for the exchange process, for example, Ca in Ca–Na, represents the final ion on the exchanger (i.e., sorbed onto the soil). Values for ΔG^0_{ex} and ΔH^0_{ex} represent ranges of values found in the literature.[b]

Exchange process	Exchanger	ΔG^0_{ex} (kJ mol^{-1})	ΔH^0_{ex} (kJ mol^{-1})
Ca–Na	Soils	2.15 to 7.77	
Ca–Na	Clay minerals	0.06 to 0.82	39.98
Ca–Mg	Calcareous soils	0.27 to 0.70	
Ca–Mg	Clay minerals	0.13	
Ca–K	Clay minerals	7.67	16.22
Ca–NH$_4$	Camp Berteau montmorillonite	8.34	23.38
Ca–Cu	Wyoming bentonite	0.11	−18.02
Na–Ca	Soils	0.49 to 4.53	
Na–Li	Clay minerals	−0.34 to −6.04	−0.47 to −23.15
Mg–Ca	Soil	1.22	
Mg–Ca	Clay minerals	1.07	6.96
Mg–Na	Soils	0.72 to 7.27	
Mg–Na	Clay minerals	−1.36	40.22
Mg–NH$_4$	Clay minerals	8.58	23.38
K–Ca	Soils	1.10 to −14.33	−3.25 to −16.28
K–Ca	Soil clay fractions	−2.84 to −6.90	−54.48
K–Ca	Clay minerals	−6.26 to −8.63	
K–Mg	Soils	−4.06	
K–Na	Soils	−3.72 to −4.54	
K–Na	Wyoming bentonite	−1.28	−2.53
K–Na	Chambers	−3.04	−4.86
K–Na	Soils	−3.72 to −4.54	
K–Na	Clay minerals	−1.28 to −3.04	−2.53 to −4.68

[a] Unless specifically noted, the exchange studies were conducted at 298 K.
[b] Mehta et al. (1983); Van Bladel and Gheyi (1980); Wild and Keay (1964); Hutcheon (1966); Laudelout et al. (1967); El-Sayed et al. (1970); Gupta et al. (1984); Gast and Klobe (1971); Jensen and Babcock (1973); Udo (1978); Deist and Talibudeen (1967a, b); Jardine and Sparks (1984b); Goulding and Talibudeen (1984); Ogwada and Sparks (1986b); Jensen (1972); Gast (1972).

BOX 6.4

Construction of a nonpreferential adsorption isotherm for a heterovalent ion exchange system

Consider the following aqueous data (from Evangelou and Lumbanraja, 2002) for the binary and heterovalent exchange of NH_4^+ and Ca^{2+} (NH_4–Ca exchange) on the surface of hydroxy-Al interlayered vermiculite (HIV). The reaction can be written as follows:

$$2NH_4^+ + Ca\text{-HIV} \rightleftharpoons Ca^{2+} + 2NH_4^+\text{-HIV} \quad \text{(B6.4a)}$$

Calculating the nonpreferential adsorption isotherm helps understand the effects that ionic strength will have on the adsorption of NH_4^+ and Ca^{2+} to HIV. It will also allow the determination of which ion has preferred selectivity for the HIV surface. In the literature several variations of the equation for nonpreferential sorption isotherms can be found (in addition to Equation 6.28, e.g., Sposito [2000], Bourg and Sposito [2011], Essington [2015]). The equation here is used in Evangelou and Lumbanraja (2002) and Sposito (1981), Bourg and Sposito, 2011:

$$\bar{E}_{NH_4} = \left[1 + \left(\frac{2}{FTN}\right)\left(\frac{1}{E_{NH_4}^2} - \frac{1}{E_{NH_4}}\right)\right]^{-\frac{1}{2}}$$

$$\text{(B6.4b)}$$

For consistency, $\bar{E}$ still represents the equivalent fraction of NH_4^+ on the exchanger phase. F is the ratio of activity coefficients $(\gamma^2 NH_4^+/\gamma ca)$, where γ is the single ion activity coefficient. TN indicates the total cation normality (or ionic strength can also be used), and E_{NH_4} (no bar) represents the equivalent fraction of NH_4^+ in solution. The equivalent fraction of NH_4^+ in solution can be calculated from the provided aqueous data, and the theoretical equivalent fraction of NH_4^+ on the exchanger ($\bar{E}$) can be calculated using the above equation.

The effects of ionic strength are also considered, and the figure below highlights how the nonpreference isotherm is affected by changes in ionic

strength. Ionic strength modifies the competition for divalent and monovalent ion selectivity for the HIV surface. Additionally, as ionic strength changes, so do the single ion activity coefficient values. The single ion activity coefficients (γ) for several cations are given in Chapter 4. However, at high ionic strength, special considerations are needed. Often, γ for NH_4^+ and K^+ are grouped together, and at high ionic strength (e.g., 0.7 M), the average γ values for K^+ and Ca^{2+} are 0.6133 and 0.2248, respectively (Kielland, 1937; Hamrouni and Dhahbi, 2001). The mole and equivalent fractions in the aqueous phases can be calculated in a similar fashion to that described for the exchanger phases in Box 6.3.

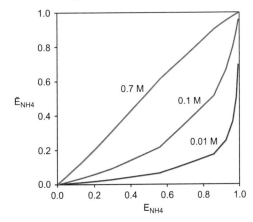

FIGURE B6.4 The influence of ionic strength on the nonpreference adsorption isotherm.

As can be seen in Figure B6.4, at lower ionic strength, there is a predicted tendency for the monovalent cation (NH_4^+) to remain in solution, i.e., the curve is closer toward the aqueous exchanger phase axis (i.e., the x-axis). However, as ionic strength increases, the exchanger phase (i.e., the mineral solid) increases in selectivity for the monovalent cation. This indicates that at

Continued

BOX 6.4 *(cont'd)*

Construction of a nonpreferential adsorption isotherm for a heterovalent ion exchange system

Equilibrium solution concentrations		Mole fractions in the aqueous phase (N)		Equivalent fraction in solution, $E_{NH_4^+}$	Nonpreference adsorption isotherm values for the exchanger phase, correcting for changes in ion activity coefficients $(\bar{E}_{NH_4^+})$		
NH_4^+ M	Ca^{2+} M	NH_4^+	Ca^{2+}	NH_4^+	0.01 M	0.1 M	0.7 M
5.00E-05	3.99E-03	0.01	0.99	0.006	0.0005	0.0016	0.0057
2.60E-04	3.89E-03	0.06	0.94	0.032	0.0025	0.0087	0.0301
4.80E-04	3.84E-03	0.11	0.89	0.059	0.0047	0.0160	0.0554
9.20E-04	3.09E-03	0.23	0.77	0.130	0.0107	0.0366	0.1260
1.62E-03	3.06E-03	0.35	0.65	0.209	0.0182	0.0619	0.2105
2.42E-03	2.95E-03	0.45	0.55	0.291	0.0267	0.0906	0.3012
5.03E-03	1.98E-03	0.72	0.28	0.560	0.0650	0.2169	0.6106
7.93E-03	6.50E-04	0.92	0.08	0.859	0.1742	0.5166	0.9024
8.76E-03	3.50E-04	0.96	0.04	0.926	0.2544	0.6678	0.9521
9.11E-03	1.70E-04	0.98	0.02	0.964	0.3656	0.8013	0.9776
9.31E-03	8.00E-05	0.99	0.01	0.983	0.5047	0.8938	0.9897
9.44E-03	3.00E-05	1.00	0.00	0.994	0.6949	0.9569	0.9962

higher ionic strength, selectivity for the divalent cation to the solid exchanger phase decreases. However, the true selectivity for the monovalent or divalent cation can only be determined when the empirical data of the equivalent fraction on the exchange phase ($\bar{E}$) are plotted along with the nonpreference isotherm. In this case if the empirical data are plotted above the nonpreference isotherm line, then the product would be preferred. The problem set for this chapter requests the calculation of this line based on empirical data to determine whether NH_4^+ or Ca^{2+} is preferred.

Note that when ternary systems are involved (e.g., NH_4^+, K^+, and Ca^{2+}), such as in other experiments by Evangelou and Lumbanraja (2002), it is important to consider a dilution factor on the exchanger phase. Typically, only two ions are considered during ion exchange reactions. In systems with three or more ions calculations for the exchanger phase and thus the mole fraction of the exchanger phase must consider dilution due to the additional ions (i.e., the third ion) in order to avoid any artifacts (Reynolds, 2021).

From the van 't Hoff equation, ΔH_{ex}^0 can be calculated:

$$d \ln K_{ex} / dT = \Delta H_{ex}^0 / RT^2 \qquad (6.32)$$

or

$$E_1 - E_{-1} = \Delta H_{ex}^0 \qquad (6.33)$$

and ΔG_{ex}^o and ΔS_{ex}^o can be determined as given in Equations. (6.21) and (6.24), respectively.

Suggested Reading

Argersinger, W.J., Davidson, A.W., Bonner, O.D., 1950. Thermodynamics and ion exchange phenomena. Trans. Kans. Acad. Sci 53, 404–410.

Babcock, K.L., 1963. Theory of the chemical properties of soil colloidal systems at equilibrium. Hilgardia 34, 417–542.

Gaines, G.L., Thomas, H.C., 1953. Adsorption studies on clay minerals. II. A formation of the thermodynamics of exchange adsorption. J. Chem. Phys. 21, 714–718.

Jensen, H.E., Babcock, K.L., 1973. Cation exchange equilibria on a Yolo loam. Hilgardia 41, 475–487.

Sposito, G., 1981. Cation exchange in soils: an historical and theoretical perspective. In: Dowdy, R.H., Ryan, J.A., Volk, V.V., Baker, D.E. (Eds.), Chemistry in the Soil Environment. Spec. Publ. 40. American Society of Agronomy/Soil Science Society of America, Madison, WI, pp. 13–30.

Sposito, G., 2000. Ion exchange phenomena. In: Sumner, M.E. (Ed.), Handbook of Soil Science. CRC Press, Boca Raton, FL, pp. 241–263.

Sumner, M.E., Miller, W.P., 1996. Cation exchange capacity and exchange coefficients. In: Sparks, D.L. (Ed.), Methods of Soil Analysis, Part 3—Chemical Methods. Soil Science Society of America Book Ser 5. Soil Science Society of America, Madison, WI, pp. 1201–1229.

Problem Set

Q1. Within the following groups or pairs of ions, which ion(s) may absorb preferentially to a soil clay mineral and why?
a) Cu^{2+}, Na^+, Pb^{2+}, and Ca^{2+}
b) Al^{3+} or Ca^{2+}
c) K^+ or Li^+

In the following questions use the data in the tables below for NH_4^+–Ca exchange on both hydroxy-Al interlayered vermiculite (HIV) and vermiculite. Data are from Evangelou and Lumbanraja (2002). The equation below can be used to summarize the reaction.

$$2NH_4^+ + Ca\text{-clay} \rightleftharpoons Ca^{2+} + 2NH_4^+\text{-clay}$$

Q2. Calculate the mole fractions of NH_4^+ and Ca^{2+} on the exchanger phases for both HIV and vermiculite for each exchanger test (i.e., $\overline{N}_{NH_4}$ and $\overline{N}_{Ca}$ for both HIV and vermiculite). $I = 0.01$ M and T =25°C. Be aware of units (molarity vs millimolarity).

HIV	Concentration of cation in solution at equilibrium (mmol liter^{-1})		Concentration of adsorbed cation at equilibrium, i.e., exchanger phase (mol kg^{-1})	
Exchanger test #	NH_4^+	Ca^{2+}	NH_4^+	Ca^{2+}
1	0.05	3.99	0.002	0.0290
2	0.26	3.89	0.007	0.0285
3	0.48	3.84	0.008	0.0290
4	0.92	3.09	0.015	0.0290
5	1.62	3.06	0.022	0.0290
6	2.42	2.95	0.029	0.0290
7	5.03	1.98	0.048	0.0290
8	7.93	0.65	0.079	0.0205
9	8.76	0.35	0.091	0.0160
10	9.11	0.17	0.100	0.0120
11	9.31	0.08	0.103	0.0085
12	9.44	0.03	0.107	0.0050

6. Ion Exchange Processes

Vermiculite	Concentration of cation in solution at equilibrium (mmol liter^{-1})		Concentration of adsorbed cation at equilibrium, i.e., exchanger phase (mol kg^{-1})	
Exchanger test #	NH$_4^+$	Ca^{2+}	NH$_4^+$	Ca^{2+}
1	0.13	3.89	0.002	0.3135
2	0.33	3.78	0.004	0.2710
3	0.52	3.65	0.006	0.2395
4	0.88	2.92	0.009	0.2010
5	1.56	2.87	0.012	0.1760
6	2.37	2.77	0.018	0.1550
7	4.81	1.89	0.020	0.1275
8	8.24	0.48	0.083	0.0925
9	8.92	0.25	0.136	0.0565
10	9.40	0.13	0.190	0.0335
11	9.49	0.07	0.224	0.0215
12	9.60	0.03	0.229	0.0120

Q3. Calculate the mole fractions of NH$_4^+$ and Ca^{2+} in the solution phases for both HIV and vermiculite for each exchanger test (i.e., $\overline{N}_{NH_4}$ and $\overline{N}_{Ca}$ for both HIV and vermiculite).

Q4. Determine K$_V$ and ln K$_V$ for exchanger tests 1–12 on both HIV and vermiculite. Use the table in Chapter 4 and values in Box 6.4 for ion activity coefficients.

Q5. Determine the equivalent fractions of NH$_4^+$ and Ca^{2+} on both exchanger phases (i.e., $\overline{E}_{NH_4^+}$ and $E_{Ca^{2+}}$ on both HIV and vermiculite) and the equivalent fractions of NH$_4^+$ and Ca^{2+} in solution for both experiments (i.e., $E_{NH_4^+}$ and $E_{Ca^{2+}}$ in solution for both HIV and vermiculite exchange experiments).

Q6. Create plots of ln K$_V$ versus $\overline{E}_{NH_4}$ and determine K$_{ex}$ by either integration or summation of the area of trapezoids under the curve. Compare and contrast the plots.

Q7. Determine ΔG_0 for the overall exchange for both HIV and vermiculite. Describe the difference between the two and which one is likely more favorable. Why might this be?

Q8. Calculate the nonpreference exchange isotherm lines for the NH$_4^+$–Ca exchange on both HIV and vermiculite using the equation in Box 6.4 at 0.01, 0.1, and 0.7 I. Then plot the cation exchange isotherms for NH$_4^+$–Ca for both HIV and vermiculite exchange experiments. Use the activity coefficients for the high ionic strength given in Box 6.4. Explain the effects of ionic strength on the nonpreference isotherm line. How do these plots relate to the ΔG values? Based on the exchange isotherm plots and the calculated Gibbs free energies of the reactions, compare NH$_4^+$–Ca exchange on vermiculite versus HIV. On which mineral is NH$_4^+$ preferred? Why?

Q9. Let us assume that 50 g soil was used for CEC analysis. An Mg(NO$_3$)$_2$ solution with a total volume of 200 mL was used to displace the NH$_4^+$. The concentration of NH$_4^+$ in solution obtained via inductively coupled plasma atomic emission spectroscopy (ICP-AES) analysis was found to be 810 mg L^{-1}. Calculate the CEC of the soil.

References

Argersinger, W.J., Davidson, A.W., Bonner, O.D., 1950. Thermodynamics and ion exchange phenomena. Trans. Kans. Acad. Sci. 53, 404–410.

Babcock, K.L., 1963. Theory of the chemical properties of soil colloidal systems at equilibrium. Hilgardia 34, 417–542.

Babcock, K.L., Doner, H.E., 1981. A macroscopic basis for soil chemistry. Soil Sci. 131 (5), 276–283.

Barrer, R.M., Klinowsk, J., 1974. Ion-exchange selectivity and electrolyte concentration. J. Chem. Soc. Faraday Trans. I 70, 2080–2091.

Bourg, I., Sposito, G., 2011. Ion exchange phenomena. In: Huang, P.M., Li, Y., Sumner, M.E. (Eds.), Handbook of Soil Sciences Properties and Processes. Second edition. CRC Press/Taylor & Francis Group, New York, NY, pp. 1–16.

Deist, J., Talibudeen, O., 1967a. Ion exchange in soils from the ion pairs K-Ca, K-Rb, and K-Na. J. Soil Sci. 18 (1), 1225–1237.

Deist, J., Talibudeen, O., 1967b. Thermodynamics of K-Ca exchange in soils. J. Soil Sci. 18 (1), 1238–1248.

Ekedahl, E., Hogfeldt, E., Sillén, L.G., 1950. Activities of the components in ion exchangers. Acta. Chem. Scand. 4, 556–558.

El-Sayed, M.H., Burau, R.G., Babcock, K.L., 1970. Thermodynamics of copper (II)-calcium exchange on bentonite clay. Soil Sci. Soc. Am. Proc. 34, 397–400.

Essington, M.E., 2015. Soil and Water Chemistry: An Integrative Approach, Second edition. CRC Press, Boca Raton, FL.

Evangelou, V.P., Lumbanraja, J., 2002. Ammonium-potassium-calcium exchange on vermiculite and hydroxy-aluminum vermiculite. Soil Sci. Soc. Am. J. 66 (2), 445–455.

Gaines, G.L., Thomas, H.C., 1953. Adsorption studies on clay minerals. II. A formation of the thermodynamics of exchange adsorption. J. Chem. Phys. 21, 714–718.

Gapon, Y.N., 1933. On the theory of exchange adsorption in soils. J. Gen. Chem. USSR Eng. Tran. 3, 144–160.

Gast, R.G., 1972. Alkali metal cation exchange on Chambers montmorillonite. Soil Sci. Soc. Am. Proc. 36, 14–19.

Gast, R.G., Klobe, W.D., 1971. Sodium-lithium exchange equilibria on vermiculite at 25° and 50°C. Clays Clay. Miner. 19, 311–319.

Glossary of Soil Science Terms, 2008. Soil Science Glossary Terms Committee and Soil Science Society of America. Soil Sci. Soc. Am., Madison, WI.

Goulding, K.W.T., 1983a. Thermodynamics and potassium exchange in soils and clay minerals. Adv. Agron. 36, 215–261.

Goulding, K.W.T., 1983b. Adsorbed ion activities and other thermodynamic parameters of ion exchange defined by mole or equivalent fractions. J. Soil Sci. 34, 69–74.

Goulding, K.W.T., Talibudeen, O., 1984. Thermodynamics of K-Ca exchange in soils. II. Effects of mineralogy, residual K and pH in soils from long-term ADAS experiments. J. Soil Sci. 35, 409–420.

Guggenheim, E.A., 1967. Thermodynamics: An Advanced Treatment for Chemists and Physicists. Netherlands: North-Holland Publishing Company, Amsterdam.

Gupta, R.K., Bhumbla, D.K., Abrol, I.P., 1984. Sodium-calcium exchange equilibria in soils as affected by calcium carbonate and organic matter. Soil Sci. 138 (2), 109–114.

Hamrouni, B., Dhahbi, M., 2001. Thermodynamic description of saline waters – Prediction of scaling limits in desalination processes. Desalination 137 (1), 275–284.

Helfferich, F., 1962a. "Ion Exchange". McGraw-Hill, New York.

Helfferich, F., 1962b. Ion exchange kinetics. III. Experimental test of the theory of particle-diffusion controlled ion exchange. J. Phys. Chem. 66, 39–44.

Hogfeldt, E., Ekedahl, E., Sillén, L.G., 1950. Activities of the components in ion exchangers with multivalent ions. Acta. Chem. Scand. 4, 828–829.

Hutcheon, A.T., 1966. Thermodynamics of cation exchange on clay: Ca-K-Montmorillonite. J. Soil Sci. 17, 339–355.

Jardine, P.M., Sparks, D.L., 1984a. Potassium-calcium exchange in a multireactive soil system. I. Kinetics. Soil Sci. Soc. Am. J. 48 (1), 39–45.

Jardine, P.M., Sparks, D.L., 1984b. Potassium-calcium exchange in a multireactive soil system. II. Thermodynamics. Soil Sci. Soc. Am. J. 48 (1), 45–50.

Jensen, H.E., 1972. Potassium-calcium exchange on a montmorillonite and a kaolinite clay. I. A test on the Argersinger thermodynamic approach. Agrochimica 17, 181–190.

Jensen, H.E., Babcock, K.L., 1973. Cation exchange equilibria on a Yolo loam. Hilgardia 41 (16), 475–487.

Kerr, H.W., 1928. The nature of base exchange and soil acidity. J. Am. Soc. Agron. 20, 309–335.

Kielland, J., 1937. Individual activity coefficients of ions in aqueous solutions. J. Am. Chem. Soc. 59, 1675–1678.

Krishnamoorthy, C., Overstreet, R., 1949. Theory of ion exchange relationships. Soil Sci. 68, 307–315.

Laudelout, H., Bladel, R.V., Bolt, G.H., Page, A.L., 1967. Thermodynamics of heterovalent cation exchange reactions in a montmorillonite clay. Trans. Faraday. Soc. 64, 1477–1488.

Laudelot, H., Thomas, H.C., 1965. The effect of water activity on ion-exchange selectivity. J. Phys. Chem. 69, 339–341.

Lewis, G.N., Randall, M., 1961. Thermodynamics, Second editon. McGraw-Hill, New York, NY.

Mehta, S.C., Poonia, S.R., Pal, R., 1983. Sodium-calcium and sodium-magnesium exchange equilibria in soil for chloride- and sulfate-dominated systems. Soil Sci. 136 (6), 339–346.

Ogwada, R.A., Sparks, D.L., 1986a. A critical evaluation on the use of kinetics for determining thermodynamics of ion exchange in soils. Soil Sci. Soc. Am. J. 50 (2), 300–305.

Ogwada, R.A., Sparks, D.L., 1986b. Use of mole or equivalent fractions in determining thermodynamic parameters for potassium exchange in soils. Soil Sci. 141 (4), 268–273.

Reynolds, J.G., 2021. Compositional constraints to identifying ternary interactions in ion-exchange equilibria. Soil Sci. Soc. Am. J. 85 (6), 1985–1989.

Siebecker, M., Li, W., Khalid, S., Sparks, D., 2014. Real-time QEXAFS spectroscopy measures rapid precipitate formation at the mineral–water interface. Nat. Commun. 5, 6003.

Siebecker, M.G., Li, W., Sparks, D.L., 2018. The important role of layered double hydroxides in soil chemical processes and remediation: what we have learned over the past 20 years. In: Sparks, D.L. (Ed.), Advances in Agronomy. Vol 147. Academic Press, Cambridge, Massachusetts, pp. 1–59.

Sparks, D.L., 1989. Kinetics of soil chemical processes. Academic Press, San Diego.

Sposito, G., 1981. Cation exchange in soils: an historical and theoretical perspective. In: Dowdy, R.H., Ryan, J.A., Volk, V.V., Baker, D.E. (Eds.), Chemistry in the Soil Environment. American Society of Agronomy/Soil Science Society of America, Madison, WI, pp. 13–30.

Sposito, G., 2000. Ion exchange phenomena. In: Sumner, M.E. (Ed.), Handbook of Soil Science. CRC Press, Boca Raton, FL, pp. B-241–B-264.

Sposito, G., Mattigod, S.V., 1979. Ideal behavior in Na-trace metal cation exchange on Camp Berteau montmorillonite. Clays Clay Miner. 27, 125–128.

Sumner, M.E., Miller, W.P., 1996. Cation exchange capacity and exchange coefficients. In: Sparks, D.L. (Ed.), "Methods of Soil Analysis, Part 3—Chemical Methods". Soil Sci. Soc. Am. Book Series 5, SSSA, Madison, WI, pp. 1201–1229.

Udo, E.J., 1978. Thermodynamics of potassium-calcium exchange reactions on a kaolinitic soil clay. Soil Sci. Soc. Am. J. 42, 556–560.

Van Bladel, R., Gheyi, H.R., 1980. Thermodynamic study of calcium-sodium and calcium-magnesium exchange in calcareous soils. Soil Sci. Soc. Am. J. 44, 938–942.

Van Bladel, R., Laudelout, H., 1967. Apparent reversibility of ion-exchange reactions in clay suspensions. Soil Sci. 104, 134–137.

Vanselow, A.P., 1932. Equilibria of the base exchange reactions of bentonites, permutites, soil colloids, and zeolites. Soil Sci. 33, 95–113.

Verburg, K., Baveye, P., 1994. Hysteresis in the binary exchange of cations on 2:1 clay minerals: a critical review. Clays Clay Miner. 42, 207–220.

Way, J.T., 1850. On the power of soils to absorb manure. J. R. Agric. Soc. Engl. 11, 313–379.

Wild, A., Keay, J., 1964. Cation-exchange equilibria with vermiculite. J. Soil Sci. 15 (2), 135–144.

CHAPTER

7

Kinetics of Soil Chemical Processes

7.1 Rate-Limiting Steps and Time Scales of Soil Chemical Reactions

Key points: Many soil chemical processes are time dependent. To fully understand the dynamic interactions of metals, oxyanions, radionuclides, pesticides, industrial chemicals, and plant nutrients with soils and to predict their fate with time, a knowledge of the kinetics of these reactions is important. An overview of this topic with applications to environmentally important reactions is presented here.

A number of transport and chemical reaction processes can affect the rate of soil chemical reactions. The slowest of these will limit the rate of a particular reaction. The actual chemical reaction at the surface, for example, adsorption, is usually very rapid and not rate limiting. Transport processes (Figure 7.1) include: (1) transport in the solution phase, which is rapid, and, in the laboratory, can be eliminated by rapid mixing; (2) transport across a liquid film at the particle–liquid interface (film diffusion); (3) transport in liquid-filled macropores (>2 nm), all of which are nonactivated diffusion processes and occur in mobile regions; (4) diffusion of a sorbate along pore wall surfaces (surface diffusion); (5) diffusion of sorbate occluded in micropores (<2 nm) (pore diffusion); and (6) diffusion processes in the bulk of the solid, all of which are activated diffusion processes. Pore and surface diffusion can be referred to as interparticle diffusion, while diffusion in the solid is intraparticle diffusion.

Soil chemical reactions occur over a wide time scale (Figure 7.2), ranging from microseconds and milliseconds for ion association (ion pairing, complexation, and chelation-type reactions in solution), ion exchange, and some sorption reactions to years for mineral solution (precipitation/dissolution reactions including discrete mineral phases) and mineral crystallization

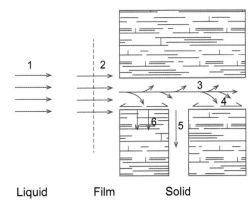

Liquid   Film   Solid

FIGURE 7.1 Transport processes in solid–liquid soil reactions: nonactivated processes, (1) transport in the soil solution, (2) transport across a liquid film at the solid–liquid interface, (3) transport in a liquid-filled macropore; activated processes, (4) diffusion of a sorbate at the surface of the solid, (5) diffusion of a sorbate occluded in a micropore, and (6) diffusion in the bulk of the solid. *From Aharoni and Sparks (1991), with permission.*

Environmental Soil Chemistry, Third Edition
https://doi.org/10.1016/B978-0-12-815880-7.00007-9

305

© 2024 Elsevier Inc. All rights reserved.

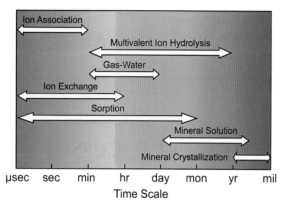

FIGURE 7.2   Time ranges required to attain equilibrium by different types of reactions in soil environments. *From Amacher (1991), with permission.*

reactions (Amacher, 1991). These reactions can occur simultaneously and consecutively.

The type of soil component can drastically affect the reaction rate. For example, sorption reactions are often more rapid on clay minerals such as kaolinite and smectites than on vermiculites and micas.

This is in large part due to the availability of sites for sorption. For example, kaolinite has readily available planar external sites and smectites have primarily interlayer sites that are also quite available for the retention of sorbates. Thus sorption reactions on these soil constituents are often quite rapid, even occurring on time scales of seconds and milliseconds (Sparks, 1989).

On the other hand, vermiculites and micas have multiple sites for the retention of metals and organics, including planar, edge, and interlayer sites, with some of the latter sites being partially to totally collapsed. Consequently, sorption and desorption reactions on these sites can be slow, tortuous, and mass transfer controlled. Often, an apparent equilibrium may not be reached even after several days or weeks. Thus with vermiculites and micas, sorption can involve two to three different reaction rates:

high rates on external sites, intermediate rates on edge sites, and low rates on interlayer sites (Jardine and Sparks, 1984; Comans and Hockley, 1992).

Metal sorption reactions on oxides, hydroxides, and humic substances depend on the type of surface and metal being studied, but the chemical reaction appears to be rapid. For example, the chemical reaction rates of metals and oxyanions on goethite occur on millisecond time scales (Sparks and Zhang, 1991; Grossl et al., 1994, 1997). Half-times for divalent Pb, Cu, and Zn sorption on peat ranged from 5 to 15 s (Bunzl et al., 1976). Siebecker et al. (2014) found that Ni surface precipitation on pyrophyllite occurred in minutes. Reaction rates of redox processes can also be very rapid. For example, Parikh et al. (2008) and Ginder-Vogel et al. (2009) found that the initial reaction process for As(III) oxidation on Mn-oxides occurred on rapid time scales of milliseconds to minutes.

A number of studies have shown that heavy metal sorption on oxides (Barrow, 1986; Bruemmer et al., 1988; Ainsworth et al., 1994; Scheidegger et al., 1997, 1998) and clay minerals (Lövgren et al., 1990) increases with longer residence times (contact time between metal and sorbent). The mechanisms for these lower reaction rates are not well understood but have been ascribed to diffusion phenomena, sites of lower reactivity, and surface nucleation/precipitation (Scheidegger and Sparks, 1996; Sparks, 1998, 1999, 2011).

Sorption and desorption of organic chemicals on soils can be very slow and may demonstrate a residence time effect, which has been attributed to diffusion into the micropores of inorganic minerals and into humic substances, retention on sites of varying reactivity, and surface nucleation/precipitation (Scheidegger and Sparks, 1996; Sparks, 1998, 1999, 2000; Strawn and Sparks, 1999; Alexander, 2000; Pignatello, 2000).

It would be instructive at this point to define two important terms: chemical kinetics and kinetics. Chemical kinetics can be defined as "the investigation of chemical reaction rates and the molecular processes by which reactions occur where transport is not limiting" (Gardiner, 1969). Kinetics is the study of time-dependent processes.

The study of chemical kinetics in homogeneous solutions is difficult, and when one studies heterogeneous systems such as soil components and, particularly, soils, the difficulties are magnified. It is extremely difficult to eliminate transport processes in soils because they are mixtures of several inorganic and organic adsorbates. Additionally, there is an array of different particle sizes and porosities in soils that enhance their heterogeneity. Thus when dealing with soils and soil components, one usually studies the kinetics of the reactions.

7.2 Rate Laws

Key point: Determining how long it takes for a reaction to reach equilibrium and gaining an understanding of chemical reaction mechanisms are the reasons for investigating the rates of soil chemical processes.

One of the most important aspects of chemical kinetics is the establishment of a rate law. By definition, a rate law is a differential equation. For the reaction (Bunnett, 1986)

$$aA + bB \rightarrow yY + zZ \qquad (7.1)$$

the rate of the reaction is proportional to some power of the concentrations of reactants A and B and/or other species (C, D, etc.) in the system. The terms a, b, y, and z are stoichiometric coefficients and are assumed to be equal to one in the following discussion. The power to which the concentration is raised may equal zero (i.e., the rate is independent of that concentration), even for reactant A or B. Rates are expressed as a decrease in reactant concentration or an increase in product concentration per unit time. Thus the rate of reactant A above, which has a concentration [A] at any time t, is $(-d[A]/(dt))$ while the rate with regard

to product Y having a concentration [Y] at time t is $(d[Y]/(dt))$.

The rate expression for Equation (7.1) is

$$\frac{d[Y]}{dt} = -\frac{d[A]}{dt} = k[A]^{\alpha}[B]^{\beta} \qquad (7.2)$$

where k is the rate constant, α is the order of the reaction with respect to reactant A and can be referred to as a partial order, and β is the order with respect to reactant B. The rate constant should only change with temperature. It is not concentration dependent. These orders are experimentally determined and are not necessarily integral numbers. The sum of all the partial orders (α, β, etc.) is the overall order (n) and may be expressed as:

$$n = \alpha + \beta \qquad (7.3)$$

Once the values of α, β, etc., are determined experimentally, the rate law is defined. Reaction order provides only information about the manner in which the rate depends on the concentration. Order does not mean the same as "molecularity," which concerns the number of reactant particles (atoms, molecules, free radicals, or ions) entering an elementary reaction. One can define an elementary reaction as one in which no reaction intermediates have been detected or need to be postulated to describe the chemical reaction on a molecular scale. An elementary reaction is assumed to occur in a single step and to pass through a single transition state (Bunnett, 1986).

To prove that a reaction is elementary, one can use experimental conditions different from those employed in determining the law. For example, if one conducted a kinetic study using a flow technique (see later discussion on this technique) and the rate of influent solution (flow rate) was 1 mL min^{-1}, one could study several other flow rates to see whether reaction rate and rate constants change. If they do, one is not determining mechanistic rate laws.

Rate laws serve three purposes: they assist one in predicting the reaction rate, mechanisms can be proposed, and reaction orders can be ascertained. There are four types of rate laws that can be determined for soil chemical processes (Skopp, 1986): mechanistic, apparent, transport with apparent, and transport with mechanistic. Mechanistic rate laws assume that only chemical kinetics are operational and transport phenomena are not occurring. Consequently, it is difficult to determine mechanistic rate laws for most soil chemical systems due to the heterogeneity of the system caused by different particle sizes, porosities, and types of retention sites. There is evidence that with some kinetic studies using relaxation and molecular-scale in situ spectroscopic techniques (see later discussion), mechanistic rate laws are determined since the agreement between equilibrium constants calculated from both kinetics and equilibrium studies are comparable (Tang and Sparks, 1993; Landrot et al., 2010). This would indicate that transport processes in the kinetics studies are severely limited (see Chapter 5). Apparent rate laws include both chemical kinetics and transport-controlled processes. Apparent rate laws and rate coefficients indicate that diffusion and other microscopic transport processes affect the reaction rate. Thus soil structure, stirring, mixing, and flow rate all would affect the kinetics. The transport with apparent rate law emphasizes transport phenomena. One often assumes first- or zero-order reactions (see discussion below on reaction order). In determining transport with mechanistic rate laws one attempts to describe simultaneously transport-controlled and chemical kinetics phenomena. One is thus trying to accurately explain both the chemistry and the physics of the system.

7.3 Determination of Reaction Order and Rate Constants

Key points: Experimental methods to determine reaction order and rate constants can

vary, but it is always critical to be able to isolate the forward reaction from any back reactions. This way, true chemical kinetics can be determined.

There are three basic ways to determine rate laws and rate constants (Bunnett, 1986; Skopp, 1986; Sparks, 1989): (1) using initial rates, (2) directly using integrated equations and graphing the data, and (3) using nonlinear least-squares analysis. The most common way of determining reaction order and rate constants is to collect time-dependent data.

Let us assume the following elementary reaction (i.e., a reaction with no intermediates) between species A, B, and Y:

$$A + B \underset{k_{-1}}{\overset{k_1}{\rightleftharpoons}} Y \qquad (7.4)$$

A forward reaction rate law can be written as

$$\frac{d[A]}{dt} = -k_1[A][B] \qquad (7.5)$$

where k_1 is the forward rate constant and α and β (see Equation (7.2)) are each assumed to be 1. Thus for this example, the overall reaction order is 2.

The reverse reaction rate law for Equation (7.4) is

$$\frac{d[A]}{dt} = +k_{-1}[Y] \qquad (7.6)$$

where k_{-1} is the reverse rate constant.

Equations (7.5) and (7.6) are only applicable far from equilibrium where back or reverse reactions are insignificant. If both these reactions are occurring, Equations (7.5) and (7.6) must be combined such that

$$\frac{d[A]}{dt} = -k_1[A][B] + k_{-1}[Y] \qquad (7.7)$$

Equation (7.7) applies the principle that the net reaction rate is the difference between the sum of all reverse reaction rates and the sum of all forward reaction rates.

One way to ensure that back reactions are not important is to measure initial rates. The initial rate is the limit of the reaction rate as time reaches zero. With an initial rate method, one plots the concentration of a reactant or product over a short reaction time period during which the concentrations of the reactants change so little that the instantaneous rate is hardly affected. Thus by measuring initial rates, one could assume that only the forward reaction in Equation (7.4) predominates. This would simplify the rate law to that given in Equation (7.5), which as written would be a second-order reaction, first-order in reactant A and first-order in reactant B. Equation (7.4), under these conditions, would represent a second-order irreversible elementary reaction. To measure initial rates, one must have available a technique that can measure rapid reactions, such as a relaxation or real-time molecular-scale technique (see later discussion), and an accurate analytical detection system for determining product concentrations.

Integrated rate equations can also be used to determine rate constants. If one assumes that reactant B in Equation (7.5) is in large excess of reactant A, which is an example of the "method of isolation" to analyze kinetic data, and $Y_0 = 0$, where Y_0 is the initial concentration of product Y, Equation (7.5) can be simplified to

$$\frac{d[A]}{dt} = -k_1[A] \qquad (7.8)$$

The first-order dependence of [A] can be evaluated using the integrated form of Equation (7.8) using the initial conditions at $t = 0$, $A = A_0$:

$$\log[A]_t = \log[A]_0 - \frac{k_1 t}{2.303} \qquad (7.9)$$

The half-time ($t_{1/2}$) for the above reaction is equal to $0.693/k_1$ and is the time required for half of reactant A to be consumed.

If a reaction is first order, a plot of $\log[A]_t$ versus t should result in a straight line with a

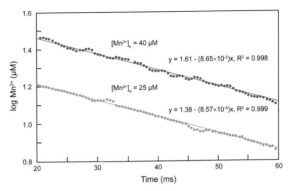

FIGURE 7.3 Initial reaction rates depicting the first-order dependence of Mn^{2+} sorption as a function of time for initial Mn^{2+} concentrations ($[Mn^{2+}]_0$) of 25 and 40 μM. *From Fendorf et al. (1993), with permission.*

slope $= -k_1/2.303$ and an intercept of $\log[A]_0$. An example of first-order plots for Mn^{2+} sorption on δ-MnO_2 at two initial Mn^{2+} concentrations, $[Mn^{2+}]_0$, 25 and 40 μM, is shown in Figure 7.3. One sees that the plots are linear at both concentrations, which would indicate that the sorption process is first order. The $[Mn^{2+}]_0$ values, obtained from the intercept of Figure 7.3, were 24 and 41 μM, in good agreement with the two $[Mn^{2+}]_0$ values. If slope $= (-k)/(2.303)$, then $-k = $ (slope) $\times$ (2.303). The slope with 40 μM Mn^{2+} is -8.65×10^{-3}. Substituting that value into the equation, $-8.65 \times 10^{-3} = (-k)/(2.303)$ and $-k = (-8.65 \times 10^{-3})(2.303)$. So at 40 μM $[Mn^{2+}]_0$, the rate constant k $= (8.65 \times 10^{-3}) \times (2.303) = 1.99 \times 10^{-2}$ ms^{-1}, and at 25 μM $[Mn^{2+}]_0$, k $= (8.57 \times 10^{-3}) \times (2.303) = 1.97 \times 10^{-2}$ ms^{-1}. The findings that the rate constants are not significantly changed with concentration is a very good indication that the reaction in Equation (7.8) is first order under the experimental conditions that were imposed. While conducting kinetic experiments, it is important to ensure that k is independent of concentration.

It is dangerous to conclude that a particular reaction order is correct based simply on the conformity of data to an integrated equation. As illustrated above, multiple initial concentrations that vary considerably should be employed to see that the rate is independent of concentration. One should also test multiple integrated equations. It may be useful to show that reaction rate is not affected by species whose concentrations do not change considerably during an experiment; these may be substances not consumed in the reaction (i.e., catalysts) or present in large excess (Bunnett, 1986; Sparks, 1989).

Least-squares analysis can also be used to determine rate constants. With this method, one fits the best straight line to a set of points linearly related as $y = mx + b$, where y is the ordinate and x is the abscissa datum point, respectively. The slope, m, and the intercept, b, can be calculated with least-squares analysis using Equations (7.10) and (7.11), respectively (Sparks, 1989), as follows:

$$m = \frac{n \sum xy - \sum x \sum y}{n \sum x^2 - (\sum x)^2} \qquad (7.10)$$

$$b = \frac{\sum y \sum x^2 - \sum x \sum xy}{n \sum x^2 - (\sum x)^2} \qquad (7.11)$$

where n is the number of data points and the summations are for all data points in the set.

Curvature may result when kinetic data are plotted. This may be due to an incorrect assumption of reaction order. If first-order kinetics is assumed and the reaction is really second order, a downward curvature is observed. If second-order kinetics is assumed but the reaction is first order, an upward curvature is observed. Curvature can also be due to a fractional, third, higher, or mixed reaction order. The nonattainment of equilibrium often results in a downward curvature. Temperature changes during the study can also cause curvature; thus it is important that the temperature be accurately controlled during a kinetic experiment.

7.4 Kinetic Models

Key points: While first-order models have been used widely to describe the kinetics of soil chemical processes, a number of other models have been employed. These include various ordered equations such as zero-, second-, and fractional-order; Elovich; power function or fractional power; and parabolic diffusion models (Table 7.1). These are the integrated rate equations from pure chemical reactions. Often, though, in soils it is difficult to obtain pure chemical kinetics.

7.4.1 Elovich Equation

The Elovich equation was originally developed to describe the kinetics of the heterogeneous chemisorption of gases on solid surfaces (Low, 1960). It seems to describe a number of reaction mechanisms, including bulk and surface diffusion and the activation and deactivation of catalytic surfaces.

In soil chemistry the Elovich equation has been used to describe the kinetics of the sorption and desorption of various inorganic materials on

TABLE 7.1 Linear forms of kinetic equations commonly used in environmental soil chemistry.[a]

Zero order[b]	$[A]_t = [A]_0 - kt$
First order[b]	$\log[A]_t = \log[A]_0 - \dfrac{kt}{2.303}$ [c]
Second order	$\dfrac{1}{[A]_t} = \dfrac{1}{[A]_0} + kt$
Elovich	$q_t = \left(\frac{1}{\beta}\right)\ln(\alpha\beta) + \left(\frac{1}{\beta}\right)\ln t$
Parabolic diffusion	$\dfrac{q_t}{q_\infty} = R_D t^{1/2}$
Power function	$\ln q = \ln k + v \ln t$

[a] Terms are defined in the text.
[b] Describing the reaction $A \rightarrow Y$.
[c] $\ln x = (2.303) \times (\log x)$ is the conversion from natural logarithms (ln) to base 10 logarithms (log).

soils (see Sparks, 1989). It can be expressed as follows (Chien and Clayton, 1980):

$$q_t = \left(\frac{1}{\beta}\right)\ln(\alpha\beta) + \left(\frac{1}{\beta}\right)\ln t \qquad (7.12)$$

where q_t is the amount of sorbate per unit mass of sorbent at time t and α and β are constants during any one experiment. A plot of q_t versus ln t should give a linear relationship if the Elovich equation is applicable with a slope of $\left(\frac{1}{\beta}\right)$ and an intercept of $\left(\frac{1}{\beta}\right)\ln(\alpha\beta)$.

An application of Equation (7.12) to phosphate sorption on soils is shown in Figure 7.4.

Some investigators have used the α and β parameters from the Elovich equation to estimate reaction rates. For example, it has been suggested that a decrease in β and/or an increase in α would increase the reaction rate. However, this is questionable. The slope of plots using Equation (7.12) changes with the concentration of the adsorptive and with the solution to soil ratio (Sharpley, 1983). Therefore the slopes are not always characteristic of the soil but may depend on various experimental conditions.

Some researchers have also suggested that "breaks" or multiple linear segments in Elovich plots could indicate a changeover from one type of binding site to another (Atkinson et al., 1970). However, such mechanistic suggestions may not be correct (Sparks, 1989).

7.4.2 Parabolic Diffusion Equation

The parabolic diffusion equation is often used to indicate that diffusion-controlled phenomena are rate limiting. It was originally derived based on radial diffusion in a cylinder where the ion concentration throughout the cylinder is uniform. It is also assumed that ion diffusion through the upper and lower faces of the cylinder is negligible. Following Crank (1976), the parabolic diffusion equation, as applied to soils, can be expressed as in Equation (7.13), where r is the average radius of the soil particle, q_t has been defined earlier, q_∞ is the corresponding quantity of sorbate at equilibrium, and D is the diffusion coefficient.

$$\frac{q_t}{q_\infty} = \frac{4}{\pi^{1/2}}\left(Dt^{1/2}\right)/r^2 - Dt/r^2 \qquad (7.13)$$

Equation (7.13) can be simply expressed as

$$\frac{q_t}{q_\infty} = R_D t^{1/2} + \text{constant} \qquad (7.14)$$

where R_D is the overall diffusion coefficient. If the parabolic diffusion law is valid, a plot of q_t/q_∞ versus $t^{1/2}$ should yield a linear relationship. If a linear relationship is found, one might assume that diffusion is taking place.

The parabolic diffusion equation has successfully described metal reactions on soils and soil constituents (Chute and Quirk, 1967; Jardine and Sparks, 1984), feldspar weathering (Wollast, 1967), and pesticide reactions (Weber and Gould, 1966).

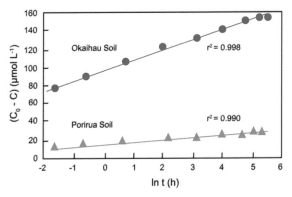

FIGURE 7.4 Plot of Elovich equation for phosphate sorption on two soils where C_0 is the initial phosphorus concentration added at time 0 and C is the phosphorus concentration in the soil solution at time t. The quantity (C_0-C) can be equated to q_t, the amount sorbed at time t. *From Chien and Clayton (1980), with permission.*

7.4.3 Fractional Power or Power Function Equation

This equation can be expressed as

$$q = kt^v \qquad (7.15)$$

where q is the amount of sorbate per unit mass of sorbent, k and v are constants, and v is positive and <1. Equation (7.15) is empirical, except for the case where v = 0.5, when Equation (7.15) is similar to the parabolic diffusion equation.

Equation (7.15) and various modified forms have been used by a number of researchers to describe the kinetics of soil chemical processes (Kuo and Lotse, 1974; Havlin and Wesfall, 1985).

7.4.4 Comparison of Kinetic Models

In a number of studies it has been shown that several kinetic models describe the rate data well, based on correlation coefficients and standard errors of the estimate (Chien and Clayton, 1980; Onken and Matheson, 1982; Sparks and Jardine, 1984). Despite this, there often is not a consistent relation between the equation that gives the best fit and the physicochemical and mineralogical properties of the soil(s) being studied. Another problem with some of the kinetic equations is that they are empirical and no meaningful rate parameters can be obtained.

Aharoni and Ungarish (1976) and Aharoni (1984) noted that some kinetic equations are approximations to which more general expressions reduce in certain limited time ranges. They suggested a generalized empirical equation by examining the applicability of power function, Elovich, and first-order equations to experimental data. By writing these as the explicit functions of the reciprocal of the rate Z, which is $(dq/dt)^{-1}$, one can show that a plot of Z versus t should be convex if the power function equation is operational (1 in Figure 7.5), linear if the Elovich equation is appropriate (2 in Figure 7.5), and concave if the first-order equation is appropriate (3 in Figure 7.5). However, Z versus t plots for soil systems (Figure 7.6) are usually S shaped, convex at small t, concave at large t, and linear at some intermediate t. These findings suggest that the reaction rate can best be described by the power function equation at small t, by the Elovich equation at intermediate t, and by a first-order

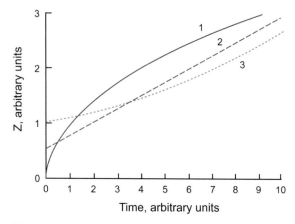

FIGURE 7.5   Plots of Z versus time implied by (1) power function model, (2) Elovich model, and (3) first-order model. The equations for the models were differentiated and expressed as explicit functions of the reciprocal of the rate, Z. *From Aharoni and Sparks (1991), with permission.*

equation at large t. Thus the S-shaped curve indicates that the above equations may be applicable, each at some limited time range.

One of the reasons a particular kinetic model appears to be applicable may be that the study is conducted during the time range when the model is most appropriate. While sorption, for example, decreases over many orders of magnitude before equilibrium is approached, with

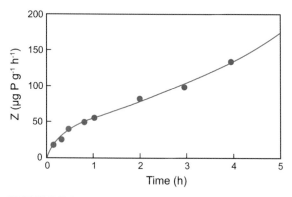

FIGURE 7.6   Sorption of phosphate by a Typic Dystrochrept soil plotted as Z versus time. The circles represent the experimental data of Polyzopoulos et al. (1986). The solid line is a curve calculated according to a homogeneous diffusion model. *From Aharoni and Sparks (1991), with permission.*

most methods and experiments, only a portion of the entire reaction is measured, and over this time range, the assumptions associated with a particular equation are valid. Aharoni and Suzin (1982a, b) showed that the S-shaped curves could be well described using homogeneous and heterogeneous diffusion models. In homogeneous diffusion situations the initial and final portions of the S-shaped curves (conforming to the power function and first-order equations, respectively) predominated (see Figure 7.6 showing data conformity to a homogeneous diffusion model), whereas in instances where the heterogeneous diffusion model was operational the linear portion of the S-shaped curve, which conformed to the Elovich equation, predominated.

The fact that diffusion models describe a number of soil chemical processes is not surprising since in most cases mass transfer and chemical kinetics phenomena are occurring simultaneously, and it is difficult to separate them.

7.5 Multiple Site Models

Key points: Soils are extremely heterogeneous systems with respect to many different factors, including mineralogy, SOM, reactive surface sites, particle size, and water content. Due to this heterogeneity, often, there are nonequilibrium factors that need to be considered when modeling soil chemical reactions. There are two main types of nonequilibrium: physical and chemical.

Based on the previous discussion, it is evident that simple kinetic models such as ordered reaction, power function, and Elovich models may not be appropriate to describe reactions in heterogeneous systems such as soils, sediments, and soil components. In these systems where there is a range of particle sizes and multiple retention sites, both chemical kinetics and transport phenomena occur simultaneously, and a fast reaction is often followed by a slower reaction(s). In such systems nonequilibrium models

that describe both chemical and physical nonequilibrium and that consider multiple components and sites are more appropriate. Physical nonequilibrium is ascribed to some rate-limiting transport mechanism such as film diffusion or interparticle diffusion, while chemical nonequilibrium is due to a rate-limiting mechanism at the particle surface (i.e., the chemical reaction). Nonequilibrium models include two-site, multiple-site, radial diffusion (pore diffusion), surface diffusion, and multiprocess models (Table 7.2). The emphasis here will be placed on the use of these models to describe sorption phenomena.

The term sites can have a number of meanings (Brusseau and Rao, 1989): (1) specific, molecular-scale reaction sites; (2) sites of differing degrees of accessibility (external, internal); (3) sites of differing sorbent type (organic matter and inorganic mineral surfaces); and (4) sites with different sorption mechanisms. With chemical nonequilibrium sorption processes, the sorbate may undergo two or more types of sorption reactions, one of which is rate limiting. For example, a metal cation may sorb to organic matter via one mechanism and to mineral surfaces via another mechanism, with one of the mechanisms being time dependent.

7.5.1 Chemical Nonequilibrium Models

Chemical nonequilibrium models describe time-dependent reactions at sorbent surfaces. The one-site model is a first-order approach that assumes that the reaction rate is limited by only one process or mechanism on a single class of sorbing sites and that all sites are of the time-dependent type. In many cases this model appears to describe soil chemical reactions quite well. However, often it does not. This model would seem not appropriate for most heterogeneous systems since multiple sorption sites exist.

The two-site (two-compartment, two-box) or bicontinuum model has been widely used to describe chemical nonequilibrium (Leenheer

TABLE 7.2 Comparison of chemical and physical nonequilibrium sorption kinetic models.[a]

Conceptual model	Fitting parameter(s)	Model limitations
Chemical nonequilibrium models		
One-site model (Coates and Elzerman, 1986) $$S \xrightarrow{k_d} C$$	k_d	Cannot describe biphasic sorption/desorption
Two-site model (Coates and Elzerman, 1986) $$S_1 \xleftarrow{X_1 K_p} C \xleftrightarrow{k_d} S_2$$	k_d, K_p^c, X_1	Cannot describe the "bleeding" or slow, reversible, nonequilibrium desorption for residual sorbed compounds (Karickhoff, 1980)
Multisite continuum compartment model (Connaughton et al., 1993) $$F(t) = 1 - \frac{M(t)}{M} = 1 - \left(\frac{\beta}{\beta + t}\right)^{\alpha}$$	α, β	Assumption of homogeneous, spherical particles and diffusion only in aqueous phase
Physical nonequilibrium models		
Radial diffusion penetration instantaneous retardation (pore diffusion) model (Wu and Gschwend, 1986) $$S' \xleftarrow{K_p} C' \xrightarrow{D_{eff}} C$$	[d]$D_{eff} = f(n, t)D_m n \,/(1-n)\rho_s K_p$	Cannot describe factor (Ball, 1989); did not describe kinetic data for times greater than 10^3 min (Wu and Gschwend, 1986)
Dual-resistance surface diffusion model (Miller and Pedit, 1992) $$S' \xleftarrow{D_s} C_s' \xrightarrow{k_b} C$$	D_s, k_b	Model calibrated with sorption data predicted more desorption than occurred in the desorption experiments (Miller and Pedit, 1992)
Pore space diffusion model (Fuller et al., 1993) $$\left(\varepsilon + \frac{S_a}{n}K_s C(r)^{(1-1/n)}\right)\frac{\partial C(r)}{\partial t} = D_e\left(\frac{\partial^2 C(r)}{\partial r^2} + \frac{2\partial C(r)}{r\partial r}\right)$$	D_e, ε, K_s, $1/n$, F_{eq}	
Multiple particle class pore diffusion model (Pedit and Miller, 1995) $$\left(\theta_p^i + \rho_a^i \frac{\partial q_r^i(r,t)}{\partial C_p^i(r,t)}\right)\frac{\partial C_p^i(r,t)}{\partial t} = \frac{\theta_p^i D_p^i}{r^2}\frac{\partial}{\partial r}\left(r^2 \frac{\partial C_p^i(r,t)}{\partial r}\right)$$ $$- \theta_p^i \lambda_p^i C_p^i(r,t) - \rho_a^i \lambda_r^i q_r^i(r,t)$$	θ_p^i, ρ_a^i, D_p^i, λ_p^i, λ_r^i	Multiple fitting parameters; variations in sorption equilibrium and rates that might occur within a particle class or an individual particle grain are not addressed

[a] Abbreviations used are as follows: C, concentration of the bulk aqueous phase contaminant ($g\ mL^{-1}$); C', concentration of contaminant free in the pore fluid ($mol\ cm^{-3}$); C_s', solution-phase solute concentration corresponding to an equilibrium with the solid-phase solute concentration at the exterior of the particle ($g\ L^{-1}$); D_{eff}, effective diffusivity of sorbate molecules or ions in the particles ($cm^2\ s^{-1}$); D_m, pore fluid diffusivity of the sorbate ($cm^2\ s^{-1}$); D_s, surface diffusion coefficient ($m\ s^{-1}$); $f(n,t)$, pore geometry factor; k_b, boundary layer mass transfer coefficient ($m\ s^{-1}$); $F(t)$, fraction of mass released through time t; k_d, first-order desorption rate coefficient (min^{-1}); K_p, sorption equilibrium partition coefficient ($mL\ g^{-1}$); M, total initial mass; $M(t)$, mass remaining after time t; n, porosity of the sorbent (cm^3 of fluid cm^{-3}); r, radius of the spherical solid particle, assumed constant (m); S_2, concentration of the sorbed contaminant that is rate limited ($g\ g^{-1}$); S, concentration of the bulk sorbed contaminant ($g\ g^{-1}$); S', concentration of contaminant in immobile bound state ($mol\ g^{-1}$); S_1, concentration of the contaminant that is in equilibrium with the bulk aqueous concentration ($g\ g^{-1}$); X_1, fraction of the bulk sorbed contaminant that is in equilibrium with the aqueous concentration; α, shape parameter; β, scale parameter necessary for the determination of mean and standard deviation of k_s; ρ, macroscopic particle density of the solid phase ($g\ m^{-3}$); ρ_s, specific gravity of the sorbent ($g\ cm^{-3}$).

[c] K_p can be determined independently.

[d] a, radius of the aggregate; $C_p^i(r,t)$, intraparticle fluid-phase solute concentration of the particle class i; $C(r)$, concentration of sorptive in the aqueous phase in the pore fluid at radial distance r; D_e, effective diffusion coefficient; D_p^i, pore diffusion coefficient for particle class i; K_p, D_m, and ρ can be determined independently; ε, internal porosity of sorbent; F_{eq}, equilibrium fraction of adsorption sites; K_s, adsorption isotherm intercept; θ_p^i, intraparticle porosity of particle class i; ρ_a^i, apparent particle density of particle class i; $q_r^i(r,t)$, intraparticle solid-phase solute concentration of particle class i; r, radial distance; S_a is the surface of sorbent per unit volume of solid; $1/n$, the adsorption isotherm slope; λ_p^i, intraparticle fluid-phase first-order reaction rate coefficient for particle class i; λ_r^i, intraparticle solid-phase first-order reaction rate coefficient for particle class i.
From Connaughton et al. (1993), with permission.

and Ahlrichs, 1971; Hamaker and Thompson, 1972; Karickhoff, 1980; McCall and Agin, 1985; Jardine et al., 1992) and physical nonequilibrium (Nkedi-Kizza et al., 1984; Lee et al., 1988; van Genuchten and Wagenet, 1989) (Table 7.2). This model assumes that there are two reactions occurring, one that is fast and reaches equilibrium quickly and a slower reaction that can continue for long time periods. The reactions can occur either in series or in parallel (Brusseau and Rao, 1989).

In describing chemical nonequilibrium with the two-site model two types of sorbent sites are assumed. One site involves an instantaneous equilibrium reaction and the other a time-dependent reaction. The former is described by an equilibrium isotherm equation while a first-order equation is usually employed for the latter.

With the two-site model, there are two adjustable or fitting parameters, the fraction of sites at local equilibrium (X_1) and the rate constant (k). A distribution (K_d) or partition coefficient (K_p) is determined independently from a sorption/ desorption isotherm.

To account for the multiple sites that may exist in heterogeneous systems, Connaughton et al. (1993) developed a multisite compartment (continuum) model (Γ) that incorporates a continuum of sites or compartments with a distribution of rate coefficients that can be described by a gamma density function. A fraction of the sorbed mass in each compartment is at equilibrium with a desorption rate coefficient or distribution coefficient for each compartment or site (Table 7.2). The multisite model has two fitting parameters, namely, α, a shape parameter, and $1/\beta$, which is a scale parameter that determines the mean standard deviation of the rate coefficients.

7.5.2 Physical Nonequilibrium Models

A number of models can be used to describe physical nonequilibrium reactions. Since transport processes in the mobile phase are not usually rate limiting, physical nonequilibrium models focus on diffusion in the immobile phase or interparticle/diffusion processes such as pore and/or surface diffusion. The transport between mobile and immobile regions is accounted for in physical nonequilibrium models in three ways (Brusseau and Rao, 1989): (1) explicitly with Fick's law to describe the physical mechanism of diffusive transfer, (2) explicitly by using an empirical first-order mass transfer expression to approximate solute transfer, and (3) implicitly by using an effective or lumped dispersion coefficient that includes the effects of sink/source differences and hydrodynamic dispersion and axial diffusion.

A pore diffusion model (Table 7.2) has been used by a number of investigators to study sorption processes using batch systems (Wu and Gschwend, 1986; Steinberg et al., 1987). The sole fitting parameter in this model is the effective diffusivity (D_e), which may be estimated *a priori* from chemical and colloidal properties. However, this estimation is only valid if the sorbent material has a narrow particle size distribution so that an accurate, average particle size can be defined. Moreover, in the pore diffusion model an average representative D_e is assumed, which means there is a continuum in properties across an entire pore size spectrum. This is not a valid assumption for micropores in which there are higher adsorption energies of sorbates, causing increased sorption. The increased sorption reduces diffusive transport rates in nonlinear isotherms for sorbents with pores less than several sorbate diameters in size. Other factors, including steric hindrance, which increases as the pore size approaches the solute size, and greatly increased surface-area-to-pore-volume ratios, which occur as pore size decreases, can cause reduced transport rates in micropores.

Another problem with the pore diffusion model is that sorption and desorption kinetics may have been measured over a narrow concentration range. This is a problem since a sorption/ desorption mechanism in micropores at one concentration may be insignificant at another concentration.

Fuller et al. (1993) used a pore space diffusion model (Table 7.2) to describe arsenate adsorption on ferrihydrite that included a subset of sites whereby sorption was at equilibrium. A Freundlich model was used to describe sorption on these sites. Intraparticle diffusion was described by Fick's second law of diffusion; homogeneous, spherical aggregates and diffusion only in the solution phase were assumed. Figure 7.7 shows the fit of the model when sorption at all sites was controlled by intraparticle diffusion. The fit was better when sites that had attained sorption equilibrium were included based on the assumption that there was an initial rapid sorption on external surface sites before intraparticle diffusion.

Pedit and Miller (1995) have developed a general multiple particle class pore diffusion model that accounts for differences in physical and sorptive properties for each particle class (Table 7.2). The model includes both instantaneous equilibrium sorption and time-dependent pore diffusion for each particle class. The pore diffusion portion of the model assumes that solute transfer between the intraparticle fluid and the solid phases is fast vis-à-vis interparticle pore diffusion processes.

Surface diffusion models, assuming a constant surface diffusion coefficient, have been used by a number of researchers (Weber and Miller, 1988; Miller and Pedit, 1992). The dual resistance model (Table 7.2) combines both pore and surface diffusion.

7.6 Kinetic Methodologies

Key points: A number of methodologies can be used to study the rates of soil chemical processes. These can be broadly classified as methods for slower reactions (>15 s), which include batch and flow techniques, and rapid techniques that can measure reactions on milli- and microsecond time scales. It should be recognized that none of these methods is a panacea for kinetic analyses. They all have advantages and disadvantages, and kinetic results can be method dependent.

7.6.1 Batch Methods

Batch methods have been the most widely used kinetic techniques. In the simplest traditional batch technique an adsorbent is placed in a series of vessels such as centrifuge tubes with a particular volume of adsorptive. The tubes are then mixed by shaking or stirring. At various times, a tube is sacrificed for analysis; that is, the suspension is either centrifuged or filtered to obtain a clear supernatant for analysis. A number of variations of batch methods exist and these are discussed by Amacher (1991).

There are a number of disadvantages to traditional batch methods. Often the reaction is complete before a measurement can be made, particularly if centrifugation is necessary, and the solid/solution ratio may be altered as the experiment proceeds. Too much mixing may cause abrasion of the adsorbent, altering the surface area, while too little mixing may enhance mass transfer and transport processes (e.g.,

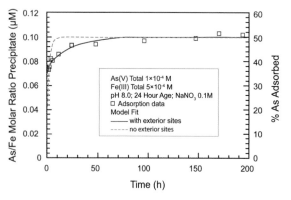

FIGURE 7.7 Comparison of pore space diffusion model fits of As(V) sorption with experimental data (dashed curve represents sorption where all surface sites are diffusion limited and the solid curve represents sorption on equilibrium sites plus diffusion-limited sites). *From Fuller et al. (1993), with permission.*

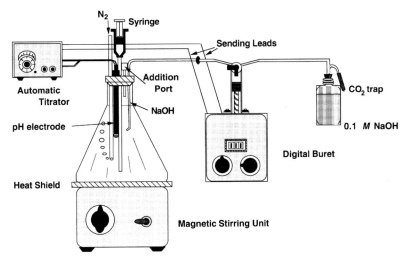

FIGURE 7.8   Schematic diagram of equipment used in the batch technique of *Zasoski and Burau (1978), with permission.*

diffusion). Another major problem with all batch techniques, unless a resin or chelate material such as sodium tetraphenylboron is used, is that released species are not removed. This can cause inhibition in further adsorbate release and the promotion of secondary precipitation in dissolution studies. Moreover, reverse reactions are not controlled, which makes the calculation of rate coefficients difficult and perhaps inaccurate.

Many of the disadvantages listed above for traditional batch techniques can be eliminated by using a method like that of Zasoski and Burau (1978), shown in Figure 7.8. In this method an adsorbent is placed in a vessel containing the adsorptive, pH and suspension volume are adjusted, and the suspension is vigorously mixed with a magnetic stirrer. At various times, suspension aliquots are withdrawn using a syringe containing N_2 gas. The N_2 gas prevents CO_2 and O_2 from entering the reaction vessel. The suspension is rapidly filtered and the filtrates are then weighed and analyzed. With this apparatus, a constant pH can be maintained, reactions can be measured at 15 s intervals, excellent mixing occurs, and a constant solid-to-solution ratio is maintained.

7.6.2 Flow Methods

Flow methods can range from continuous flow techniques (Figure 7.9), which are similar to liquid-phase chromatography, to stirred-flow methods (Figure 7.10) that combine aspects of both batch and flow methods. Important attributes of flow techniques are that one can conduct studies at realistic soil-to-solution ratios that better simulate field conditions, the adsorbent is exposed to a greater mass of ions than

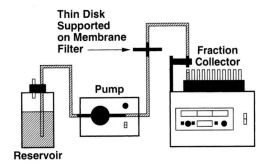

FIGURE 7.9   Thin-disk flow (continuous flow) method experimental setup. Background solution and solute are pumped from the reservoir through the thin disk and are collected as aliquots by the fraction collector. *From Amacher (1991), with permission.*

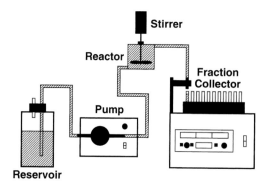

FIGURE 7.10   Stirred-flow reactor method experimental setup. Background solution and solute are pumped from the reservoir through the stirred reactor containing the solid phase and are collected as aliquots by the fraction collector. The separation of solid and liquid phases is accomplished by a membrane filter at the outlet end of the stirred reactor. *From Amacher (1991), with permission.*

in a static batch system, and the flowing solution removes desorbed and detached species.

With continuous flow methods, samples can be injected as suspensions or spread dry on a membrane filter. The filter is attached to its holder by securely capping it, and the filter holder is connected to a fraction collector and peristaltic pump, the latter maintaining a constant flow rate. The influent solution then passes through the filter and reacts with the adsorbent, and at various times, effluents are collected for analysis. Depending on flow rate and the amount of effluent needed for analysis, samples can be collected about every 30–60 s. One of the major problems with this method is that the colloidal particles may not be dispersed; that is, the time necessary for an adsorptive to travel through a thin layer of colloidal particles is not equal at all locations of the layer. This plus minimal mixing promotes significant transport effects. Thus apparent rate laws and rate coefficients are measured, with the rate coefficients changing with flow rate. There can also be dilution of the incoming adsorptive solution by the liquid used to load the adsorbent on the filter, particularly if the adsorbent is placed on the filter as a

suspension, or if there is washing out of the remaining adsorptive solution during desorption. This can cause concentration changes that are not due to adsorption or desorption.

A more preferred method for measuring soil chemical reaction rates is the stirred-flow method. The experimental setup is similar to the continuous flow method (Figure 7.9), except there is a stirred-flow reaction chamber rather than a membrane filter. A schematic of this method is shown in Figure 7.10. The sorbent is placed into the reaction chamber, where a magnetic stir bar or an overhead stirrer keeps it suspended during the experiment. There is a filter placed at the top of the chamber that keeps the solids in the reaction chamber. A peristaltic pump maintains a constant flow rate, and a fraction collector is used to collect the leachates. The stirrer effects perfect mixing; that is, the concentration of the adsorptive in the chamber is equal to the effluent concentration.

This method has several advantages over the continuous flow technique and other kinetic methods. Reaction rates are independent of the physical properties of the porous media, the same apparatus can be used for adsorption and desorption experiments, desorbed species are removed, continuous measurements allow for monitoring reaction progress, experimental factors such as flow rate and adsorbent mass can be easily altered, a variety of solids can be used (however, sometimes fine particles can clog the filter, causing a buildup in pressure, which results in a nonconstant flow rate) with the technique, the adsorbent is dispersed, and dilution errors can be measured. With this method, one can also use stopped-flow tests and vary influent concentrations and flow rates to elucidate possible reaction mechanisms (Bar-Tal et al., 1990).

7.6.3 Relaxation Techniques

As noted earlier, many soil chemical processes are very rapid, occurring on milli- and microsecond time scales. These include metal and organic sorption–desorption reactions, ion exchange processes, and ion association reactions. Batch and flow techniques, which measure reaction rates of >15 s, cannot be employed to measure these reactions. Chemical relaxation methods can be used to measure very rapid reactions. These include pressure-jump (p-jump), electric field pulse, temperature-jump (t-jump), and concentration-jump (c-jump) methods. These methods are fully outlined in other sources (Sparks, 1989; Sparks and Zhang, 1991). Only a brief discussion of the theory of chemical relaxation and a description of p-jump methods will be given here. The theory of chemical relaxation can be found in a number of sources (Eigen, 1954; Takahashi and Alberty, 1969; Bernasconi, 1976). It should be noted that relaxation techniques are best used with soil components such as oxides and phyllosilicates and not whole soils. Soils are heterogeneous, which complicates the analyses of the relaxation data.

All chemical relaxation methods are based on the theory that the equilibrium of a system can be rapidly perturbed by some external factor such as pressure, temperature, or electric field strength. Rate information can then be obtained by measuring the approach from the perturbed equilibrium to the final equilibrium by measuring the relaxation time, τ (the time that it takes for the system to relax from one equilibrium state to another, after the perturbation pulse), by using a detection system such as conductivity. The relaxation time is related to the specific rates of the elementary reactions involved. Since the perturbation is small, all rate expressions reduce to first-order equations regardless of reaction order or molecularity (Bernasconi, 1976). The rate equations are then linearized such that

$$\tau^{-1} = k_1(C_A + C_B) + k_{-1} \qquad (7.16)$$

where k_1 and k_{-1} are the forward and backward rate constants, respectively, and C_A and C_B are the concentrations of reactants A and B at

equilibrium, respectively. From a linear plot of τ^{-1} versus (C_A+C_B), one could calculate k_1 and k_{-1} from the slope and intercept, respectively. Pressure-jump relaxation is based on the principle that chemical equilibria depend on the pressure, as shown below (Bernasconi, 1976):

$$\left(\frac{\partial \ln K^{\circ}}{\partial \ln p}\right)_T = -\Delta V/RT \qquad (7.17)$$

where K° is the equilibrium constant, ΔV is the standard molar volume change of the reaction, p is pressure, and R and T were defined earlier. For a small perturbation,

$$\frac{\Delta K^{\circ}}{K^{\circ}} = \frac{-\Delta V \Delta p}{RT} \qquad (7.18)$$

Details on the experimental protocol for a p-jump study can be found in several sources (Sparks, 1989; Zhang and Sparks, 1989; Grossl et al., 1994). Chemical relaxation techniques such as p-jump and c-jump (e.g., stopped-flow) allow rapid initial rate data collection on time scales of milliseconds, which minimizes back reactions. They are predicated on the assumption that the reactions are reversible, which often is not the case with sorption processes. A major disadvantage of most chemical relaxation techniques is that the rate "constants" are calculated from linearized rate equations that often include parameters that were determined from equilibrium and modeling studies. Consequently, the rate "constants" are not directly determined (Ginder-Vogel et al., 2009; Ginder-Vogel and Sparks, 2010; Sparks, 2013).

One example of a chemical relaxation method that avoids the above pitfalls is an electron paramagnetic resonance stopped-flow (EPR-SF) method (an example of a c-jump method) used by Fendorf et al. (1993) to study reactions in colloidal suspensions in situ on millisecond time scales. With the EPR-SF method of Fendorf et al. (1993), the mixing can be done in <10 ms and EPR digitized within a few microseconds. A diagram of the EPR-SF instrument is shown

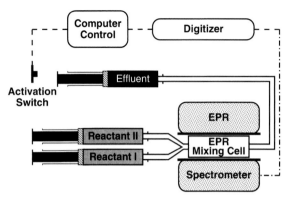

FIGURE 7.11   Schematic diagram of the electron paramagnetic resonance monitored stopped-flow kinetic apparatus. *From Fendorf et al. (1993), with permission.*

in Figure 7.11. Dual 2 mL in-port syringes feed a mixing cell located in the EPR spectrometer. This allows for the EPR detection of the cell contents. A single outflow port is fitted with a 2 mL effluent collection syringe equipped with a triggering switch. The switch activates the data acquisition system. Each run consists of filling the in-port syringes with the desired reactants, flushing the system with the reactants several times, and initiating and monitoring the reaction. Fendorf et al. (1993) used this system to study the kinetics of Mn^{2+} sorption on δ-MnO_2. The sorption reaction was complete in 200 ms. Data were taken every 50 μs and 100 points were averaged to give the time-dependent sorption of Mn(II).

Another example of a stopped-flow technique is the use of UV-Vis to determine the reduction kinetics of polymeric MnO_2 via Fe^{2+}. This redox couple, which is prevalent at oxic/suboxic interfaces in sediments, was analyzed by Siebecker et al. (2015), and the experiments employed a pneumatic stopped-flow apparatus coupled to a UV-Vis spectrophotometer. Similar to other stopped-flow techniques, solutions from two syringes are flushed together through a flow cell where the reaction is monitored. Rapid scanning mode on the UV-Vis enabled scans over a 70 nm

range every 100 ms, and the decrease in a broad ligand to metal charge transfer (LMCT) peak of the MnO_2 was analyzed to determine initial reaction rates and rate constants. As can be seen in Figure 7.12, the reactions were very fast as electron transfer from Fe^{2+} to MnO_2 took place. In this redox reaction Fe oxidizes, and Mn becomes reduced. The initial slopes of three exemplary reactions are labeled with brackets and are per 0.2 s. The slope is -2.29 μM MnO_2 reduced per 0.2 s. As noted in Table 7.3, the reaction is exceedingly fast and can be 50% complete within less than 1 s, depending on reaction conditions.

7.6.4 Real-Time Molecular-Scale Kinetic Methods

Major advances have been made in the employment of real-time molecular-scale kinetic techniques where initial reaction rates can be measured on time scales of milliseconds. This minimizes back reactions and transport, enabling the measurement of chemical kinetics and the determination of reaction mechanisms.

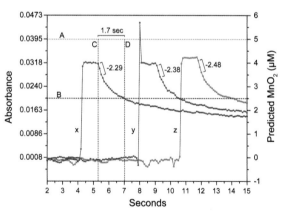

FIGURE 7.12   Reduction kinetics of polymeric MnO_2 by Fe^{2+}. The sharp increase in absorbance at x, y, and z occurs when the pneumatic fast mixer is fired and the MnO_2 and Fe^{2+} are flushed together in the flow cell. The plateau is the time it takes for the solutions to pump through the cell. When flow stops, the reaction continues and absorbance decreases. *From Siebecker et al. (2015), with permission.*

TABLE 7.3 Reduction kinetics of polymeric MnO_2 by Fe^{2+}. In these experiments MnO_2 and Fe^{2+} were reacted under pseudo-first-order reaction conditions, where an excess (10x-20x) of Fe^{2+} was used to reduce MnO_2. Observed pseudo-first-order rate constants (s^{-1}) for 5 μM MnO_2 reacted with 50 and 100 μM Fe^{2+} were obtained from $\ln[MnO_2]$ versus time plots.

Fe/Mn ratio	$-MnO_2$ (μM)/0.2 s	$-MnO_2$ (μM)/s	$-MnO_2$ (μM)/min	k (s^{-1})	R^2	Time (s) to reach 50% abs
10.5	2.35	11.8	706	0.649	0.995	1.77
20.5	3.21	16.1	964	0.988	0.997	0.70

Data are averages from Siebecker et al. (2015), with permission.

Real-time molecular-scale ATR-FTIR and Q-XAS techniques have been employed to elucidate the rapid redox kinetics and mechanisms of As(III) oxidation (Ginder-Vogel et al., 2009; Parikh et al. 2008), Cr(III) oxidation (Landrot et al. 2010) on Mn-oxides, and Ni−Al layered double hydroxide (LDH) phase precipitate formation on pyrophyllite (Siebecker et al. 2014).

Figure 7.13 shows Q-XAS spectra collected from batch experiments for As(III) oxidation on hydrous Mn-oxide (HMO). An initial concentration of 5 mM of As(III) was used. After 1 s of reaction, the As(V) concentration, indicating the oxidation of As(III), reached 0.37 mM and continued to increase rapidly for 45 s, eventually reaching a concentration of 1 mM (Ginder-Vogel et al. 2009). A complete XANES spectrum was collected in 980 ms (Figure 7.13). The oxidation states in both the solution and at the mineral/water interface were probed using Q-XAS. The average oxidation state was then determined by fitting each individual XANES spectra with a linear combination of As(III) and As(V) standards (in this experiment, 5 mM aqueous As standards were used). The fits provide the molar ratio of As in solution, from which the concentrations of As(III) and As(V) in the system can be calculated based on the initial As concentration. In Figure 7.13 one sees the XANES spectra for the reaction of As(III) with HMO over time scales ranging from 0.98 to 298.9 s and the rapid transformation of As(III) to As(V) on the HMO surface. Rate constants were determined from first-order plots for both the Q-XAS and batch

experiments based on biphasic kinetics (Table 7.4) and show, independent of the method, that rate constants for As(III) depletion are about an order of magnitude larger during the initial portion of the reaction than in the later segment. However, the initial rate constant of As(III) depletion based on Q-XAS was nearly twice as

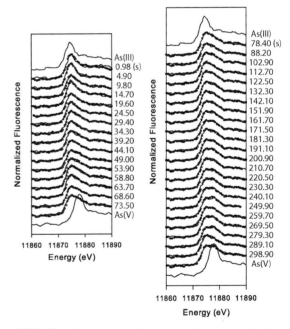

FIGURE 7.13 Data (solid lines) and linear combination fits (dots) of As K-edge XANES spectra collected via Q-XAS. Data were used to determine changes in As(III) and As(V) concentrations during the batch reactions. Each spectrum was collected in approximately 980 ms. *From Ginder-Vogel et al. (2009), with permission.*

TABLE 7.4 Apparent first-order rate constants of arsenite oxidation by HMO determined from batch reactions and Q-XAS. The rate constants of As(III) depletion were determined by using linear regression analysis of the noted time periods for the plots in Figure 7.13.

Experiment type	Time period (s)	# of data points	k (s^{-1})	R^2
As(III)–Batch	1–60	4	$2.5\,(3) \times 10^{-3}$	0.96
As(III)–Batch	135–300	14	$6.1\,(4) \times 10^{-4}$	0.82
As(III)–Q-XAS	1–30	30	$4.7\,(4) \times 10^{-3}$	0.91
As(III)–Q-XAS	135–300	168	$4.9\,(4) \times 10^{-4}$	0.74

From Ginder-Vogel et al. (2009), with permission.

large (4.7×10^{-3} s^{-1}) as the rate constant measured with the batch method (2.5×10^{-3} s^{-1}).

Landrot et al. (2010) employed the same Q-XAS technique as Ginder-Vogel et al. (2009) to measure the rapid initial kinetics of Cr(III) oxidation on HMO. A complete XANES spectrum was collected in 0.75 s. Data were collected at pH values of 2.5, 3.0, and 3.5 and at Cr(III) concentrations of 40–100 mM. Chromium(VI), resulting from the oxidation of Cr(III) on the HMO surface, displays a prominent preedge feature in the XANES spectrum, which is related to the proportion of Cr(VI) in the system. Therefore one only needs to measure the height of the preedge feature and compare it to a set of Cr(III)/Cr(VI) mixtures to determine the amount of Cr(VI) present in the system (Landrot et al. 2010; Peterson et al. 1997). One sees in Figure 7.14 that the preedge feature increases in intensity with time, indicating the oxidation of Cr(III) to Cr(VI). Chromium(III) oxidation is quite rapid during the first 120 s of the reaction, when about 35 mM of chromate was produced. Initial reaction rates were determined by quantifying Cr(VI) from the preedge feature height for each experiment during the first 60 s of the reaction using the kinetic model below.

With this approach, one of the reactants is fixed at a constant value (but not necessarily in excess) in a set of experiments, while the concentration of the other reactant is successively varied. Therefore both the reactants must be considered in the overall rate equation. During the early stage of the reaction, a linear relationship relates [Cr(VI)] and time. The slope of this linear trend is the initial rate i of the reaction, which is defined as:

$$i = \frac{d[Cr(VI)]}{dt} = -k[Cr(III)]_0^{\alpha}[MnO_2]_0^{\beta} \quad (7.19)$$

where k, α, and β were defined earlier, and 0 represents the initial concentrations. To measure the partial rate coefficient of chromium α at a

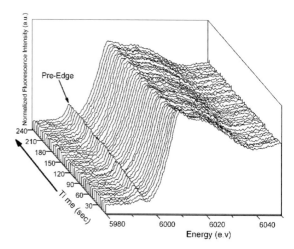

FIGURE 7.14 Cr(III) oxidation kinetics using Q-XAS at pH 2.5. [Cr(III)] = 100 mM, hydrous manganese oxide [HMO] = 20 g L^{-1}, and time 0–240 s. Each XANES spectrum represents 3 s of the reaction and is an average of four spectra collected over 0.75 s each. *From Landrot et al. (2010), with permission.*

given pH value for the experiments (where [MnO$_2$] was fixed and [Cr(III)] was varied), Equation (7.22) was used, which is a simplified expression of Equation (7.20), where exp = experiment:

$$\frac{i_{exp1} = -k[Cr(III)_{exp1}]^{\alpha}[MnO_{2\,fixed}]^{\beta}}{i_{exp2} = -k[Cr(III)_{exp2}]^{\alpha}[MnO_{2\,fixed}]^{\beta}} \qquad (7.20)$$

Hence:

$$\frac{i_{exp1}}{i_{exp2}} = \left(\frac{[Cr(III)_{exp1}]}{[Cr(III)_{exp2}]}\right)^{\alpha} \qquad (7.21)$$

Therefore:

$$\alpha = \frac{\log\left(\dfrac{i_{exp1}}{i_{exp2}}\right)}{\log\left(\dfrac{[Cr(III)_{exp1}]}{[Cr(III)_{exp2}]}\right)} \qquad (7.22)$$

To measure the partial rate coefficient of manganese β at a given pH value for the experiments (where [Cr(III)] was fixed and [MnO$_2$] was varied), Equation (7.25) was used, which is a simplification of Equation (7.23):

$$\frac{i_{exp3} = -k[Cr(III)_{fixed}]^{\alpha}[MnO_{2\,exp3}]^{\beta}}{i_{exp4} = -k[Cr(III)_{fixed}]^{\alpha}[MnO_{2\,exp4}]^{\beta}} \qquad (7.23)$$

Thus:

$$\frac{i_{exp3}}{i_{exp4}} = \left(\frac{[MnO_{2\,exp3}]}{[MnO_{2\,exp4}]}\right)^{\beta} \qquad (7.24)$$

Therefore:

$$\beta = \frac{\log\left(\dfrac{i_{exp3}}{i_{exp4}}\right)}{\log\left(\dfrac{[MnO_{2\,exp3}]}{[MnO_{2\,exp4}]}\right)} \qquad (7.25)$$

The rate constant k was calculated for each experiment using the i values experimentally measured, and the rate coefficients β and α calculated for each pH value, using Equation (7.26), which is a rearrangement of Equation (7.19):

$$k = -\frac{i}{[Cr(III)]_0^{\alpha}[MnO_2]_0^{\beta}} \qquad (7.26)$$

A plot of one initial rate experiment is shown in Figure 7.15.

Landrot et al. (2010) found that at a given pH and for varying Cr(III) and HMO concentrations, the rate constants were similar (Table 7.5). This strongly suggests that the k values are chemical rate constants since only temperature should impact the magnitude of the k values. Therefore, the use of a method like Q-XAS not only provides valuable information on rapid, initial reaction processes, but an additional benefit is that one can determine chemical kinetics rate constants.

7.6.5 Choice of Kinetic Method

The method that one chooses to study the kinetics of soil chemical reactions depends on several factors. The reaction rate will certainly dictate the choice of method. With batch and

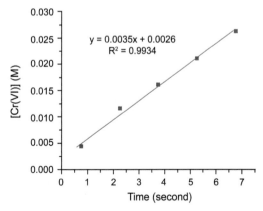

FIGURE 7.15 An initial rate measurement for Cr(III) oxidation during an experiment at pH 2.5, where [Cr(III)] = 80 mM and hydrous manganese oxide [HMO] = 20 g L^{-1}. *From Landrot et al (2010), with permission.*

TABLE 7.5　Initial reaction conditions, initial rates, and rate constants of Cr(III) oxidation by hydrous manganese oxide (HMO). The partial rate coefficients (α, β) at pH 2.5, pH 3, and pH 3.5 are averaged values from α and β calculated using Equations (7.22) and (7.25) and by combining the initial rates measured at different experimental conditions. The rate constant for each experiment was measured with Equation (7.26).

[Cr(III)] (mM)	[HMO] (g L^{-1})	pH 2.5	pH 3.0	pH 3.5
		Initial rates (averaged) (mol L^{-1} s^{-1})		
100	20	0.0049	0.0045	0.0046
40	20	0.0019	0.0017	0.0012
		Rate constant, k (averaged) (s^{-1})		
40 to 100	5 to 20	0.201	0.242	0.322
Cr partial rate coefficient α		1.09	1.09	1.48
Mn partial rate coefficient β		0.94	1.02	0.55

From Landrot et al. (2010), with permission.

flow methods, the most rapid measurements one can make require about 15 s. For more rapid reactions, one must use relaxation and certain molecular-scale real-time techniques where millisecond time scales can be measured.

Another factor in deciding on a kinetic method is the objective of one's experiments. If one wishes to measure the chemical kinetics of a reaction where transport is minimal, most batch and flow techniques are unsuitable, and a relaxation or real-time molecular-scale technique should be employed. On the other hand, if one wants to simulate time-dependent reactions in the field, perhaps a flow technique would be more realistic than a batch method. For comprehensive discussions on kinetic methodologies, one should consult Sparks (1989), Amacher (1991), Sparks and Zhang (1991), and Sparks et al. (1996).

7.7　Effect of Temperature on Reaction Rates

Key points: Temperature is one of the key factors that controls the rates of chemical reactions.

In environmental systems temperature often is the only variable that will change a reaction rate constant. Thus it can be an extremely valuable parameter to better understand reaction mechanisms.

Temperature has a marked effect on reaction rate. Arrhenius noted the following relationship between k and T:

$$k = A_f e^{-E_a/RT} \qquad (7.27)$$

where A_f is a frequency factor and E_a is the energy of activation. Converting Equation (7.27) to a linear form results in

$$\ln k = \ln A_f - E_a/RT \qquad (7.28)$$

A plot of ln k versus 1/T would yield a linear relationship with the slope equal to $-E_a/R$ and the intercept equal to ln A_f. Thus by measuring k values at several temperatures, one could determine the E_a value.

The magnitude of E_a can reveal important information on the reaction process or mechanism. Some E_a values for various reactions or processes are given in Table 7.6. Low E_a values usually indicate diffusion-controlled transport and

TABLE 7.6 Energy of activation (E_a) values for various reactions or processes.

Reaction or process	Typical range of E_a values (kJ mol^{-1})
Physical adsorption	8 to 25
Aqueous diffusion	<21
Pore diffusion	20 to 40
Cellular and life-related reactions	21 to 84
Mineral dissolution or precipitation	34 to 151
Mineral dissolution via surface reaction control	42 to 84
Polymer diffusion	60 to 70
Ion exchange	>84
Isotopic exchange in solution	75 to 201
Solid-state diffusion in minerals at low temperatures	84 to 503

From Langmuir (1997), with permission.

physical adsorption processes, whereas higher E_a values would indicate a chemical reaction or surface-controlled processes (Sparks, 1985).

For example, E_a values of 6.7–26.4 kJ mol^{-1} were observed for pesticide sorption on soils and soil components, which appeared to be diffusion controlled (Haque et al., 1968; Leenheer and Ahlrichs, 1971; Khan, 1973), while gibbsite dissolution in acid solutions, which appeared to be a surface-controlled reaction, was characterized by E_a values ranging from 59 ± 4.3 to 67 ± 0.6 kJ mol^{-1} (Bloom and Erich, 1987).

Kinetic experiments at different temperatures also shed important light on the understanding of surface-induced precipitation mechanisms at mineral–water interfaces, particularly with respect to the determination of activation energies for trace metals (e.g., Mo et al., 2021). Surface chemical reactions dictate metal mobility and transport in soils. For example, the sorption of Ni on palygorskite at pH 6 was found to have an E_a value of 35.1 kJ mol^{-1}, which was

significantly lower than that of the E_a at pH 7.5, which was 102.1 kJ mol^{-1}. The E_a values indicate an adsorption reaction at low pH and a precipitation reaction at high pH. Those conclusions based on the thermodynamic data were confirmed spectroscopically using EXAFS analysis, which identified outer-sphere surface complexation at low ionic strength and low pH and the formation of surface precipitates (specifically Ni phyllosilicates) at high pH. Based on this study, a new process of "continuous nucleation" has been proposed, which differs from the step-by-step "adsorption-to-polymerization" process that has been observed on other metal oxide surfaces. As seen in Figure 7.16, both pH and temperature played critical roles in the behavior of Ni at the mineral–water interface. At low pH, less Ni sorbed to the surface. As the temperature increased from 15 to 35°C, the rate of Ni sorption also increased.

7.8 Kinetics of Important Soil Chemical Processes

Key points: Soil chemical reactions and their kinetics include those of organic and inorganic contaminants, surface precipitation, ion exchange, and mineral dissolution. Often those processes are not mutually exclusive. For example, mineral dissolution has been shown to impact surface precipitation. Additionally, the reaction rates are impacted by both the solution chemistry (i.e., sorptive ions/molecules) and solids (i.e., sorbents).

7.8.1 Sorption–Desorption Reactions

(a) Organic Contaminants

There have been a number of studies on the kinetics of organic chemical sorption/desorption with soils and soil components. Similar to metals and oxyanions, many of these investigations have shown that sorption/desorption is characterized by a rapid, reversible stage followed by

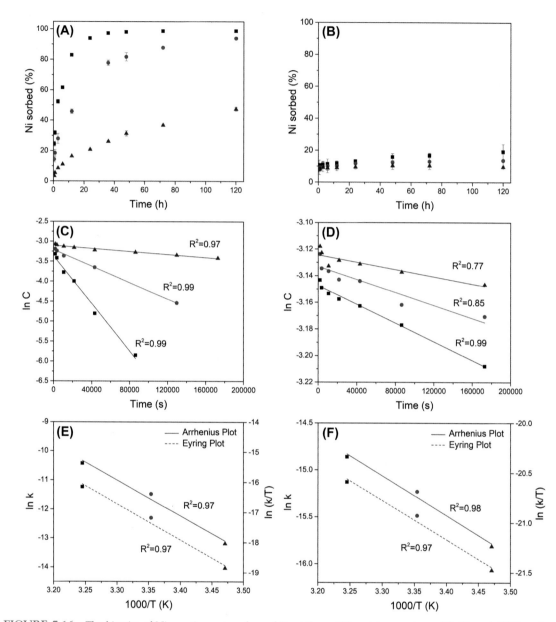

FIGURE 7.16   The kinetics of Ni sorption onto palygorskite at three different temperatures (15, 25, and 35°C) and two different pH values (pH 7.5 [A, C, E] and 6.0 [B, D, F]) versus time. Apparent first-order kinetic plots are shown in C and D. The Arrhenius and Eyring plots were used to determine energies of activation. *From Mo et al. (2021), with permission.*

a much slower, nonreversible stage (Karickhoff et al., 1979; DiToro and Horzempa, 1982; Karickhoff and Morris, 1985) or biphasic kinetics.

The rapid phase has been ascribed to the retention of the organic chemical in a labile form that is easily desorbed. The labile form of the chemical is also available for microbial attack. However, the much slower reaction phase involves the entrapment of the chemical in a nonlabile form that is difficult to desorb and is resistant to biodegradation. This slower sorption/desorption reaction has been ascribed to the diffusion of the chemical into the micropores of organic matter and inorganic soil components (Wu and Gschwend, 1986; Steinberg et al., 1987; Ball and Roberts, 1991).

Recent theories (Weber and Huang, 1996; Pignatello, 2000; Xing and Pignatello, 1997), for example, the dual mode model, have explained the slow diffusion of the organic chemical in SOM by considering SOM as a combination of "rubbery" and "glassy" polymers (Figure 7.17). The rubbery-like phases are characterized by an expanded, flexible, and highly solvated structure with pores of subnanometer dimensions (holes) (Pignatello, 1998, 2000; Xing and Pignatello, 1997). Sorption in the rubbery phase results in linear, noncompetitive, and reversible behavior. The glassy phases have pores that are of subnanometer size and sorption in this phase

is characterized by nonlinearity and is competitive (Xing and Pignatello, 1997).

The above theories, relating slow diffusion into organic matter to diffusion in polymers, are partially validated in recent studies measuring E_a values for organic chemical sorption. Cornelissen et al. (1997) studied the temperature dependence of the slow adsorption and desorption kinetics of some chlorobenzenes, polychlorinated biphenyls (PCBs), and polycyclic aromatic hydrocarbons (PAHs) in laboratory and field contaminated sediments. E_a values of $60-70$ kJ mol^{-1}, which are in the range for diffusion in polymers (Table 7.6), were determined. These values are much higher than those for pore diffusion ($20-40$ kJ mol^{-1}), suggesting that intraorganic matter diffusion may be a more important mechanism for slow organic chemical sorption than interparticle pore diffusion.

An example of the biphasic kinetics observed for many organic chemical reactions in soils/sediments is shown in Figure 7.18. In this study 55% of the labile PCBs were desorbed from sediments in a 24 h period, while little of the remaining 45% nonlabile fraction was desorbed in 170 h (Figure 7.18A). Over another 1 year period, about 50% of the remaining nonlabile fraction is desorbed (Figure 7.18B).

In another study with volatile organic compounds (VOCs) Pavlostathis and Mathavan (1992) observed a biphasic desorption process for field soils contaminated with trichloroethylene (TCE), tetrachloroethylene (PCE), toluene (TOL), and xylene (XYL). A fast desorption reaction occurred within 24 h, followed by a much slower desorption reaction beyond 24 h. In 24 h $9-29$, $14-48$, $9-40$, and $4-37\%$ of the TCE, PCE, TOL, and XYL, respectively, were released.

A number of studies have also shown that with increased residence time, the nonlabile portion of the organic chemical in the soil/sediment becomes more resistant to release (McCall and Agin, 1985; Steinberg et al., 1987; Pignatello and Huang, 1991; Pavlostathis and Mathavan,

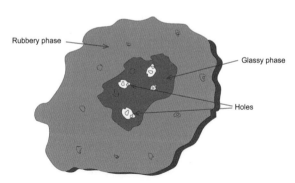

FIGURE 7.17 Rubbery—glassy polymer concept (dual mode model) of soil organic matter. The perspective is intended to be three-dimensional. The rubbery and glassy phases both have dissolution domains in which sorption is linear and noncompetitive. The glassy phase, in addition, has pores of subnanometer dimension ("holes") in which adsorption-like interactions occur with the walls, giving rise to nonlinearity and competitive sorption. *From Xing and Pignatello (1997), with permission.*

(A)

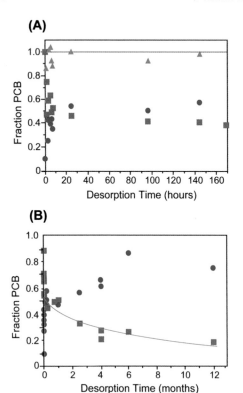

(B)

FIGURE 7.18   (a) Short-term PCB desorption in hours (h) from Hudson River sediment contaminated with 25 mg kg^{-1} PCB. The distribution of the PCB between the sediment (■) and XAD-4 resin (●) is shown, as well as the overall mass balance (▲). The resin acts as a sink to retain the PCB that is desorbed. (b) Long-term PCB desorption in months (mo) from Hudson River sediment contaminated with 25 mg kg^{-1} PCB. The distribution of the PCB between the sediment (■) and XAD-4 resin (●) is shown. The line represents a nonlinear regression of the data by the two-box (site) model. *From Carroll et al. (1994), with permission.*

1992; Alexander, 1995). One of the first studies to clearly demonstrate this aging or residence time effect was the research of Steinberg et al. (1987). They showed that 1,2-dibromoethane (EDB) release from soils reacted in the laboratory over a short period of time was much more rapid than EDB release from field soils that had been contaminated with EDB for many years. This difference in release was related to a greater

diffusion into micropores of clay minerals and humic components that occurred at longer times (Steinberg et al., 1987).

One way to gauge the effect of time on organic contaminant retention in soils is to compare K_d (sorption distribution coefficient) values for "freshly aged" and "aged" soil samples (see Chapter 5 for a discussion of these coefficients). In most studies K_d values are measured based on a 24 h equilibration between the soil and the organic chemical. When these values are compared to K_d values for field soils previously reacted with the organic chemical (aged samples), the latter have much higher K_d values, indicating that much more of the organic chemical is in a sorbed state. For example, Pignatello and Huang (1991) measured K_d values in freshly aged (K_d) and aged soils (K_{app}, apparent sorption distribution coefficient) reacted with atrazine and metolachlor, two widely used herbicides. The aged soils had been treated with the herbicides 15–62 months before sampling. The K_{app} values were 2.3–42 times higher than the K_d values (Table 7.7).

TABLE 7.7   Sorption distribution coefficients for herbicides in "freshly aged" and "aged" soils.[a]

Herbicide	Soil	K_d[b]	K_{app}[c]
Metolachlor	CVa	2.96	39
	CVb	1.46	27
	W1	1.28	49
	W2	0.77	33
Atrazine	CVa	2.17	28
	CVb	1.32	29
	W3	1.75	4

[a] Herbicides had been added to soils 31 months prior to sampling for CVa and CVb soils, 15 months for the W1 and W2 soils, and 62 months for the W3 soil.
[b] Sorption distribution coefficient (L kg^{-1}) of "freshly aged" soil based on a 24 h equilibration period.
[c] Apparent sorption distribution coefficient (L kg^{-1}) in contaminated soil ("aged" soil) determined using a 24 h equilibration period.
Adapted from Pignatello and Huang (1991), with permission.

Scribner et al. (1992), studying simazine (a widely used triazine herbicide for broadleaf and grass control in crops) desorption and bioavailability in aged soils, found that K_{app} values were 15 times higher than K_d values. Scribner et al. (1992) also showed that 48% of the simazine added to the freshly aged soils was biodegradable over a 34-day incubation period while none of the simazine in the aged soil was biodegraded.

One of the implications of these results is that while many transport and degradation models for organic contaminants in soils and waters assume that the sorption process is an equilibrium process, the above studies clearly show that kinetic reactions must be considered when making predictions about the mobility and fate of organic chemicals. Moreover, K_d values determined based on a 24 h equilibration period and commonly used in fate and risk assessment models can be inaccurate since 24 h K_d values often overestimate the amount of organic chemical in the solution phase.

The finding that many organic chemicals are quite persistent in the soil environment has both good and bad features. The beneficial aspect is that the organic chemicals are less mobile and may not be readily transported in groundwater supplies. The negative aspect is that their persistence and inaccessibility to microbes may make decontamination more difficult, particularly if in situ remediation techniques such as biodegradation are employed.

The previously established sorption models, such as the dual mode model, based on the glassy and rubbery states of soil organic matter (SOM), have been widely used in various sorption systems of hydrophobic organic contaminants (HOCs). With the increasing research concern of personal care products and pharmaceuticals, the sorption of ionic organic contaminants (IOCs) has attracted special attention. The cation bridging, ligand exchange, and electrostatic interactions are all emphasized in IOCs sorption (Pan et al., 2009, Li et al., 2015 and Zhao et al., 2018).

However, because most IOCs, SOM, and carbonaceous particles (including black carbon) are negatively charged in environmentally relevant pH values, charge-assisted hydrogen bonding (CAHB) was found to be a key mechanism for IOCs sorption (Li et al., 2015) and among SOM molecules (Zhao et al., 2018). CAHB may greatly facilitate IOCs immobilization and carbon sequestration in soil. For positively charged IOCs, cation–π interaction with carbonaceous particles is clearly a major force (Zhao et al., 2017).

(b) Heavy Metals and Oxyanions

The chemical reaction rate rate of heavy metal (e.g., Cu^{2+}, Pb^{2+}) and oxyanion (e.g., arsenate, chromate, and phosphate) sorption on soil components is rapid, occurring on millisecond time scales. For such rapid reactions, chemical relaxation and real-time molecular-scale techniques, for example, pressure-jump relaxation and QXAS, must be employed (Sparks and Zhang, 1991; Sparks, 2000; Landrot et al., 2010).

Zhang and Sparks (1990) studied the kinetics of selenate SeO_4^{2-} adsorption on goethite using pressure-jump relaxation and found that adsorption occurred mainly under acidic conditions. The dominant species was SeO_4^{2-}. As pH increased SeO_4^{2-} adsorption decreased. Selenate was described using the modified triple-layer model (see Chapter 5). A single relaxation was observed and the mechanism proposed was where XOH is 1 mol of reactive surface hydroxyl bound to an Fe ion in goethite.

$$XOH + H^+ + SeO_4^{2-} \rightleftharpoons XOH_2^+ - SeO_4^{2-} \quad (7.29)$$

A linearized rate equation given below was developed and tested:

$$\tau^{-1} = k_1 \left([XOH][SeO_4^{2-}] + [XOH][H^+] + [SeO_4^{2-}][H^+] \right) + k_{-1} \quad (7.30)$$

where the terms in the brackets are the concentrations of species at equilibrium. Since the reaction was conducted at the solid/liquid interface, the electrostatic effect must be considered to calculate the intrinsic rate constants (k_1^{int} and k_{-1}^{int}). Using the modified triple-layer model to obtain electrostatic parameters, a first-order reaction was derived (Zhang and Sparks, 1990):

$$\tau^{-1} \exp\left(\frac{-F(\Psi_\alpha - 2\Psi_\beta)}{2RT}\right)$$

$$= k_1^{int}\left[\exp\left(\frac{-F(\Psi_\alpha - 2\Psi_\beta)}{RT}\right)\right. \qquad (7.31)$$

$$\times \left([XOH]\left[SeO_4^{2-}\right] + [XOH]\left[H^+\right]\right.$$

$$\left.\left. + \left[SeO_4^{2-}\right]\left[H^+\right]\right)\right] + k_{-1}^{int}$$

A plot of the left side of Equation (7.31) versus the terms in brackets on the right side was linear and the k_1^{int} and k_{-1}^{int} values were calculated from the slope and intercept, respectively (Figure 7.19). The linear relationship would indicate that the outer-sphere complexation mechanism proposed in Equation (7.29) was plausible. Of course, one would need to use spectroscopic approaches to definitively determine the mechanism. This was done earlier with X-ray absorption fine structure (XAFS) spectroscopy to prove that selenate is adsorbed as an outer-sphere complex on goethite (Hayes et al., 1987).

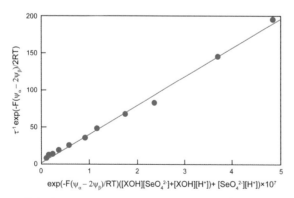

FIGURE 7.19 Plot of relationship between τ^{-1} with exponential and concentration terms in Equation (7.31). *From Zhang and Sparks (1990), with permission.*

The kinetics of heavy metal and oxyanion sorption on soil components and soils is typically characterized by a biphasic process in which sorption is initially rapid, followed by slow reactions ((Figure 7.20a). The rapid step, which occurs over milliseconds to hours, can be ascribed to chemical reaction and film diffusion processes. During this rapid reaction process, a large portion of the sorption may occur. For example, in Figure 7.20a one sees that ~90% of the total Ni sorbed on kaolinite and pyrophyllite occurred within the first 24 h. For Pb sorption on a Matapeake soil, 78% of the total Pb sorption occurred in 8 min. Following the initial fast reaction, slow sorption continued, but only about 1% additional Pb was sorbed after 800 h (Figure 7.20b). Figure 7.20c shows a biphasic reaction for As(V) sorption on ferrihydrite. Within 5 min, a majority of the total sorption had occurred. Slow sorption continued for at least 192 h.

However, in some cases the magnitude of sorption can greatly increase with longer reaction times. For example, Bruemmer et al. (1988) studied Ni^{2+}, Zn^{2+}, and Cd^{2+} sorption on goethite, a porous Fe oxide that has defect structures in which metals can be incorporated to satisfy charge imbalances. It was found at pH 6 that as the reaction time increased from 2 h to 42 days (at 293 K), adsorbed Ni^{2+} increased from 12 to 70% of total sorption, and total Zn^{2+} and Cd^{2+} sorption over the same time increased by 33 and 21%, respectively. Metal uptake was hypothesized to occur via a three-step mechanism: (1) adsorption of metals on external surfaces, (2) solid-state diffusion of metals from external to internal sites, and (3) metal binding and fixation at positions inside the goethite particle.

The amount of contact time between metals and soil sorbents (residence time) can dramatically affect the degree of desorption, depending on the metal and sorbent. Examples of this are shown in Figure 7.21 and Table 7.8. In Figure 7.21 the effect of residence time on Pb^{2+} and Co^{2+} desorption from hydrous Fe oxide (HFO) was studied. With Pb, between pH 3 and 5.5, there

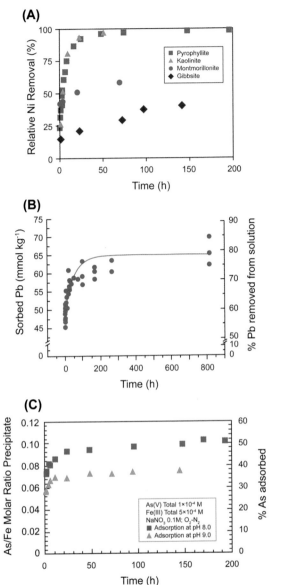

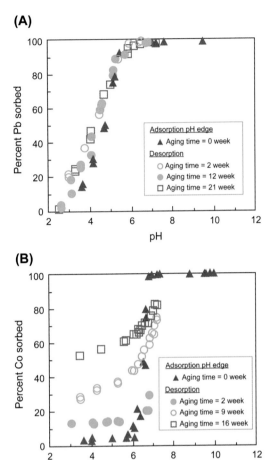

FIGURE 7.21 Fractional sorption–desorption of (a) Pb^{2+} and (b) Co^{2+} to hydrous Fe(III) oxide (HFO) as a function of pH and HFO–Pb^{2+} (a) and HFO–Co^{2+} (b) aging time. *From Ainsworth et al. (1994), with permission.*

FIGURE 7.20 Kinetics of metal and oxyanion sorption on soil minerals and soil. (a) Kinetics of Ni sorption (%) on pyrophyllite, kaolinite, gibbsite, and montmorillonite from a 3 mM Ni solution, an ionic strength I = 0.1 M $NaNO_3$, and a pH of 7.5 (*From Scheidegger et al. (1997), with permission*); (b) kinetics of Pb sorption on a Matapeake soil from a 12.25 mM Pb solution, an ionic strength I = 0.05 M, and a pH of 5.5 (*From Strawn and Sparks (2000), with permission*); (c) kinetics of As(V) sorption on ferrihydrite at pH 8.0 and 9.0 *From Fuller et al. (1993), with permission.*

TABLE 7.8 Effect of residence time on Pb desorption from Matapeake soils,

Residence time (d)	Sorbed Pb (mmol kg^{-1})	Desorbed Pb (mmol kg^{-1})	Percentage Pb desorbed
1	54.9	27.9	50.8
10	60.0	28.7	47.
32	66.1	30.5	46.1

From Strawn and Sparks (2000), with permission.

was a minor effect of residence time (over 21 weeks) on Pb^{2+} desorption, with only minor hysteresis occurring (hysteresis varied from <2% difference between sorption and desorption to ~10%). At pH 2.5, Pb^{2+} desorption was complete within a 16 h period and was not affected by residence time (Figure 7.21A). In a soil where 2.1% SOM was present, residence time had little effect on the amount of Pb desorbed, but marked hysteresis was observed at all residence times (Table 7.8). This could be ascribed to the strong metal–soil complexes that occur and perhaps to diffusion processes.

With Co^{2+}, extensive hysteresis was observed over a 16-week residence time (Figure 7.21B), and the hysteresis increased with residence time. After a 16-week residence time, 53% of the Co^{2+} was not desorbed, and even at pH 2.5, hysteresis was observed. The extent of Co reversibility with residence time was attributed to Co incorporation into a recrystallizing solid via isomorphic substitution and not to micropore diffusion. Similar residence time effects on Co desorption have been observed both with Fe and Mn oxides (Backes et al., 1995) and with soil clays (McLaren et al., 1998).

Srivastava et al. (2008) also observed a significant decrease in the desorption of Pb and Cd from kaolinite at two temperatures (20 and 40°C) with different trends for the two metals. For Pb at a loading of 43.44 μmol g^{-1}, about 23% of the adsorbed Pb was desorbed from the "no aged" sample, which decreased to ~17 and 3% for the samples aged for 30 and 60 days at 20°C, respectively. At 40°C, ~13% of the adsorbed Pb was desorbed from the sample aged for 30 days and very little (~0.2%) from the sample aged for 60 days at high surface loading. Based on the EXAFS spectroscopy, the significant decrease in Pb desorption was attributed to the formation of polynuclear Pb-hydroxide complexes, which with an increasing aging period stabilized at the kaolinite surface.

In the case of Cd a significant decrease in Cd desorption was observed only for the sample

aged for 60 days at 40°C, where 40% Cd was desorbed (compared to 92% desorption from the "no aged" sample) at a surface loading of 19.33 μmol g^{-1} (Srivastava et al. 2008). EXAFS spectroscopy showed no detectable change in the surface speciation (outer-sphere complexation) of Cd with aging for 60 days at 40°C. The authors attributed the decrease in Cd desorption to inter- and intraparticle diffusion of Cd ions, due to a reduction in the solvated ion radius with aging at elevated temperatures.

The aging effect on Cd desorption from goethite was different from kaolinite, with a much greater decrease in Cd desorption from goethite with aging at 20°C (Mustafa et al., 2006). As can be seen in Figure 7.22, the cumulative Cd desorption at pH 6.0 decreased from 71% to 19.8% at 20°C and the corresponding decrease ranged from 50.6% to 0.1% at 40°C.

Structural parameters obtained from the EXAFS (Figure 7.23) data revealed that Cd ions form single-edge-sharing surface complexes on goethite after 16 h, regardless of equilibrium temperature, which changed to double-edge-sharing

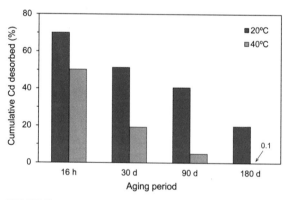

FIGURE 7.22 Effects of equilibrium temperature (20 and 40°C) and aging (16 h, 30, 90, and 180 days) on Cd desorption from goethite. An initial Cd concentration of 180 mM was used to equilibrate Cd and goethite at pH 6 for 2 h prior to centrifugation and aging as a moist paste. The surface density of Cd on goethite was 0.06 mmol m^{-2} for the samples, which corresponds to ~73% surface coverage. *From Mustafa et al., 2006, with permission.*

(A)   **(B)**

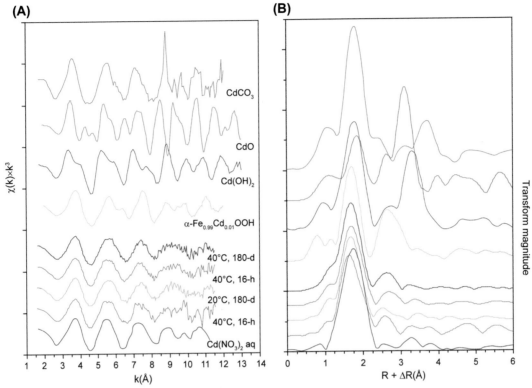

FIGURE 7.23   (a) Raw k^3-weighted w(k) spectra of reference and experimental samples and (b) their corresponding radial structure functions. Sample and experimental details are given in Figure 7.22 caption. Cadmium-adsorbed goethite samples from an initial Cd concentration of 180 mM at pH 6 and moist paste was aged for 6 h, 30, 90, and 180 days at 20 and 40°C. *From Gräfe et al. (2008), with permission.*

surface complexes after 180 days both at 20 and 40°C (Gräfe et al., 2008). The effect of increased temperature on Cd desorption from goethite was suggested to be due to the increased electrostatic attraction of Cd ions to the goethite surface at 40°C as a result of a decreasing point of zero charge with temperature

7.8.2 Kinetics of Metal Hydroxide Surface Precipitation/Dissolution

In addition to diffusion processes, the formation of metal hydroxide surface precipitates (these were discussed in Chapter 5) and subsequent residence time effects on soil sorbents can greatly affect metal release and hysteresis.

Metal hydroxide precipitates can form on phyllosilicates, metal oxides, and soils. It has generally been thought that the kinetics of formation of surface precipitates was slow. However, recent studies have shown that metal hydroxide precipitates can form on time scales of minutes. In Figure 7.24 one sees that mixed Ni−Al hydroxide precipitates formed on pyrophyllite within 15 min, and they grew in intensity as time increased. Similar results have been observed with other soil components and with soils (Scheidegger et al., 1998; Roberts et al., 1999).

The formation and subsequent "aging" of the metal hydroxide surface precipitate can have a significant effect on metal release. In Figure 7.25

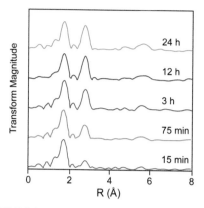

FIGURE 7.24 Radial structure functions (derived from XAFS analyses) for Ni sorption on pyrophyllite for reaction times up to 24 h, demonstrating the appearance and growth of the second shell (peak at ~2.8 Å) contributions due to surface precipitation and growth of a mixed Ni–Al hydroxide phase. *From Scheidegger et al. (1998), with permission.*

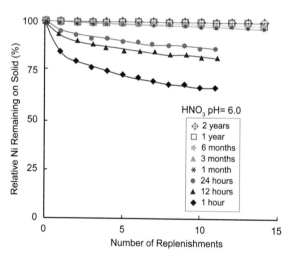

FIGURE 7.25 Dissolution of Ni from surface precipitates formed on pyrophyllite at residence times of 1 h to 2 years. The figure shows the relative amount of Ni^{2+} remaining on the pyrophyllite surface following extraction for 24 h periods (each replenishment represents a 24 h extraction) with HNO_3 at pH 6.0. *From Scheckel et al. (2001), with permission.*

one sees that as residence time (aging) increases from 1 h to 2 years, Ni release from pyrophyllite, as a percentage of total Ni sorption, decreased

from 23 to ~0%, when HNO_3 (pH 6.0) was employed as a dissolution agent for 14 days. This enhanced stability is due to the transformation of the metal–Al hydroxide precipitates to a metal–Al phyllosilicate precursor phase as residence time increases. This transformation occurs via a number of steps (Figure 7.26). There is diffusion of Si originating from the weathering of the sorbent into the interlayer space of the LDH, replacing the anions such as NO_3^-. The polymerization and condensation of the interlayer Si slowly transforms the LDH into a precursor metal–Al phyllosilicate. Peltier et al. (2006) proved that silication of the interlayer space of LDH minerals is the most thermodynamically stable phase. Silica in the interlayer space is more stable than nitrate, sulfate, or carbonate.

The metal stabilization that occurs in surface precipitates on Al-free sorbents (e.g., talc) may be due to Ostwald ripening, resulting in increased crystallization (Scheckel and Sparks, 2001). Thus with time, one sees that metal (e.g., Co, Ni, and Zn) sorption on soil minerals often results in a continuum of processes from adsorption to precipitation to solid phase transformation (Figure 7.27).

This continuum of adsorption to surface precipitation is highlighted in a set of experiments by Siebecker et al. (2014), who employed a Q-XAS flow technique, for the first time, to measure the rapid formation of Ni LDH precipitate phases on pyrophyllite in situ and in real time. While a Ni solution was pumped through the flow cell, the Ni that sorbed to the pyrophyllite surface provided time-resolved EXAFS spectra in real-time.

In Figure 7.28a at 5.3 Å^{-1}, a shoulder can be seen forming over time (arrow 1). This shoulder feature is not present at the beginning of the experiment (6–15 min). The shoulder is caused by focused, linear multiple scattering paths from both Ni and Al in the newly formed LDH sheets. Additionally, between 7 and 9 Å^{-1}, one peak separates into two peaks, and the peak

(A) Exchange

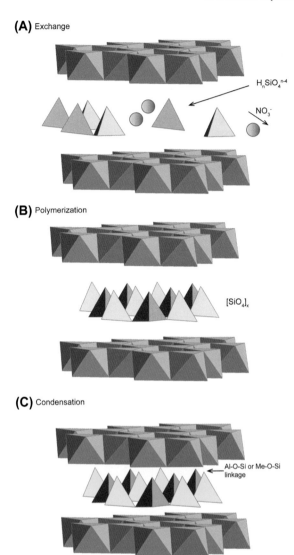

(B) Polymerization

(C) Condensation

FIGURE 7.26 Hypothetical reaction process illustrating the transformation of an initially precipitated Ni–Al LDH into a phyllosilicate-like phase during aging. The initial step involves the exchange of dissolved silica for nitrate within the LDH interlayer (A) followed by the polymerization (B) and condensation of silica onto the octahedral Ni–Al layer (C) The resultant solid possesses structural features common to 1:1 and 2:1 phyllosilicates. *From Ford et al. (2001), with permission.*

height at 7.5 Å^{-1} (arrow 2) becomes equal to that of the peak at 8.2 Å^{-1} (arrow 3) starting at 31–40 min. This is caused by a beat pattern of destructive overlapping Ni–Ni and Ni–Al photoelectric waves and is indicative of Ni–Al LDH.

With respect to the FT shells in Figure 7.28b, the amplitude of the first shell remained constant, which is the oxygen shell. The amplitude of the second shell increased over time, indicating an increase in the coordination number of the second nearest neighbor (e.g., Ni, Al, and/or Si). Analysis of the second shell is complicated due to the contribution of different backscattering atoms that surround the Ni atom. Al substitution for Ni in the octahedral sheet and the presence of Si in a tetrahedral sheet simultaneously produce backscattered photoelectric waves that are partially destructive and constructive with the wave produced by Ni in the octahedral layer. The fits of each shell are described in detail in Siebecker et al. (2014).

These results suggest that adsorption and precipitation occur on similar time scales and perhaps concurrently. In a dynamic flow environment where the readsorption of dissolved products is limited (as compared to batch-type experiments), the incorporation of dissolved Al into newly formed solids was rapid. The mineral (adsorbent) often cannot be considered a stable surface even over short time scales because the cycling of metals between solution and solid phases is very rapid. Thus there can be a continuum in sorption processes, which dictates that more robust sorption models need development that capture multiple sorption processes over reaction time. The formation of metal surface precipitates could be an important mechanism for sequestering metals in soils such that they are less mobile and bioavailable. Such products must be considered when modeling the fate and mobility of metals such as Co^{2+}, Mn^{2+}, Ni^{2+}, and Zn^{2+} in soil and water environments.

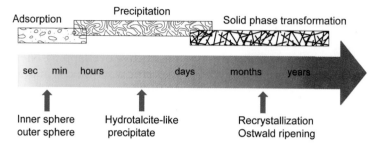

FIGURE 7.27 Changes in sorption processes with time showing a continuum from adsorption to precipitation to solid phase transformation.

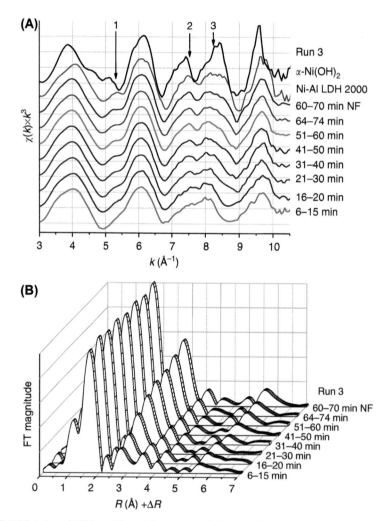

FIGURE 7.28 Q-XAS data and FT from Ni sorption to pyrophyllite in a flow cell. In panel (a) the experiment "Run 3" is compared with two isostructural compounds, α-Ni(OH)$_2$ and Ni—Al LDH. There is a decrease in the peak at 8.2 Å^{-1}, indicating the formation of LDH in as little as 31—40 min in flow conditions. In panel (b) the FT plots show an increase in the first, second, and third metal coordination shell amplitudes throughout the reaction, which also indicates the formation of three-dimensional surface precipitates. *From Siebecker et al. (2014), with permission.*

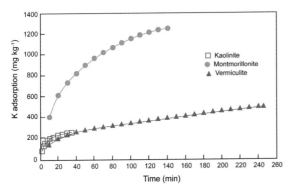

FIGURE 7.29   Potassium adsorption versus time for clay minerals. *From Sparks and Jardine (1984), with permission.*

7.8.3 Ion Exchange Kinetics

Ion exchange kinetics are greatly dependent on the adsorbent and the ion. Figure 7.29 clearly shows how the type of clay mineral affects K–Ca exchange. Reaction rates are much more rapid on kaolinite, where the exchange sites are external, than on vermiculite, which contains both external and internal exchange sites. The internal sites may be fully expanded or partially collapsed.

The type of ion also has a pronounced effect on the rate of exchange. The exchange of ions like K^+, NH_4^+, and Cs^+ is often slower than that of ions such as Ca^{2+} and Mg^{2+}. This is related to the smaller hydrated radius of the former ions. The smaller ions fit well in the interlayer spaces of clay minerals, which causes partial or total interlayer space collapse. The exchange is thus slow and interparticle diffusion-controlled. However, with the exception of K^+, NH_4^+, and Cs^+ exchange on 2:1 clay minerals like vermiculite and mica, ion exchange kinetics are usually very rapid, occurring in millisecond time scales (Tang and Sparks, 1993). Figure 7.30 shows that Ca–Na exchange on montmorillonite was complete in <100 ms.

7.8.4 Kinetics of Mineral Dissolution

(a) Rate-Limiting Steps

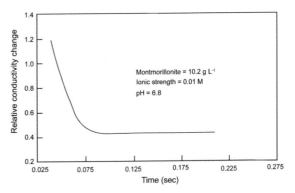

FIGURE 7.30   Typical pressure-jump relaxation curve for Ca–Na exchange on montmorillonite showing relative change in conductivity versus time. *From Tang and Sparks (1993), with permission.*

The dissolution of minerals involves several steps (Stumm and Wollast, 1990): (1) mass transfer of dissolved reactants from the bulk solution to the mineral surface, (2) adsorption of solutes, (3) interlattice transfer of reacting species, (4) surface chemical reactions, (5) removal of reactants from the surface, and (6) mass transfer of products into the bulk solution. Under field conditions, mineral dissolution is slow and mass transfer of reactants or products in the aqueous phase (Steps 1 and 6) is not rate limiting. Thus the rate-limiting steps are either transport of reactants and products in the solid phase (Step 3) or surface chemical reactions (Step 4) and removal of reactants from the surface (Step 5).

Transport-controlled dissolution reactions or those controlled by mass transfer or diffusion can be described using the parabolic rate law (Stumm and Wollast, 1990):

$$r = \frac{dC}{dt} = kt^{-1/2} \qquad (7.32)$$

where r is the reaction rate, C is the concentration in solution, t is time, and k is the reaction rate constant. Integrating, C increases with $t^{1/2}$:

$$C = C_0 + 2kt^{1/2} \qquad (7.33)$$

where C_0 is the initial concentration in solution.

If the surface reactions are slow compared to the transport reactions, dissolution is surface controlled, which is the case for most dissolution reactions of silicates and oxides. In surface-controlled reactions the concentrations of solutes next to the surface are equal to the bulk solution concentrations and the dissolution kinetics are zero order if steady-state conditions are operational on the surface. Thus the dissolution rate, r, is

$$r = \frac{dC}{dt} = kA \qquad (7.34)$$

and r is proportional to the mineral's surface area, A. Thus for a surface-controlled reaction, the relationship between time and C should be linear. Figure 7.31 compares transport- and surface-controlled dissolution mechanisms.

Intense arguments concerning the mechanism for mineral dissolution have ensued over the years. Those that supported a transport-controlled mechanism believed that a leached layer formed as mineral dissolution proceeded and that subsequent dissolution took place via diffusion through the leached layer (Petrovic et al., 1976). Advocates of this theory found that dissolution was described by the parabolic rate law (Equation [7.32]). However, the "apparent" transport-controlled kinetics may be an artifact caused by the dissolution of hyper-fine particles formed on the mineral surfaces after grinding, which are highly reactive sites, or by the use of batch methods that cause reaction products to accumulate, causing precipitation of secondary minerals. These experimental arti-facts can cause incongruent reactions and pseudoparabolic kinetics. Studies employing surface spectroscopies such as X-ray photoelectron spectroscopy and nuclear resonance profiling (Schott and Petit, 1987; Casey et al., 1989) have demonstrated that although some incongruency may occur in the initial dissolution process, which may be diffusion controlled, the overall reaction is surface controlled. An illustration of the surface-controlled dissolution of δ-Al_2O_3 resulting in a linear release of Al^{3+} with time is shown in Figure 7.32. The dissolution rate, r, can be obtained from the slope of Figure 7.32.

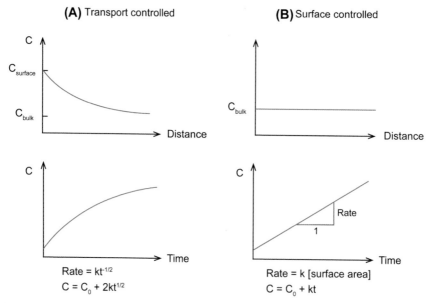

(A) Transport controlled

(B) Surface controlled

Rate = $kt^{-1/2}$
$C = C_0 + 2kt^{1/2}$

Rate = k [surface area]
$C = C_0 + kt$

FIGURE 7.31   (A) Transport- versus (B) surface-controlled dissolution. Schematic representation of concentration in solution, C, as a function of distance from the surface of the dissolving mineral. In the lower part of the figure the change in concentration is given as a function of time. *From Stumm (1992), with permission.*

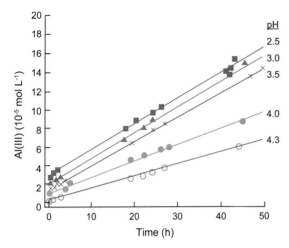

FIGURE 7.32 Linear dissolution kinetics observed for the dissolution of δ-Al$_2$O$_3$. Representative of processes whose rates are controlled by a surface reaction and not by transport. *From Furrer and Stumm (1986), with permission.*

(b) Surface-Controlled Dissolution Mechanisms

The dissolution of oxide minerals via a surface-controlled reaction by ligand- and proton-promoted processes has been described by Stumm and coworkers (Furrer and Stumm, 1986; Zinder et al., 1986; Stumm and Furrer, 1987) using a surface coordination approach. The important reactants in these processes are H$_2$O, H$^+$, OH$^-$, ligands, and reductants and oxidants (see definitions in Chapter 8). The reaction mechanism occurs in two steps (Stumm and Wollast, 1990):

Surface sites + reactants $\left(H^+, \ OH^-, \ or \ ligands \right)$

$\xrightarrow{\text{fast}}$ Surface species

$$(7.35)$$

Surface species $\xrightarrow[\text{Detachment of metal (M)}]{\text{Slow}}$ M(aq)

$$(7.36)$$

Thus the attachment of the reactants to the surface sites is fast and the detachment of metal species from the surface into solution is slow and rate limiting.

(c) Ligand-Promoted Dissolution

Figure 7.33 shows how the surface chemistry of the mineral affects dissolution. One sees that surface protonation of the surface ligand increases dissolution by polarizing interatomic bonds close to the central surface ions, which promotes the release of a cation surface group into solution. Hydroxyls that bind to surface groups at higher pH values can ease the release of an anionic surface group into the solution phase.

Ligands that form surface complexes via ligand exchange with a surface hydroxyl add negative charge to the Lewis acid center coordination sphere and lower the Lewis acid acidity. This polarizes the M—oxygen bonds, causing the detachment of the metal cation into the solution phase. Thus inner-sphere surface complexation plays an important role in mineral dissolution. Ligands such as oxalate, salicylate, F$^-$, EDTA, and NTA increase dissolution, but others, for example, sulfate (SO$_4^{2-}$), chromate (CrO$_4^{2-}$), and benzoate, inhibit dissolution. Phosphate and arsenate enhance dissolution at low pH and dissolution is inhibited at pH > 4 (Stumm, 1992).

The reason for these differences may be that bidentate species that are mononuclear promote dissolution while binuclear bidentate species inhibit dissolution. With binuclear bidentate complexes, more energy may be needed to remove two central atoms from the crystal structure. With phosphate and arsenate, at low pH, mononuclear species are formed, while at higher pH (around pH 7), binuclear or trinuclear surface complexes form. Mononuclear bidentate complexes are formed with oxalate while binuclear bidentate complexes form with CrO$_4^{2-}$. Additionally, the electron donor properties of CrO$_4^{2-}$ and oxalate are also different. With CrO$_4^{2-}$, a high redox potential is maintained at the oxide surface, which restricts reductive dissolution (Stumm and Wollast, 1990; Stumm, 1992).

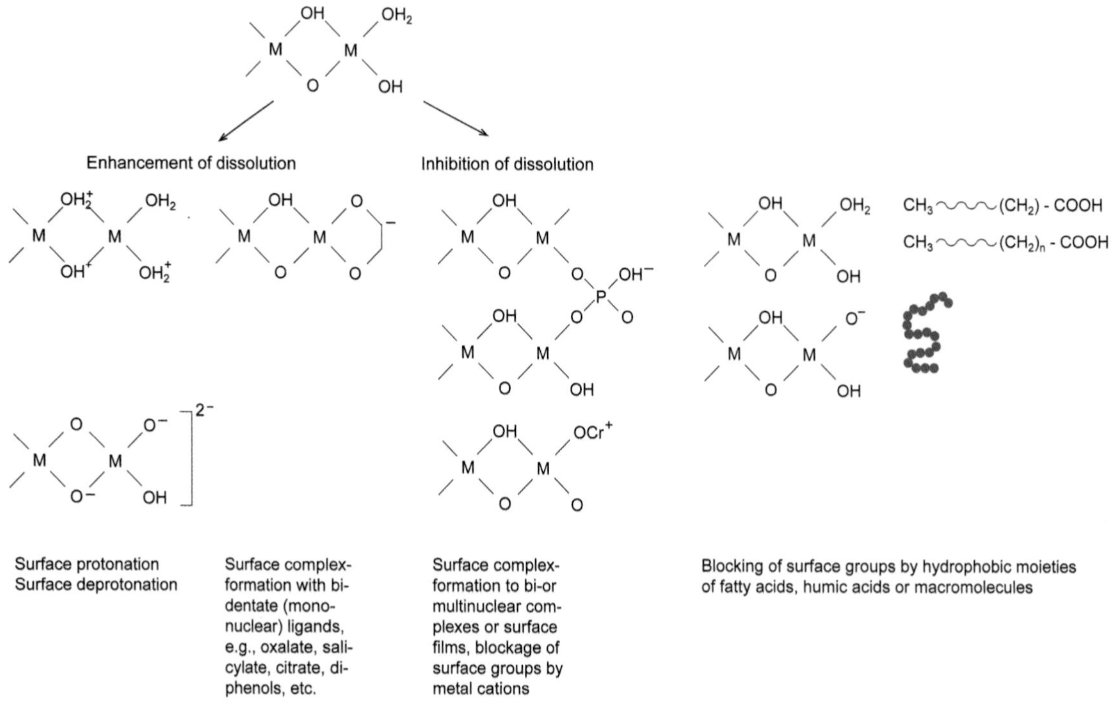

FIGURE 7.33   The dependence of surface reactivity and of kinetic mechanisms on the coordinative environment of the surface groups. *From Stumm and Wollast (1990), with permission.*

Dissolution can also be inhibited by cations such as VO^{2+}, Cr(III), and Al(III) that block surface functional groups.

One can express the rate of the ligand-promoted dissolution, R_L, as

$$R_L = k'_L(\equiv ML) = k'_L\, C_L^S \qquad (7.37)$$

where k_L' is the rate constant for ligand-promoted dissolution (time^{-1}), $\equiv ML$ is the metal–ligand complex, and C_L^S is the surface concentration of the ligand complex (mol m^{-2}). Figure 7.34 shows that Equation (7.37) adequately described the ligand-promoted dissolution of δ-Al$_2$O$_3$.

(d) Proton-Promoted Dissolution

Under acid conditions, protons can promote mineral dissolution by binding to surface oxide ions, causing bonds to weaken. This is followed by the detachment of metal species into solution. The proton-promoted dissolution rate, R_H, can be expressed as (Stumm, 1992)

$$R_H = k_{H'}(\equiv MOH_2^+)^j = k'_H(C_H^s)^j \qquad (7.38)$$

where k_H' is the rate constant for proton-promoted dissolution, $\equiv MOH_2^+$ is the metal–proton complex, C_H^s is the concentration of the surface-adsorbed proton complex (mol m^{-2}), and j corresponds to the oxidation state of the central metal ion in the oxide structure (i.e., $j = 3$ for Al(III) and Fe(III) in simple cases). If dissolution occurs through only one mechanism, j is an integer. Figure 7.35 shows an application of Equation (7.38) for the proton-promoted dissolution of δ-Al$_2$O$_3$.

(e) Overall Dissolution Mechanisms

The rate of mineral dissolution, which is the sum of the ligand-, proton-, and deprotonation-

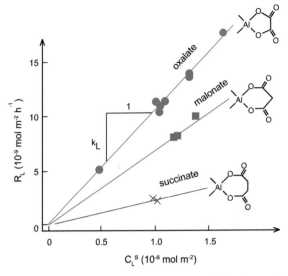

FIGURE 7.34 The rate of ligand-catalyzed dissolution of δ-Al_2O_3 by the aliphatic ligands oxalate, malonate, and succinate, R_L (nmol m^{-2} h^{-1}), can be interpreted as a linear dependence on the surface concentrations of the ligand complexes, C_L^s. In each case the individual values for C_L^s were determined experimentally. *From Furrer and Stumm (1986), with permission.*

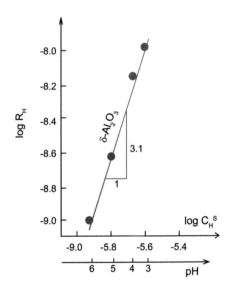

FIGURE 7.35 The dependence of the rate of proton-promoted dissolution of δ-Al_2O_3, RH (mol m^{-2} h^{-1}) on the surface concentration of the proton complexes. CH^s (mol m^{-2}). *From Furrer and Stumm (1986), with permission.*

promoted (or bonding of OH$^-$ ligands) dissociation ($R_{OH} = (k'_{OH})(C_{OH}^s)^i$) rates, along with the pH-independent portion of the dissolution rate (k'_{H_2O}), which is due to hydration, can be expressed as (Stumm, 1992)

$$R = k'_L(C_L^s) + k'_H(C_H^s)^j \\ + k'_{OH}(C_{OH}^s)^i + k'_{H_2O} \qquad (7.39)$$

Equation (7.39) is valid if dissolution occurs in parallel at varying metal centers (Furrer and Stumm, 1986).

Suggested Reading

Alexander, M., 2000. Aging, bioavailability, and overestimation of risk from environmental pollutants. Environ. Sci. Technol. 34, 4259–4265.

Borda, M.J., Sparks, D.L., 2008. Kinetics and mechanisms of sorption-desorption in soils: a multiscale assessment. In: Violante, A., Huang, P.M., Gadd, G.M. (Eds.), Biophysico-Chemical Processes of Heavy Metals and Metalloids in Soil Environments". John Wiley & Sons, Inc., New York, pp. 75–124.

Lasaga, A.C., 1998. Kinetic Theory in the Earth Sciences. Princeton University Press, Princeton, NJ.

Lasaga, A.C., Kirkpatrick, R.J. (Eds.), 1981. Kinetics of Geochemical Processes. Mineralogical Society of America, Washington, DC.

Liberti, L., Helfferich, F.G. (Eds.), 1983. Mass Transfer and Kinetics of Ion Exchange," NATO ASI Ser. E, No. 71. Nijhoff, The Hague, The Netherlands.

Khaokaew, S., Landrot, G., Sparks, D.L., 2016. Speciation and release kinetics of cadmium and zinc in paddy soils. In: Rinklebe, J., Knox, A., Paller, M. (Eds.), Trace Elements in Waterlogged Soils and Sediments". Taylor & Francis, Abingdon, UK, pp. 75–99.

Pignatello, J.J., 2000. The measurement and interpretation of sorption and desorption rates for organic compounds in soil media. Adv. Agron. 69, 1–73.

Sparks, D.L., 1985. Kinetics of ion reactions in clay minerals and soils. Adv. Agron. 38, 231–266.

Sparks, D.L., 1989. Kinetics of Soil Chemical Processes. Academic Press, San Diego, CA.

Sparks, D.L., 1992. Soil kinetics. In: Nirenberg, W.A. (Ed.), Encyclopedia of Earth Systems Science, vol. 4. Academic Press, San Diego, CA, pp. 219–229.

Sparks, D.L., 1999. Kinetics of reactions in pure and mixed systems. In: Sparks, D.L. (Ed.), Soil Physical Chemistry", second ed. CRC Press, Boca Raton, FL, pp. 83–145.

Sparks, D.L., Suarez, D.L. (Eds.), 1991. Rates of Soil Chemical Processes. SSSA Spec. Publ. 27. Soil Science Society of America, Madison, WI.

Sparks, D.L., 2005. Sorption-desorption, kinetics. In: Hillel, D., Hatfield, J.L., Powlson, D.S., Rosenweig, C., Scow, K.M., Singer, M.J., Sparks, D.L. (Eds.), Encyclopedia of Soils in the Environment". Elsevier Ltd., Oxford UK, pp. 556–561.

Sparks, D.L., 2011. Kinetics and mechanisms of soil chemical reactions. In: Huang, P.M., Li, Y., Sumner, M.E. (Eds.), Handbook of Soil Sciences: Properties and Processes, second ed. CRC Press, Boca Raton, FL, pp. 13-1–13-30.

Sparks, D.L., 2018. Kinetics of geochemical processes. In: White, W.M. (Ed.), Encyclopedia of Geochemistry: A Comprehensive Reference Source on the Chemistry of the Earth". Springer International Publishing, Cham, Switzerland, pp. 775–784.

Sposito, G., 1994. Chemical Equilibria and Kinetics in Soils. Wiley, New York.

Stumm, W. (Ed.), 1990. Aquatic Chemical Kinetics. Wiley, New York.

Problem Set

Q1. To assess the binding mechanism of zinc to aluminum oxides, a sorption study is carried out. The main objective is to determine whether zinc binds to the surface of the aluminum oxide or if it is incorporated into a surface precipitate. Gibbsite is the aluminum (hydr)oxide used in the experiment. The temperature of the experiment and the pH are varied at 15, 25, and 35°C and pH 6.2 and pH 7.6. Based on the following experimental data, create a series of plots that demonstrate the % Zn that is adsorbed to the surface of gibbsite over time, at each pH value, and at different temperatures. Describe the differences in the sorption reactions. Then, determine the energies of activation for the reaction at pH 6.2 versus at 7.6 and discuss your findings. The starting concentration of Zn in solution is 56 μM. The Zn concentrations in the table are the amounts of Zn remaining in solution at the noted equilibration time.

pH 7.6					
15°C		25°C		35°C	
Time (h)	Zn (μM)	Time (h)	Zn (μM)	Time (h)	Zn (μM)
0.5	46.5	0.5	41.0	0.5	37.0
1.0	45.5	1.0	39.0	1.0	33.0
3.0	44.0	3.0	34.2	3.0	24.0
6.0	42.6	12.0	26.0	6.0	19.0
12.0	42.0	36.0	10.4	12.0	8.0
24.0	38.0			24.0	2.5
36.0	35.5				
48.0	33.0				

pH 6.2					
15°C		25°C		35°C	
Time (h)	Zn (μM)	Time (h)	Zn (μM)	Time (h)	Zn (μM)
0.5	48.2	0.5	48.0	0.5	47.1
1.0	48.0	1.0	47.8	1.0	47.0
3.0	47.9	3.0	47.7	3.0	46.8
6.0	47.8	6.0	47.5	6.0	46.6
12.0	47.7	12.0	47.4	12.0	46.5
24.0	47.6	24.0	47.3	24.0	46.3
48.0	47.4	48.0	47.0	48.0	46.1

Q2. Given the following data for two sorption reactions of aqueous Mn^{2+} onto the surface of manganese oxide, determine the order of each reaction and the rate constants. The data were acquired using a stopped-flow kinetic apparatus coupled with electron paramagnetic resonance (EPR). Data were collected extremely quickly (about every 50 μs). What are the approximate initial concentrations for both reactions? Make a plot to illustrate.

				Reaction 1								Reaction 2	
time (ms)	Mn²⁺ (µM)	time (ms, cont.)	Mn²⁺ (µM) (cont.)	time (ms, cont.)	Mn²⁺ (µM) (cont.)	time (ms, cont.)	Mn²⁺ (µM) (cont.)	time (ms)	Mn²⁺ (µM)	time (ms, cont.)	Mn²⁺ (µM) (cont.)	time (ms, cont.)	Mn²⁺ (µM) (cont.)
66.12	18.20	99.09	14.45	133.68	10.96	166.59	8.71	79.47	9.12	112.53	7.24	145.38	5.75
67.83	17.78	100.59	14.13	135.21	10.72	168.18	8.71	81.09	8.91	114.12	7.24	147.15	5.62
69.27	17.38	102	14.13	136.59	10.96	169.65	8.71	82.56	8.91	115.41	7.08	148.59	5.62
70.77	16.98	103.5	14.13	138.06	11.22	171.06	8.71	84.09	8.71	116.88	7.08	150.03	5.62
72.18	16.60	104.94	13.80	139.59	10.96	172.62	8.51	85.56	8.51	118.41	6.92	151.5	5.50
73.59	16.60	108.12	13.49	141.06	10.96	174.36	8.32	87.03	8.32	119.82	6.92	152.88	5.50
75.12	16.22	109.5	13.18	142.62	10.47	175.8	8.32	88.44	8.32	121.65	6.76	154.41	5.50
76.59	16.22	111.03	13.18	144	10.47	177.27	8.13	89.91	8.51	123.24	6.76	155.82	5.37
78.03	15.85	112.59	12.88	145.47	10.23	178.68	7.94	91.41	8.32	124.62	6.61	157.32	5.37
79.47	15.49	114.06	12.59	147.18	10.23			92.7	8.32	126.18	6.61	158.85	5.37
81.27	15.85	115.56	12.30	148.68	10.00			94.56	8.51	127.68	6.46	160.62	5.25
82.68	15.85	116.88	12.02	150.18	10.00			96.15	8.51	129.09	6.46	161.97	5.25
84.12	15.85	118.47	12.02	151.65	10.00			97.62	8.32	130.56	6.31	163.5	5.13
85.62	15.85	119.94	12.02	153.12	10.00			99.09	7.94	131.97	6.31	165.06	5.01
87.15	15.85	121.71	12.02	154.56	10.00			100.56	7.76	133.71	6.17	166.53	5.01
88.41	15.14	123.15	12.02	156.03	10.00			102	7.76	135.12	6.03	168.09	4.90
89.97	14.79	124.71	11.75	157.44	9.55			103.44	7.76	136.59	5.89	169.56	4.90
91.41	14.79	126.18	11.22	158.91	9.33			104.82	7.59	137.97	5.75	171.03	4.90
92.82	14.45	127.65	11.22	160.68	9.33			106.35	7.59	139.5	5.75	172.5	4.79
94.65	14.45	129.12	11.22	162.21	9.12			108.03	7.41	141	5.62	174.24	4.79
96.21	14.13	130.56	11.48	163.62	9.12			109.56	7.41	142.47	5.62	175.74	4.68
97.56	14.13	131.91	11.22	165.12	8.91			111.03	7.41	143.88	5.75	177.21	4.68

References

Aharoni, C., 1984. Kinetics of adsorption: the S-shaped Z(t) plot. Adsorpt. Sci. Technol. 1, 1—29.

Aharoni, C., Sparks, D.L., 1991. Kinetics of soil chemical reactions: a theoretical treatment. In: Sparks, D.L., Suarez, D.L. (Eds.), Rates of Soil Chemical Processes. SSSA Spec. Publ. No. 27. Soil Science Society of America, Madison, WI, pp. 1—18.

Aharoni, C., Suzin, Y., 1982a. Application of the Elovich equation to the kinetics of occlusion: part 1. Homogenous microporosity. J. Chem. Soc. Faraday Trans. 1 78, 2313—2320.

Aharoni, C., Suzin, Y., 1982b. Application of the Elovich equation to the kinetics of occlusion: part 3. Heterogenous microporosity. J. Chem. Soc. Faraday Trans. 1 78, 2329–2336.

Aharoni, C., Ungarish, M., 1976. Kinetics of activated chemisorption. Part 1. The non-elovichian part of the isotherm. J. Chem. Soc. Faraday Trans. 172, 400–408.

Ainsworth, C.C., Pilou, J.L., Gassman, P.L., Van Der Sluys, W.G., 1994. Cobalt, cadmium, and lead sorption to hydrous iron oxide: residence time effect. Soil. Sci. Soc. Am. J. 58, 1615–1623.

Alexander, M., 1995. How toxic are toxic chemicals in soils? Environ. Sci. Technol. 29, 2713–2717.

Alexander, M., 2000. Aging, bioavailability, and overestimation of risk from environmental pollutants. Environ. Sci. Technol 34, 4259–4265.

Amacher, M.C., 1991. Methods of obtaining and analyzing kinetic data. In: Sparks, D.L., Suarez, D.L. (Eds.), Rates of Soil Chemical Processes. SSSA Spec. Publ. No. 27. Soil Science Society of America, Madison, WI, pp. 19–59.

Atkinson, R.J., Hingston, F.J., Posner, A.M., Quirk, J.P., 1970. Elovich equation for the kinetics of isotope exchange reactions at solid-liquid interfaces. Nature (London) 226, 148–149.

Backes, C.A., McLaren, R.G., Rate, A.W., Swift, R.S., 1995. Kinetics of cadmium and cobalt desorption from iron and manganese oxides. Soil. Sci. Soc. Am. J 59, 778–785.

Ball, W.P., 1989. Equilibrium Sorption and Diffusion Rate Studies With Halogenated Organic Chemicals and Sandy Aquifer Materials. Ph.D. Dissertation, Stanford University, Palo Alto, CA.

Ball, W.P., Roberts, P.V., 1991. Long-term sorption of halogenated organic chemicals by aquifer material: 1. Equilibrium. Environ. Sci. Technol 25 (7), 1223–1236.

Barrow, N.J., 1986. Testing a mechanistic model: II. the effects of time and temperature on the reaction of zinc with a soil. J. Soil. Sci 37, 277–286.

Bar-Tal, A., Sparks, D.L., Pesek, J.D., Feigenbaum, S., 1990. Analyses of adsorption kinetics using a stirred-flow chamber: I. theory and critical tests. Soil. Sci. Soc. Am. J 54, 1273–1278.

Bernasconi, E.C., 1976. Relaxation Kinetics. Academic Press, New York, NY.

Bloom, P.R., Erich, M.S., 1987. Effect of solution composition on the rate and mechanism of gibbsite dissolution in acid solutions. Soil. Sci. Soc. Am. J 51, 1131–1136.

Bruemmer, G.W., Gerth, J., Tiller, K.G., 1988. Reaction kinetics of the adsorption and desorption of nickel, zinc and cadmium by goethite: I. Adsorption and diffusion of metals. J. Soil. Sci 39, 37–52.

Brusseau, M.L., Rao, P.S.C., 1989. Sorption nonideality during organic contaminant transport in porous media. CRC Crit. Rev. Environ. Control 19, 33–99.

Bunnett, J.F., 1986. Kinetics in solution. In: Bernasconi, C.F. (Ed.), Investigations of Rates and Mechanisms of Reactions. Fourth edition. John Wiley & Sons, New York, NY, pp. 171–250.

Bunzl, K., Schmidt, W., Sansoni, B., 1976. Kinetics of ion exchange in soil organic matter. IV. Adsorption and desorption of Pb^{2+}, Cu^{2+}, Zn^{2+}, and Ca^{2+} by peat. J. Soil. Sci 27, 32–41.

Carroll, K.M., Harkness, M.R., Bracco, A.A., Balcarcel, R.B., 1994. Application of a permeant/polymer diffusional model to the desorption of polychlorinated biphenyls from Hudson River sediments. Environ. Sci. Technol 28, 253–258.

Casey, W.H., Westrich, H.R., Arnold, G.W., Banfield, J.F., 1989. The surface chemistry of dissolving labradorite feldspar. Geochim. Cosmochim. Acta 53, 821–832.

Chien, S.H., Clayton, W.R., 1980. Application of Elovich equation to the kinetics of phosphate release and sorption in soils. Soil. Sci. Soc. Am. J 44, 265–268.

Chute, J.H., Quirk, J.P., 1967. Diffusion of potassium from mica-like materials. Nature (London). 213, 1156–1157.

Coates, J.T., Elzerman, A.W., 1986. Desorption kinetics for selected PCB congeners from river sediments. J. Contam. Hydrol 1, 191–210.

Comans, R.N.J., Hockley, D.E., 1992. Kinetics of cesium sorption on illite. Geochim. Cosmochim. Acta 56, 1157–1164.

Connaughton, D.F., Stedinger, J.R., Lion, L.W., Shuler, M.L., 1993. Description of time-varying desorption kinetics: release of naphthalene from contaminated soils. Environ. Sci. Technol 27, 2397–2403.

Cornelissen, G., van Noort, P.C.M., Parsons, J.R., Govers, H.A.J., 1997. The temperature dependence of slow adsorption and desorption kinetics of organic compounds in sediments. Environ. Sci. Technol 31, 454–460.

Crank, J., 1976. The Mathematics of Diffusion, Second edition. Clarendon Press, Oxford, England.

DiToro, D.M., Horzempa, L.M., 1982. Reversible and resistant components of PCB adsorption-desorption: Isotherms. Environ. Sci. Technol 16, 594–602.

Eigen, M., 1954. Ionic reactions in aqueous solutions with half-times as short as 10^{-9} second. Applications to neutralization and hydrolysis reactions. Discuss. Faraday Soc. 17, 194–205.

Fendorf, S.E., Sparks, D.L., Franz, J.A., Camaioni, D.M., 1993. Electron paramagnetic resonance stopped-flow kinetic study of manganese (II) sorption-desorption on birnessite. Soil Sci. Soc. Am. J. 57, 57–62.

Ford, R.G., Scheinost, A.C., Sparks, D.L., 2001. Frontiers in metal sorption/precipitation mechanisms on soil mineral surfaces. Adv. Agron 74, 41–62.

Fuller, C.C., Davis, J.A., Waychunas, G.A., 1993. Surface chemistry of ferrihydride: part 2. Kinetics of arsenate adsorption and coprecipitation. Geochim. Cosmochim. Acta 57, 2271–2282.

Furrer, G., Stumm, W., 1986. The coordination chemistry of weathering. I. Dissolution kinetics of δ-Al_2O_3 and BeO. Geochim. Cosmochim. Acta 50, 1847−1860.

Gardiner Jr., W.C., 1969. Rates and Mechanisms of Chemical Reactions. Benjamin, New York, NY.

Ginder-Vogel, M., Landrot, G., Fischel, J.S., Sparks, D.L., 2009. Quantification of rapid environmental redox processes with quick-scanning X-ray absorption spectroscopy (Q-XAS). Proc. Natl. Acad. Sci. U.S.A. 106, 16124−16128.

Ginder-Vogel, M., Sparks, D.L., 2010. The impacts of X-ray absorption spectroscopy on understanding soil processes and reaction mechanisms. In: Singh, B., Gräfe, M. (Eds.), Synchrotron-Based Techniques in Soils and Sediments. Developments in Soil Science 34. Elsevier, New York, NY, pp. 1−26.

Gräfe, M., Mustafa, G., Singh, B., Kookana, R.S., 2008. Temperature and aging effects on the surface speciation of Cd(II) at the goethite-water interface. In: Barnett, M.O., Kent, D.B. (Eds.), Adsorption of Metals by Geomedia II: Variables, Mechanisms, and Model Applications. Developments in Earth and Environmental Sciences, 7. Elsevier, Amsterdam, pp. 187−204.

Grossl, P.R., Eick, M.J., Sparks, D.L., Goldberg, S., Ainsworth, C.C., 1997. Arsenate and chromate retention mechanisms on goethite. 2. Kinetic evaluation using a pressure-jump relaxation technique. Environ. Sci. Technol 31, 321−326.

Grossl, P.R., Sparks, D.L., Ainsworth, C.C., 1994. Rapid kinetics of Cu(II) adsorption/desorption on goethite. Environ. Sci. Technol 28, 1422−1429.

Hamaker, J.W., Thompson, J.M., 1972. Adsorption. In: Goring, C.A.I., Hamaker, J.W. (Eds.), Organic Chemicals in the Environment. Dekker, New York, NY, pp. 39−151.

Haque, R., Lindstrom, F.T., Freed, V.H., Sexton, R., 1968. Kinetic study of the sorption of 2,4-D on some clays. Environ. Sci. Technol 2, 207−211.

Havlin, J.L., Westfall, D.G., 1985. Potassium release kinetics and plant response in calcareous soils. Soil Sci. Soc. Am. J 49, 366−370.

Hayes, K.F., Roe, A.L., Brown Jr., G.E., Hodgson, K.O., Leckie, J.O., Parks, G.A., 1987. In situ X-ray absorption study of surface complexes: selenium oxyanions on α-FeOOH. Science 238, 783−786.

Jardine, P.M., Dunnivant, F.M., Selim, H.M., McCarthy, J.F., 1992. Comparison of models for describing the transport of dissolved organic carbon in aquifer columns. Soil Sci. Soc. Am. J 56, 393−401.

Jardine, P.M., Sparks, D.L., 1984. Potassium-calcium exchange in a multireactive soil system. I. Kinetics. Soil Sci. Soc. Am. J 48, 39−45.

Karickhoff, S.W., 1980. Sorption kinetics of hydrophobic pollutants in natural sediments. In: Baker, R.A. (Ed.), Contaminants and Sediments. Ann Arbor Science Publishers, Ann Arbor, MI, pp. 193−205.

Karickhoff, S.W., Brown, D.S., Scott, T.A., 1979. Sorption of hydrophobic pollutants on natural sediments. Water Res 13, 241−248.

Karickhoff, S.W., Morris, K.R., 1985. Sorption dynamics of hydrophobic pollutants in sediment suspensions. Environ. Toxicol. Chem 4, 469−479.

Khan, S.U., 1973. Equilibrium and kinetic studies on the adsorption of 2,4-D and picloram on humic acid. Can. J. Soil. Sci 53, 429−434.

Kuo, S., Lotse, E.G., 1974. Kinetics of phosphate adsorption and desorption by lake sediments. Soil. Sci. Soc. Am. Proc 38, 50−54.

Landrot, G., Ginder-Vogel, M., Sparks, D.L., 2010. Kinetics of chromium(III) oxidation by manganese(IV) oxides using quick scanning X-ray absorption fine structure spectroscopy (Q-XAFS). Environ. Sci. Technol 44, 143−149.

Langmuir, D., 1997. Aqueous Environmental Geochemistry. Prentice-Hall, Englewood Cliffs, NJ.

Lee, L.S., Rao, P.S.C., Brusseau, M.L., Ogwada, R.A., 1988. Nonequilibrium sorption of organic contaminants during flow through columns of aquifer materials. Environ. Toxicol. Chem 7, 779−793.

Leenheer, J.A., Ahlrichs, J.L., 1971. A kinetic and equilibrium study of the adsorption of carbaryl and parathion upon soil organic matter surfaces. Soil Sci. Soc. Am. Proc 35, 700−704.

Li, X.Y., Gamiz, B., Wang, Y.Q., Pignatello, J.J., Xing, B.S., 2015. Competitive sorption used to probe strong hydrogen bonding sites for weak organic acids on carbon nanotubes. Environ. Sci. Technol 49, 1409−1417.

Lövgren, L., Sjöberg, S., Schindler, P.W., 1990. Acid/base reactions and Al(III) complexation at the surface of goethite. Geochim. Cosmochim. Acta 54, 1301−1306.

Low, M.J.D., 1960. Kinetics of chemisorption of gases on solids. Chem. Rev 60, 267−312.

McCall, P.J., Agin, G.L., 1985. Desorption kinetics of picloram as affected by residence time in the soil. Environ. Toxicol. Chem 4, 37−44.

McLaren, R.G., Backes, C.A., Rate, A.W., Swift, R.S., 1998. Cadmium and cobalt desorption kinetics from soil clays: effect of sorption period. Soil Sci. Soc. Am. J 62, 332−337.

Miller, C.T., Pedit, J., 1992. Use of a reactive surface-diffusion model to describe apparent sorption-desorption hysteresis and abiotic degradation of lindane in a subsurface material. Environ. Sci. Technol 26 (7), 1417−1427.

Mo, X., Siebecker, M.G., Gou, W., Li, W., 2021. EXAFS investigation of Ni(II) sorption at the palygorskite-solution interface: new insights into surface-induced precipitation phenomena. Geochim. Cosmochim. Acta 314, 85−107.

Mustafa, G., Kookana, R.S., Singh, B., 2006. Desorption of cadmium from goethite: effects of pH, temperature and aging. Chemosphere 64, 856–865.

Nkedi-Kizza, P., Biggar, J.W., Selim, H.M., van Genuchten, M.Th., Wierenga, P.J., Davidson, J.M., et al., 1984. On the equivalence of two conceptual models for describing ion exchange during transport through an aggregated Oxisol. Water Resour. Res 20, 1123–1130.

Onken, A.B., Matheson, R.L., 1982. Dissolution rate of EDTA-extractable phosphate from soils. Soil Sci. Soc. Am. J 46, 276–279.

Pan, B., Ning, P., Xing, B., 2009. Part V—sorption of pharmaceuticals and personal care products. Environ. Sci. Pollut. Res. 16, 106–116.

Parikh, S.J., Lafferty, B.J., Sparks, D.L., 2008. An ATR-FTIR spectroscopic approach for measuring rapid kinetics at the mineral/water interface. J. Colloid Interface. Sci 320, 177–185.

Pavlostathis, S.G., Mathavan, G.N., 1992. Desorption kinetics of selected volatile organic compounds from field contaminated soils. Environ. Sci. Technol 26, 532–538.

Pedit, J.A., Miller, C.T., 1995. Heterogenous sorption processes in subsurface systems. 2. Diffusion modeling approaches. Environ. Sci. Technol 29, 1766–1772.

Peltier, E., Allada, R., Navrotsky, A., Sparks, D.L., 2006. Nickel solubility and precipitation in soils: a thermodynamic study. Clays Clay Miner 54, 153–164.

Peterson, M.L., Brown, G.E., Parks, G.A., Stein, C.L., 1997. Differential redox and sorption of Cr(III/VI) on natural silicate and oxide minerals: EXAFS and XANES results. Geochim. Cosmochim. Acta 61, 3399–4413.

Petrovic, R., Berner, R.A., Goldhaber, M.B., 1976. Rate control in dissolution of alkali feldspars. I. Study of residual feldspar grains by X-ray photoelectron spectroscopy. Geochim. Cosmochim. Acta 40, 537–548.

Pignatello, J.J., 1998. Soil organic matter as a nanoporous sorbent of organic pollutants. Adv. Colloid Interface Sci. 76–77, 445–467.

Pignatello, J.J., 2000. The measurement and interpretation of sorption and desorption rates for organic compounds in soil media. Adv. Agron 69, 1–73.

Pignatello, J.J., Huang, L.Q., 1991. Sorptive reversibility of atrazine and metolachlor residues in field soil samples. J. Environ. Qual 20, 222–228.

Polyzopoulos, N.A., Keramidas, V.Z., Pavlatou, A., 1986. On the limitations of the simplified Elovich equation in describing the kinetics of phosphate sorption and release from soils. J. Soil Sci. 37, 81–87.

Roberts, D.R., Scheidegger, A.M., Sparks, D.L., 1999. Kinetics of mixed Ni-Al precipitate formation on a soil clay fraction. Environ. Sci. Technol 33, 3749–3754.

Scheckel, K.G., Sparks, D.L., 2001. Dissolution kinetics of nickel surface precipitates on clay mineral and oxide surfaces. Soil. Sci. Soc. Am. J 65, 685–694.

Scheidegger, A.M., Lamble, G.M., Sparks, D.L., 1997. Spectroscopic evidence for the formation of mixed-cation, hydroxide phases upon metal sorption on clays and aluminum oxides. J. Colloid Interface. Sci 186, 118–128.

Scheidegger, A.M., Sparks, D.L., 1996. Kinetics of the formation and the dissolution of nickel surface precipitates on pyrophyllite. Chem. Geol 132, 157–164.

Scheidegger, A.M., Strawn, D.G., Lamble, G.M., Sparks, D.L., 1998. The kinetics of mixed Ni-Al hydroxide formation on clay and aluminum oxide minerals: a time-resolved XAFS study. Geochim. Cosmochim. Acta 62, 2233–2245.

Schott, J., Petit, J.C., 1987. New evidence for the mechanisms of dissolution of silicate minerals. In: Stumm, W. (Ed.), Aquatic Surface Chemistry. John Wiley & Sons (Interscience), New York, NY, pp. 293–312.

Scribner, S.L., Benzing, T.R., Sun, S., Boyd, S.A., 1992. Desorption and bioavailability of aged simazine residues in soil from a continuous corn field. J. Environ. Qual 21, 115–120.

Sharpley, A.N., 1983. Effect of soil properties on the kinetics of phosphorus desorption. Soil. Sci. Soc. Am. J 47, 462–467.

Siebecker, M., Li, W., Khalid, S., Sparks, D., 2014. Real-time QEXAFS spectroscopy measures rapid precipitate formation at the mineral–water interface. Nat. Commun 5, 6003.

Siebecker, M.G., Madison, A.S., Luther, G.W., 2015. Reduction kinetics of polymeric (soluble) manganese (IV) oxide (MnO_2) by ferrous iron (Fe^{2+}). Aquat. Geochem 21, 143–158.

Skopp, J., 1986. Analysis of time dependent chemical processes in soils. J. Environ. Qual 15, 205–213.

Sparks, D.L., 1985. Kinetics of ionic reactions in clay minerals and soils. Adv. Agron 38, 231–266.

Sparks, D.L., 1989. Kinetics of Soil Chemical Processes. Academic Press, San Diego, CA.

Sparks, D.L., 1998. Kinetics of sorption/release reactions on natural particles. In: Huang, P.M., Senesi, N., Buffle, J. (Eds.), Structure and Surface Reactions of Soil Particles. John Wiley & Sons, New York, NY, pp. 413–448.

Sparks, D.L., 1999. Kinetics and mechanisms of chemical reactions at the soil mineral/water interface. In: Sparks, D.L. (Ed.), Soil Physical Chemistry. Second edition. CRC Press, Boca Raton, FL, pp. 135–191.

Sparks, D.L., 1999. Kinetics of reactions in pure and mixed systems. In: Sparks, D.L. (Ed.), Soil Physical Chemistry, second ed. CRC Press, Boca Raton, FL, pp. 83–145.

Sparks, D.L., 2000. Kinetics and mechanisms of soil chemical reactions. In: Sumner, M.E. (Ed.), Handbook of Soil Science. CRC Press, Boca Raton, FL, pp. B-123—167.

Sparks, D.L., 2011. Kinetics and Mechanisms of Soil Chemical Reactions. In: Huang, P.M., Li, Y., Sumner, M.E. (Eds.), Handbook of Soil Sciences: Properties and Processes, second ed. CRC Press, Boca Raton, FL, pp. 13-1—13-30.

Sparks, D.L., 2013. Advances in the use of synchrotron radiation to elucidate environmental interfacial reaction processes and mechanisms in the Earth's critical zone. In: Xu, J., Sparks, D.L. (Eds.), Molecular Environmental Soil Science: Progress in Soil Science. Springer, Dordrecht, pp. 93—114.

Sparks, D.L., Fendorf, S.E., Toner, C.V., Carski, T.H., 1996. Kinetic methods and measurements. In: Sparks, D.L. (Ed.), Methods of Soil Analysis: Part 3 — Chemical Methods. SSSA Book Ser. 5. Soil Science Society of America, Madison, WI, pp. 1275—1307.

Sparks, D.L., Jardine, P.M., 1984. Comparison of kinetic equations to describe K-Ca exchange in pure and in mixed systems. Soil Sci. 138, 115—122.

Sparks, D.L., Suarez, D.L. (Eds.), 1991. Rates of Soil Chemical Processes. SSSA Spec. Publ. No. 27. Soil Science Society of America, Madison, WI.

Sparks, D.L., Zhang, P.C., 1991. Relaxation methods for studying kinetics of soil chemical phenomena. In: Sparks, D.L., Suarez, D.L. (Eds.), Rates of Soil Chemical Processes. SSSA Spec. Publ. No. 27. Soil Science Society of America, Madison, WI, pp. 61—94.

Sposito, G., 1994. Chemical Equilibria and Kinetics in Soils. John Wiley & Sons, New York, NY.

Srivastava, P., Gräfe, M., Singh, B., Balasubramanian, M., 2008. Cadmium and lead desorption from kaolinite. In: Barnett, M.O., Kent, D.B. (Eds.), Adsorption of Metals by Geomedia II: Variables, Mechanisms, and Model Applications. Developments in Earth and Environmental Sciences, 7. Elsevier, Amsterdam, pp. 205—233.

Steinberg, S.M., Pignatello, J.J., Sawhney, B.L., 1987. Persistence of 1,2-dibromoethane in soils: Entrapment in intraparticle micropores. Environ. Sci. Technol 21, 1201—1208.

Strawn, D.G., Sparks, D.L., 1999. The use of XAFS to distinguish between inner- and outer-sphere lead adsorption complexes on montmorillonite. J. Colloid Interf. Sci 216, 257—269.

Strawn, D.G., Sparks, D.L., 2000. Effects of soil organic matter on the kinetics and mechanisms of Pb(II) sorption and desorption in soil. Soil Sci. Soc. Am. J 64, 145—156.

Stumm, W. (Ed.), 1992. Chemistry of the Solid-Water Interface. John Wiley & Sons, New York, NY.

Stumm, W., Furrer, G., 1987. The dissolution of oxides and aluminum silicates: examples of surface-coordination-controlled kinetics. In: Stumm, W. (Ed.), Aquatic Surface Chemistry. John Wiley & Sons (Interscience), New York, NY, pp. 197—219.

Stumm, W., Wollast, R., 1990. Coordination chemistry of weathering. Kinetics of the surface-controlled dissolution of oxide minerals. Rev. Geophys 28, 53—69.

Takahashi, M.T., Alberty, R.A., 1969. The pressure-jump methods. In: Kustin, K. (Ed.), Methods in Enzymology, Vol. 16. Academic Press, New York, NY, pp. 31—55.

Tang, L., Sparks, D.L., 1993. Cation exchange kinetics on montmorillonite using pressure-jump relaxation. Soil Sci. Soc. Am. J 57, 42—46.

van Genuchten, M.Th., Wagenet, R.J., 1989. Two-site/two-region models for pesticide transport and degradation: Theoretical development and analytical solutions. Soil Sci. Soc. Am. J 53, 1303—1310.

Weber Jr., W.J., Gould, J.P., 1966. Sorption of organic pesticides from aqueous solution. Adv. Chem. Ser 60, 280—305.

Weber Jr., W.J., Huang, W., 1996. A distributed reactivity model for sorption by soils and sediments. 4. Intraparticle heterogeneity and phase-distribution relationships under nonequilibrium conditions. Environ. Sci. Technol 30, 881—888.

Weber Jr., W.J., Miller, C.T., 1988. Modeling the sorption of hydrophobic contaminants by aquifer materials. 1. Rates and equilibria. Water Res 22, 457—464.

Wollast, R., 1967. Kinetics of the alteration of K-feldspar in buffered solutions at low temperature. Geochim. Cosmochim. Acta 31, 635—648.

Wu, S., Gschwend, P.M., 1986. Sorption kinetics of hydrophobic organic compounds to natural sediments and soils. Environ. Sci. Technol 20, 717—725.

Xing, B., Pignatello, J.J., 1997. Dual-mode sorption of low-polarity compounds in glassy poly(vinyl chloride) and soil organic matter. Environ. Sci. Technol 31 (3), 792—799.

Zasoski, R.G., Burau, R.G., 1978. A technique for studying the kinetics of adsorption in suspensions. Soil Sci. Soc. Am. J 42, 372—374.

Zhang, P.C., Sparks, D.L., 1989. Kinetics and mechanisms of molybdate adsorption/desorption at the goethite/water interface using pressure-jump relaxation. Soil Sci. Soc. Am. J 53, 1028—1034.

Zhang, P.C., Sparks, D.L., 1990. Kinetics of selenate and selenite adsorption/desorption at the goethite/water interface. Environ. Sci. Technol 24, 1848—1856.

Zhao, Q., Zhang, S., Zhang, X., Lei, L., Ma, W., et al., 2017. Cation—Pi Interaction: a key force for sorption of fluoroquinolone antibiotics on pyrogenic carbonaceous materials. Environ. Sci. Technol 51, 13659–13667.

Zhao, Q., Ma, W., Pan, B., Zhang, Q., Zhang, X., et al., 2018. Wrinkle-induced high sorption makes few-layered black phosphorus a superior adsorbent for ionic organic compounds. Environ. Sci. Nano 5, 1454.

Zinder, B., Furrer, G., Stumm, W., 1986. The coordination chemistry of weathering. II. Dissolution of Fe(III) oxides. Geochim. Cosmochim. Acta 50, 1861–1869.

Redox Chemistry of Soils

8.1 Oxidation—Reduction Reactions and Potentials

Key points: Oxidation and reduction reactions greatly impact the cycling of elements in soils. This is because solid-phase minerals, for example, iron and manganese oxides, which are toxic metal and nutrient scavengers, are stable under oxidizing soil conditions yet become reduced and can dissolve under reducing conditions.

Soil chemical reactions involve some combination of proton and/or electron transfer. Oxidation occurs when there is a loss of electrons from an element or compound during the transfer process, and reduction occurs when an element or compound gains those electrons. Oxidation is always partnered with reduction: if an element or compound is oxidized, there must always be another element or compound to receive those electrons and it becomes reduced. The element or compound that gains the electrons (i.e., it was previously in the oxidized state) is the electron acceptor, and it can also be termed the "oxidant" because it oxidizes its counterpart. Oppositely, the element or compound that loses electrons (i.e., it was previously in the reduced state) can also be termed as the "reductant" because it is the electron donor, and it reduces its counterpart.

Table 8.1 lists oxidants and reductants found in natural environments. The electrons are not free in the soil solution; thus the oxidant must be in close contact with the reductant. Both oxidation and reduction must be considered to completely describe oxidation—reduction (redox) reactions (Bartlett and James, 1993; Patrick et al., 1996).

To determine if a particular reaction will occur (i.e., the Gibbs free energy for the reaction, $\Delta G_r < 0$), one can write reduction and oxidation half-reactions (a half-reaction or half-cell reaction can be referred to as a redox couple) and calculate equilibrium constants for the half-reactions. The redox reactions of soil oxidants can be defined conventionally by using the general half-reduction reaction (Patrick et al., 1996, Sidebar 8.1).

$$Ox + mH^+ + ne^- \rightarrow Red \qquad (8.1)$$

Sidebar 8.1

Half reactions are always written as reduction reactions, with the electron on the left side of the equation. They can be written on the basis of 1 electron ($1e^-$) transfer or several electron transfer (ne^-). In Table 8.1 all the stoichiometry is adjusted so that the reactions are based on $1e^-$ transfer.

In Equation 8.1 Ox is the oxidized component and eventual electron acceptor, Red is the reduced component and eventual electron

Environmental Soil Chemistry, Third Edition
https://doi.org/10.1016/B978-0-12-815880-7.00008-0

8. Redox Chemistry of Soils

TABLE 8.1 Selected reduction half-reactions pertinent to soil, natural water, plant, and microbial systems.[a]

Half-reaction	$\log K^{ob}$	pe[c] pH 5	pe[c] pH 7
Nitrogen species			
$\frac{1}{2} N_2O + e^- + H^+ = \frac{1}{2} N_2 + \frac{1}{2} H_2O$	29.8	22.9	20.9
$NO + e^- + H^+ = \frac{1}{2} N_2O + \frac{1}{2} H_2O$	26.8	19.8	17.8
$\frac{1}{2} NO_2^- + e^- + \frac{3}{2} H^+ = \frac{1}{4} N_2O + \frac{3}{4} H_2O$	23.6	15.1	12.1
$\frac{1}{5} NO_3^- + e^- + \frac{6}{5} H^+ = \frac{1}{10} N_2 + \frac{3}{5} H_2O$	21.1	14.3	11.9
$NO_2^- + e^- + 2H^+ = NO + H_2O$	19.8	9.8	5.8
$\frac{1}{4} NO_3^- + e^- + \frac{5}{4} H^+ = \frac{1}{8} N_2O + \frac{5}{8} H_2O$	18.9	12.1	9.6
$\frac{1}{6} NO_2^- + e^- + \frac{4}{3} H^+ = \frac{1}{6} NH_4^+ + \frac{1}{3} H_2O$	15.1	8.4	5.7
$\frac{1}{8} NO_3^- + e^- + \frac{5}{4} H^+ = \frac{1}{8} NH_4^+ + \frac{3}{8} H_2O$	14.9	8.6	6.1
$\frac{1}{2} NO_3^- + e^- + H^+ = \frac{1}{2} NO_2^- + \frac{1}{2} H_2O$	14.1	9.1	7.1
$\frac{1}{6} NO_3^- + e^- + \frac{7}{6} H^+ = \frac{1}{6} NH_2OH + \frac{1}{3} H_2O$	11.3	5.4	3.1
$\frac{1}{6} N_2 + e^- + \frac{4}{3} H^+ = \frac{1}{3} NH_4^+$	4.6	−0.7	−3.3
Oxygen species			
$\frac{1}{2} O_3 + e^- + H^+ = \frac{1}{2} O_2 + \frac{1}{2} H_2O$	35.1	28.4	26.4
$OH\cdot + e^- = OH^-$	33.6	33.6	33.6
$O_2^- + e^- + 2H^+ = H_2O_2$	32.6	22.6	18.6
$\frac{1}{2} H_2O_2 + e^- + H^+ = H_2O$	30.0	23.0	21.0
$\frac{1}{4} O_2 + e^- + H^+ = \frac{1}{2} H_2O$	20.8	15.6	13.6
$\frac{1}{2} O_2 + e^- + H^+ = \frac{1}{2} H_2O_2$	11.6	8.2	6.2
$O_2 + e^- = O_2^-$	−9.5	−6.2	−6.2
Sulfur species			
$\frac{1}{8} SO_4^{2-} + e^- + \frac{5}{4} H^+ = \frac{1}{8} H_2S + \frac{1}{2} H_2O$	5.2	−1.0	−3.5
$\frac{1}{2} SO_4^{2-} + e^- + 2H^+ = \frac{1}{2} SO_2 + H_2O$	2.9	−7.1	−11.1
Iron and manganese compounds			
$\frac{1}{2} Mn_3O_4 + e^- + 4H^+ = \frac{3}{2} Mn^{2+} + 2H_2O$	30.7	16.7	8.7
$\frac{1}{2} Mn_2O_3 + e^- + 3H^+ = Mn^{2+} + \frac{3}{2} H_2O$	25.7	14.7	8.7
$Mn^{3+} + e^- = Mn^{2+}$	25.5	25.5	25.5
$\gamma MnOOH + e^- + 3H^+ = Mn^{2+} + 2H_2O$	25.4	14.4	8.4
$0.62MnO_{1.8} + e^- + 2.2H^+ = 0.62Mn^{2+} + 1.1H_2O$	22.1	13.4	8.9
$\frac{1}{2} Fe_3(OH)_8 + e^- + 4H^+ = \frac{3}{2} Fe^{2+} + 4H_2O$	21.9	7.9	−0.1

TABLE 8.1   Selected reduction half-reactions pertinent to soil, natural water, plant, and microbial systems.[a]—cont'd

Half-reaction	log K^{ob}	pe[c] pH 5	pe[c] pH 7
		pe[c]	
		pH 5	pH 7
Iron and manganese compounds			
$\frac{1}{2}MnO_2 + e^- + 2H^+ = \frac{1}{2}Mn^{2+} + H_2O$	20.8	12.8	8.8
$[Mn^{3+}(PO_4)_2]^{3-} + e^- = [Mn^{2+}(PO_4)_2]^{4-}$	20.7	20.7	20.7
$Fe(OH)_2^+ + e^- + 2H^+ = Fe^{2+} + 2H_2O$	20.2	10.2	6.2
$\frac{1}{2}Fe_3O_4 + e^- + 4H^+ = \frac{3}{2}Fe^{2+} + 2H_2O$	17.8	3.9	−4.1
$MnO_2 + e^- + 4H^+ = Mn^{3+} + 2H_2O$	16.5	0.54	−7.5
$Fe(OH)_3 + e^- + 3H^+ = Fe^{2+} + 3H_2O$	15.8	4.8	−1.2
$Fe(OH)^{2+} + e^- + H^+ = Fe^{2+} + H_2O$	15.2	10.2	8.2
$\frac{1}{2}Fe_2O_3 + e^- + 3H^+ = Fe^{2+} + \frac{3}{2}H_2O$	13.4	2.4	−3.6
$FeOOH + e^- + 3H^+ = Fe^{2+} + 2H_2O$	13.0	2.0	−4.0
$Fe^{3+} + e^- = Fe^{2+}$phenanthroline	18.0	—[d]	—
$Fe^{3+} + e^- = Fe^{2+}$	13.0	13.0	13.0
$Fe^{3+} + e^- = Fe^{2+}$acetate	—	5.8	—
$Fe^{3+} + e^- = Fe^{2+}$malonate	—	4.4 (pH 4)	—
$Fe^{3+} + e^- = Fe^{2+}$salicylate	—	4.4 (pH 4)	—
$Fe^{3+} + e^- = Fe^{2+}$hemoglobin	—	—	2.4
$Fe^{3+} + e^- = Fe^{2+}$cyt b_3 (plants)	—	—	0.68
$Fe^{3+} + e^- = Fe^{2+}$oxalate	—	—	0.034
$Fe^{3+} + e^- = Fe^{2+}$pyrophosphate	−2.4	—	—
$Fe^{3+} + e^- = Fe^{2+}$peroxidase	—	—	−4.6
$Fe^{3+} + e^- = Fe^{2+}$ferredoxin (spinach)	—	—	−7.3
$\frac{1}{3}KFe_3(SO_4)_2(OH)_6 + e^- + 2H^+ = Fe^{2+} + 2H_2O + \frac{2}{3}SO_4^{2-} + \frac{1}{3}K^+$	8.9	6.9	2.9
$[Fe(CN)_6]^{3-} + e^- = [Fe(CN)_6]^{4-}$	—	—	6.1
Carbon species			
$\frac{1}{2}CH_3OH + e^- + H^+ = \frac{1}{2}CH_4 + \frac{1}{2}H_2O$	9.9	4.9	2.9
$\frac{1}{2}$o-quinone $+ e^- + H^+ = \frac{1}{2}$diphenol	—	—	5.9
$\frac{1}{2}$p-quinone $+ e^- + H^+ = \frac{1}{2}$hydroquinone	—	—	4.7
$\frac{1}{12}C_6H_{12}O_6 + e^- + H^+ = \frac{1}{4}C_2H_5OH + \frac{1}{4}H_2O$	4.4	0.1	−1.9

(Continued)

TABLE 8.1   Selected reduction half-reactions pertinent to soil, natural water, plant, and microbial systems.[a]—cont'd

Half-reaction	log K^{ob}	pe[c] pH 5	pH 7
Carbon species			
Pyruvate $+\, e^- + H^+ =$ lactate	—	—	−3.1
$\frac{1}{8} CO_2 + e^- + H^+ = \frac{1}{8} CH_4 + \frac{1}{4} H_2O$	2.9	−2.1	−4.1
$\frac{1}{2} CH_2O + e^- + H^+ = \frac{1}{2} CH_3OH$	2.1	−2.9	−4.9
$\frac{1}{2} HCOOH + e^- + H^+ = \frac{1}{2} CH_2O + \frac{1}{2} H_2O$	1.5	−3.5	−5.5
$\frac{1}{4} CO_2 + e^- + H^+ = \frac{1}{24} C_6H_{12}O_6 + \frac{1}{4} H_2O$	−0.21	−5.9	−7.9
$\frac{1}{2}$ deasc $+\, e^- + H^+ = \frac{1}{2}$ asc	1.0	−3.5	−5.5
$\frac{1}{4} CO_2 + e^- + H^+ = \frac{1}{4} CH_2O + \frac{1}{4} H_2O$	−1.2	−6.1	−8.1
$\frac{1}{2} CO_2 + e^- + H^+ = \frac{1}{2} HCOOH$	−1.9	−6.7	−8.7
Pollutant/nutrient group			
$Co^{3+} + e^- = Co^{2+}$	30.6	30.6	30.6
$\frac{1}{2} NiO_2 + e^- + 2H^+ = \frac{1}{2} Ni^{2+} + H_2O$	29.8	21.8	17.8
$PuO_2^+ + e^- = PuO_2$	26.0	22.0	22.0
$\frac{1}{2} PbO_2 + e^- + 2H^+ = \frac{1}{2} Pb^{2+} + H_2O$	24.8	16.8	12.8
$PuO_2 + e^- + 4H^+ = Pu^{3+} + 2H_2O$	9.9	−6.1	−14.1
$\frac{1}{3} HCrO_4^- + e^- + \frac{4}{3} H^+ = \frac{1}{3} Cr(OH)_3 + \frac{1}{3} H_2O$	18.9	10.9	8.2
$\frac{1}{2} AsO_4^{3-} + e^- + 2H^+ = \frac{1}{2} AsO_2^- + H_2O$	16.5	6.5	2.5
$Hg^{2+} + e^- = \frac{1}{2} Hg_2^{2+}$	15.4	13.4	13.4
$\frac{1}{2} MoO_4^{2-} + e^- + 2H^+ = \frac{1}{2} MoO_2 + H_2O$	15.0	3.0	−1.0
$\frac{1}{2} SeO_4^{2-} + e^- + H^+ = \frac{1}{2} SeO_3^{2-} + \frac{1}{2} H_2O$	14.9	9.9	7.9
$\frac{1}{4} SeO_3^{2-} + e^- + \frac{3}{2} H^+ = \frac{1}{4} Se + \frac{3}{4} H_2O$	14.8	6.3	3.3
$\frac{1}{6} SeO_3^{2-} + e^- + \frac{4}{3} H^+ = \frac{1}{6} H_2Se + \frac{1}{2} H_2O$	7.62	1.0	−1.7
$\frac{1}{2} VO_2^+ + e^- + \frac{1}{2} H^+ + \frac{1}{2} H_2O = \frac{1}{2} V(OH)_3$	6.9	2.4	1.4
$Cu^{2+} + e^- = Cu^+$	2.6	2.6	2.6
$PuO_2 + e^- + 3H^+ = PuOH^{2+} + H_2O$	2.9	−8.1	−14.1
Analytical couples			
$CeO_2 + e^- + 4H^+ = Ce^{3+} + 2H_2O$	47.6	31.6	23.6
$\frac{1}{2} ClO^- + e^- + H^+ = \frac{1}{2} Cl^- + \frac{1}{2} H_2O$	29.0	24.0	22.0
$HClO + e^- = \frac{1}{2} Cl_2 + H_2O$	27.6	20.6	18.6
$\frac{1}{2} Cl_2 + e^- = Cl^-$	23.0	25.0	25.0
$\frac{1}{6} IO_3^- + e^- + H^+ = \frac{1}{6} I^- + \frac{1}{2} H_2O$	18.6	13.6	11.6

TABLE 8.1   Selected reduction half-reactions pertinent to soil, natural water, plant, and microbial systems.[a]—cont'd

Half-reaction	log K^{ob}	pe[c] pH 5	pe[c] pH 7
Analytical couples			
$\frac{1}{2} Pt(OH)_2 + e^- + H^+ = \frac{1}{2} Pt + H_2O$	16.6	11.6	9.6
$\frac{1}{2} I_2 + e^- = I^-$	9.1	11.1	11.1
$\frac{1}{2} Hg_2Cl_2 + e^- = Hg + Cl^-$	4.5	3.9	3.9
$e^- + H^+ = \frac{1}{2} H_2$	0	−5.0	−7
$\frac{1}{2} PtS + e^- + H^+ = \frac{1}{2} Pt + \frac{1}{2} H_2S$	−5.0	−10.0	−12.0

[a] From Bartlett and James (1993), with permission.
[b] Calculated for reaction as written according to Equation 8.14. Free energy of formation data were taken from Lindsay (1979) as a primary source and, when not available from that source, from Garrels and Christ (1965) and Loach (1976).
[c] Calculated using tabulated log $K°$ values, reductant and oxidant $= 10^{-4}$ M soluble ions and molecules, and activities of solid phases $= 1$; partial pressures for gases that are pertinent to soils: 1.01×10^{-4} MPa for trace gases, 2.12×10^{-2} MPa for O_2, 7.78×10^{-2} MPa for N_2, and 3.23×10^{-5} MPa for CO_2.
[d] Values not listed by Loach (1976).

donor, m is the number of hydrogen ions participating in the reaction, and n is the number of electrons involved in the reaction. The electrons in Equation 8.1 must be supplied by an accompanying oxidation half-reaction. For example, in soils soil organic matter (SOM) is a primary source of electrons. Thus to completely describe a redox reaction, an oxidation reaction must balance the reduction reaction. Let us illustrate these concepts for the redox reaction of $Fe(OH)_3$ reduction (Patrick et al., 1996):

$$4Fe(OH)_3(s) + 12H^+ + 4e^- \rightarrow 4Fe^{2+}(aq)$$
$$+ 12H_2O \text{ (reduction)} \qquad (8.2)$$

$$CH_2O + H_2O \rightarrow CO_2 + 4H^+ + 4e^- \text{ (oxidation)}$$
$$(8.3)$$

$$4Fe(OH)_3(s) + CH_2O + 8H^+ \rightarrow 4Fe^{2+}(aq)$$
$$+ CO_2(g) + 11H_2O \text{ (net reduction)}$$
$$(8.4)$$

where CH_2O is SOM. Equation 8.2 represents the reduction half-reaction and Equation 8.3 represents the oxidation half-reaction. Note that the complete redox reaction (Equation 8.4) must be balanced with stoichiometry and charge. All the

atoms on each side of the equation must have an equal number. Also, it can be seen that the eight positive charges from H^+ are balanced by the eight positive charges of $4Fe^{2+}$.

The reduction reaction (Equation 8.2) can also be described by calculating ΔG_r, the Gibbs free energy for the reaction:

$$\Delta G_r = \Delta G_r^o + RT \ln\left(\frac{Red}{(Ox)(H^+)^m}\right) \qquad (8.5)$$

where ΔG_r^o is the standard free energy change for the reaction. The Nernst equation can be employed to express the reduction reaction in terms of electrochemical energy (millivolts) using the expression $\Delta G_r = -nFE$ such that (Patrick et al., 1996):

$$Eh = E° - \frac{RT}{nF} \ln\left(\frac{Red}{Ox}\right) + \frac{mRT}{nF} \ln(H^+) \qquad (8.6)$$

where Eh is the electrode potential or, in the case of the reduction half-reaction in Equation 8.2, a reduction potential, $E°$ is the standard half-reaction reduction potential (with each half-reaction, for example, Equations 8.2 and 8.3, there is a standard potential; the standard potential means the activities of all reactants and products are unity [i.e., one]), F is the Faraday constant, n is the number of electrons exchanged in the half-cell reaction, m is the number of protons exchanged, and the activities of the oxidized and reduced species are in parentheses.

The determination of Eh will provide quantitative information on electron availability and can be either an oxidation or reduction potential depending on how the reaction is written (see Equations 8.2–8.3). Oxidation potentials are more often used in chemistry, while in soil chemistry, reduction potentials are more frequently used to describe soil and other natural systems (Patrick et al., 1996). It should also be pointed out that the Nernst equation is valid for predicting the activity of oxidized and reduced species only if the system is at equilibrium, which is seldom the case for soils and sediments. As noted in Chapter 7, the heterogeneity of soils that promotes transport processes causes many soil chemical reactions to be very slow. Thus it is difficult to use Eh values to quantitatively measure the activities of oxidized and reduced species for such heterogeneous systems (Bohn, 1968).

Using the values of 8.31 J K^{-1} mol^{-1} for R, 9.65×10^4 C mol^{-1} for F, and 298 K for T and the relationship $\ln(x) = 2.303 \log(x)$, Equation 8.6 becomes the following, which can be used to relate Eh to pH:

$$Eh \, (mV) = E^\circ - \frac{59}{n} \log \left(\frac{Red}{Ox} \right) - 59 \frac{m}{n} pH$$

$$(8.7)$$

From Equations 8.6–8.7, it can be seen that Eh increases as the activity (concentration) of the oxidized species increases; Eh decreases with an increase in the activity of the reduced species, and Eh increases as H$^+$ activity increases or pH decreases. If the ratio of protons to electrons is 1 (i.e., $m/n = 1$), one would predict that Eh would change by 59 mV for every unit change in pH. Thus one could predict the Eh at various pH values by using the 59 mV factor. However, this relationship assumes that redox controls the pH of the system. This assumption is valid for solutions, but in soils pH buffering is affected

by soil components such as silicates, carbonates, and oxides, which are not involved in redox reactions. Thus it may be inappropriate to apply the 59 mV factor (Patrick et al., 1996).

Eh is positive and high in strongly oxidizing systems, while it is negative and low in strongly reducing systems. There is not a neutral point, as one observes with pH. Eh, like pH, is an intensity factor. The lower the Eh, the higher the activity of the electrons. Soil pe can be calculated from Eh (in mV) using Equation 8.7a:

$$(pe)(59) = Eh \qquad (8.7a)$$

Redox zones for different elements help describe electron activity conditions. For example, the oxygen–nitrogen range has been defined by Eh values of +250 to +100 mV, the iron range as +100 to 0.0 mV, the sulfate range as 0.0 to −200 mV, and the methane–hydrogen range as < −200 mV (Liu and Narasimhan, 1989).

8.1.1 Eh Versus pH and pe Versus pH Diagrams

Key points: The pe–pH relationship can be used to determine whether an oxidation–reduction reaction can occur spontaneously. Eh versus pH diagrams are useful in determining how oxidized or reduced a soil is as pH varies.

Diagrams of the activities of Eh versus pH can be very useful in delineating the redox status of a system. Figure 8.1 shows such a diagram for soils. The pH range was narrower in reduced soils (negative Eh) than in oxidized soils (positive Eh). Based on these results, Baas Becking et al. (1960) divided the soils into three categories: normal (oxidized), wet (seasonally saturated), and waterlogged (semipermanently saturated) (Figure 8.1).

The reduction half-reaction given in Equation 8.1 can also be expressed in terms of an equilibrium constant K$^\circ$ (Patrick et al., 1996).

$$K^\circ = \frac{(Red)}{(Ox)(e^-)^n (H^+)^m} \qquad (8.8)$$

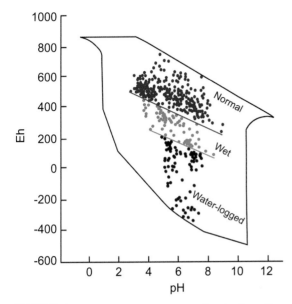

FIGURE 8.1   Eh–pH characteristics of soils. *From Baas Becking et al. (1960), with permission.*

Expressed in log form, Equation 8.8 becomes

$$\log K^o = \log (Red) - \log (Ox) - n \log (e^-)$$
$$- m \log (H^+) \qquad (8.9)$$

The $-\log (e^-)$ term in Equation 8.9 is defined as pe in a similar way that pH is expressed as $-\log (H^+)$. The pe is an intensity factor as it is an index of the electron free energy level per mole of electrons (Ponnamperuma, 1972). Thus pe and pH are master variables of a soil and must be known to completely understand the equilibrium state of a soil. Moreover, to fully determine the redox status of a soil, pe and pH cannot be separated (Bartlett and James, 1993). In strongly oxidizing systems the e^- activity is low and pe is large and positive. In reducing systems pe is small and negative. Sposito (1989) proposed "oxic" (oxidized) soils as those with pe > 7, "suboxic" soils as those in the pe range between +2 and +7, and "anoxic" (reduced) soils as those with pe < +2, all at pH 7. These ranges

are consistent with redox control by oxygen–nitrogen, manganese–iron, and sulfur couples (James and Bartlett, 2000). The pe range of most soils is −6 to +12 (Lindsay, 1979). Rearranging Equation 8.9, one arrives at an expression that relates pe to pH:

$$pe = \frac{\log K^o - \log (Red) + \log (Ox)}{n} - \frac{m}{n} pH$$
$$(8.10)$$

which represents a straight line with a slope of m/n and an intercept given in brackets. The intercept is a function of $\log K^o$ for the half-reaction and the activities of the oxidized and reduced species. When there is a one-electron transfer (i.e., $n = 1$) and consumption of one proton (i.e., $m = 1$), and when $(Red) = (Ox)$, Equation 8.10 is simplified to

$$pe + pH = \log K^o \qquad (8.11)$$

At pH = 0,

$$pe = \log K^o \qquad (8.12)$$

One can relate $\log K^o$ to ΔG_r^o using the equation

$$\Delta G_r = - RT \ln K^o \qquad (8.13)$$

At 298 K and converting to log,

$$\frac{-G_r^o}{5.71} = \log K^o \qquad (8.14)$$

where 5.71 is derived from the product of (RT)(2.303), R is 0.008314 kJ mol^{-1} K^{-1}, and T = 298.15 K. Therefore $\log K^o$ could be estimated by knowing the free energies of formation (ΔG_f^o) of H_2O and the Red and Ox species since those for H^+ and e^- are zero by convention (Bartlett and James, 1993).

Information in Box 8.1 shows how one would calculate $\log K^o$ and pe for a reduction half-reaction at pH 5 and 7 using Equations 8.11–8.14.

The values of $\log K^o$ can be used to predict whether a reduction and oxidation reaction will

---

BOX 8.1

Calculation of log K° and pe

The reduction half-reaction below (see Table 8.1) shows the reduction of Fe^{3+} to Fe^{2+}.

$$Fe(OH)^{2+} + e^- + H^+ = Fe^{2+} + H_2O \quad \text{(B8.1a)}$$

In this reaction there is one electron transfer, i.e., n in (Equation 8.8) is 1, there is consumption of one proton, i.e., $m = 1$, and $(Fe^{3+}) = (Fe^{2+})$ is an imposed condition. Thus (Equation 8.10) reduces to (Equation 8.11) and at pH 0, (Equation 8.12) results. Relating ln K° to ΔG_r^o, one can employ (Equation 8.13):

$$\Delta G_r^o = -RT \ln K^o$$

We know from (Equation 4.7) that

$$\Delta G_r^o = \sum \Delta G_f^o \text{ products} - \sum \Delta G_f^o \text{ reactants}$$

Solving ΔG_r^o for (Equation B8.1a) above,

$$\Delta G_r^o = [(-91.342 \text{ kJ mol}^{-1}) + (-237.52 \text{ kJ mol}^{-1})] - [(-241.85 \text{ kJ mol}^{-1}) + (0)]$$

$$\Delta G_r^o = \left[-328.86 \text{ kJ mol}^{-1} + 241.85 \text{ kJ mol}^{-1} \right] \quad \text{(B8.1b)}$$

$$\Delta G_r^o = -87.01 \text{ kJ mol}^{-1}$$

Using (Equation 8.14),

$$\log K^o = \frac{87.01}{5.71} = 15.2 \quad \text{(B8.1c)}$$

This value for log K° is the one shown in Table 8.1 for the reaction in (Equation B8.1a).

To calculate pe at pH 5 and pH 7, one would use (Equation 8.11). For pH 5 and substituting in the value of 15.2 for log K°,

$$pe = \log K^o - pH$$

$$pe = 15.2 - 5 = 10.2$$

For pH 7,

$$pe = 15.2 - 7 = 8.2$$

These are the pe values shown in Table 8.1 for the reduction half-reaction in (Equation B8.1a).

---

combine to allow the transfer of electrons from the reductant to the oxidant. Table 8.1 lists a number of reduction half-reactions important in natural systems. The log K° values are given in descending order and are pe values at pH 0, when the activities of oxidant and reductant are 1. The log K° values are the standard reference pe values for the reactions. The larger the value of log K° or pe, the greater the tendency for an oxidant (left side of the half-reaction equation) to be reduced (converted to the right side of the half-reaction equation). Conversely, lower log K° and pe values in Table 8.1 indicate reactions that tend to produce electrons (i.e., proceed in the opposite direction in which they are written).

Therefore an oxidant in a given reduction half-reaction can oxidize the reductant in another half-reaction with a lower pe, at a particular pH. Alternatively stated, elements on the right side of the equations will donate electrons to elements on the left side of the equations if the right side has a lower pe value.

As an example, Mn(III,IV) oxides could oxidize Cr(III) to Cr(VI) at pH 5 because the pe value for the reduction half-reaction of Mn (12.8) is higher than that for the Cr(VI) reduction half-reaction (10.9) (Bartlett and James, 1993). In field moist soils over a pH range of 4–7 it has indeed been observed that Mn(III,IV) oxides can oxidize Cr(III) to Cr(VI) (Bartlett and

James, 1979; James and Bartlett, 1993). Alternatively stated, Cr(III) reduces MnO_2.

A practical understanding of predicting a redox reaction based on pe values is as follows: comparing the log $K°$ values from two different reduction half-reactions will predict whether a full redox reaction will take place when the half-reactions are coupled together. By using environmentally relevant pH values (e.g., 5 or 7), the comparison to determine if a full redox reaction is favorable or not will be between the corresponding pe values.

Low pe values mean that the activity of the electron (e^-) is high. An analogy can be made to pH, where a very low pH indicates a very large concentration (activity) of protons. A clear example is the following: pH $= -\log(H^+)$ just as pe $= -\log(e^-)$. At low pH, there is a large concentration (activity) of protons, just as with low pe there is a large activity of electrons. Thus in systems that are very reduced the activity of electrons is very high (and pe is very low) and there is a large tendency for electron transfer to occur. This is the case with soils saturated with water for extended periods of time.

When evaluating if a hypothetical redox reaction will take place, compare the pe values of the corresponding reduction half-reactions. The reduction half-reaction with the smaller pe value will be the one that releases electrons because the activity of the electrons is greater; that is, there is a higher tendency of the electron to be released from the reduced species. This can often be confusing because this means that the reduction half-reaction that releases the electron will take place in the direction opposite in which it is written. The electron that is released will reduce the oxidized species; it will be used in the reduction half-rection when the two half-reactions are coupled together. The half-reaction with the lowest pe will be the half-reaction that supplies the electron.

An alternative way to state this is that a reduction half-reaction with a lower pe value will proceed in the direction opposite to which it is

written when that half-reaction is coupled with a half-reaction with a higher pe value. The half-reaction with the higher pe value will take place as written because the oxidant becomes reduced (accepts the electron). It is important to keep in mind that while some of these reactions may be thermodynamically favorable, the kinetics of the reaction need to be understood to know if the reaction will indeed take place over relevant timescales. For example, Fe^{2+} is not a stable species in most oxygenated or oxidized environments (except at low pH), so it will start to lose electrons very quickly in a matter of seconds.

The pe—pH relationship expressed in (Equation 8.10) can be used to determine whether an oxidation—reduction reaction can occur spontaneously, that is, $\Delta G_r < 0$. Figure 8.2 shows pe versus pH stability lines between oxidized and reduced species for several redox couples. If thermodynamic equilibrium is present, the oxidized form of the couple would be preferred if the pe and pH region was above a given line and the reduced form would be favored below a given line (Bartlett, 1986). The line for Fe is often considered the dividing point between an aerobic (oxidized) and an anaerobic (reduced) soil. In aerobic soils oxidized species stay oxidized even though the thermodynamic tendency is toward reduction, as indicated by the high pe. Below the iron line, reduced species are prevalent, even though the thermodynamic tendency is toward oxidation. Sulfide is easily oxidized, and nitrite is easily reduced (Bartlett and James, 1993).

Figure 8.3 illustrates a pe—pH diagram for several Mn species. One sees that Mn oxides can oxidize Pu(III) to Pu(IV), V(III) to V(V), As(III) to As(V), Se(IV) to Se(VI), and Cr(III) to Cr(VI), because the pe for each of these couples is below the pe for Mn oxides. It has been shown that Mn oxides in soils can indeed effect the oxidation of Pu(III), As(III), Se(IV), and, as noted earlier, Cr(III) (Bartlett and James, 1979; Bartlett, 1981; Amacher and Baker, 1982; Moore et al., 1990). The environmental aspects of some of these oxidation processes are discussed later in this chapter.

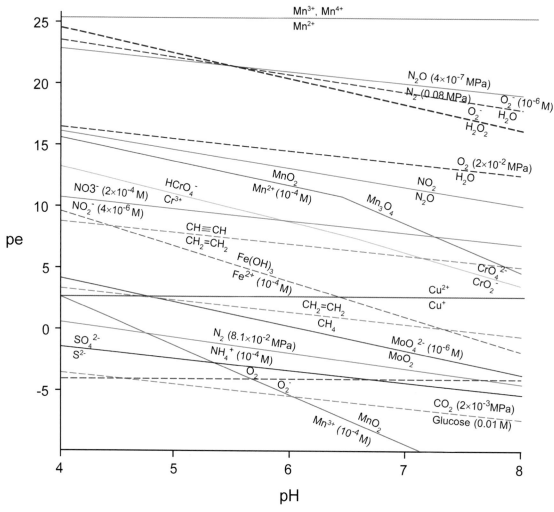

FIGURE 8.2   Stability lines between oxidized and reduced species for several redox couples. Solid phases of Mn and Fe are at unit activity and the activities of other species are designated if not equal within a couple. *From Bartlett (1986), with permission.*

Another term often used in studying the redox chemistry of soils is poise. The poise of a redox system is the resistance to change in redox potential with the addition of small amounts of oxidant or reductant. Poise increases with the total concentration of oxidant plus reductant, and for a fixed total concentration, it reaches a maximum when the ratio of oxidant to reductant is 1 (Ponnamperuma, 1955).

8.2 Measurement and Use of Redox Potentials

Key point: Redox potential can be very useful in characterizing the redox status of a soil.

The measurement of redox potential in soils is usually done with a platinum electrode. This electrode will transfer electrons to or from the soil pore water, but it should not react with the pore

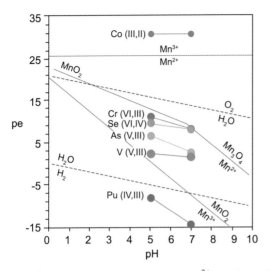

FIGURE 8.3 A pe–pH diagram for Mn^{3+}, MnO_2, and Mn_3O_4; as compared with reduction between pH 5 and 7 for Co, Cr, Se, As, V, and Pu. Activity for ionic species is 10^{-4} M. *From Bartlett and James (1993), with permission.*

water. Once the platinum electrode is combined with a half-cell of known potential, reducing systems will transfer electrons to the electrode while oxidizing systems will remove electrons from the electrode. When experimental redox potential measurements are done, there is no electron flow, and the potential between the half-cell composed of the platinum in contact with the soil pore water and the known potential of the reference electrode half-cell is determined with a meter that measures the electromotive force (i.e., potential) (Patrick et al., 1996). A number of investigators have noted that the measurement of redox potentials in aerated soils is questionable due to the lack of poising of reduction–oxidation systems that are well aerated and have plentiful quantities of oxygen (Ponnamperuma, 1955). Soil atmospheric oxygen measurements are preferred to characterize well-aerated soils. Thus redox measurements are most reliable for flooded soils and sediments.

Redox potentials can be very useful in characterizing the oxidation–reduction status of a soil. Oxidized soils have redox potentials of +400 to +700 mV. Seasonally saturated soils have redox

potentials of +400 to +700 mV (oxidized) to highly reduced (−250 to −300 mV) (Patrick et al., 1996). Redox potentials can help one predict when reducing conditions will begin due to the depletion of oxidants such as oxygen and nitrate and the initiation of oxidizing conditions when oxygen is reintroduced into the soil. Redox potentials can also provide information on conditions that are favorable for the increased bioavailability of heavy metals (Gambrell et al., 1977; Reddy and Patrick, 1977), changes in plant metabolism (Mendelssohn et al., 1981), distribution of plant species (Josselyn et al., 1990), and location of wetlands (Faulkner et al., 1989).

If redox potential data are combined with other information such as depth to the water table and oxygen content of the soil, even more accurate information can be gleaned about the wetness of an environment. In nonwetland environments the Eh and oxygen content do not change much during the year. Transitional areas may be either oxidized or reduced as the water table rises and falls. The redox potentials are low until after the water is drained and oxygen moves through the soil. Wetland sites that have low redox have had long periods of flooding and soil saturation (Patrick et al., 1996).

Redox data are also useful in understanding the morphology and genesis of the soil. The color of a soil and the degree of mottling (spots or blotches of different colors or shades of color interspersed with the dominant color; Glossary of Soil Science Terms, 2008) can reveal much about the soil's moisture status. Both color and mottling depend on the redox chemistry of Fe in the soil. When the soil is saturated for long times, Fe oxides are reduced under low redox potentials, and the soil will exhibit a gray color. Soils that undergo alternate oxidation and reduction cycles are usually mottled (Patrick et al., 1996).

8.3 Submerged Soils

Key points: Submerged soils are reduced, and they have a low redox potential. Variables such as the presence of oxygen, SOM, and nitrate affect the redox state of a submerged soil.

Submerged soils are reduced, and they have a low oxidation–reduction potential (Ponnamperuma, 1972). When an aerobic soil is submerged, the Eh decreases for the first few days and reaches a minimum; then it increases, reaches a maximum, and then decreases again to a value characteristic of the soil, usually after 8–12 weeks of submergence (Ponnamperuma, 1955, 1965). The magnitude and rate of the Eh decrease depend on the properties and amount of SOM, the nature and content of e^- acceptors, temperature, and period of submergence. Native or added SOM enhances the first Eh minimum, while nitrate causes the minimum to disappear. Temperatures above and below 298 K slow the Eh decrease but the retardation varies with the soil. It is greatest in acid soils and not observable in neutral soils with high SOM.

Ponnamperuma and Castro (1964) and Ponnamperuma (1965) have listed the effects of soil properties on Eh changes in submerged soils as follows: soils high in nitrate, with more than 275 mg kg^{-1} NO$_3^-$, exhibit positive potentials for several weeks after submergence; soils low in SOM (<1.5%) or high in Mn (>0.2%) have positive potentials even 6 months after submergence; soils low in active Mn and Fe (sandy soils) with more than 3% SOM reach Eh values of -0.2 to -0.3 V within 2 weeks of submergence; and stable potentials of 0.2 to -0.3 V are attained after several weeks of submergence.

8.4 Redox Reactions Involving Inorganic and Organic Pollutants

Key points: Redox processes impact the speciation (form), solubility, and mobility of inorganic and organic contaminants in soils. Significant research has been carried out specifically for the Fe, Mn, Cr, and As redox couples.

As noted in Chapter 2, metal oxides, for example, Fe(III) and Mn(III/IV), are quite common in soils and sediments as suspended particles and as coatings on clay minerals. Manganese(III/IV), Fe(III), Co(III), and Pb(IV) oxides/hydroxides are thermodynamically stable in oxygenated systems at neutral pH. However, under anoxic conditions, the reductive dissolution of oxides by reducing agents occurs as shown below for MnOOH and MnO$_2$ (Stone, 1991):

$$Mn(III)OOH(s) + 3H^+ + e^- = Mn^{2+} + 2H_2O$$

$$(8.15)$$

$$E^o = +1.50 \text{ V}$$

$$Mn(IV)O_2(s) + 4H^+ + 2e^- = Mn^{2+} + 2H_2O$$

$$(8.16)$$

$$E^o = +1.23 \text{ V}$$

Changes in the oxidation state of the metals associated with the oxides above can greatly affect their solubility and mobility in soil and aqueous environments. The reductants can be either inorganic or organic.

There are a number of natural and xenobiotic organic functional groups that are good reducers of oxides and hydroxides. These include carboxyl, carbonyl, phenolic, and alcoholic functional groups of SOM. Microorganisms in soils and sediments are also examples of organic reductants. Stone (1987a) showed that oxalate and pyruvate, two microbial metabolites, could reduce and dissolve Mn(III/IV) oxide particles. Inorganic reductants include As(III), Cr(III), and Pu(III).

Table 8.2 gives standard reduction potentials (E^o) for oxide/hydroxide minerals at a metal concentration of 1.0 M and potentials ($E^{o'}$) determined under more normal environmental conditions of pH 7 and a metal concentration of 1×10^{-6} M. Oxidant strength decreases in the order Mn(III,IV) oxides > Co(III) oxides > Fe(III) oxides. The Fe oxides are more difficult to reduce than Mn(III,IV) oxides. Thus Fe(II) is easier to oxidize than Mn(II). Reduction potentials for the oxidation of some important organic half-reduction reactions are provided in Table 8.3.

If the potential of the oxidant half-reaction is greater than the reductant half-reaction potential, then the overall reaction is thermodynamically favored. Thus comparing $E^{o\prime}$ values in Tables 8.2 and 8.3, one observes that Mn(III/IV) oxides can oxidize hydroquinone, ascorbate, bisulfide, and oxalate under the specified conditions. Goethite and hematite could only oxidize oxalate; they are not strong enough oxidants to oxidize hydrogen sulfide, ascorbate, or hydroquinone. At pH 4, the Fe(III) oxides can oxidize all of the reductants listed in Table 8.3 (Stone, 1991). Manganese (III/IV) oxides can react with many organic compounds over a wide pH range while reactions with Fe(III) oxides are not thermodynamically stable at neutral and alkaline pHs and with organic compounds that are slightly reduced (Stone, 1991).

8.4.1 Mechanisms for the Reductive Dissolution of Metal Oxides/Hydroxides

Key point: The reductive dissolution of metal oxides occurs via a number of sequential rate processes and inner- and outer-sphere complex mechanisms.

The reductive dissolution of metal oxides/hydroxides occurs in the following sequential steps (Stone, 1986, 1991): (1) diffusion of the reductant molecules to the oxide surface, (2) surface chemical reaction, and (3) the release of reaction products and diffusion away from the oxide surface. Steps (1) and (3) are transport steps. The rate-controlling step in the reductive dissolution of oxides appears to be surface chemical reaction control. Reductive dissolution can be described by both inner- and outer-sphere complex mechanisms that involve (A) precursor complex formation, (B) electron transfer, and (C) the breakdown of the successor complex (Figure 8.4). Inner- and

TABLE 8.2 Reduction half-reactions along with standard reduction potentials (E°) and reduction potentials calculated under more realistic environmental conditions ($E^{o\prime}$) for mineral phases containing Mn(IV), Mn(III), Fe(III), and Co(III).[a]

	$E^{\circ b}$(V)	$E^{o\prime c}$ (V)
Vernadite (Bricker, 1965)		
$\frac{1}{2}$ Mn(IV)O$_2$(s) + 2H$^+$ + e$^-$ = $\frac{1}{2}$ Mn^{2+} + H$_2$O	+1.29	+0.64
Manganite (Bricker, 1965)		
Mn(III)OOH(s) + 3H$^+$ + e$^-$ = Mn^{2+} + 2H$_2$O	+1.50	+0.61
Goethite (Robie et al., 1978)		
Fe(III)OOH(s) + 3H$^+$ + e$^-$ = Fe^{2+} + 2H$_2$O	+0.67	−0.22
Hematite (Robie et al., 1978)		
$\frac{1}{2}$ Fe(III)O$_3$(s) + 3H$^+$ + e$^-$ = Fe^{2+} + $\frac{3}{2}$ H$_2$O	+0.66	−0.23
Magnetite (Robie et al., 1978)		
$\frac{1}{2}$ Fe$_2$(III)Fe(II)O$_4$(s) + 4H$^+$ + e$^-$ = $\frac{3}{2}$ Fe^{2+} + 2H$_2$O	+0.90	−0.23
Cobalt hydroxide oxide (crystalline) (Hem et al., 1985)		
Co(III)OOH(s) + 3H$^+$ + e$^-$ = Co^{2+} + 2H$_2$O	+1.48	+0.23

[a] From Stone (1991), with permission.
[b] E° = standard reduction potential ([I] = 1.0 M).
[c] $E^{o\prime}$ = reduction potential under the following conditions: [H$^+$] = 1.0 × 10^{-7} M; [M^{2+}] = 1.0 × 10^{-6} M.

TABLE 8.3   Reduction half-reactions along with standard reduction potentials ($E°$) and reduction potentials calculated under more realistic environmental conditions ($E°'$) for several organic compounds.[a,b]

	$E°$[c] (V)	$E°'$[d] (V)
Hydroquinone		
p-Benzoquinone + $2H^+$ + $2e^-$ = hydroquinone	0.699	0.196
Ascorbate		
Dehydroascorbate + $2H^+$ + $2e^-$ = ascorbate	0.40	−0.103
Hydrogen sulfide		
$S°(s) + H^+ + e^-$ = HS^-	−0.062	−0.17
Oxalate		
$2HCO_3^- + 2H^+ + 2e^- = {}^-OOC - COO^- + 2H_2O$	−0.18	−0.69

[a] From Stone (1991), with permission.
[b] Using thermodynamic data compiled in Latimer (1952) and Stone and Morgan (1984b).
[c] $E°$ = standard reduction potential ($[I] = 1.0$ M).
[d] $E°'$ = Reduction potential under the following conditions: $[H^+] = 1.0 \times 10^{-7}$ M; reductant concentration = 1.0×10^{-3} M; oxidant concentration = 1.0×10^{-6} M; C_T (total dissolved carbonate) = 1.0×10^{-3} M.

outer-sphere precursor complex formations are adsorption reactions that increase the density of reductant molecules at the oxide surface, which promotes electron transfer (Stone, 1991). In the inner-sphere mechanism the reductant enters the inner coordination sphere via ligand exchange and bonds directly to the metal center prior to electron transfer. With the outer-sphere complex, the inner coordination sphere is left intact and electron transfer is enhanced by an outer-sphere precursor complex (Stone, 1986). Kinetic studies have shown that high rates of reductive dissolution are favored by high rates of precursor complex formation, that is, large

k_1 and k_{-1} values; high electron transfer rates, that is, large k_2; and high rates of product release, that is, high k_3 (Figure 8.4).

Specifically adsorbed cations and anions may reduce reductive dissolution rates by blocking oxide surface sites or by preventing the release of Mn(II) into solution. Stone and Morgan (1984a) showed that PO_4^{3-} inhibited the reductive dissolution of Mn(III/IV) oxides by hydroquinone. The addition of 10^{-2} M PO_4^{3-} at pH 7.68 caused the dissolution rate to be only 25% of the rate when PO_4^{3-} was not present. Phosphate had a greater effect than Ca^{2+}.

8.4.2 Oxidation of Inorganic Pollutants

Key point: Manganese oxides and bacteria play important roles in oxidizing elements such as As and Cr.

As mentioned earlier, Mn oxides can oxidize a number of environmentally important ions that can be toxic to humans and animals. Chromium and plutonium are similar in their chemical behavior in aqueous settings (Rai and Serne, 1977; Bartlett and James, 1979). They can exist in multiple oxidation states and as both cationic and anionic species. Chromium(III) is quite stable and innocuous, which occurs as Cr^{3+} and its hydrolysis products or as CrO_2^-. Chromium(III) can be oxidized to Cr(VI) by Mn(III/IV) oxides (Bartlett and James, 1979; Fendorf and Zasoski, 1992). Chromium(VI) is mobile in the soil environment and is a suspected carcinogen. It occurs as the dichromate, $Cr_2O_7^{2-}$, or chromate, $HCrO_4^-$ and CrO_4^{2-}, anions (Huang, 1991).

Figure 8.5 shows the oxidation kinetics of Cr(III) to Cr(VI) in a soil. Most of the oxidation occurred during the first hour. At higher temperatures, there was a rapid oxidation rate, followed by a slower rate. The decrease in the rate of Cr(III) oxidation has been ascribed to a number of factors, including the formation of a surface

	Inner-sphere	Outer-sphere

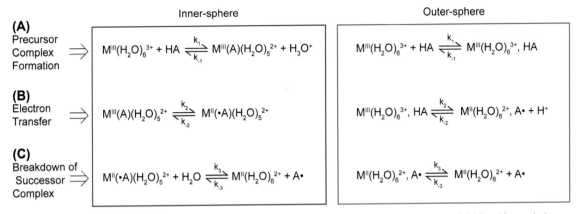

(A)
Precursor
Complex
Formation
$\Rightarrow$

Inner-sphere:
$$M^{III}(H_2O)_6^{3+} + HA \underset{k_{-1}}{\overset{k_1}{\rightleftharpoons}} M^{III}(A)(H_2O)_5^{2+} + H_3O^+$$

Outer-sphere:
$$M^{III}(H_2O)_6^{3+} + HA \underset{k_{-1}}{\overset{k_1}{\rightleftharpoons}} M^{III}(H_2O)_6^{3+}, HA$$

(B)
Electron
Transfer
$\Rightarrow$

Inner-sphere:
$$M^{III}(A)(H_2O)_5^{2+} \underset{k_{-2}}{\overset{k_2}{\rightleftharpoons}} M^{II}(\cdot A)(H_2O)_5^{2+}$$

Outer-sphere:
$$M^{III}(H_2O)_6^{3+}, HA \underset{k_{-2}}{\overset{k_2}{\rightleftharpoons}} M^{II}(H_2O)_6^{2+}, A\cdot + H^+$$

(C)
Breakdown of
Successor
Complex
$\Rightarrow$

Inner-sphere:
$$M^{II}(\cdot A)(H_2O)_5^{2+} + H_2O \underset{k_{-3}}{\overset{k_3}{\rightleftharpoons}} M^{II}(H_2O)_6^{2+} + A\cdot$$

Outer-sphere:
$$M^{II}(H_2O)_6^{2+}, A\cdot \underset{k_{-3}}{\overset{k_3}{\rightleftharpoons}} M^{II}(H_2O)_6^{2+} + A\cdot$$

FIGURE 8.4   Reduction of $M(H_2O)_6^{3+}$ by phenol (HA) in homogeneous solution. *From Stone (1986), with permission.*

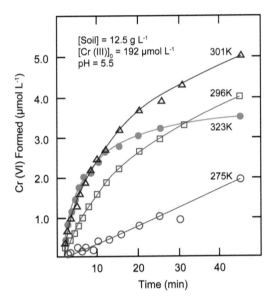

[Soil] = 12.5 g L^{-1}
[Cr (III)]$_0$ = 192 μmol L^{-1}
pH = 5.5

301K
296K
323K
275K

FIGURE 8.5   Effect of temperature on the kinetics of Cr(III) oxidation in moist Hagerstown silt loam soil. *From Amacher and Baker (1982), with permission.*

precipitate that effectively inhibits further oxidation (Fendorf et al., 1992).

Plutonium can exist in the +3 to +6 oxidation states as Pu^{3+}, Pu^{4+}, PuO_2^+, and PuO_2^{2+} in strongly acid solutions (Huang, 1991). Plutonium(VI), which can result from the oxidation of Pu(III/IV) by Mn(III/IV) oxides (Amacher

and Baker, 1982), is highly toxic and mobile in soils and waters.

Arsenic (As) can exist in several oxidation states and forms in soils and waters. In waters As can exist in the +5, +3, 0, and −3 oxidation states. Arsenite (As(III)) and arsine (AsH$_3$, where the oxidation state of As is −3) are much more toxic to humans than arsenate, As(V). Manganese(III/IV) oxides can oxidize As(III) to As(V) as shown below, where As(III) as HAsO$_2$ is added to MnO$_2$ to produce As(V) as H$_3$AsO$_4$ (Oscarson et al., 1983):

$$HAsO_2 + MnO_2 = MnO_2 \cdot HAsO_2 \quad (8.17)$$

$$MnO_2 \cdot HAsO_2 + H_2O = H_3AsO_4 + MnO \quad (8.18)$$

$$H_3AsO_4 = H_2AsO_4^- + H^+ \quad (8.19)$$

$$H_2AsO_4 = HAsO_4^{2-} + H^+ \quad (8.20)$$

$$MnO_2 \cdot HAsO_2 + 2H^+ = H_3AsO_4 + Mn^{2+} \quad (8.21)$$

Equation 8.18 involves the formation of an adsorbed layer. Oxygen transfer occurs and HAsO$_2$ is oxidized to H$_3$AsO$_4$ Equation 8.18. At pH $\leq$ 7, the predominant As(III) species is arsenious acid (HAsO$_2$), but the oxidation product, arsenic acid (H$_3$AsO$_4$), will dissociate and form the same quantities of dihydrogen arsenate (H$_2$AsO$_4^-$) and hydrogen arsenate (HAsO$_4^{2-}$)

with little H_3AsO_4 present at equilibrium (Equations 8.19 and 8.20). Each mole of As(III) oxidized releases about 1.5 mol H^+. The H^+ produced after H_3AsO_4 dissociation reacts with the adsorbed $HAsO_2$ on MnO_2, forming H_3AsO_4, and leads to the reduction and dissolution of Mn(IV) (Equation 8.21). Thus every mole of As(III) oxidized to As(V) results in 1 mol of Mn(IV) in the solid phase being reduced to Mn(II) and partially dissolved in solution (Oscarson et al., 1981). Oscarson et al. (1980) studied the oxidation of As(III) to As(V) in sediments from five lakes in Saskatchewan, Canada. The oxidation of As(III) to As(V) occurred within 48 h. In general, >90% of the added As was sorbed on the sediments within 72 h.

In addition to the As/Mn redox couple, the Cr/Mn redox couple has also received significant attention. The Cr oxidation reaction with Mn oxides proceeds in three phases: first (1) Cr sorbs to the Mn surface, then (2) Cr exchanges electrons with Mn, and finally (3) the reaction products are desorbed. An overall equation for the oxidation of Cr(III) is shown in Equation 8.22 (Eary and Rai, 1987; Fendorf et al., 1992 Nico and Zasoski, 2000;) and a schematic of the multimodal processes is shown in Figure 8.6.

$$Cr(OH)^{2+} + 1.5MnO_2 + H_2O \rightarrow HCrO_4^-$$
$$+ 1.5Mn^{2+} + H^+$$

$$(8.22)$$

The pH of the solution can greatly impact Cr oxidation kinetics. As pH rises, the Mn oxide surface becomes more negative, further attracting positively charged Cr(III). This increases the electrostatic interaction between the Mn mineral surface and Cr(III). Accordingly, more Cr(III) is sorbed to the Mn surface, leading to initially greater rates of Cr(VI) production (Rai et al., 1987). However, not all of the Cr(III) is oxidized because Mn(IV) and Cr(III) share the same structural environment. Therefore some of the Cr(III) can be sorbed to the surface of the manganese and remain

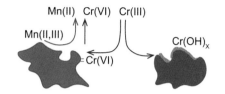

FIGURE 8.6   Chromium(III) reactivity with MnO_2.

unoxidized. As Cr(III) is oxidized, Mn(IV) in the mineral structure is reduced to Mn(II) and Mn(III), which can go into solution or become sorbed on the mineral surface (Figure 8.6).

In addition, Cr(III) can become sorbed onto the Mn oxide surface as an inner-sphere complex, with subsequent oxidation to Cr(VI); Cr(VI) is primarily held as an outer-sphere complex (Fendorf and Zasoski, 1992). At higher pH and Cr(III) concentrations, Cr(III) oxidation decreases, and Cr(III) can precipitate as a Cr hydroxide on the mineral surface. The formation of Cr hydroxides on the Mn mineral surface decreases the availability (i.e., accessibility) of surface reactive sites, which in turn decreases Cr(III) sorption and oxidation. Additionally, Cr hydroxide can precipitate from bulk solution onto the surface of Mn minerals at concentrations below its saturation index. Examples of the pH effect on Cr(III) oxidation kinetics on δ-MnO_2 and birnessite are shown in Figure 8.7. Greater oxidation of Cr(III) is observed at more acidic pH values, within the pH range of 2.5 to 3.5 (Figure 8.7). The combination of higher pH values and higher Cr concentrations promotes the precipitation of Cr (oxyhydr)oxide phases (Fendorf and Zasoski, 1992; Landrot et al., 2012).

(a) A Special Case: Arsenic (As)

Arsenic (As) is a toxic element in the environment, originating from both geogenic and anthropogenic sources. Its toxicity and mobility are determined by its speciation. Arsenite is generally considered to be more toxic than As(V), although both are hazardous to human

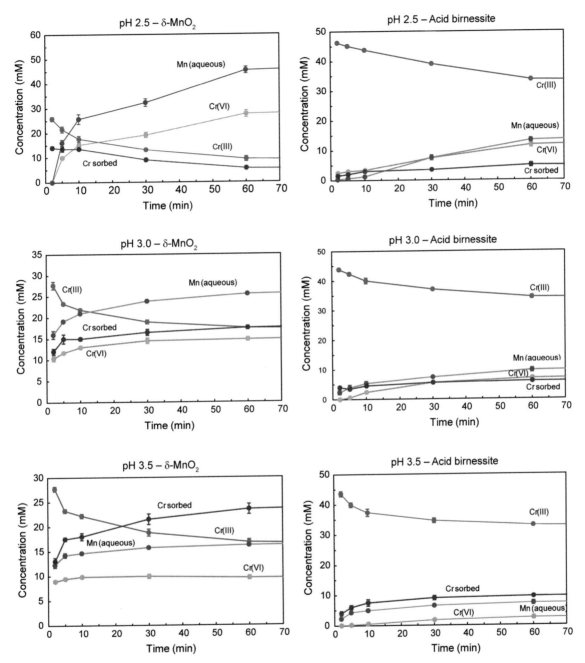

FIGURE 8.7 Mn oxides react rapidly with Cr. The reaction is largely complete within the first 20–60 min. Lower pH values result in greater oxidation of Cr(VI), with less overall surface sorption. *Reproduced from Landrot et al., (2012), with permission.*

health. Arsenic(V) is more mobile in soils than As(III) because of the pK_a values at which it deprotonates. The sorption of As on oxide minerals also depends on pH and speciation. Manganese oxides, including phyllomanganates, can oxidize As(III) to As(V). Additionally, As(III) oxidation by Mn oxides can decrease the mobility of As (Oscarson et al., 1983; Scott and Morgan, 1995; Raven et al. 1998; Dixit and Hering, 2003; Lafferty et al., 2010a, b).

The toxic impacts of As contamination are a global issue that affects the health and wellbeing of many diverse populations. It is estimated that over 150 million people worldwide are endangered by As contamination (Ravenscroft et al., 2009). Arsenite has the greatest impact on human health in a few critical developing countries where natural geological formations contain hazardous levels of As. The World Health Organization deemed Bangladesh "the largest mass poisoning of a population in history," due to the installation of wells in naturally As-contaminated areas (Argos et al., 2010). These shallow wells often tap into alluvial sediments that are nearly devoid of oxidizing agents such as Mn and Fe and are rich in As(III) accumulated from the weathering of upstream rocks. The pH of these soils also impacts As transport, with older deposits ranging from pH 6.0 to 6.5 and more recent alluvium having a pH of 7.0–8.5 (Gerrard, 1992).

The impacts of climate change, particularly related to flooding and sea level rise in coastal areas, is a major concern in terms of cycling of As, as well as other toxic elements. In addition to geogenic sources of As, many coastal soils are contaminated with As from anthropogenic sources, for example, legacy industrial operations. Flooding and sea water intrusion will alter soil redox conditions and introduce salinity and reducing zones in areas that have long remained oxic or less brackish. This could impact water quality and human health (LeMonte et al., 2017; Izaditame et al., 2022).

Sidebar 8.2

Comproportionation (sometimes written conproportionation) and disproportionation are two terms that are used to describe redox changes of an element during a reaction. They are opposite to each other. Conproportionation takes place when two ions (species) of the same element have different oxidation states and react to create a product that has an oxidation state between the two (e.g., $Mn^{4+} + Mn^{2+} \rightarrow 2Mn^{3+}$). Disproportionation takes place in a reaction where two ions (species) of the same oxidation state react to create a product with different oxidant states (e.g., $2Mn^{3+} \rightarrow Mn^{4+} + Mn^{2+}$).

With respect to As reactions with Mn oxides, there are four possible pathways for the oxidation of As(III) by δ-MnO_2. Each pathway employs a different combination of Mn reactive sites on the δ-MnO_2 surface for As(III) oxidation (Shiller and Stephens, 2005). In the first pathway one As(III) molecule reacts with one Mn(IV) reactive site (represented by >Mn(IV)—OH in (Equations 8.23–8.26). The Mn surface is protonated (Peacock and Sherman, 2007), and the reaction produces one As(V) molecule and one Mn(II) molecule (Equation 8.23). One As(III) molecule could also react with two Mn(IV) reactive sites producing one As(V) molecule and two Mn(III) reactive sites (represented by >Mn(III)—OH$_2$ (Equation 8.24); protonation state based on Ramstedt et al. (2004). This proposed mechanism for Mn(III) formation results directly through As(III) oxidation (Nesbitt et al. 1998). The third and fourth possible As(III) oxidation pathways necessitate that the δ-MnO_2 surface first contain Mn(III) reactive sites.

$$> Mn(IV)\text{-}OH + H_3As(III)O_3(aq)$$

$$\rightarrow Mn(II)(aq) + HAs(V)O_4^{2-}(aq) + 3H^+$$

$$(8.23)$$

$$2(> Mn(IV)\text{-}OH) + H_3As(III)O_3(aq)$$

$$+ H_2O \rightarrow 2(> Mn(III)\text{-}OH_2)$$

$$+ HAs(V)O_4^{2-}(aq) + 2H^+ \quad (8.24)$$

Manganese(III) can also form by the comproportionation of Mn(II) and Mn(IV) at the δ-MnO₂ surface (see Sidebar 8.2). In this reaction Mn(II) is oxidized, Mn(IV) is reduced, and the resulting product is Mn(III) (Morgan and Stumm, 1964; Tan et al., 2008). Mn(III) resulting from the comproportionation of Mn(II) and Mn(IV) could produce soluble Mn(III); however, if there is no ligand to stabilize Mn(III) in the reactions, then soluble Mn(III) should rapidly disproportionate back into Mn(II) and Mn(IV) (Lafferty et al., 2010b). Thus one can assume that all Mn(III) in these reactions is associated with the δ-MnO₂ surface.

Another possibility for As(III) to be oxidized by Mn(III) is via the reaction of one As(III) molecule with one Mn(III) reactive site and one Mn(IV) reactive site, resulting in one As(V) molecule, one Mn(II) molecule, and one Mn(III) reactive site Equation 8.25. The other possible pathway for As(III) oxidation by δ-MnO₂ is for one As(III) molecule to react with two Mn(III) reactive sites, creating one As(V) molecule and two Mn(II) molecules (Equation 8.26) (Tani et al., 2004; Shiller and Stephens 2005):

$$> Mn(IV)\text{-}OH + > Mn(III)\text{-}OH_2$$

$$+ H_3As(III)O_3(aq) \rightarrow Mn(II)(aq) +$$

$$> Mn(III)\text{-}OH_2 + HAs(V)O_4^{2-}(aq) + 3H^+$$

$$(8.25)$$

$$2(> Mn(III)\text{-}OH_2) + H_3As(III)O_3(aq)$$

$$\rightarrow 2Mn(II)(aq) + HAs(V)O_4^{2-}(aq)$$

$$+ H_2O + 4H^+ \quad (8.26)$$

Lafferty et al. (2010a, b), using a combined kinetic and molecular scale approach, determined the rates and mechanisms of As(III) oxidation on hydrous manganese oxide (HMO), a high-surface-area, amorphous Mn oxide. A stirred-flow technique was employed to follow the kinetics of As(III) oxidation over a time scale of days (Figure 8.8).

During the first 0.4 h, no As or Mn appeared in the effluent solutions, indicating that all of the As(III) was oxidized to As(V). Complementary EXAFS data did not show any sorption of As(III) on the HMO. Up until 6.4 h, only As(V) appeared in the effluent solution, suggesting that all the As(III) was oxidized to As(V) and subsequently bound to the HMO or desorbed into the solution. After 6.4 h, As(V), as well as Mn(II) and As(III), appeared in the effluent. This was also when As(III) oxidation decreased. Synchrotron-based XRD and XAS analyses of Mn revealed that initially Mn was sorbed into the vacant sites of the HMO as both Mn^{2+} and Mn^{3+}, and between 4 and 10 h, the vacancy sites were filled.

Subsequently, Mn^{2+} began to compete with As(V) for edge sites, which caused a decrease in As(III) oxidation. The data also indicated that Mn^{3+} could be sorbed onto edge sites via comproportionation. This would occur when Mn^{2+} is oxidized and Mn^{4+} is reduced. The passivation of the HMO surface was significantly affected not only by As(V) but, most importantly, by Mn^{2+} and Mn^{3+}. Thus even under flow conditions, a buildup of products could inhibit the reaction. As time proceeded, the passivation of the surface increased. The XAS of As showed that only As(V) species were present on the surface; however, the type of As(V) bonding changed as reaction time proceeded (Figure 8.10). At up to 4 h of reaction time, the predominant inner-sphere adsorption complexes were bidentate binuclear and monodentate mononuclear, but between 4 and 10 h, when passivation was more pronounced, the two predominant species were: bidentate binuclear and a bidentate mononuclear species. A schematic showing the overall mechanism of As(III) oxidation on HMO is shown in Figures 8.9

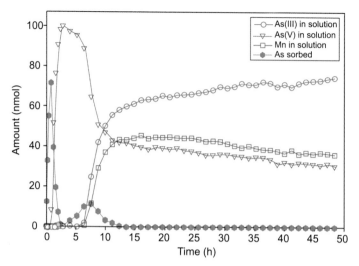

FIGURE 8.8   The amount (nmol) of As sorbed as well as amounts (nmol) of As(III), As(V), and Mn(II) in the effluent of a stirred-flow experiment reacting $1 \, g \, L^{-1}$ δ-MnO$_2$ with 100 μM As(III) flowing at 1 mL min^{-1} for 48 h. *From Lafferty et al., (2010a), with permission.*

and 8.10. The data clearly show, once again, the importance of following reaction processes over time and at the molecular scale.

The addition of coions and bacteria can also have competitive effects on the kinetics of the As(III) oxidation on Mn oxides (Parikh et al., 2010; Jones et al. 2012). Ion competition impacts the As(V) sorption behavior on minerals (Violante and Pigna, 2002: Parikh et al., 2010), but only a few studies have appeared on the effect of ion competition on As(III) oxidation kinetics on Mn oxides. Parikh et al. (2010) studied the effects of Fe and phosphate competition on the rapid, initial kinetics of As(III) oxidation on δ-Mn oxide. Phosphate caused decreased total As sorption on Mn oxide. This could be ascribed to the similar chemical structure and reactivity of As(V) and phosphate (both are tetrahedral oxyanions). There was also decreased retention of both anions. The decrease in As sorption in the presence of phosphate is advantageous in environments where animal waste containing both As and P is applied on land.

Jones et al. (2012) studied the kinetics of As(III) oxidation by using two heterotrophic soil bacteria (*Agrobacterium tumefaciens* and *Pseudomonas fluorescens*) and poorly crystalline manganese oxide (δ-Mn oxide) using batch experiments. The apparent rate of As(V) production was higher for the combined batch experiments in which bacteria and δ-Mn oxide were oxidizing As(III) at the same time than for either component alone. The additive effects of the mixed bacteria and δ-Mn oxide system was observed in both short- (<1 h) and long-term (24 h) coincubation experiments, illustrating that mineral surface inhibition by cells had little effect on the As(III) oxidation rate. Interactions between cells and mineral surfaces were indicated by sorption and pH-induced desorption data. Total sorption of As on the mineral was lower with bacteria present (16.1 ± 0.8% As sorbed) and higher with δ-Mn oxide alone (23.4 ± 1%). Arsenic was more easily desorbed from the cell–δ-Mn oxide system than from δ-Mn oxide alone. Therefore the presence

(A) δ-MnO$_2$

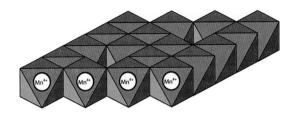

(B) As(III) reaction with δ-MnO$_2$: 0-4 h

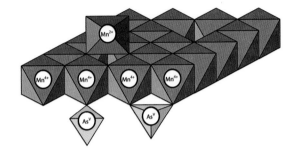

(C) As(III) reaction with δ-MnO$_2$: 4- ~6 h

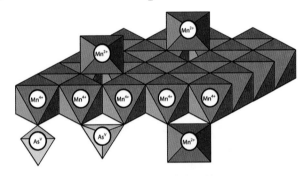

(D) As(III) reaction with δ-MnO$_2$: ~6-48 h

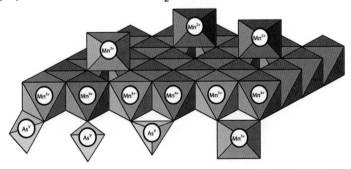

FIGURE 8.9   Proposed reaction mechanism for As(III) oxidation by δ-MnO$_2$ over 48 h in a stirred-flow reactor. Throughout the reaction, As(III) is oxidized by Mn(IV) at δ-MnO$_2$ edge sites, producing Mn(II) and As(V). (a) Unreacted δ-MnO$_2$ octahedral layers consist primarily of Mn^{4+} and have reaction sites at layer edges (edge sites) and vacancy sites. (b) During the first 4 h of As(III) oxidation, Mn(II) sorbs at δ-MnO$_2$ vacancy sites, and As(V) sorbs at edge sites in bidentate–binuclear and monodentate–mononuclear complexes. Also, a portion of sorbed Mn(II) reacts with Mn(IV) at vacancy sites to form Mn(III). (c) Between 4 and 6 h of reaction, vacancy sites become filled with Mn(II/III), Mn(II) begins to sorb at δ-MnO$_2$ edge sites, and As(V) sorption continues in the same sorption complexes. (d) Beyond 6 h of reaction, Mn(II) at edge sites (and probably vacancy sites) reacts with Mn(III) in δ-MnO$_2$ octahedral layers to form Mn(III). The resulting Mn(III) changes the bonding environment of As(V), which begins to sorb in bidentate–mononuclear complexes, and the As–Mn distance in As(V) bidentate–binuclear complexes increases slightly. *From Lafferty et al. (2010b), with permission.*

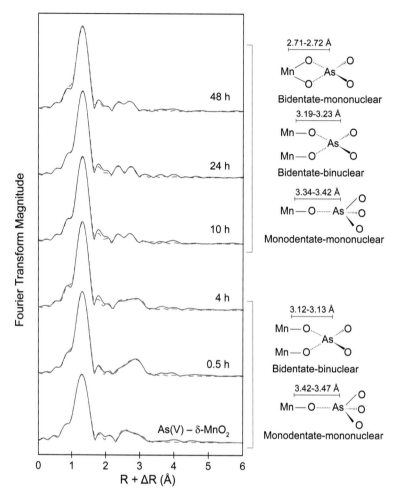

FIGURE 8.10   Reaction mechanisms for As(III) adsorption on δ-MnO₂ as a function of time. At time 0 through 4 h As(V) is bound as bidentate–binuclear and monodentate–mononuclear complexes. This is the time when it is primarily sorbing at vacancy sites. Time 10 through 48 h As(V) is bound in bidentate–binuclear, monodentate–mononuclear, and bidentate–mononuclear complexes. The As–Mn length increases. Mn(III) is present when As(V) bonding changes. This is the time when Mn has filled all vacancy sites and begins to compete with As for edge sites. *From Lafferty et al.(2010b), with permission.*

of bacteria inhibited As sorption and decreased the stability of sorbed As on δ-Mn oxide. This was true even though As(III) was oxidized most rapidly in the mixed cell/mineral system. The additive effect of biotic (As-oxidizing bacteria) and abiotic (δ-Mn oxide mineral) oxidation processes in a system containing both oxidants suggests that mineral-only data

may underestimate the oxidative capacity of natural systems where both biotic and abiotic As(III) oxidation pathways operate (Jones et al., 2012).

(b) A Special Case: MnO₂

Manganese oxides, for example, Mn_3O_4 and MnOOH, may also catalyze the oxidation of other trace metals such as Co^{2+}, Co^{3+}, Cu^{2+}, Ni^{2+}, Ni^{3+}, Pb^{2+}, and Se^{4+} by disproportionation to Mn^{2+} and MnO_2 (Hem, 1978; Scott and Morgan, 1996). The disproportionation results in vacancies in the Mn oxide structure. Since the Mn^{2+} and Mn^{3+} in the oxides have physical sizes similar to those of Co^{2+}, Co^{3+}, Cu^{2+}, Ni^{2+}, Ni^{3+}, and Pb^{2+}, these metals can occupy the vacancies in the Mn oxide and become part of the structure. With disproportionation or with other redox processes involving the Mn oxides, the solubility of the metals can be affected. For example, if, during the disproportionation process, Co_3O_4, the oxidized form of the metal, forms from Co^{2+}, the reaction can be expressed as (Hem, 1978):

$$2Mn_3O_4(s) + 3Co^{2+} + 4H^+ \rightarrow MnO_2(s)$$

$$+ Co_3O_4(s) + 5Mn^{2+} + 2H_2O \quad (8.27)$$

and the equilibrium constant ($K°$) is (Hem, 1978)

$$K° = \frac{(Mn^{2+})^5}{(Co^{2+})^3(H^+)^4} = 10^{18.73} \quad (8.28)$$

Thus the oxidation of Co(II) to Co(III) reduces its solubility and mobility in the environment. Using X-ray photoelectron spectroscopic analyses (Murray and Dillard, 1979), this reaction has been shown to occur.

Scott and Morgan (1996) studied the oxidation of Se(IV) by synthetic birnessite. Se(IV) was oxidized to Se(VI) with Se(VI) first appearing in the aqueous suspension after 12 h and was produced at a constant rate over the duration of the experiment (28 days). The following oxidation mechanism was suggested: (i) birnessite directly oxidized Se(IV) through a surface complex mechanism; (ii) the rate-limiting step in the production of Se(VI) was the electron transfer step involving a transfer of two electrons from the anion to the metal ion, breaking of two Mn–O bonds, and addition of an O from water to Se(VI); and (iii)

the reaction products Se(VI) and Mn(II) were released from the surface by different steps.

Scott and Morgan (1996) compared their results to those of Eary and Rai (1987), who studied Cr(III) oxidation by pyrolusite (β-MnO_2) between pH 3.0 and 4.7, and Scott and Morgan (1995), who studied As(III) oxidation by birnessite (δ-MnO_2) (Table 8.4). The Cr(III) redox transformation on pyrolusite was slowest, which was attributed to unfavorable adsorption on both a positively charged surface and aqueous species and the small thermodynamic driving force. Also, the transfer of three electrons from Cr(III) to Mn(IV) requires the involvement of more than one Mn(IV) atom per Cr(III) atom.

Manganese oxides also appear to play an important role in ligand-facilitated metal transport. Using soil columns that consisted of fractured saprolite coated with amorphous Fe– and Mn–oxides, Jardine et al. (1993) studied the transport of Co(II) $EDTA^{2-}$, a mixture of Co(II) $EDTA^{2-}$ and Co(III) $EDTA^-$ and Sr $EDTA^{2-}$. The Mn oxides oxidized Co(II) $EDTA^{2-}$ into Co(III) $EDTA^-$, a very stable complex (log $K°$ value of 41.4, Xue and Traina, 1996). The formation of this complex resulted in the enhanced transport of Co.

Xue and Traina (1996) found that an aerobic goethite suspension catalyzed the oxidation of Co(II) $EDTA^{2-}$ to Co(III) $EDTA^-$ by dissolved O_2. The kinetics were described using a pseudo-first-order rate constant, k_1, of 0.0078 ± 0.0002 h^{-1} at pH 5 and a goethite concentration of 3.09 g L^{-1}.

8.4.3 Reductive Dissolution of Mn Oxides by Organic Pollutants

Key point: Organic pollutants such as hydroquinone and phenols can cause the reductive dissolution of Mn oxides.

A number of investigators have studied the reductive dissolution of Mn oxides by organic

TABLE 8.4   Inorganic redox reactions with manganese oxides.[a]

System	Time to oxidize 50%	Driving force at pH 4 Δe^o (V)[b]	Source
		δ-MnO$_2$: As(III) $\rightarrow$ As(V)	
pH 4, 298 K, 14 m^2 L^{-1}	10 min	+0.529	Scott and Morgan (1995)
		δ-MnO$_2$: Se(IV) $\rightarrow$ Se(VI)	
pH 4, 308 K, 14 m^2 L^{-1}	10 days		
pH 4, 298 K, 28 m^2 L^{-1}	16 days	+0.092	
pH 4, 298 K, 14 m^2 L^{-1}	30 days		
		β-MnO$_2$: Cr(III) $\rightarrow$ Cr(VI)	
pH 4, 298 K, 71 m^2 L^{-1}	95 days	+0.011	Eary and Rai (1987)

[a] From Scott and Morgan (1996), with permission.
[b] The activity ratio for each oxidant/reductant pair is taken as unity.

pollutants such as hydroquinone (Stone and Morgan, 1984a), substituted phenols (Stone, 1987b), and other organic compounds (Stone and Morgan, 1984b). With substituted phenols, the rate of dissolution was proportional to substituted phenol concentration and the rate increased as pH decreased (Stone, 1987b). Phenols containing alkyl, alkoxy, or other electron-donating substituents were more slowly degraded; p-nitrophenol reacted slowly with Mn(III/IV) oxides. The increased rate of reductive dissolution at lower pH may be due to more protonation reactions that enhance the creation of precursor complexes or increases in the protonation level of the surface precursor complexes that increase electron transfer rates (Stone, 1987b).

8.4.4 Reduction of Contaminants by Iron and Microbes

Key point: Iron(II) oxides and microbes are effective at reducing inorganic (e.g., Cr(VI)) and organic contaminants.

Iron(II)-containing oxides, such as magnetite ((Fe^{2+}Fe$_2^{3+}$)O$_4$) and ilmenite ((Fe^{2+}Ti)O$_3$), and microbes (microbially mediated processes) can play a significant role in the reduction of inorganic and organic contaminants. The chemical reduction of Cr(VI) in Cr$_2$O$_7^{2-}$ to Cr(III) by a generic Fe^{2+} oxide is shown below (White and Peterson, 1998):

$$9Fe^{2+}_{oxide} + Cr_2O_7^{2-} + 14H^+ \rightarrow 6Fe^{3+}_{oxide}$$

$$+ 3Fe^{2+} + 2Cr^{3+} + 7H_2O \qquad (8.29)$$

Figure 8.11 shows changes that occur in the aqueous concentrations of Fe, Cr, and V species with time after reaction with ilmenite at pH 3. As time increases, there is a linear increase in the transformation of the oxidized species to the reduced species. That is, Fe(III) is reduced to Fe(II), Cr(VI) is reduced to Cr(III), and V(V) is reduced to V(IV).

However, under field conditions and at longer times, the effectiveness of Fe(II)-containing oxides in the reduction of contaminants is affected by the reductive capacity of the oxide minerals, the impact of surface passivation (surface layer that forms, which is different from the bulk oxide structure), and competition effects and poisoning by other aqueous species.

Several classes of organic pollutants can also undergo abiotic reduction in anoxic aqueous environments when ferrous iron and Fe(III) oxides are present (Haderlein and Pecher, 1998). Figure 8.12A shows the reduction of 4-chloro-

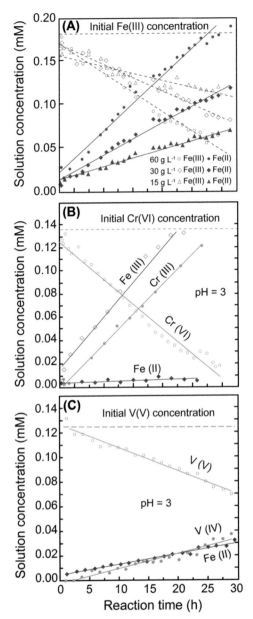

FIGURE 8.11 Time trends showing the rates of ferric (A), chromate (B), and vanadate (C) reduction in ilmenite suspensions at pH 3 at 298 K. Rates of ferric reduction increase with increasing suspended ilmenite (g L⁻¹). Solid lines are linear regression fits to data. Dashed lines are initial solution concentrations. *From White and Peterson (1996), with permission.*

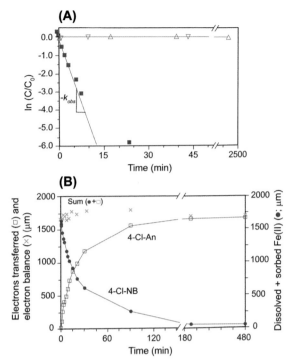

FIGURE 8.12 (A) Reduction of 50 μM 4-chloro-nitrobenzene (4-Cl-NB) in the presence of 17 m² L⁻¹ magnetite and an initial concentration of 2.3 mM Fe(II) at pH 7.0 and 298 K (■). The rate law deviates from pseudo-first-order behavior for longer observation times. 4-Cl-NB was not reduced significantly in suspensions of magnetite without dissolved Fe(II) (▽) or in solutions of Fe(II) without magnetite (△). (B) Electron balance (×) for the reduction of 4-Cl-NB repeatedly added at times t = 0, 6, 17 min to a suspension containing 11.2 m² L⁻¹ magnetite and an initial concentration of dissolved Fe(II) of 1.6 mM at pH 7.75 and T = 298 K. *Adapted from Klaussen et al. (1995), with permission.*

nitrobenzene (4-Cl-NB) with time when both magnetite and Fe(II) in solution were present and when only magnetite or solution Fe(II) was included. Only when both magnetite and solution Fe(II) were present did rapid reduction in 4-Cl-NB occur (Figure 8.12A). This suggests that the continued reduction of the organic contaminant is dependent on the continual replacement of Fe(II) on the mineral surface that has been consumed by oxidation of the organic pollutant, as shown in Figure 8.12B. During the entire period of 4-Cl-NB reduction, the number of electrons transferred to the 4-Cl-NB was similar to the consumption of aqueous Fe(II).

While chemical reduction, for example, Cr(VI) reduction by Fe(II) or S(−II), is a major pathway for the reduction of contaminants in anaerobic environments, the presence of reductant pools of Fe(II) and S(−II) is dependent on microbial activity (Wielinga et al., 2001). Fe(III) and SO_4^{2-} reduction occurs predominantly through dissimilatory reduction pathways by dissimilatory metal-reducing bacteria (DMRB) (Lovely, 1991). DMRB can reduce solid-phase Fe(III) oxides

and oxyhydroxides such as ferrihydrite, goethite, hematite, and magnetite (Kostka and Nealson, 1995).

For example, one sees in the following two-step biotic−abiotic reaction the role of DMRB on the reduction of Fe(III) oxide and subsequent reduction of Cr(VI) (Wielinga et al., 2001):

$$\frac{3}{4}C_3H_5O_3^- + 3Fe(OH)_3 \rightarrow \frac{3}{4}C_2H_3O_2^-$$
$$+ 3Fe^{2+} + \frac{3}{4}HCO_3^- + 2H_2O + 5\frac{1}{4}\ OH^-$$

(8.30)

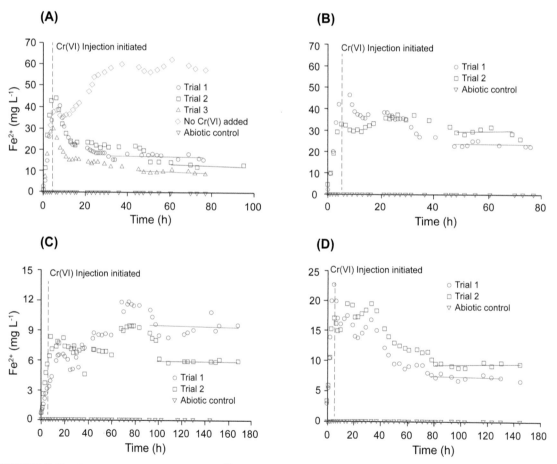

FIGURE 8.13   Temporal changes in aqueous Fe^{2+} from a flow reactor during experiments with Fe-reducing bacteria and different iron minerals: (A) goethite, (B) hematite, (C) ferrihydrite, (D) hydrous ferric oxide. In these experiments 30 mM HEPES at pH 7 were used, as well as 10 mM lactate as an electron donor. *From Wielinga et al. (2001), with permission.*

$$3Fe^{2+} + HCrO_4^- + 8H_2O \rightarrow 3Fe(OH)_3$$
$$+ Cr(OH)_3 + 5H^+ \qquad (8.31)$$

Ferrous iron produced in Equation 8.30 is cycled back to Fe(III) in Equation 8.31, thus serving as an electron shuttle between the bacteria and Cr.

Figure 8.13 shows microbially induced, ferrous iron–mediated reduction of Cr(VI). *Shewanella alga* strain BrY was the model DIRB. BrY is a facultative anaerobic bacterium that can couple the oxidation of organic acids and H_2 to the reduction of Fe(III), Mn(IV), and U(VI) under anaerobic conditions (Caccavo et al., 1992). In Figure 8.13 one sees the ability of BrY to reduce Cr(VI) via Fe(II) production during iron respiration. When Cr(VI) is introduced, the concentration of $Fe(II)_{(aq)}$ decreases dramatically. Without the bacterium, little Fe^{2+} is produced.

Suggested Reading

Baas Becking, L.G.M., Kaplan, I.R., Moore, D., 1960. Limits of the natural environment in terms of plant and oxidation-reduction potentials. J. Geol. 68, 243–284.

Bartlett, R.J., 1986. Soil redox behavior. In: Sparks, D.L. (Ed.), Soil Physical Chemistry. CRC Press, Boca Raton, FL, pp. 179–207.

Bartlett, R.J., James, B.R., 1993. Redox chemistry of soils. Adv. Agron. 50, 151–208.

DeLaune, R.D., Reddy, K.R., Richardson, C.J., Megonigal, J.P. (Eds.), 2013. Methods in Biogeocheistry of Wetlands. Soil Science Society of America Book Series 10. Soil Science Society of America, Madison, WI.

Haderlein, S.B., Pecher, K., 1998. Pollutant reduction in heterogeneous Fe(II)-Fe(III) systems. In: Sparks, D.L., Grundl, T.J. (Eds.), Mineral-Water Interfacial Reactions. ACS Symposium Series 715. American Chemical Society, Washington, DC, pp. 342–357.

Huang, P.M., 1991. Kinetics of redox reactions on manganese oxides and its impact on environmental quality. In: Sparks, D.L., Suarez, D.L. (Eds.), Rates of Soil Chemical Processes. SSSA Special Publications No. 27. Soil Science Society of America, Madison, WI, pp. 191–230.

James, B.R., Bartlett, R.J., 2000. Redox phenomena. In: Sumner, M.E. (Ed.), Handbook of Soil Science. CRC Press, Boca Raton, FL, pp. B169–B194.

James, B.R., Brose, D.A., 2012. Oxidation-reduction phenomena. In: Huang, P.M., Li, Y., Sumner, M.E. (Eds.), Handbook of Soil Science. CRC Press, Boca Raton, FL, pp. 14-1–14-24.

Patrick Jr., W.H., Gambrell, R.P., Faulkner, S.P., 1996. Redox measurements of soils. In: Sparks, D.L. (Ed.), Methods of Soil Analysis: Part 3 – Chemical Methods. SSSA Book Series 5. Soil Science Society of America, Madison, WI, pp. 1255–1273.

Ponnamperuma, F.N., 1972. The chemistry of submerged soils. Adv. Agron. 24, 29–96.

Reddy, K.R., DeLaune, R.D., 2008. Biogeochemistry of Wetlands. Science and Applications. CRC Press, Boca Raton, FL.

Stone, A.T., 1986. Adsorption of organic reductants and subsequent electron transfer on metal oxide surfaces. In: Sparks, D.L., Suarez, D.L. (Eds.), Rates of Soil Chemical Processes. SSSA Special Publications 27. Soil Science Society of America, Madison, WI, pp. 231–254.

Stumm, W., Morgan, J.J., 1981. Aquatic Chemistry. Wiley, New York, NY.

White, A.F., Peterson, M.L., 1998. The reduction of aqueous metal species on the surfaces of Fe(II)-containing oxides: the role of surface passivation. In: Sparks, D.L., Grundl, T.J. (Eds.), Mineral-water Interfacial Reactions. ACS Symposium Series 715. American Chemical Society, Washington, DC, pp. 323–341.

Problem Set

Q.1: Write the balanced half-reactions for each reactant as well as a balanced (full) redox reaction.

a) Arsenite and MnO_2 are reacted to form Mn^{2+} and arsenate.

b) Cr^{6+} and Fe^{2+} are reacted to form Cr^{3+} and $Fe(OH)_{3(s)}$

Q.2: List three common reductants and oxidants in the soil environment.

Q.3: Give the standard reduction potentials for the following reactants (as reduction half-reactions) at pH 7 and describe whether or not the full reaction will occur as stated.

a) Reaction in question: MnO_2 can oxidize Cr^{3+} to Cr^{6+}

b) Reaction in question: MnO_2 can oxidize Fe^{2+} to Fe^{3+}

c) Reaction in question: Fe^{2+} can reduce Cr^{6+} to Cr^{3+}

d) Reaction in question: Fe^{2+} can reduce NO_3^- to $N_{2(g)}$

e) Reaction in question: $O_{2(g)}$ can oxidize As^{3+} to As^{5+}

f) Reaction in question: CO_2 can oxidize Fe^{2+} to $Fe(OH)_3$

g) Reaction in question: $Fe^{3+}_{(aq)}$ can oxide As^{3+} to As^{5+}

h) Reaction in question: $Fe(OH)_{3(s)}$ can oxide As^{3+} to As^{5+}

i) Reaction in question: $O_{2(g)}$ can oxidize SeO_3^{2-} to SeO_4^{2-}

Q.4: Describe how to measure the redox potentials in soils. What are some of the common redox potentials found on oxidized and reduced soils?

Q.5: Explain the mechanisms for the reductive dissolution of metal oxides/hydroxides. What is a potential mechanism that could impede reductive dissolution?

Q.6: Using Visual MINTEQ, plot the distribution of As(V) and As(III) species over pH 2–12 at 0.1 increments. Start with 100 μM As and use ionic strength of 0.1. At pH 7 and pH 10, what are the dominant As(V) and As(III) species?

Q.7: Often As(III) is described as being the more mobile and toxic form of As versus As(V). Based on your response to Question 6, describe why As(III) might be more mobile than As(V). Use an example mineral, such as aluminum oxide or iron oxide, to describe this mobility.

Q.8: In a groundwater system contaminated with arsenic you need to determine what the major predicted arsenic species will be. The pH of the system is 7, and groundwater data indicate

that Fe^{2+} is present. What is the dominant phase of arsenic going to be As(III) or As(V)? Why? (Problem Credit: Dr. Amrika Deonarine).

References

Amacher, M.C., Baker, D.E., 1982. Redox Reactions Involving Chromium, Plutonium, and Manganese in Soils. DOE/DP/04515-1. Pennsylvania State University, University Park, PA.

Argos, M., Kalra, T., Rathouz, P.J., Chen, Y., Pierce, B., Parvez, F., et al., 2010. Arsenic exposure from drinking water, and all-cause and chronic-disease mortalities in Bangladesh (HEALS): a prospective cohort study. Lancet. 376, 252–258.

Baas Becking, L.G.M., Kaplan, I.R., Moore, D., 1960. Limits of the natural environment in terms of plant and oxidation-reduction potentials. J. Geol. 68, 243–284.

Bartlett, R.J., 1981. Nonmicrobial nitrite-to-nitrate transformation in soils. Soil. Sci. Soc. Am. J. 45, 1054–1058.

Bartlett, R.J., 1986. Soil redox behavior. In: Sparks, D.L. (Ed.), Soil Physical Chemistry. CRC Press, Boca Raton, FL, pp. 179–207.

Bartlett, R.J., James, B.R., 1979. Behavior of chromium in soils. III. Oxidation. J. Environ. Qual. 8, 31–35.

Bartlett, R.J., James, B.R., 1993. Redox chemistry of soils. Adv. Agron. 50, 151–208.

Bohn, H.L., 1968. Electromotive force of inert electrodes in soil suspensions. Soil. Sci. Soc. Am. Proc. 32, 211–215.

Bricker, O.P., 1965. Some stability relationships in the system MnO_2-H_2O at 25°C and 1 atm total pressure. Am. Mineral. 50, 1296–1354.

Caccavo Jr., F., Blakemore, R.P., Lovely, D.R., 1992. A hydrogen-oxidizing, Fe(III)-reducing micro-organism from the great bay estuary. Appl. Environ. Microbiol. 58, 3211–3216.

Dixit, S., Hering, J.G., 2003. Comparison of arsenic(V) and arsenic(III) sorption onto iron oxide minerals: implications for arsenic mobility. Environ. Sci. Technol. 37 (18), 4182–4189.

Eary, E., Rai, D., 1987. Kinetics of chromium(III) oxidation to chromium(VI) by reaction with manganese dioxide. Environ. Sci. Technol. 21, 1187–1193.

Faulkner, S.P., Patrick Jr., W.H., Gambrell, R.P., 1989. Field techniques for measuring wetland soil parameters. Soil Sci. Soc. Am. J. 53, 883–890.

Fendorf, S.E., Fendorf, M., Sparks, D.L., Gronsky, R., 1992. Inhibitory mechanisms of Cr(III) oxidation by δ-MnO_2. J. Colloid Interf. Sci. 153, 37–54.

Fendorf, S.E., Zasoski, R.J., 1992. Chromium(III) oxidation by δ-MnO_2 1. Characterization. Environ. Sci. Technol. 26, 79–85.

Gambrell, R.P., Khalid, R.A., Verlow, M.G., Patrick Jr., W.H., 1977. Transformation of Heavy Metals and Plant Nutrients in Dredged Sediments as Affected by Oxidation-Reduction Potential and pH. Part II. Materials and Methods, Results and Discussion, DACW-39-74-C-0076. U.S. Army Engineer Waterways Experiment Station, Vicksburg, MS.

Garrels, R.M., Christ, C.L., 1965. Solutions, Minerals, and Equilibria. Freeman, Cooper, San Francisco.

Gerrard, J., 1992. Soil Geomorphology. Chapman & Hall, London.

Glossary of Soil Science Terms, 2008. Soil Science Glossary Terms Committee and Soil Science Society of America. Soil Sci. Soc. Am, Madison, WI.

Haderlein, S.B., Pecher, K., 1998. Pollutant reduction in heterogeneous Fe(II)-Fe(III) systems. In: Sparks, D.L., Grundl, T.J. (Eds.), Mineral-Water Interfacial Reactions. ACS Symposium Series 715. American Chemical Society, Washington, DC, pp. 323–341.

Hem, J.D., 1978. Redox processes at the surface of manganese oxide and their effects on aqueous metal ions. Chem. Geol. 21, 199–218.

Hem, J.D., Roberson, C.E., Lind, C.J., 1985. Thermodynamic stability of CoOOH and its coprecipitation with manganese. Geochim. Cosmochim. Acta. 49, 801–810.

Huang, P.M., 1991. Kinetics of redox reactions on manganese oxides and its impact on environmental quality. In: Sparks, D.L., Suarez, D.L. (Eds.), Rates of Soil Chemical Processes. SSSA Special Publications No. 27. Soil Science Society of America, Madison, WI, pp. 191–230.

Izaditame, F., Siebecker, M.G., Sparks, D.L., 2022. Sea-level-rise-induced flooding drives arsenic release from coastal sediments. J. Hazard Mater. 423, 127161.

James, B.R., Bartlett, R.J., 2000. Redox phenomena. In: Sumner, M.E. (Ed.), Handbook of Soil Science. CRC Press, Boca Raton, FL, pp. B169–B194.

Jardine, P.M., Jacobs, G.K., O'Dell, J.D., 1993. Unsaturated transport processes in undisturbed heterogeneous porous media. II. Co-contaminants. Soil. Sci. Soc. Am. J. 57, 954–962.

Jones, L.C., Lafferty, B.J., Sparks, D.L., 2012. Additive and competitive effects of bacteria and Mn oxides on arsenite oxidation kinetics. Environ. Sci. Technol. 46 (12), 6548–6555.

Josselyn, M.N., Faulkner, S.P., Patrick Jr., W.H., 1990. Relationships between seasonally wet soils and occurrence of wetland plants in California. Wetlands 10, 7–26.

Klausen, J., Tröber, S.P., Haderlein, S.B., Schwarzenbach, R.P., 1995. Reduction of substituted nitrobenzenes by Fe(II) in aqueous mineral suspensions. Environ. Sci. Technol 29, 2396–2404.

Kostka, J.E., Nealson, K.H., 1995. Dissolution and reduction of magnetite by bacteria. Environ. Sci. Technol. 29, 2535–2540.

Lafferty, B.J., Ginder-Vogel, M., Sparks, D.L., 2010a. Arsenite oxidation by a poorly crystalline manganese-oxide 1. Stirred-flow experiments. Environ. Sci. Technol. 44 (22), 8460–8466.

Lafferty, B.J., Ginder-Vogel, M., Zhu, M., Livi, K.J.T., Sparks, D.L., 2010b. Arsenite oxidation by a poorly crystalline manganese-oxide. 2. Results from X-ray absorption spectroscopy and X-ray diffraction. Environ. Sci. Technol. 44 (22), 8467–8472.

Landrot, G., Ginder-Vogel, M.A., Livi, K.J., Fitts, J.P., Sparks, D.L., 2012. Chromium(III) oxidation by three poorly-crystalline manganese(IV) oxides 1. Chromium(III)-oxidizing capacity. Environ. Sci. Technol. 46 (21), 11594–11600.

LeMonte, J.J., Stuckey, J.W., Sanchez, J.Z., Tappero, R., Rinklebe, J., Sparks, D.L., 2017. Sea level rise induced arsenic release from historically contaminated coastal soils. Environ. Sci. Technol. 51, 5913–5922.

Lindsay, W.L., 1979. Chemical Equilibria in Soils. John Wiley & Sons, New York, NY.

Liu, C.W., Narasimhan, T.N., 1989. Redox-controlled multiple-species reactive chemical transport. 1. Model development. Water Resour. Res. 25, 869–882.

Loach, P.A., 1976. Oxidation-reduction potentials, absorbance bonds, and molar absorbance of compounds used in biochemical studies. In: Handbook of Biochemistry and Molecular Biology: Physical and Chemical Data. In: Fasman, G.D. (Ed.), Third edition, vol. 1. Chemical Rubber, Cleveland, OH, pp. 122–130.

Lovely, D.R., 1991. Dissimilatory Fe(III) and Mn(IV) reduction. Microbiol. Rev. 55, 259–287.

Mendelssohn, I.A., McKee, K.L., Patrick Jr., W.H., 1981. Oxygen deficiency in Spartina alterniflora roots: metabolic adaptation to anions. Science 214, 439–441.

Moore, J.N., Walker, J.R., Hayes, T.H., 1990. Reaction scheme for the oxidation of As(III) to As(V) by birnessite. Clays. Clay. Miner. 38, 549–555.

Morgan, J.J., Stumm, W., 1964. Colloid-chemical properties of manganese dioxide. J. Colloid Sci. 19 (4), 347–359.

Murray, J.W., Dillard, J.G., 1979. The oxidation of cobalt(II) adsorbed on manganese dioxide. Geochim. Cosmochim. Acta 43, 781–787.

Nesbitt, H.W., Canning, G.W., Bancroft, G.M., 1998. XPS study of reductive dissolution of 7 angstrom-birnessite by H_3AsO_3, with constraints on reaction mechanism. Geochim. Cosmochim. Acta 62 (12), 2097–2110.

Nico, P.S., Zasoski, R.J., 2000. Importance of Mn(III) availability on the rate of Cr(III) oxidation on delta-MnO_2. Environ. Sci. Technol. 34, 3363–3367.

Oscarson, D.W., Huang, P.M., Defosse, C., Herbillon, A., 1981. In: The oxidative power of Mn(IV) and Fe(III) oxides with respect to As(III) in terrestrial and aquatic environments, Nature 291, pp. 50–51.

Oscarson, D.W., Huang, P.M., Hammer, U.T., Liaw, W.K., 1983. Oxidation and sorption of arsenite by manganese dioxide as influenced by surface coatings of iron and aluminum oxides and calcium carbonate. Water. Air. Soil. Pollut. 20 (2), 233–244.

Oscarson, D.W., Huang, P.M., Liaw, W.K., 1980. The oxidation of arsenite by aquatic sediments. J. Environ. Qual. 9, 700–703.

Oscarson, D.W., Huang, P.M., Liaw, W.K., 1981. Role of manganese in the oxidation of arsenite by freshwater lake sediments. Clay. Miner. 29 (3), 219–225.

Parikh, S.J., Lafferty, B.J., Meade, T.G., Sparks, D.L., 2010. Evaluating environmental influences on AsIII oxidation kinetics by a poorly crystalline Mn-oxide. Environ. Sci. Technol. 44 (10), 3772–3778.

Patrick Jr., W.H., Gambrell, R.P., Faulkner, S.P., 1996. Redox measurements of soils. In: Sparks, D.L. (Ed.), Methods of Soil Analysis: Part 3 – Chemical Methods. SSSA Book Series 5. Soil Science Society of America, Madison, WI, pp. 1255–1273.

Peacock, C.L., Sherman, D.M., 2007. Sorption of Ni by birnessite: equilibrium controls on Ni in seawater. Chem. Geol. 238 (1–2), 94–106.

Ponnamperuma, F.N., 1955. The Chemistry of Submerged Soils in Relation to the Yield of Rice. Ph.D. Dissertation, Cornell University, Ithaca, NY.

Ponnamperuma, F.N., 1965. Dynamic aspects of flooded soils and the nutrition of the rice plant. In: Chandler, R.F. (Ed.), The Mineral Nutrition of the Rice Plant. Johns Hopkins Press, Baltimore, pp. 295–328.

Ponnamperuma, F.N., 1972. The chemistry of submerged soils. Adv. Agron. 24, 29–96.

Ponnamperuma, F.N., Castro, R.U., 1964. Redox systems in submerged soils. Trans. 8th International Soil Science, Vol. III. Bucharest, Romania, pp. 379–386.

Postel, 1989. Water for Agriculture: Facing the Limits, Worldwatch Paper 93. Worldwatch Institute, Washington, DC.

Rai, D., Sass, B., Moore, D., 1987. Chromium(III) hydrolysis constants and solubility of chromium(III) hydroxide. Inorg. Chem. 26, 345–349.

Rai, E., Serne, R.J., 1977. Plutonium activities in soil solutions and the stability and formation of selected plutonium minerals. J. Environ. Qual. 6, 89–95.

Ramstedt, M., Andersson, B.M., Shchukarev, A., Sjöberg, S., 2004. Surface properties of hydrous manganite (γ-MnOOH). A potentiometric, electroacoustic, and X-ray photoelectron spectroscopy study. Langmuir 20, 8224–8229.

Raven, K.P., Jain, A., Loeppert, R.H., 1998. Arsenite and arsenate adsorption on ferrihydrite: kinetics, equilibrium, and adsorption envelopes. Environ. Sci. Technol. 32 (3), 344–349.

Ravenscroft, P., Brammer, H., Richards, K., 2009. Arsenic Pollution: A Global Synthesis. Wiley, West Sussex.

Reddy, C.N., Patrick Jr., W.H., 1977. Effect of redox potential and pH on the uptake of Cd and Pb by rice plants. J. Environ. Qual. 6, 259–262.

Robie, R.A., Hemingway, B.S., Fisher, J.R., 1978. Thermodynamic Properties of Minerals and Related Substances at 298.15 K and 1 Bar (10^5 Pascals) Pressure and at Higher Temperatures. U.S. Geological Survey Bulletin 1452. U.S. Govt. Printing Office, Washington, DC.

Scott, M.J., Morgan, J.J., 1995. Reactions at oxide surfaces. 1. Oxidation of As(III) by synthetic birnessite. Environ. Sci. Technol. 29 (8), 1898–1905.

Scott, M.J., Morgan, J.J., 1996. Reactions at oxide surfaces. 2. Oxidation of Se(IV) by synthetic birnessite. Environ. Sci. Technol. 30, 1990–1996.

Shiller, A.M., Stephens, T.H., 2005. Microbial manganese oxidation in the lower Mississippi river: methods and evidence. Geomicrobiol. J. 22 (3), 117–125.

Sposito, G., 1989. The Chemistry of Soils. Oxford University Press, New York, NY.

Stone, A.T., 1986. Adsorption of organic reductants and subsequent electron transfer on metal oxide surfaces. In: Davis, J.A., Hayes, K.F. (Eds.), Geochemical Processes at Mineral Surfaces. ACS Symposium Series 323. American Chemical Society, Washington, DC, pp. 446–461.

Stone, A.T., 1987a. Microbial metabolites and the reductive dissolution of manganese oxides: oxalate and pyruvate. Geochim. Cosmochim. Acta 51, 919–925.

Stone, A.T., 1987b. Reductive dissolution of manganese(III/IV) oxides by substituted phenols. Environ. Sci. Technol. 21, 979–988.

Stone, A.T., 1991. Oxidation and hydrolysis of ionizable organic pollutants at hydrous metal oxide surfaces. In: Sparks, D.L., Suarez, D.L. (Eds.), Rates of Soil Chemical Processes. SSSA Special Publications No. 27. Soil Science Society of America, Madison, WI, pp. 231–254.

Stone, A.T., Morgan, J.J., 1984a. Reduction and dissolution of manganese(III) and manganese(IV) oxides by organics. 1. Reaction with hydroquinone. Environ. Sci. Technol. 18, 450–456.

Stone, A.T., Morgan, J.J., 1984b. Reduction and dissolution of manganese (III) and manganese (IV) oxides by organics. 2. Survey of the reactivity of organics. Environ. Sci. Technol. 18, 617–624.

Tan, W.-F., Lu, S.-J., Liu, F., Feng, X.-H., He, J.-Z., Koopal, L.K., 2008. Determination of the point-of-zero charge of manganese oxides with different methods including an improved salt titration method. Soil. Sci. 173 (4), 277–286.

Tani, Y., Miyata, N., Ohashi, M., Ohnuki, T., Seyama, H., Iwahori, K., et al., 2004. Interaction of inorganic arsenic with biogenic manganese oxide produced by a Mn-oxidizing fungus, strain KR21-2. Environ. Sci. Technol. 38 (24), 6618–6624.

Violante, A., Pigna, M., 2002. Competitive sorption of arsenate and phosphate on different clay minerals and soils. Soil Sci. Soc. Am. J. 66 (6), 1788–1796.

White, A.F., Peterson, M.L., 1996. Reduction of aqueous transition metal species on the surfaces of Fe(II)-containing oxides. Geochim. Cosmochim. Acta 60, 3799–3814.

White, A.F., Peterson, M.L., 1998. The reduction of aqueous metal species on the surfaces of Fe(II)-containing oxides: the role of surface passivation. In: Sparks, D.L., Grundl, T.J. (Eds.), Mineral-water Interfacial Reactions. ACS Symposium Series 715. American Chemical Society, Washington, DC, pp. 323–341.

Wielinga, B., Mizuba, M.M., Hansel, C., Fendorf, S., 2001. Iron promoted reduction of chromate by dissimilatory iron-reducing bacteria. Environ. Sci. Technol. 35, 522–527.

Xue, Y., Traina, S.J., 1996. Oxidation kinetics of Co(II)-EDTA in aqueous goethite suspensions. Environ. Sci. Technol. 30, 1975–1981.

The Chemistry of Soil Acidity

9.1 Introduction

Key points: Globally, about 30% of the land area is occupied by acid soils. Soil acidification causes a significant loss in crop productivity and adversely affects the health of terrestrial and aquatic ecosystems.

Soil pH has often been called the master variable of soils and greatly affects numerous soil chemical reactions and processes. It is an important measurement in deciding how acid a soil is and can be expressed as $pH = -log\ (H^+)$, where H^+ is the hydrogen ion activity. Soils that have a $pH < 7$ are acid, those with a $pH > 7$ are considered alkaline, and those with a pH of 7 are assumed to be neutral. Soil pH ranges can be classified as given in Table 9.1. The most important culprits of soil acidity in mineral–organic soils are H and Al, with Al being more important in soils except for those with extremely low pH values (<4).

Although soil acidification is a natural process, it has been exacerbated by farming practices in many parts of the world. Globally, about 4 billion ha of land surface is occupied by acid soils, which make up 30% of the total land area (von Uexküll and Mutert, 1995). About 75% of the soils with topsoil acidity also exhibit significant subsoil acidity, indicating a strong relationship between surface and subsurface soil acidity under natural conditions.

Soil acidity is a major constraint to crop production, particularly in tropical and subtropical regions of the world. In some parts of the world subsoil acidity exists, which is difficult to ameliorate and has become worse with continued cropping. In developing and low-resource countries farmers cannot afford to apply adequate lime to ameliorate soil acidity, resulting in substantial yield losses, particularly under rainfed conditions. The global distribution of acid soils and the distribution of acid soils in major soil types are presented in Figure 9.1 and Table 9.2, respectively.

TABLE 9.1   Descriptive terms and proposed buffering mechanisms for various soil pH ranges.[a]

Descriptive terms[b]	pH range	Buffering mechanism[c]
Extremely acid	<4.5	Iron range (pH 2.4–3.8)
Very strongly acid	4.5–5.0	Aluminum/iron range (pH 3.0–4.8)
Strongly acid	5.1–5.5	Aluminum range (pH 3.0–4.8)
Moderately acid	5.6–6.0	Cation exchange (pH 4.2–5.0)
Slightly acid to neutral	6.1–7.3	Silicate buffers (all pH values typically >5)
Slightly alkaline	7.4–7.8	Carbonate (pH 6.5–8.3)

[a] From Robarge and Johnson (1992), with permission.
[b] Glossary of Soil Science Terms (2008)
[c] Schwertmann et al. (1987), Ulrich (1991), Reuss and Walthall (1989).

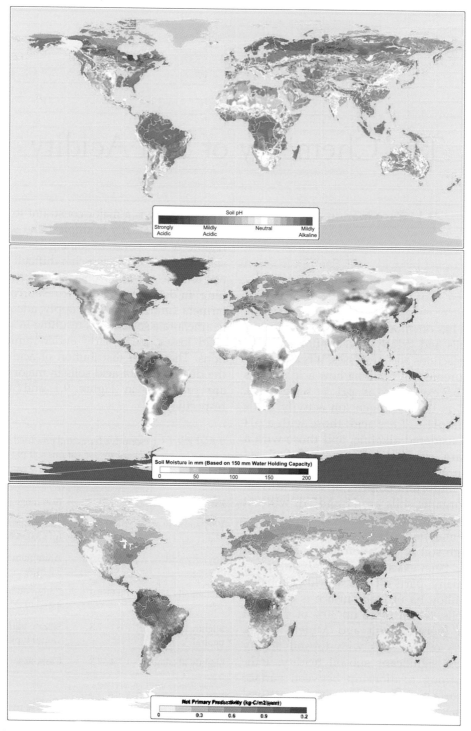

FIGURE 9.1   Soil pH, soil moisture distribution, and net primary plant production (NPP) maps of the world. Maps sourced from http://www.sage.wisc.edu/atlas with permission from The Center for Sustainability and the Global Environment, Nelson Institute for Environmental Studies, University of Wisconsin-Madison, USA.

TABLE 9.2 Global distribution of acid soils based on soil types.

WRB soil group	Area (million ha)	Soil taxonomy order	Area (million ha)	(%)
Regosols	293	Entisols	824	20.9
Arenosols	280			
Fluvisols	50			
Gleysols	201			
Cambisols	299	Inceptisols	561	14.2
Gleysols	201			
Leptosols	61			
Luvisols	128	Alfisols	255	6.5
Planosols	127			
Andosols	34	Andisols	34	0.9
Podzols	415	Spodosols	415	10.5
Acrisols	731	Ultisols	864	21.8
Nitosols	118			
Planosols	15			
Ferralsols	727	Oxisols	727	18.4
Histosols	270	Histosols	270	6.8
Total	3950		3950	100.0

From von Uexküll and Mutert (1995), with permission.

9.2 Historical Perspective of Soil Acidity

Key points: Soil acidification and the beneficial effects of liming on crops have been recognized for well over a century. However, the toxicity of Al to plants in acid soils was confirmed a few decades later.

As previously noted in Chapter 1, one of the great debates in soil chemistry has been the cause of soil acidity. This debate went on for over five decades and there were heated arguments over whether the culprit in soil acidity was H or Al. The history of this debate was described in a lively manner by Thomas (1977), and the discussion below is largely taken from this review.

As noted in the discussion on the history of soil chemistry in Chapter 1, Edmund Ruffin was the first person to lime soils for the proper reason, to neutralize acidity, when he applied oyster shells to his soils. It was 70 years after Ruffin's work before research on soil acidity was initiated again. F.P. Veitch (1902) found that the titration of soils that had been equilibrated with $Ca(OH)_2$ to a pink endpoint with phenolphthalein was a good test for predicting whether lime (e.g., $CaCO_3$) was needed to neutralize acidity that would be detrimental to crop growth (Thomas, 1977). Hopkins et al. (1903) developed a lime requirement (LR) test based on the titration of a soil equilibrated with 1 M NaCl. Veitch (1904) showed that a 1 M NaCl extract, while not replacing all the soil's acidity, was a good LR test. A very important finding by Veitch (1904) not recognized as such at the time was that the acidity replaced by 1 M NaCl was $AlCl_3$, not HCl. After Veitch's work, a number of soil chemists started to study soil acidity and debate whether acidity was caused by Al or H. Bradfield (1923, 1925) titrated clays and observed that their pK_a values were similar to those found for weak acids. Kelley and Brown (1926) and Page (1926) hypothesized that "exchangeable Al" was dissolved by exchangeable H^+ during the extraction with salt. Paver and Marshall (1934) believed that the exchangeable H^+ dissolved the clay structure, releasing Al, which in turn became a counterion on the exchange complex. This was indeed an important discovery that was not definitively proved and accepted until the 1950s and early 1960s.

Chernov (1947) had shown that electrodialyzed clays and naturally acid clays were primarily Al saturated. Shortly thereafter, Coleman and Harward (1953) found that H resin-treated clays or clays leached rapidly with 1 M HCl had properties quite different from those of clays that were slowly leached, leached with dilute acid solutions, or electrodialyzed. They concluded that hydrogen clays were strongly acidic. Low (1955), employing potentiometric and conductometric titration analyses (these are discussed later in this chapter), proved that an electrodialyzed clay was Al saturated. Coleman and Craig (1961) confirmed the earlier finding of Coleman and Harward (1953) that H clays are unstable and rapidly convert to Al clay, with temperature having a dramatic effect on the transformation rate. The research on H versus Al clays was very important in that it showed that Al is more important in soil acidity than H^+.

Also, in the 1950s and 1960s there were some important discoveries made about the types of Al found in soils. Rich and Obenshain (1955) showed that in some Virginia soils, formed from mica schist, there was not only exchangeable Al^{3+} but also nonexchangeable Al, with the latter blocking exchange sites and thus lowering the cation exchange capacity (CEC) of the soils. The nonexchangeable Al also kept vermiculite from collapsing (Rich, 1964; Rich and Black, 1964) and was referred to as interlayer hydroxy-Al.

9.3 Soil Acidification Processes

Key points: Natural acidification of soils occurs with weathering processes and a large area of acid soils that are highly weathered exists in the tropical and subtropical regions of the world. Leaching of nitrate (derived from NH_4^+ fertilizers and biological N_2 fixation), removal of products, and organic matter build-up are the main soil acidification processes. Deposition of acid rain and oxidation of sulfide minerals can also cause soil acidification.

9.3.1 Natural Soil Weathering

Soil weathering is a natural acidifying process that is driven by precipitation and vegetation. Precipitation is generally acidic due to the presence of H_2CO_3, which reacts with the soil solum and removes basic cations (i.e., Ca, Mg, K, and Na; these cations are also referred to as "nonacid cations" because they do not cause soil acidity;

however, they do not cause the formation of OH^- either) and increases the proportion of acidic cations, such as Al and Fe, over a long period of time. The acidification rate by weathering is dependent on the nature of the parent rock, precipitation, and temperature (Sumner and Noble, 2003). Mafic parent rocks weather more rapidly than highly siliceous and acidic rocks, and weathering intensity increases with increasing precipitation and temperature. The northern boreal forests and the humid tropics have a high proportion of acid soils, which have formed naturally. This region has old soils and is not impacted by recent glaciation. The assimilation of carbon dioxide into carboxylic acids in higher plants indirectly acidifies the soil explored by their roots (Bolan and Hedley, 2003).

The cumulative effect of natural soil weathering and net plant productivity on soil pH is evident in Figure 9.1, where low pH soils are present in the areas where there is excess soil moisture and high net productivity. Consequently, Ultisols and Oxisols, present in the tropical and subtropical regions of the world, are more weathered and naturally acidic (Table 9.1). In natural ecosystems soils become more acid over a long period of time; however, a recent study comprising a large dataset shows that soil pH can change abruptly from alkaline to acid when the mean annual precipitation begins to exceed the mean annual potential evapotranspiration (Slessarev et al., 2016).

Sidebar 9.1 — Spodosols (Podzols)

Spodosols or Podzols have a characteristic spodic subsurface horizon (from Greek "spodos," meaning wood ash) that contains an illuvial horizon composed of organic matter and Al and Fe oxides. The illuvial horizon is overlain by an ash-gray eluvial horizon. These are acid soils that often occur under coniferous forests in cool and moist climates.

Some soils, which occur in the Spodosol (Podzols), Histosol (Histosols), and Entisol

(Fluvisols, Gleysols, Regosols, Arenosols) orders, derived from basic, cation poor, unbuffered, and coarse-textured parent material, become acid via the translocation of basic cations, iron, and organic matter from the surface horizons (Sumner and Noble, 2003). Spodosols or Podzols (Sidebar 9.1) are sandy acid soils that exist mainly in the forested areas of temperate regions. Histosols have very high organic matter contents and occur in cool mountain areas or in poorly drained soil; these soils are usually acidic to strongly acidic.

9.3.2 Fertilizers and Legume Pastures

Sidebar 9.2 — Chemical Reactions of Biologically Fixed N, Urea, and Ammonium Fertilizers

(i) *Enzymatic deaminization of amino compounds*

$$R - C - H - COOH + H_2O + 2H^+ + 2e^- \xrightarrow[\text{deamination}]{\text{Enzymatic}}$$
$$\underset{NH_2}{|}$$

$$R - CH_2 - COOH + NH_4^+ + OH^-$$

(ii) *Hydrolysis of urea*

$$CO(NH_2)_2 \text{ (urea)} + 3H_2O \xrightarrow{\text{urease}} 2NH_4^+ + 2OH^- + CO_2$$

(iii) *Nitrification of ammonium fertilizer*

$$(NH_4)_2SO_4 \text{ (ammonium sulfate)} + 2O_2 \rightarrow 2NO_3^- + 4H^+ + SO_4^{2-} + 2H_2O$$

(iv) *Generalized Nitrification Reaction (for all three NH_4 sources):*

$$NH_4 + 2O_2 \rightarrow NO_3^- + 2H^+ + H_2O$$

In the nitrification of NH_4^+ fertilizers 2 mol of H^+ are generated for every mole of N, whereas in the transformation of organic N compounds (from biological N fixation) and urea fertilizer one net mole of H^+ is generated for every mole of N transformed.

The application of ammonium fertilizers and biological N_2 fixation causes the acidification of soils. Nitrification (microbially mediated) of NH_4^+ applied through fertilizers produces 2 mol of H^+ for every mole of N, and the transformation of organic N (from biological N_2 fixation) and urea fertilizer result in the production of one net mole of H^+ for every mole of N (see Sidebar 9.2). If the NO_3^- produced through these transformations is taken up by plants (or remains in the soil), there is no acidification as NO_3^- uptake occurs via proton–anion cotransport, that is, H^+ uptake with NO_3^- ions (Marschner, 2012). However, under field conditions, NO_3^- is frequently leached and soil acidification occurs. Additionally, there is excess uptake of cations over anions by legumes in the process of N_2 fixation that vary from 0.2 to 0.7 mole H^+ per mole of fixed N (Bolan and Hedley, 2003).

The effect of NH_4 fertilizer on soil acidification is evident in the data from long-term studies from Rothamsted Experimental Station in the U.K. (Figure 9.2; Johnston et al., 1986). Surface soil pH decreased by up to 1.6 pH units with the application of $(NH_4)_2SO_4$ fertilizer, whereas the pH remained about the same with nitrate fertilizer ($NaNO_3$). Soil acidification rate increased with the increasing rate of $(NH_4)_2SO_4$ as the surface soil pH decreased 0.9, 1.2, and 1.6 pH units with the annual addition of 48, 96, and 145 kg N ha^{-1}, respectively. Soil pH has been reported to decrease by 1 pH unit after growing legume pasture continuously for more than 30 years in Australia (Bolan and Hedley, 2003). The application of S fertilizer also leads to soil acidification, with the elemental S fertilizer oxidizing to produce sulfuric acid, similar to the oxidation of sulfides as described in Section 9.3.5.

9.3.3 Removal of Crop Products

Most plant materials are slightly alkaline and their removal by grazing or the removal of produce, such as hay, grain, or animal products

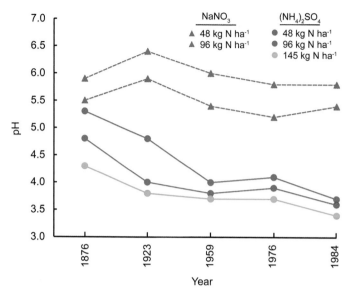

FIGURE 9.2   The effects of different rates of ammonium and nitrate fertilizers over 100 years on the surface soil (0–23 cm) pH. From the Park Grass Experiment, Rothamsted in the U.K. The fertilizer doses given in the figure were applied annually *drawn using data from Johnston et al., 1986.*

(e.g., wool, milk, meat), from the land leaves residual H^+ ions in the soil. Generally, soil acidification is greater under legume pastures/crops than under other crop species, which is attributed to greater excretion of protons from the excess uptake of cations during N_2 fixation (Tang and Rengel, 2003). The extent of soil acidification also depends on the soil type, crop rotation, and management of crop residues.

Over time, the removal of most agricultural produce causes the acidification of agricultural soils (Table 9.3). Hay, corn, soybean, banana, and tobacco are some of the major contributors to acidity in agricultural soils.

9.3.4 Accumulation and Decomposition of Organic Matter

The accumulation and subsequent decomposition of organic matter is considered to be a major cause of soil acidification. The role of organic matter in soil acidification can be considered in two separate processes (Bolan and Hedley 2003). Microbial decomposition of organic matter and root respiration can increase the CO_2 concentration by an order to two orders of magnitude in the soil environment as compared to air. The dissolved CO_2 forms H_2CO_3 and creates slight acidity as in acid rain. The decomposition of plant residues in soils produces humic substances, which have functional groups that dissociate to produce hydrogen ions. The release of H^+ ions from humic substances is pH dependent and varies with both the soil pH and the dissociation constant of functional groups of humic substances. This type of acidity is most prominent in Histosols.

In contrast, there may be increases in pH from the mineralization of organic anions to CO_2 and water (thereby removing H^+) or because of the "alkaline" nature of the organic plant residues. Also, humic substances can form complexes with Al^{3+} and thus may help in reducing the deleterious effects on plant growth (as discussed in Chapter 3).

9.3.5 Oxidation of Sulfide Minerals

Sidebar 9.3 — Acid Sulfate Soils

Acid sulfate soils are the soils that contain soil materials consisting of sulfide minerals or are affected by transformations of iron sulfide minerals, predominantly pyrite (FeS). In the natural state, i.e., under water-logged conditions, acid sulfate soils are benign; however, once drained, the microbial oxidation of pyrite and sulfidic material produce extreme acidity in the soils with a pH of ≤ 4 and often <3. The chemolithotroph iron bacteria are "rock eating" bacteria that can grow on a carbon-free medium and obtain their energy from the oxidation of ferrous (Fe^{2+}) iron compounds.

Mine spoil and acid sulfate soils have very low pH values due to the oxidation of sulfide minerals (Sidebar 9.3), with pyrite being the most predominant. Mine spoil soils are common in surface-mined coal areas, and acid sulfate soils occur in coastal areas near mangrove forests, salt marshes, and floodplains in temperate and tropical areas. When they are drained, pyrite oxidation occurs, and extreme acidity is produced. The complete oxidation of pyrite (FeS_2), which is mediated by chemolithotrophic iron bacteria, typically Acidithiobacillus ferrooxidans, can be expressed as:

$$FeS + \frac{15}{4} O_2 + \frac{7}{2} H_2O \rightarrow Fe(OH)_3 + 2H_2SO_4$$

$$(9.1)$$

The high concentrations of sulfuric acid cause pH values as low as 2 in mine spoil soils (McFee et al., 1981) and <4 in acid sulfate soils. The extreme acid produced moves into drainage and floodwaters, corrodes steel and concrete, and causes the dissolution of clay minerals, releasing soluble Al. The drainage waters can also contain heavy metals and arsenic, which can have profound effects on animal, plant, and human health (Ritsema et al., 2000).

TABLE 9.3   Amount of lime ($CaCO_3$) required to balance the acidity generated with the removal of base cations by various crops and pasture species.

Crop or Pasture species	Yield (t ha^{-1})	CaCO$_3$ (kg ha^{-1})	Source
Lucerne or alfalfa hay, Medicago sativa L.	5	350	[a]
Mixed grass hay	5	125	[a]
Subterranean clover, Trifolium subterraneum	5	200	[a]
Lupin, Lupinus sp.	2	40	[a]
Wheat grain, Triticum aestivum	2	18	[a]
Corn, Zea mays — whole plant	13	264	[b]
Corn — grain	5.4	29	[b]
Soybean, Glycine max L. — whole plant	6.4	324	[b]
Soybean — grain	1.9	42	[b]
Small grain cereals (wheat, barley, oats)	5.3	101	[b]
Tobacco (Nicotiana sp.)	2.24	258	[c]
Banana	8.0	200	[a]
Wool (per sheep)	5 kg	0.07	[a]
Meat (per lamb)	1	0.02	[a]
Milk (1000 L)	1	4	[a]

[a] Fenton and Helyar (2000) [b] Robarge (2008) [c] Nyatsanga and Pierre (1973).

Acid sulfate soils are estimated to occupy an area of 50 million ha worldwide (Sullivan et al. 2012). These soils generally are grouped in the Inceptisol (Gleysols, Cambisols, Leptosols) and Entisol (Fluvisols, Gleysols, Regosols, Arenosols) orders of soil taxonomy (Sumner and Noble, 2003).

9.3.6 Acid Deposition

Acid vapors, primarily sulfuric (H_2SO_4) and nitric (HNO_3), form in the atmosphere as a result of the emission of sulfur dioxide (SO_2) and nitrogen oxides (Sidebar 9.4) from natural and anthropogenic sources. Natural sources include volcanic emission of SO_2 and HCl, lightning (NO), decaying vegetation, wildfires, and biological processes (e.g., dimethyl sulfide, Sidebar 9.5). The largest anthropogenic sources of these gases are the burning of fossil fuels (source of sulfur gases) and the exhaust from motor vehicles (source of nitrogen oxides). These vapors condense to form aerosol particles and, along with basic materials in the atmospheric water, determine the pH of precipitation. Major cations in the precipitation water are H^+, NH_4^+, Na^+, Ca^{2+}, Mg^{2+}, and K^+ while the major anions are SO_4^{2-}, NO_3^-, and Cl^- (Meszaros, 1992).

Sidebar 9.4 — Nitrogen Oxides

"Nitrogen oxides" (NOx) is a collective term used to refer to nitrogen monoxide (nitric oxide or NO) and nitrogen dioxide (NO_2). Nitrogen oxides are mostly produced in the combustion processes of fossil fuels and small amounts are produced naturally by lightning.

Sidebar 9.5 — Dimethyl Sulfide

Dimethyl sulfide, $(CH_3)_2S$, is the most abundant biological sulfur compound emitted into the atmosphere. The compound is a major metabolite in some marine algae, and thus, its emission mostly occurs over the oceans.

There has been great concern about the increased acidity of rainfall or acid rain and dry acid deposition. An average value for the amount of H^+ produced per year from acid precipitation falling on industrialized areas is 1 kmol H^+ ha^{-1} $year^{-1}$, but depending on the proximity to the pollution source, it may vary from 0.1 to 6.4 kmol H^+ ha^{-1} $year^{-1}$ (Ulrich, 1991). The long-term data from Rothamsted show that soil pH at 0–23, 24–46, and 47–69 cm decreased by 2.0, 2.5, and 1.4 pH units, respectively, over 100 years (Figure 9.3) in the Geescroft Wilderness (Johnston et al., 1986). The total (dry and wet) deposition of 1.1, 1.7, and 3.9 kmol H^+ ha^{-1} $year^{-1}$ was estimated in the 1890s, 1930s, and 1980s, respectively.

In the United States 60%–70% of the acidity in precipitation comes from H_2SO_4 and the

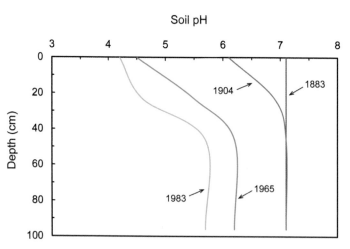

FIGURE 9.3   Soil pH (in water) at three depths from Geescroft Wilderness at Rothamsted over a 100-year period, from 1883 to 1983. The soil in the wilderness area (regenerated woodland, mainly deciduous species), which has not been limed or fertilized since 1883, has been acidified from dry and wet acid deposition *From Johnston et al., (1986), with permission.*

remaining 30–40% is derived from HNO_3. Acid rain can deleteriously affect aquatic life by significantly lowering the pH of lakes and streams and can cause damage to buildings, monuments, and plants. Severe forest decline from acid rain has been observed in Europe and North America, particularly on coarse-textured, poorly buffered soils (Sumner and Noble, 2003).

In most cases the amount of soil acidification (Sidebar 9.6) that occurs naturally or results from agronomic practices is significantly greater than that resulting from the deposition of acid rain. For example, if one assumes annual fertilizer application rates of 50–200 kg N ha^{-1} to soils being cropped, soil acidification from ammonium fertilizer (due to the reaction NH_4^+ $+ 2O_2 \rightarrow NO_3^- + H_2O + 2H^+$) would be 4–16 times greater than the acidification from acid rain in highly industrialized areas (Sumner, 1991). Thus in most soils used in agriculture acid rain does not appear to be a problem. This is particularly true for soils that are limed periodically and that have appreciable buffering capacities due to significant clay and organic matter contents. However, on poorly buffered soils, such as many sandy soils, acid rain could increase their acidity over time.

Sidebar 9.6 – Soil Acidification Quantification

Soil acidification is generally quantified as the amount of H^+ added or generated in the soil per hectare per year.

It is frequently expressed in kmol (H^+) ha^{-1} $year^{-1}$ units. A reaction describing the exchange of H^+ for Ca^{2+} is as follows:

Soil-$2H^+$ + $CaCO_3$ $\rightarrow$ Soil-Ca^{2+} + H_2O + CO_2

On a charge basis, 1 mol H^+ = 0.5 mol Ca^{2+}, and if Ca is added in the form of lime, $CaCO_3$ (molecular weight = 100.09 g mol^{-1}), to neutralize 1 kmol (H^+), 50 kg $CaCO_3$ will be needed.

In forests and grasslands acid rain can have a significant effect not only on the trees but also on the chemistry of the soils. Liming of forests is seldom done and acid rain can cause leaching of nutrient cations such as Ca^{2+}, Mg^{2+}, and K^+ from the soil, resulting in low pH values and the solubilization of toxic metals such as Al^{3+} and Mn^{2+}. This can cause a decrease in soil biological activity, such as ammonification (conversion of NH_4^+ to NO_3^-), decrease the fixation of atmospheric N_2 by leguminous plants (such as soybeans), and also decrease nutrient cycling. Over time, the productivity of forests and grasslands is decreased due to fewer nutrients and higher levels of toxic metals.

9.4 Subsoil Acidity

Key point: Subsoil acidity occurs in soils that have acid surface soils and in certain land use systems, where a spatial separation of acidity and alkalinity (produced in the N and C cycles) occurs.

Subsoil acidification is widespread, and the causes of subsoil acidification may be the same as that for surface soil, and/or separate acidification processes may occur only in the subsoil. A large proportion of highly weathered acid soils in the tropical and subtropical areas of South America, Africa, and Asia also exhibit marked subsoil acidity of similar intensity, indicating the strong relationship between top- and subsoil acidity under natural conditions (Sumner and Noble, 2003). In some soils acidification occurs at a greater rate in the subsurface soil than in the topsoil and can extend to a depth of more than 80 cm layers. Dolling and Porter (1994) showed an increased rate of subsoil acidification with time of clearing in deep yellow sands, which were under a cereal-annual pasture rotation in Western Australia (Figure 9.4). The soil pH increased with depth in uncleared soil. Subsoil acidity has a significant effect on crop and

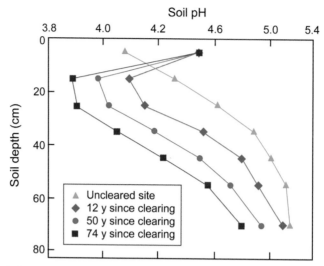

FIGURE 9.4   Soil pH at different depths in deep sandy soils (in Western Australia) at an uncleared site and other sites after 12, 50, and 74 years since clearing. *Redrawn with permission from Dolling and Porter (1994).*

pasture productivity in many soils, particularly in low rainfall areas, where, due to the lack of root penetration, plants are unable to access water from the subsoil.

Although the cause of accelerated subsoil acidification is not fully understood, Tang et al. (2013) attributed this to the spatial separation of acidity and alkalinity produced by the N and C cycles. The acid produced by pasture roots due to the excess uptake of cations over anions is distributed through the rooting zone, whereas plant residues are mainly oxidized in the surface soil. The oxidation of organic anions in plant residues is an acid-consuming process that neutralizes the acidity generated by roots in the topsoil, but acidity persists in the subsoil.

9.5 Impacts of Acidification in the Soil

Key points: Soil acidification adversely affects the availability of many plant nutrients and some elements, such as Al and Mn, which may be present in high concentrations and are toxic to plants.

Sidebar 9.7 — Percent Base Saturation

Percent base saturation (% BS) or base saturation percentage is the percentage of the cation exchange capacity (CEC) or effective cation exchange capacity (ECEC, sum of all exchangeable cations) that is occupied by base forming cations, i.e., Ca^{2+}, Mg^{2+}, Na^+ and K^+.

$$BS\,(\%) = \frac{Exchangeable\,(Ca+Mg+Na+K)}{CEC} \times 100$$

Percent base saturation reflects the exchangeable acidity of a soil; as the concentration of base forming cations decreases, the exchangeable acidity of the soil increases.

Soil pH has direct and indirect effects on the availability and solubility of plant nutrients as well as toxic plant elements (Figure 9.5). The availability of plant nutrients in soils is generally at an optimum level between pH 5.5 and 7.0. At low pH, one sees that Al, Fe, and Mn become more soluble and can be toxic to plants. As pH increases, their solubility decreases, and precipitation occurs

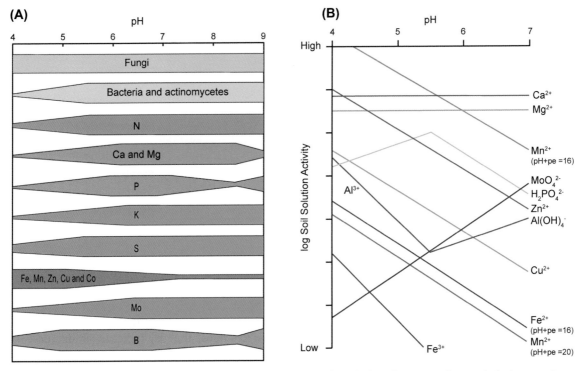

FIGURE 9.5 The effects of pH on the (A) availability of essential (and toxic) plant elements to plants and of microorganisms in soils and (B) soil solution activity of elements. In figure (A), as the band for a particular nutrient or microbe widens, the availability of the nutrient or activity of the microbes is greater. For example, with K, the greatest availability is from pH ~6–9. The activities of various ions as a function of pH are based on thermodynamic stability considerations, as described in Lindsay (1980). Figure (A) *from Brady (1984) and (B) From Sumner et al. (1991), with permission.*

(often as metal hydroxide species). Plants may suffer deficiencies as pH rises above neutrality. In contrast to Al, Fe, and Mn, there is decreased availability of other nutrients, particularly Ca, Mg, P, and Mo, which may become deficient in acid soils. The contribution of the variable charge to soil CEC decreases at low pH, and percent base saturation (Sidebar 9.7) decreases in acid soils.

In addition to the solubility and availability of nutrients, low pH has a negative effect on beneficial soil microorganisms, particularly bacteria. Microbial processes, such as mineralization of organic matter, nitrification, and biological N fixation, are adversely affected at low pH caused by soil acidification.

Plant toxicity of Al in acid soils, particularly monomeric Al^{3+} $(Al(H_2O)_6^{3+})$ species, is

considered to be the major impediment to plant growth (Foy, 1984). Increased soluble Al in the soil solution causes reduced cell division in root apices that results in stunted, club-shaped roots; consequently, the plant's ability to extract moisture from deep in the soil is reduced. The extent of Al toxicity varies substantially with the type of plant and the Al species. Figure 9.6 summarizes the sensitivity of common field crops and forage species to soil pH (measured in 0.01 M $CaCl_2$) and Al (Slattery et al., 1999). While the toxicity of monomeric Al^{3+} $(Al(H_2O)_6^{3+})$ species is well recognized, some studies have shown that polymeric Al species (especially triskaideka aluminum; $AlO_4Al_{12}(OH)_{24}(H_2O)_{12}^{7+}$ or Al_{13}, discussed in Section 9.7) are also toxic to soybeans and wheat (Parker et al., 1988, 1989).

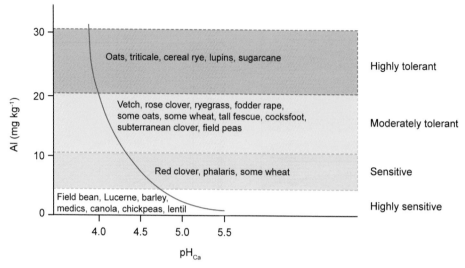

FIGURE 9.6   Relationship between exchangeable Al and soil pH (in 0.01 M CaCl₂) and Al and soil pH sensitivity (tolerance) of some crop and pasture plants. *From Slattery et al. [1999], with permission.*

9.6 Soil Buffering Mechanisms

Key points: Different soil components provide buffering to counter acid inputs. Buffering in soils results from the dissolution of minerals or adsorption of protons on exchange sites or surface functional groups of organic matter and variable charge minerals.

Buffering capacity is the ability of a soil to resist changes, that is, increase or decrease in pH. Soils are buffered against acidification by a series of buffering mechanisms that operate within different and some overlapping pH ranges (Table 9.1; Figure 9.7).

The dissolution of carbonates (predominantly $CaCO_3$) and weathering of primary

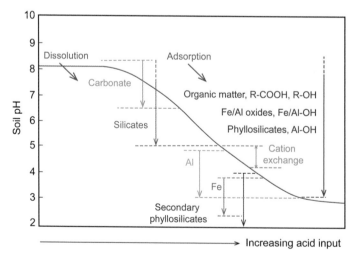

FIGURE 9.7   A hypothetical titration curve with acid input showing different buffering mechanisms and their pH ranges.

silicate minerals buffer acid input simultaneously in many soils. Carbonate minerals are present in some sedimentary and metamorphic rocks, neutralizing acidity and maintaining soil pH at around 8.0. Once the carbonate minerals have dissolved, the soil pH will decline below 7.0 and the dissolution of primary silicates becomes the dominant weathering reaction. The weathering of $CaCO_3$ can be expressed as follows:

$$CaCO_3 + H_2CO_3 \rightarrow Ca^{2+} + 2HCO_3^- \quad (9.2)$$

The suspension of $CaCO_3$ in distilled water in the ambient atmosphere will have a pH of about 8.3 and two $2HCO_3^-$ ions are present for every Ca^{2+} ion in the solution. The pH buffer range for $CaCO_3$ varies widely (Figure 9.7) in soils because of the influence of various factors, such as the CO_2 concentration that is often much greater in the soil environment than in the ambient air, the presence of other Ca salts (e.g., gypsum) and other carbonate salts (e.g., $MgCO_3$), and the nonuniform distribution of carbonate in the soils.

The role of silicate (primary) minerals in acid buffering is demonstrated by the following reaction where anorthite (Ca-feldspar) weathers to form kaolinite and two protons are consumed in this process:

$$CaAl_2Si_2O_8 (anorthite) + H_2O + 2H^+ \rightarrow$$
$$Al_2Si_2O_5(OH)_4 (kaolinite) + Ca^{2+}$$
$$(9.3)$$

As seen in the above reaction, the hydrolysis of the silicate mineral anorthite results in the release of a base cation (Ca^{2+}) from the mineral structure with the consumption of protons equal to the equivalents of base cation released. Similar acid buffering reactions can be written for other silicates.

The exchange reaction for acid buffering can be expressed as follows:

$$X\text{-}Ca^{2+} + 2H^+ \rightarrow X\text{-}2H^+ + Ca^{2+} \quad (9.4)$$

where X represents soil exchange sites and the Ca^{2+} ion is replaced by two H^+ ions. Similar to the cation exchange reaction, Al-hydroxy polymers carrying positive charge can exchange base cations from the interlayer of 2:1 clay minerals in the pH range between 5 and 7.

The dissolution of Al and Fe oxides buffers soil pH in a highly acidic pH range and the chemical reactions are as follows:

$$Al(OH)_3 (gibbsite) + 3H^+ \rightarrow Al^{3+} + 3H_2O$$
$$(9.5)$$

$$FeOOH (goethite) + 3H^+ \rightarrow Fe^{3+} + 2H_2O$$
$$(9.6)$$

In addition to Fe and Al oxides, at pH below 5, the dissolution of Mn oxides and secondary phyllosilicates also buffer soil acidity. Variable charge surfaces of organic matter, Fe and Al oxides, allophane and imogolite, and the edges of phyllosilicate minerals can adsorb (and desorb) H^+ over the whole soil pH range.

It is apparent from the above that soils are buffered against the acid input by different soil components, processes, and properties. The main properties that govern the soil buffering capacity include clay content, clay mineral type, organic matter content, carbonate and primary mineral content, and initial soil pH. Soil buffering processes are complex and overlapping, and it may not be possible to predict the soil buffering capacity based on measured soil properties (Singh et al., 2003).

9.7 Solution Chemistry of Aluminum

Key points: Aluminum exists in different monomeric forms depending on solution pH; the Al^{3+} form is highly phytotoxic and the dominant species at pH < 5.0. With increasing solution pH, other Al monomeric species and Al-hydroxy polymers form. Al-hydroxy polymers may be preferentially adsorbed in the interlayer of 2:1 clay minerals and decrease the CEC of soils.

9.7.1 Monomeric Al Species

Sidebar 9.8 − Hydrolysis

Hydrolysis is a chemical reaction whereby a substance is split or decomposed by water (Baes and Mesmer, 1976). The positive charge on metal ions, such as Al^{3+} and Fe^{3+}, draws electron density from the O−H bond in the water, which increases the bond's polarity, making it easier to break the water molecule.

Aluminum in aqueous solution rapidly and reversibly hydrolyzes (Sidebar 9.8) in dilute solutions (<0.001 M), which leads to the progressive loss of hydration shell protons to water molecules in the solution to maintain dissociation equilibrium. Hydrolysis reactions of Al, by omitting water of hydration, can be expressed as follows:

$$Al^{3+} + H_2O \leftrightarrow Al(OH)^{2+} + H^+ \qquad (9.7)$$

$$Al^{3+} + 2H_2O \leftrightarrow Al(OH)_2^+ + 2H^+ \qquad (9.8)$$

TABLE 9.4   A summary of Al monomeric hydrolysis products at zero ionic strength (infinite dilution) and 298.15 K.

Equation	Equilibrium constant expression	Negative log of equilibrium constant
9.7	$^*K_1 = \dfrac{\left(AlOH^{2+}\right)\left(H^+\right)}{\left(Al^{3+}\right)}$	4.95
9.8	$^*\beta_2 = \dfrac{\left(Al(OH)_2^+\right)\left(H^+\right)^2}{\left(Al^{3+}\right)}$	10.01
9.9	$^*\beta_3 = \dfrac{\left(Al(OH)_3^0\right)\left(H^+\right)^3}{\left(Al^{3+}\right)}$	16.8
9.10	$^*\beta_4 = \dfrac{\left(Al(OH)_4^-\right)\left(H^+\right)^4}{\left(Al^{3+}\right)}$	22.87

In general, 'K' is used to denote the consecutive or stepwise formation constant, and 'β' is used to denote the cumulative or overall formation constant. Additionally, when the formation constants of metal ion complexes are determined by adding a ligand in its protonated form to a metal ion solution, the complex formation reactions involve a deprotonation reaction of the ligand. In such cases, an asterisk (*) is added to the equilibrium constant.
From Nordstrom and May, (1996), with permission.

$$Al^{3+} + 3H_2O \leftrightarrow Al(OH)_3^0 + 3H^+ \qquad (9.9)$$

$$Al^{3+} + 4H_2O \leftrightarrow Al(OH)_4^- + 4H^+ \qquad (9.10)$$

The formation quotients for the monomeric hydrolysis products are shown in Table 9.4. The formation quotients for the +2 and −1 products are best known. One of the important aspects of the reactions in Equations (9.7)–(9.10) is that H_3O^+ or H^+ is produced, resulting in a decrease in pH or increased acidity. The magnitude of the pH decrease depends on the Al concentration in the solution.

The form of monomeric Al in the soil solution depends on the pH. One can see the effect of pH on the solubilities of Al in water solutions in Figure 9.8. At pH values below 5.0, Al^{3+} dominates; between pH 5.0 and 5.25, the $Al(OH)^{2+}$ species dominates; between pH of 5.25 and 6.50, the $Al(OH)_2^+$ species dominates; and at pH 6.5 and above, $Al(OH)_4^-$ is the primary species. At a pH of about 6.4, three species, namely, $Al(OH)_2^+$, $Al(OH)_3^0$, and $Al(OH)_4^-$ are present in nearly equal proportions. From pH 4.7 to 7.5, the solubility of Al is low. This is the pH range where Al is precipitated and remains as $Al(OH)_3$ (s). Below pH 4.7 and above 7.5, the concentration of Al in solution increases rapidly.

The structure of the free aqueous Al^{3+} ion is shown in Figure 9.9. It is coordinated by six water molecules in an octahedral coordination, $Al(H_2O)_6^{3+}$. Due to the high positive charge of the Al^{3+} ion, the water molecules form a tightly bound primary hydration shell (Nordstrom and May, 1996).

It should be noted that free Al^{3+} may comprise a small fraction of the total soil solution Al. Much of the Al may be complexed with inorganic species such as F^- and SO_4^{2-} or with organic species such as humic substances and organic acids. For example, Wolt (1981) found that free Al^{3+} comprised 2−61% of the total Al in the soil solutions from acid soils where SO_4^{2-} was a major complexing ligand. David and Driscoll (1984) found that 6%−28% of the total Al in soil

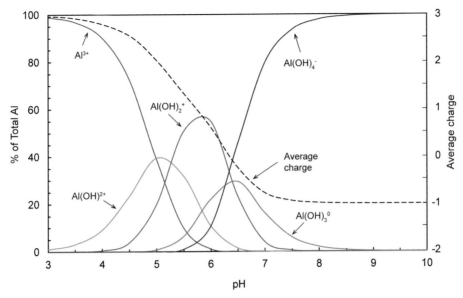

FIGURE 9.8 Relationship between pH and the distribution and average charge of soluble aluminum species. The data for various Al species were obtained using Visual MINTEQ with a total Al concentration of 1×10^{-4} M.

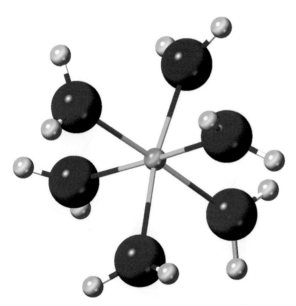

FIGURE 9.9 Structure of the free aqueous aluminum $[Al(H_2O)_6^{3+}]$ ion.

solutions occurred as free Al^{3+}. Most of the soil solution Al was complexed with organic species and with F^-.

9.7.2 Polymeric Al Species

In addition to monomeric Al species, polymeric Al species can also form through hydrolysis reactions in aqueous solutions. The presence of Al polymers in soil solutions has not been proven, and thermodynamic data necessary to calculate stability constants for Al polymeric species are also lacking. One of the reasons it has been difficult to determine the significance of Al polymers in the soil solution is that they are preferentially adsorbed onto clay minerals and organic matter and are usually difficult to exchange.

A number of polymeric Al species have been proposed based on solution experiments in the laboratory. The hydrolysis of Al forming polymeric Al species can be represented as (Bertsch and Parker, 1996):

$$xAl^{3+} + yH_2O \leftrightarrow Al_x(OH)_y^{(3x-y)+} + yH^+ \quad (9.11)$$

where $Al_x(OH)_y^{(3x-y)+}$ represents the polymeric Al species.

Jardine and Zelazny (1996) have noted that the polymeric species are transient, metastable intermediates formed prior to the precipitation

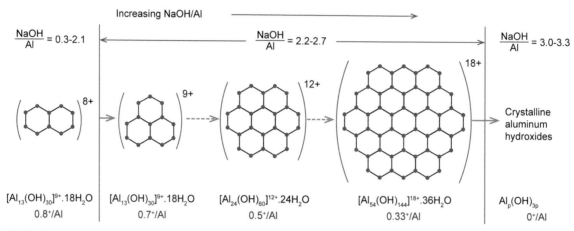

of crystalline $Al(OH)_3$. The nature and distribution of polymeric species depend on the ionic strength, total Al concentration, total OH added, pH, temperature, types of anions present, time, and method of preparation (Smith, 1971; Jardine and Zelazny, 1996).

Hsu (1977) proposed a polymerization scheme (Figure 9.10) that consists of single or double gibbsite-like rings at $\bar{n} \leq 2.1$. With an increase in pH ($\bar{n} = 2.2-2.7$), large polymers that have a reduced net positive charge per Al atom form, with the ionic charge being constant until $\bar{n} = 3$. This positive charge is balanced by counteranions in solution or the negative charge on the clay minerals.

The use of ^{27}Al NMR spectroscopy and of colorimetric methods has verified a number of polymeric species in aqueous solutions. Figure 9.10 shows a series of positively charged Al-OH polymers. Johansson (1960) proposed the Al_{13} polymer or $[Al_{13}O_4(OH)_{24}(H_2O)_{12}]^{7+}$ (Figure 9.11). It has one Al^{3+} at the center, tetrahedrally coordinated to four oxygens and surrounded by 12 Al^{3+}, each coordinated to 6 OH^-, H_2O, or the O^{2-} shared with the Al^{3+} at

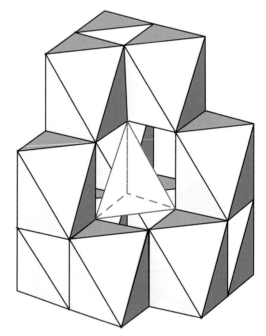

FIGURE 9.11 The $[Al_{13}O_4(OH)_{24}(H_2O)_{12}]^{7+}$ species. The drawing shows how the 12 AlO_6 octahedra are joined together by common edges. The tetrahedra of oxygen atoms in the center of the structure contain one 4-coordinate Al atom. *From Johansson (1960), with permission.*

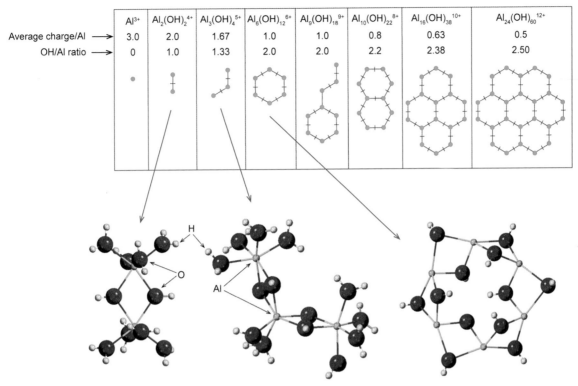

	Al^{3+}	$Al_2(OH)_2^{4+}$	$Al_3(OH)_4^{5+}$	$Al_6(OH)_{12}^{6+}$	$Al_9(OH)_{18}^{9+}$	$Al_{10}(OH)_{22}^{8+}$	$Al_{16}(OH)_{38}^{10+}$	$Al_{24}(OH)_{60}^{12+}$
Average charge/Al $\rightarrow$	3.0	2.0	1.67	1.0	1.0	0.8	0.63	0.5
OH/Al ratio $\rightarrow$	0	1.0	1.33	2.0	2.0	2.2	2.38	2.50

FIGURE 9.12 Summary of Al species with progressive polymerization, which can be deduced from the solid-state structure. OH^- and H_2O are not shown in the polymerization scheme in the top part of the figure for the sake of clarity. Detailed structures of Al–dimer, trimer, and hexamer are presented with OH^- and H_2O. The OH/Al ratio refers to the molar ratio in the complexes and not the OH/Al ratio of the system. *From Stol et al. (1976), with permission.*

the center. Later studies using ^{27}Al NMR (Denney and Hsu, 1986) indicate that the Al_{13} polymer is present under only limited conditions, is transient, and does not represent all of the polymers that are present. Bertsch (1987) has shown that the quantity of Al_{13} polymers depends on the Al concentration, the OH/Al ratio, and the rate of base additions to the Al solutions. It is not known whether Al_{13} polymers form in soils, but the alkalinization of the microenvironment at root apexes could cause their formation (Kinraide, 1991). However, Al_{13} polymers may not be stable in natural systems over long time periods (Bertsch and Parker, 1996).

In general, one can say that as polymerization increases, the number of Al atoms increases, the average charge/Al decreases, and the OH/Al ratio increases (Figure 9.12).

Hsu (1989) has noted that some investigators have found that only polymeric or monomeric species appear in partially neutralized solutions. Whether monomeric or polymeric hydrolysis products result may depend on how the Al solutions were prepared (Hsu, 1989). If the Al solutions were prepared by dissolving an Al salt in water, monomeric species would predominate. Polymeric species would tend to predominate if the solutions were prepared via the addition of base.

9.7.3 Exchangeable and Nonexchangeable Aluminum

Exchangeable Al in soils is primarily associated with the monomeric hexaaqua ion, $[Al(H_2O)_6]^{3+}$. Exchangeable Al^{3+} is bound to the negatively charged surfaces of clay minerals and organic matter in the soil. It is readily displaced with a neutral, unbuffered salt such as 1 M KCl, CaCl$_2$, or BaCl$_2$. Unbuffered KCl is the most commonly used extractant. The extracting solution should be fairly concentrated to remove the Al^{3+} and at a low pH to maintain the Al in a soluble form.

As pH increases, the amount of the monomeric Al^{3+} ion decreases and Al-hydroxy species, both monomeric and polymeric hydroxy species, predominate. These species are complicated and can be grouped collectively into the noncrystalline Al fraction (Bertsch and Bloom, 1996). This fraction consists of Al-hydroxy polymers in the interlayers of smectite and vermiculites, organically complexed Al-hydroxy species, monomeric and polymeric Al-hydroxy species of various sizes and basicities, and metastable forms present as Al oxyhydroxide coatings and as discrete amorphous to poorly structured phases (Bertsch and Bloom, 1996). These species are considered nonexchangeable forms of Al since they are not readily replaced with an unbuffered salt like 1 M KCl. However, it has been shown that KCl can cause the hydrolysis of nonexchangeable Al, which is then measured as exchangeable Al^{3+}. Nonexchangeable Al is titratable with a base such as NaOH and is a major source of variable charge in soils (Thomas and Hargrove, 1984).

Since the polymers are large and positive, they remain in the clay interlayer space and are nonexchangeable. In fact, the polymeric Al species are preferred over monomeric Al species in the interlayers. The polymeric Al species may exist as uniformly distributed islands Figure 9.13a) in the interlayer space (Grim and Johns, 1954) or be concentrated at the edges of the clay mineral surfaces (Chakravarti and Talibudeen, 1961; Dixon and Jackson, 1962) as "atolls" (Frink, 1965), as seen in Figure 9.13b. If the polymers are at the edge, they can block exchange sites in the center of the interlayer space. Moreover, the clay mineral structure will collapse to 1 nm when K-saturated if the polymers are at the clay edges. However, if the polymers occur as islands in the interlayer space, they can serve as props to keep the interlayer space open and enhance the exchange of ions.

9.7.4 Effect of Adsorbed Aluminum on Soil Chemical Properties

As noted earlier, nonexchangeable Al-hydroxy polymers can decrease the CEC of clays and soils (Rich and Obenshain, 1955). The removal of the interlayer Al increases the CEC. The polymers may cause a lower CEC by occupying exchange sites, physically blocking the exchange reaction, and preventing the exchangeable ion from getting in contact with the exchange site.

Since the Al-hydroxy polymers can keep the clay interlayer space open, K$^+$ fixation can be decreased. The fact that the interlayer space is open may also enhance the exchange of ions like K$^+$ or NH$_4^+$ by ions of similar size, for example, NH$_4^+$ or H$_3$O$^+$ exchanging for K$^+$, and K$^+$ and H$_3$O$^+$ exchanging for NH$_4^+$, from "wedge zones" that are partially collapsed regions in the interlayer spaces of weathered micas and vermiculites (Rich, 1964; Rich and Black, 1964). Larger hydrated cations such as Ca^{2+} and Mg^{2+} cannot fit into the wedge zones (Figure 9.14). However, the position of the wedge zone in the interlayer space is important in affecting exchange and selectivity phenomena. If it is at the edge of the interlayer space, the ion selectivity of K$^+$, for example, is less, but if it is deep within the interlayer space, the selectivity is greater (Rich, 1968). Monomeric Al^{3+} may also decrease fixation when KCl solutions are added since Al^{3+} is difficult to displace, particularly at low KCl concentrations.

(A)

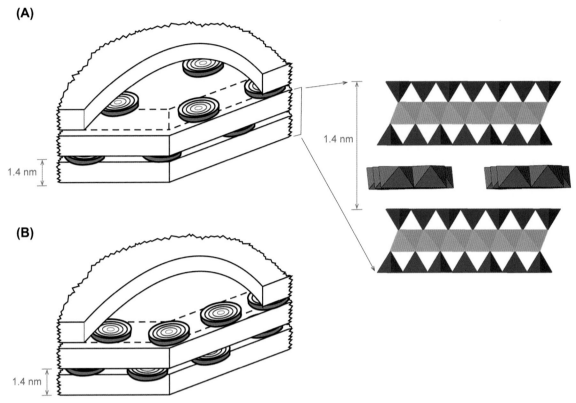

(B)

1.4 nm

1.4 nm

1.4 nm

FIGURE 9.13 Illustration of the distribution of aluminum-hydroxy polymers in the interlayer space of 2:1 clay minerals: (A) a uniform distribution and (B) an "atoll" arrangement. The structure of a 2:1 layer clay mineral with Al-hydroxy polymers in the interlayer is also depicted. *Redrawn with permission from Dixon and Jackson (1962).*

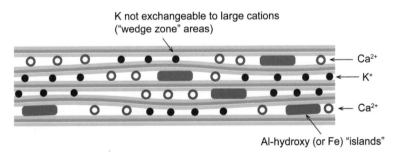

K not exchangeable to large cations ("wedge zone" areas)

Ca²⁺

K⁺

Ca²⁺

Al-hydroxy (or Fe) "islands"

FIGURE 9.14 Effect of Al-hydroxy polymers in an expansible clay mineral on cation fixation. *Redrawn with permission from Rich (1968).*

9.8 Soil Acidity

9.8.1 Forms of Soil Acidity

Key points: Soil acidity exists in three pools, namely, active, exchangeable, and reserve acidity. Active acidity, though easily measured by the pH of the soil solution, forms a small portion of the total acidity. Exchangeable and reserve pools constitute the bulk of soil acidity, and the amount in these pools depends on clay and organic matter contents and the composition of the soil clay fraction.

The main forms of acidity in mineral soils are associated with Al, which can be exchangeable (extractable is a preferred term since the technique, kind, and concentration of extractant and pH can all affect the removal Jardine and Zelazny [1996]); nonexchangeable, or precipitated as an array of solid phases such as bayerite, gibbsite, or nordstrandite. Only in very acidic soils with a pH <4 and in soils high in organic matter does one find major quantities of exchangeable H^+.

Soil acidity is considered to exist in three different pools: active acidity, exchangeable acidity, and residual (nonexchangeable) acidity, as illustrated in Figure 9.15.

The active acidity is all titratable acidity in solution phase, mainly free Al^{3+} ($[Al(H_2O)_6]^{3+}$) and H^+ (H_3O^+) ions. The active acidity is indicated by the solution pH; this is a very small component of the total acidity in soils. Exchangeable acidity is the amount of the total CEC due to H^+ and primarily Al^{3+} that are easily exchangeable by other cations in a simple unbuffered salt solution, such as KCl (as described in Section 9.7.3). As a proportion of the total acidity, its quantity depends on the type of soil (e.g., type and quantity of clay minerals) and the

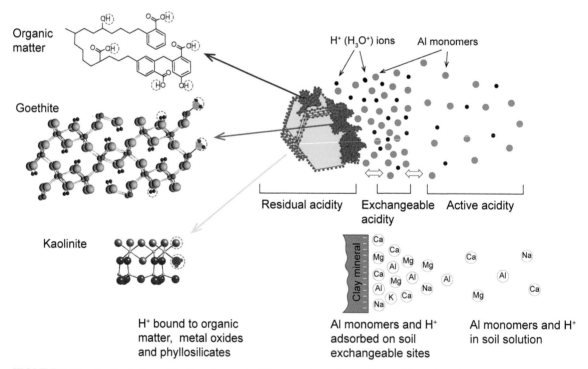

FIGURE 9.15   An illustration of the three forms of acidity, namely, active, exchangeable, and residual (or nonexchangeable) acidity, in soils.

percentage of the CEC composed of exchangeable bases such as Ca^{2+}, Mg^{2+}, K^+, and Na^+, or the percentage base saturation (Thomas and Hargrove, 1984). The ratio of exchangeable acidity to total acidity is greatest for smectites, intermediate for dioctahedral vermiculites, and the least for kaolinite clay minerals. This is related to the higher CEC of smectites, which is predominantly composed of constant charge sites, as opposed to kaolinite, which has a low CEC and is variably charged. With dioctahedral vermiculites, a significant portion of the total acidity is nonexchangeable, which occurs in the form of Al-hydroxy polymeric material in the interlayer space and that is pH dependent. When Al and Fe oxides coat clay surfaces, the ratio of exchangeable acidity to total acidity is decreased since the oxide coatings reduce the quantity of exchangeable acidity that can be extracted (Coleman et al., 1964; Thomas and Hargrove, 1984). Aluminum oxides appear to be more effective in reducing the quantity of exchangeable acidity (Coleman et al., 1964) than Fe oxides.

Coleman et al. (1959) showed that soils high in montmorillonite (a smectite) had greater amounts of exchangeable acidity than soils high in kaolinite due to the higher CEC of the montmorillonitic soils and the predominance of constant charge. Evans and Kamprath (1970) found less exchangeable Al in organic soils than in mineral soils, even when the pH of the organic soil was lower. This can be ascribed to hydroxy-metal–organic complexes involving Al^{3+} and Fe^{3+}. The extraction of Al from these complexes is difficult (Mortensen, 1963; Schnitzer and Skinner, 1963; White and Thomas, 1981). Another reason that organic soils often have lower exchangeable Al content than mineral soils is that due to their lower pH, a large portion of the exchangeable acidity appears to be H^+ but there may be significant amounts of Al that are difficult to exchange from the organic matter. Moreover, much of the H^+ that appears to be exchangeable in organic soils may be due to the hydrolysis of Al on the organic matter resulting in H^+.

Residual acidity consists of all bound Al and H in soil minerals and organic matter. Residual acidity can be neutralized by limestone or other alkaline materials or a buffered salt solution to raise the soil pH to a specified pH (generally 7.0); however, it cannot be readily replaced by an unbuffered salt solution. Total acidity is the sum of active, exchangeable, and residual acidity in soils. Though less common than the three forms of acidity described earlier, in some acid sulfate soils another acidity pool exists, called potential acidity, which results from the oxidation of sulfur compounds.

9.8.2 Titration Analyses

Key points: Potentiometric and conductometric titrations are useful tools for characterizing H^+ and Al^{3+} acidity associated with minerals and organic matter in soils.

With potentiometric titration analysis, one investigates the relationship between pH and the amount of base added (Figure 9.16). The first inflection point or break in the titration curve is ascribed to the end of the neutralization of

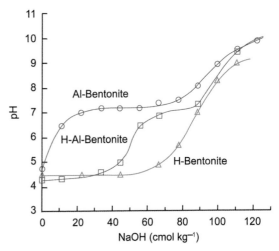

FIGURE 9.16 Potentiometric titration curve of H– and Al–bentonite (a smectitic clay). *From Coleman and Harward (1953), with permission.*

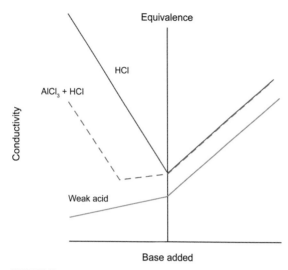

FIGURE 9.17   Conductimetric titration curves of a strong (HCl) and a weak (AlCl₃) acid and of a mixture of HCl and AlCl₃.

exchangeable H^+ and the beginning of the titration of exchangeable Al^{3+}.

Another important tool for differentiating between H^+ and Al^{3+} is the use of conductometric titration curves. Their use in differentiating exchangeable Al^{3+} from nonexchangeable Al is limited. Figure 9.17 shows a conductometric titration curve for various acids where conductivity is plotted versus base added. With HCl, one notes a sharp drop in conductivity as base is added and then a sharp rise in conductivity after the acidity has been neutralized or after the equivalence point is reached. When a strong acid like HCl is titrated with a strong base (e.g., NaOH), the conductivity of the solution at first decreases, due to the replacement of H^+ ions that have a high mobility ($36.30 \times 10^{-8} \ m^2 \ s^{-1} \ V^{-1}$) by slower-moving cations, for example, Na^+ (the mobility is $5.19 \times 10^{-8} \ m^2 \ s^{-1} \ V^{-1}$). At the equivalence point, the decrease in conductance stops, and as more base is added, the conductance increases since the OH^- ions, which have a high mobility ($20.50 \times 10^{-8} \ m^2 \ s^{-1} \ V^{-1}$), are no longer

neutralized but remain free to carry electricity (Moore, 1972). One sees in Figure 9.17 that with HCl, the negative slope is sharp before the equivalence point is reached, indicating H^+, whereas, with a combination of AlCl₃ and HCl, there is a negative slope due to H^+ and a slightly positive slope due to exchangeable Al^{3+}. With a weak acid, the negative slope, diagnostic for exchangeable H^+, is missing, but there is a slightly positive slope due to exchangeable Al^{3+}.

Thus titration curves can be used to measure both the type and quantity of acidity. Soils and clays titrate as weak acids. The titration of SOM is a good example of weak acid behavior. While carboxylic acids, which are the main source of acidity in SOM, give acid dissociation constants (K_a) of 10^{-4} to 10^{-5}, titration curves of SOM give values of 10^{-6}. Martin and Reeve (1958) showed this was due to exchangeable Al^{3+} and Fe^{3+} on the exchange sites. The removal of the Fe^{3+} and Al^{3+} from the organic matter resulted in a K_a of 10^{-4}. One can see the effect of Al on the titration of SOM in Figure 9.18.

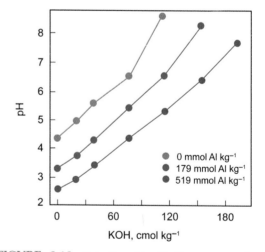

FIGURE 9.18   Potentiometric titration curves of soil organic matter of varying Al content (from a muck soil, e.g., wetland soil). *From Hargrove and Thomas (1982), with permission.*

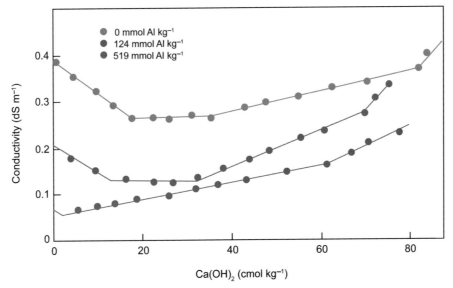

FIGURE 9.19   Conductimetric titration curves of Michigan muck soil (e.g., wetland soil) of various Al contents. *From Hargrove and Thomas (1982), with permission.*

With muck soil (e.g., wetland soil), Hargrove and Thomas (1982) used $Ca(OH)_2$ to differentiate between exchangeable H^+, Al^{3+}, and nonexchangeable Al (Figure 9.19). The negative slope is strong acidity due to exchangeable H^+, the horizontal slope is weak acidity due to exchangeable Al^{3+}, and the third segment, which has a small positive slope, is due to very weak acidity. As Al was added, the amount of acidity decreased, while the amount of very weak acidity increased or did not change. The total amount of acidity titrated decreased as the Al content increased. Thus it appears that the Al that complexed with SOM is not exchangeable and apparently not titratable at pH values below 8 (Thomas and Hargrove, 1984). Another practical ramification of these results is that since the Al^{3+} blocks sites and limits the dissociation of functional groups, the SOM in acid soils containing large quantities of SOM complexed with Al^{3+} may not contribute much to the soil's CEC, except at low pH where Al is more labile.

9.9 Amelioration of Soil Acidity

Key points: Lime is commonly applied to ameliorate surface acid soils. Both laboratory and field methods have been used to determine the amount of lime required to obtain the desired soil pH. The amelioration of subsoil acidity is achieved by the application of gypsum to the surface soil.

9.9.1 Lime Requirement

Sidebar 9.9 — Soil pH and Lime Requirement

McLean (1982) considered soil pH as the intensity factor (index) and the lime requirement as the capacity factor of soil acidity. Soil pH reflects the amount of acidity in soil solution, whereas the lime requirement represents the fraction of total soil acidity (soluble and exchangeable) that must be neutralized to achieve a desired soil pH.

The amelioration of an acid soil is done by the application of lime (i.e., liming) to raise the soil pH that is optimum for a given crop on a given soil type. The amount of lime required to increase the initial soil pH to the desired soil pH is called the lime requirement (LR). Soil pH and LR are regarded as the intensity and capacity factors, respectively, of soil acidity (Sidebar 9.9).

LR of a soil depends on a number of complex and interrelated factors, such as the nature and extent of soil acidity, soil properties (e.g., clay content, organic matter content, clay mineralogy), land use, and fertilizer application regime. Field experiments are the most accurate way to determine the LR of a soil; however, such experiments are time consuming and costly. Thus various rapid laboratory procedures have been proposed to determine the LR of soils. The most common laboratory-based methods for estimating LR are soil-lime incubations, soil-buffer equilibrations followed by the titration of the leachate or estimating acidity by pH change, soil-base titrations, and exchangeable acidity or Al (Sims, 1996; van Lierop, 1990). In addition to these, LR is estimated based on soil pH and some readily measured soil properties, such as clay and organic carbon content, which are well correlated with the soil buffer capacity.

Among the laboratory-based methods, soil-buffer equilibrations (e.g., Adams and Evans, Shoemaker, McLean, and Pratt (SMP), Mehlich, Woodruff) are most widely used for LR estimations (van Lierop, 1990). A soil sample is equilibrated with a buffered chemical solution for a short time (15–30 min), and then the pH of the soil-buffer mixture is measured. The buffer solution is a mixture of two or more buffers and resists significant changes in pH (mimicking a soil system) when equilibrated with an acid soil. The decrease in the pH of a soil-buffer mixture as compared to the original buffer pH provides an estimate of the amount of soil acidity, which must be neutralized by the application of lime to increase from the current pH to the desired pH. LR is calculated from calibration equations, some of which are

derived based on field trials (van Lierop, 1990; Sims, 1996).

Some researchers are critical of the approach of LR estimation using laboratory-based methods and, in particular, the choice of the desired pH based on the soil acidity concepts of exchangeable acidity or base saturation that are not appropriate, especially in relation to variable charge soils (Edmeades and Ridley, 2003). Biological lime requirement (BLR), which is the amount of lime required to eliminate restrictions to plant growth, has been suggested to be a better approach to optimizing soil pH. The LR using the BLR approach has been successfully used to determine the optimum soil pH for pastures in New Zealand and cereals in Australia, and in both cases the laboratory-based methods underestimated the LR (Edmeades et al., 1984; Slattery and Coventry, 1993). Although the BLR approach is sound for the estimation of LR, it requires site- and soil-specific field trials that are time consuming and expensive and may not be feasible for routine recommendations.

The application of lime is most commonly used to overcome the impact of soil acidification on field crops. However, an integrated approach, combining liming with management practices to minimize soil acidification and the use of tolerant species of crop species and genotypes, should be practiced, particularly in highly weathered naturally acid soils.

9.9.2 Lime — Characteristics and Chemical Reactions

Calcium carbonate (calcite, $CaCO_3$) and calcium–magnesium carbonate (dolomite, $CaMg(CO_3)_2$) derived from natural sources (e.g., sedimentary rocks) and industrial by-products are used for liming acid soils. Hydrated lime $(Ca(OH)_2)$ and burnt or quick lime (CaO) have greater solubility, but they are more expensive than calcite; however, they are preferred when a quick response for high-value crops is needed to ameliorate soil acidity. The acid-neutralizing efficiency of liming materials on a weight basis is expressed as the calcium carbonate equivalent (Sidebar 9.10). The application rate of a liming material must be adjusted based on its neutralizing value in comparison to pure lime. Additionally, the application rate should be adjusted based on the presence of any impurities in the liming material. The effectiveness of lime in the field depends on several factors, such as the fineness of the liming material, uniform application and mixing of lime in the soil, and the presence of soil moisture. Because of the poor efficiency of lime under field conditions, the amount of lime applied in the field is often doubled.

Sidebar 9.10 — Neutralizing Value of Liming Materials

The neutralizing value (NV) represents the effectiveness of liming compounds as compared to pure calcium carbonate (molecular weight ~ 100 g mol^{-1}) that has an NV of 100. For example, CaO has a molecular weight of ~ 56 g mol^{-1} and contains 1 mol of Ca as $CaCO_3$, so the NV of CaO in % of $CaCO_3$ is equal to $100 \times (100/56) = 179$. Similarly, one can compute the NV of any other liming material.

Rounded values for molecular masses have been used here, as these values are for field application rates and absolute accuracy is not needed. Highly accurate NV values for any liming material can also be calculated based on this concept.

The general reaction that explains the interaction of a liming material such as $CaCO_3$ with water to form OH^- ions is (Thomas and Hargrove, 1984):

$$CaCO_3 + H_2O \rightarrow Ca^{2+} + HCO_3^- + OH^- \tag{9.12}$$

The OH^- reacts with indigenous H^+ or H^+ formed from the hydrolysis of Al^{3+}. The overall reaction of lime with an acid soil can be expressed as:

$$2Al\text{-soil} + 3CaCO_3 + 3H_2O \rightarrow \\ 3Ca\text{-soil} + 2Al(OH)_3 + 3CO_2 \tag{9.13}$$

One sees that the products are exchangeable Ca^{2+} and $Al(OH)_3$. Assuming that all of the acidity was completely neutralized, one would find that the soil pH would be close to 8.3 and the soil would be completely saturated. However, usually soils are limed to bring the soil pH at a level so that the toxicity of Al is minimized and any restrictions to a particular crop are eliminated, as discussed earlier.

9.9.3 Amelioration of Subsoil Acidity

The amelioration of acid subsoil is particularly difficult because lime application in the surface layer is not effective in overcoming subsoil acidity (Tang et al., 2013). The application of lime at high rates on the soil surface can be partially effective, but this can create deficiency of some nutrients in the surface soil. The direct application of lime into the acidic subsoil layer is costly and may not be feasible due to the requirement of suitable machinery and high energy costs (Sumner et al., 1986).

The application of gypsum, which has a much greater solubility than lime, has been found to be an economically viable option for ameliorating subsoil acidity (Tang et al., 2013). Gypsum has a small neutralizing value, and the effectiveness of gypsum has been attributed to several different processes. Some of the processes are (i) the displacement of OH^- by SO_4^{2-} from oxide surfaces and the displaced OH^- ions reacting with free Al^{3+}; (ii) the formation of less phytotoxic $AlSO_4^+$ complexes or precipitation of minerals such as jurbanite, basaluminite, or alunite in acid media containing Al^{3+} and SO_4^{2-} ions; (iii) decreased Al^{3+} activity in soil solution from increased ionic strength from gypsum addition; and (iv) exchange of Al from soil colloids and its subsequent leaching. It is possible that more than one process may operate in soil environments after gypsum application.

Suggested Reading

Adams, F. (Ed.), 1984. The Chemistry of Soil Acidity, sec-ondnd ed. Agronomy Monograph 12. American Society of Agronomy, Madison, WI.

Rengel, Z. (Ed.), 2003. Handbook of Soil Acidity. Marcel Dek-ker, New York.

Robson, A.D. (Ed.), 1998. Soil Acidity and Plant Growth. Academic Press, Sydney.

Sposito, G. (Ed.), 1996. The Environmental Chemistry of Aluminum, 2nd ed. CRC Press (Lewis), Boca Raton, FL.

Ulrich, B., Summer, M.E. (Eds.), 1991. Soil Acidity. Springer-Verlag, Berlin.

Problem Set

Q. 1. The table below provides laboratory data for some acid soils (0–10 cm) from Australia and Nigeria. Calculate the base saturation percentage for each of the soils.

Soil type, country	pH	Exchangeable cations (cmol$_c$ kg^{-1})						
		Ca	Mg	Na	K	Al	Mn	H
Australia	(1:5 CaCl$_2$)							
Solodic Planosol	4.73	3.37	0.75	0.03	0.90	0.17	0.63	–
Calcic Luvisol	4.70	2.74	0.84	0.04	0.92	0.17	0.51	–
Nigeria	(1:1 H$_2$O)							
Alfisol	4.50	1.90	0.40	–	0.20	0.70	0.10	0.30
Alfisol	4.80	3.60	0.70	–	0.30	0.40	0.10	0.20

From Slattery and Coventry (1993) and Juo et al. (1996), with permission.

Q. 2. Using the above data, calculate the lime requirement (kg ha^{-1}) to neutralize the exchangeable acidity of the soils. Assume the bulk density of the surface soil (0–10 cm) to be 1.3 g cm^{-3} or 1.3 Mg m^{-3}.

Q. 3. An agricultural field that had been used for growing corn received ammonium sulfate fertilizer application at a rate of 150 kg N ha^{-1} annually in the surface (0–15 cm) soil. The field is situated in a high-rainfall region and 30% of the applied fertilizer was lost through nitrate leaching from the surface soil. Calculate the acidification rate (kmol ha^{-1} year^{-1}) of the surface soil that would result from the nitrate leaching, and also determine the amount of lime needed annually to neutralize the acidity.

Q. 4. Aluminum in aqueous solution reversibly hydrolyzes and the proportion of Al monomeric species depends on solution pH. Al hydrolysis reactions can be written as:

$$Al^{3+} + H_2O \leftrightarrow Al(OH)^{2+} + H^+$$
$$p*K_1 = 4.95$$

$$Al^{3+} + 2H_2O \leftrightarrow Al(OH)_2 + 2H^+$$
$$p*\beta_2 = 10.01$$

$$Al^{3+} + 3H_2O \leftrightarrow Al(OH)_3^0 + 3H^+$$
$$p*\beta_3 = 16.8$$

$$Al^{3+} + 4H_2O \leftrightarrow Al(OH)_4^+ + 4H^+$$
$$p*\beta_4 = 22.87$$

Calculate the percentage of total Al in solution that remains as free Al^{3+} if the H$^+$ concentration is 0.00005 M (assume activities are equal to concentrations).

Q. 5. The dissolution constant $\left(K_{dis}^\circ\right)$ for diaspore is given by the following reaction:
AlOOH (*diaspore*) + 3H$^+$ → Al^{3+} + 2H$_2$O;
log K_{dis}° = 7.92. What is the activity of Al^{3+} in a pH 4.15 solution in equilibrium with AlOOH$_{(s)}$?

Q. 6. To determine the removal of excess inorganic cations over anions in a lucerne hay pasture, 2 g dried herbage was weighed into a 50 mL porcelain crucible, heated slowly, and held at 500°C for 4 h in a muffle furnace. The ashed sample was dissolved in 20 mL of 1 M HCl and 5 mL of the aliquot was then titrated against a 0.25 M NaOH that gave the endpoint value of 15.5 mL. A blank titration of 5 mL of HCl was also done that consumed 19.9 mL of 0.25 M NaOH. Calculate the lime requirement to maintain the soil pH if the 4 t ha^{-1} year^{-1} of lucerne hay is removed from the field.

Q. 7. To determine the lime requirement using the two-point titration method, 10 g (<2 mm) of oven-dry acid soil was weighed into each of two polythene tubes. Then, 25 mL distilled water was added to Tube 1 and 25 mL of 0.01 M Ca(OH)$_2$ was added to Tube 2. The tubes were equilibrated on a rotary shaker for 2 h and allowed to settle for 15 min, and then the pH was measured. The pH of soil suspension in Tube 1 (or soil pH [1:2.5 H$_2$O]) was 4.5 and the pH of soil suspension in Tube 2 (soil after equilibration of lime) was 5.3. Based on this data, calculate the amount of lime (kg ha^{-1}) required to raise the pH of the surface soil (0—15 cm) to 6.0. The bulk density of the surface soil (0—15 cm) is 1.2 g cm^{-3} or 1.2 Mg m^{-3}.

Q. 8. The figure below shows the pH trend over 150 years in soil from three long-term experiment sites at Rothamsted in the United Kingdom. Discuss various buffering mechanisms that might operate at each of these sites. Plot A represents soil that is under Broadbalk Wilderness, Plot B is soil from the Park Grass that has received no fertilizer or manure application, and Plot C is soil from the Park Grass where ammonium sulfate fertilizer has been applied regularly.

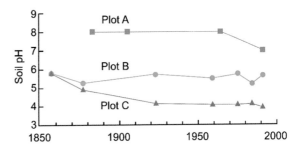

From Goulding (2016), with permission.

References

Baes, C.F., Jr., Mesmer, R.E., 1976. The Hydrolysis of Cations. John Wiley & Sons, New York.

Bertsch, P.M., 1987. Conditions for Al$_{13}$ polymer formation in partially neutralized Al solutions. Soil Sci. Soc. Am. J. 51, 825—828.

Bertsch, P.M., 1989. Aqueous polynuclear aluminum species. In: Sposito, G. (Ed.), The Environmental Chemistry of Aluminum. CRC Press, Boca Raton, FL, pp. 87—115.

Bertsch, P.M., Bloom, P.R., 1996. Aluminum. In: Sparks, D.L. (Ed.), Methods of Soil Analysis, Part 3 — Chemical Methods. Soil Science Society of America Book Series 5. SSSA, Madison, WI, pp. 517—550.

Bertsch, P.M., Parker, D.R., 1996. Aqueous polynuclear aluminum species. In: Sposito, G. (Ed.), The Environmental Chemistry of Aluminum. Second ed. CRC Press, Boca Raton, FL, pp. 117—168.

Bolan, N.S., Hedley, M.J., 2003. Role of carbon, nitrogen and sulfur cycles in soil acidification. In: Rengel, Z. (Ed.), Handbook of Soil Acidity. Marcel Dekker, New York, NY, pp. 29—52.

Bradfield, R., 1923. The nature of the acidity of the colloidal clay of acid soils. J. Am. Chem. Soc. 45, 2669—2678.

Bradfield, R., 1925. The chemical nature of colloidal clay. J. Am. Soc. Agron. 17, 253—370.

Brady, N.C., 1984. The Nature and Properties of Soils, Nineth edition. Macmillan, New York, NY.

Chakravarti, S.N., Talibudeen, O., 1961. Phosphate interaction with clay minerals. Soil Sci. 92 (4), 232—242.

Chernov, V.A., 1947. The Nature of Soil Acidity (English Translation Furnished by Hans Jenny—Translator Unknown). Press of Academy of Sciences, Moscow. Soil Sci. Soc. Am. (1964).

Coleman, N.T., Craig, D., 1961. The spontaneous alteration of hydrogen clay. Soil Sci. 91, 14—18.

Coleman, N.T., Harward, M.E., 1953. The heats of neutralization of acid clays and cation exchange resins. J. Am. Chem. Soc. 75, 6045–6046.

Coleman, N.T., Kamprath, E.J., Weed, S.B., 1959. Liming. Adv. Agron. 10, 475–522.

Coleman, N.T., Thomas, G.W., LeRoux, F.H., Bredell, G., 1964. Salt-exchangeable and titratable acidity in bentonite-sesquioxide mixtures. Soil Sci. Soc. Am. Proc. 28, 35–37.

David, M.B., Driscoll, C.T., 1984. Aluminum speciation and equilibria in soil solutions of a Haplorthod in the Adirondack Mountains (New York, U.S.A.). Geoderma 33, 297–318.

Denney, D.Z., Hsu, P.H., 1986. 27Al nuclear magnetic resonance and ferron kinetic studies of partially neutralized AlCl3 solutions. Clays Clay Miner. 34, 604–607.

Dixon, J.B., Jackson, M.L., 1962. Properties of intergradient chlorite-expansible layer silicates of soils. Soil Sci. Soc. Am. Proc. 26, 358–362.

Dolling, P.J., Porter, W.M., 1994. Acidification rates in the central wheatbelt of Western Australia. I. On a deep yellow sand. Aust. J. Exp. Agric. 34, 1155–1164.

Edmeades, D.C., Pringle, R.M., Shannon, P.W., Mansel, G.P., 1984. Effects of lime on pasture production on soils in the North Island of New Zealand. 4. Predicting lime responses. NZ. J. Agric. Res. 27, 371–382.

Edmeades, D.C., Ridley, A.M., 2003. Using lime to ameliorate topsoil and subsoil acidity. In: Rengel, Z. (Ed.), Handbook of Soil Acidity. Marcel Dekker Inc., New York, NY, pp. 297–336.

Evans, C.E., Kamprath, E.J., 1970. Lime response as related to percent Al saturation, solution Al and organic matter response. Soil Sci. Soc. Am. Proc. 34, 893–896.

Fenton, G., Helyar, K.R., 2000. Soil acidification. In: Charman, P.E.V., Murphy, B.W. (Eds.), Soils: Their Properties and Management. Oxford University Press, Melbourne, pp. 223–237.

Foy, C.D., 1984. Physiological effects of hydrogen, aluminum, and manganese toxicities in acid soil. In: Adams, F. (Ed.), Soil Acidity and Liming. Agronomy 12. Second edition. American Society of Agronomy, Madison, WI, pp. 57–97.

Frink, C.R., 1965. Characterization of aluminum interlayers in soil clays. Soil Sci. Soc. Am. Proc. 29, 379–382.

Glossary of Soil Science Terms, 2008. Soil Science Glossary Terms Committee and Soil Science Society of America. Soil Sci. Soc. Am., Madison, WI.

Goulding, K.W.T., 2016. Soil acidification and the importance of liming agricultural soils with particular reference to the United Kingdom. Soil Use Manag. 32, 390–399.

Grim, R.E., Johns, W.D., 1954. Clay mineral investigation of sediments in the northern Gulf of Mexico. Clays Clay Miner. 2, 81–103.

Hargrove, W.L., Thomas, G.W., 1982. Titration properties of Al-organic matter. Soil Sci. 134, 216–225.

Hopkins, C.G., Knox, W.H., Pettit, J.H., 1903. A quantitative method for determining the acidity of soils. USDA Bureau Chem. Bull. 73, 114–119.

Hsu, P.H., 1977. Aluminum hydroxides and oxyhydroxides. In: Dixon, J.B., Weed, S.B. (Eds.), Minerals in Soil Environments. Soil Science Society of America, Madison, WI, pp. 99–143.

Hsu, P.H., 1989. Aluminum oxides and oxyhydroxides. In: Dixon, J.B., Weed, S.B. (Eds.), Minerals in Soil Environments. SSSA Book Ser. 1. Soil Science Society of America, Madison, WI, pp. 331–378.

Hsu, P.H., Bates, T.E., 1964. Formation of X-ray amorphous and crystalline aluminum hydroxides. Mineral. Mag. 33, 749–768.

Jardine, P.M., Zelazny, L.W., 1996. Surface reactions of aqueous aluminum species. In: Sposito, G. (Ed.), The Environmental Chemistry of Aluminum, Second edition. CRC Press (Lewis Publishers), Boca Raton, FL, pp. 221–270.

Johansson, G., 1960. On the crystal structure of some basic aluminum salts. Acta. Chem. Scand. 14, 771–773.

Johnston, A.E., Goulding, K.W.T., Poulton, P.R., 1986. Soil acidification during more than 100 years under permanent grassland and woodland at Rothamsted. Soil Use Manag. 2, 3–10.

Juo, A.S.R, Franzluebbers, K., Dabiri, A., Ikhile, B., 1996. Soil properties and crop performance on a kaolinitic Alfisol after 15 years of fallow and continuous cultivation. Plant Soil. 180, 209–217.

Kelley, W.P., Brown, S.M., 1926. Ion exchange in relation to soil acidity. Soil Sci. 21, 289–302.

Kinraide, T.B., 1991. Identity of the rhizotoxic aluminum species. Plant Soil. 134, 167–178.

Lindsay, W.L., 1980. Chemical Equilibria in Soils. John Wiley & Sons, New York, NY.

Low, P.F., 1955. The role of aluminum in the titration of bentonite. Soil Sci. Soc. Am. Proc. 19, 135–139.

Martin, A.E., Reeve, R., 1958. Chemical studies of podzolic illuvial horizons: III. Titration curves of organic matter suspensions. J. Soil Sci. 9, 89–100.

McFee, W.W., Byrnes, W.R., Stockton, J.G., 1981. Characteristics of coal mine overburden important to plant growth. J. Environ. Qual. 10, 300–308.

McLean, E.O., 1982. Soil pH and lime requirement. In: Page, A.L., Miller, R.H., Keeney, D.R. (Eds.), Methods of Soil Analysis. Part 2 - Chemical and Microbiological Properties. Agronomy No. 9, Second edition. ASA and SSSA Publ., Madison, WI, pp. 199–223.

Mesarzos, E., 1992. Occurrence of atmospheric acidity. In: Radojevic, M., Harrison, R.M. (Eds.), Atmospheric Acidity: Sources, Consequences and Abatement. Elsevier Applied Science, London, pp. 1–37.

Moore, W.J., 1972. Physical Chemistry, Fourth edition. Prentice-Hall, Englewood Cliffs, NJ.

Mortensen, J.L., 1963. Complexing of metals by soil organic matter. Soil Sci. Soc. Am. Proc. 27, 179–186.

Nordstrom, D.K., May, H.M., 1996. Aqueous equilibrium data for mononuclear aluminum species. In: Sposito, G. (Ed.), The Environmental Chemistry of Aluminum, Second edition. CRC Press (Lewis Publishers), Boca Raton, FL, pp. 39–80.

Nyatsanga, T., Pierre, W.H., 1973. Effect of nitrogen fixation by legumes on soil acidity. Agron. J. 65, 936–940.

Page, H.J., 1926. The nature of soil acidity. Int. Soc. Soil Sci. Trans. Comm. 2A, 232–244.

Parker, D.R., Kinraide, T.B., Zelazny, L.W., 1988. Aluminum speciation and phytotoxicity in dilute hydroxy-aluminum solutions. Soil Sci. Soc. Am. J 52, 438–444.

Parker, D.R., Kinraide, T.B., Zelazny, L.W., 1989. On the phytotoxicity of polynuclear hydroxy-aluminum complexes. Soil Sci. Soc. Am. J 53, 789–796.

Paver, H., Marshall, C.E., 1934. The role of aluminum in the reactions of the clays. Chem. Ind. (London) 53, 750–760.

Reuss, J.O., Walthall, P.M., 1989. Soil reaction and acidic deposition. In: Norton, S.A., Lindberg, S.E., Page, A.L. (Eds.), Acidic Precipitation, Vol. 4. Springer-Verlag, Berlin, pp. 1–33.

Rich, C.I., 1964. Effect of cation size and pH on potassium exchange in Nason soil. Soil Sci. 98, 100–106.

Rich, C.I., 1968. Mineralogy of soil potassium. In: Kilmer, V.J., Younts, S.E., Brady, N.C. (Eds.), The Role of Potassium in Agriculture. American Society of Agronomy, Madison, WI, pp. 79–96.

Rich, C.I., Black, W.R., 1964. Potassium exchange as affected by cation size, pH, and mineral structure. Soil Sci. 97, 384–390.

Rich, C.I., Obenshain, S.S., 1955. Chemical and clay mineral properties of a red-yellow podzolic soil derived from muscovite schist. Soil Sci. Soc. Am. Proc. 19, 334–339.

Ritsema, C.J., van Meusvoot, M.E.F., Dent, D.L., Tau, Y., vanden Bosch, H., van Wijk, A.L.M., 2000. Acid sulfate soils. In: Sumner, M.E. (Ed.), Handbook of Soil Science. CRC Press, Boca Raton, FL, pp. G121–G154.

Robarge, W.P., 2008. Acidity. In: Chesworth, W (Ed.), Encyclopedia of Soil Science. Encyclopedia of Earth Sciences Series. Springer, Dordrecht, pp. 10–20.

Robarge, W.P., Johnson, D.W., 1992. The effects of acidic deposition on forested soils. Adv. Agron. 47, 1–83.

Schnitzer, M., Skinner, S.J.M., 1963. Organo-metallic interactions in soils: 1. Reactions between a number of metal ions and the organic matter of a podzol Bh horizon. Soil Sci. 96, 86–93.

Schwertmann, U., Susser, P., Natscher, L., 1987. Protonenpuffersubstanzen in Boden. Z. Pflanzernähr. Düng. Bodenkunde 150, 174–178.

Sims, J.T., 1996. Lime requirement. In: Sparks, D.L. (Ed.), Methods of Soil Analysis. Part 2: Chemical Properties. Third edition. ASA, SSSA, CSSA, Madison, WI, pp. 491–515.

Singh, B., Odeh, I.O.A., McBratney, A.B., 2003. Acid buffering capacity and potential acidification of cotton soils in northern New South Wales. Aust. J. Soil Res. 41, 875–888.

Slattery, W.J., Conyers, M.K., Aitken, R.L., 1999. Soil pH, aluminium, manganese and lime requirement. In: Peverill, K.I., Sparrow, L.A., Reuter, D.J. (Eds.), Soil Analysis: An Interpretation Manual. CSIRO Publishing, Melbourne, pp. 103–128.

Slattery, W.J., Coventry, D.R., 1993. Response of wheat, triticale, barley and canola to lime on four soil types in the north-eastern Victoria. Aust. J. Exp. Agric. 33, 609–618.

Slessarev, E., Lin, Y., Bingham, N.J., Johnson, E., Dai, Y., Schimel, J.P., et al., 2016. Water balance creates a threshold in soil pH at the global scale. Nature. 540, 567–569.

Smith, R.W., 1971. Relations among equilibrium and nonequilibrium aqueous species of aluminum hydroxy complexes. Adv. Chem. Ser. 106, 250–279.

Stol, R.J., van Helden, A.K., de Bruyn, P.L., 1976. Hydrolysis precipitation studies of aluminum (III) solutions. 2. A kinetic study and model. J. Colloid Interf. Sci. 57, 115–131.

Sullivan, L.A., Bush, R.T., Burton, E.D., Ritsema, C.J., van Mensvoort, M.E.F., 2012. Acid sulfate soils. In: Huang, P.M., Li, Y.C., Sumner, M.E. (Eds.), Handbook of Soil Science, Volume II: Resource Management and Environmental Impacts. Second ed. Taylor and Francis, Boca Raton, FL, pp. 21-21–21-26.

Sumner, M.E., 1991. Soil acidity control under the impact of industrial society. In: Bolt, G.H., Boodt, M.F.D., Hayes, M.H.B., McBride, M.B. (Eds.), Interactions at the Soil Colloid-Soil Solution Interface. NATO ASI, Ser. E, 190. Kluwer Academic Publ., Dordrecht, The Netherlands, pp. 517–541.

Sumner, M.E., Fey, M.V., Noble, A.D., 1991. Nutrient status and toxicity problems in acid soils. In: Ulrich, B., Sumner, M.E. (Eds.), Soil Acidity. Springer-Verlag, Berlin, pp. 149–182.

Sumner, M.E., Noble, A.D., 2003. Soil acidification: the world story. In: Rengel, Z. (Ed.), Handbook of Soil Acidity. Marcel Dekker, New York, NY, pp. 1–28.

Sumner, M.E., Shahandesh, H., Bonton, J., Hammel, J., 1986. Amelioration of acid soil profile through deep liming and surface application of gypsum. Soil Sci. Soc. Am. J. 50, 1254–1258.

Tang, C., Rengel, Z., 2003. Role of plant cation/anion uptake ratio in soil acidification. In: Rengel, Z. (Ed.), Handbook of Soil Acidity. Marcel Dekker, New York, NY, pp. 57–81.

Tang, C., Weligama, C., Sale, P., 2013. Subsurface soil acidification in farming systems: its possible causes and management options. In: Xu, J., Sparks, D.L. (Eds.),

Molecular Environmental Soil Science. Springer, Dordrecht, pp. 389–412.

Thomas, G.W., 1977. Historical developments in soil chemistry: ion exchange. Soil Sci. Soc. Am. J. 41, 230–237.

Thomas, G.W., Hargrove, W.L., 1984. The chemistry of soil acidity. In: Adams, F. (Ed.), Soil Acidity and Liming. Agronomy Monograph 12. Second edition. American Society of Agronomy, Madison, WI, pp. 3–56.

Ulrich, B., 1991. An ecosystem approach to soil acidification. In: Ulrich, B., Summer, M.E. (Eds.), Soil Acidity. Springer-Verlag, Berlin, pp. 28–79.

van Lierop, W., 1990. Testing soils for pH and lime requirement. In: Westerman, R.L. (Ed.), Soil Testing and Plant Analysis, Third edition. SSSA, Madison, WI, pp. 73–126.

Veitch, F.P., 1902. The estimation of soil acidity and the lime requirement of soils. J. Am. Chem. Soc. 24, 1120–1128.

Veitch, F.P., 1904. Comparison of methods for the estimation of soil acidity. J. Am. Chem. Soc. 26, 637–662.

von Uexküll, H.R., Mutert, E., 1995. Global extent, development and economic impact of acid soils. Plant Soil. 171, 1–15.

White, R.E., Thomas, G.W., 1981. Hydrolysis of aluminum on weakly acid organic exchangers. Implications for phosphorus adsorption. Fert. Inst. 2, 159–167.

Wolt, J.D., 1981. Sulfate retention by acid sulfate-polluted soils in the Cooper Basin area of Tennessee. Soil Sci. Soc. Am. J 45, 283–287.

The Chemistry of Saline and Sodic Soils

10.1 Introduction

Key points: Globally, approximately 58% of cultivated lands are salt affected, that is, saline or sodic. Salt-affected soils mostly occur in arid and semiarid regions where the dominance of evaporation and transpiration over precipitation causes the accumulation of salts in the root zone. Irrigated cropped lands contribute one-third of the food supply and about 20% of the irrigated lands are salt affected. Saline soils are plagued by high levels of soluble salts, whereas sodic soils have high levels of exchangeable sodium.

Oceans contain about 97.5% of the earth's water, continents about 2.5% water, that is, fresh water, and the atmosphere about 0.001% (Shiklomanov, 2000). About 68.7% of the fresh water occurs in ice caps and glaciers, about 29.9% is fresh groundwater, and 0.26% of the fresh water occurs as surface water that is present in lakes, reservoirs, and rivers. The land area of the earth is 13.0 billion ha; of this area, 1.6 billion ha (12%) is currently used for growing agricultural crops, 4.6 billion ha (35%) is under grasslands and woodland ecosystems, and 3.7 billion ha (28%) is under forest (FAO, 2021). Of the cultivated lands, approximately 0.35 billion ha (22%) are saline, that is, containing excessive amounts of soluble salts, and another 0.58 billion ha (36%) are sodic, that is, containing excessive levels of exchangeable Na^+. About 17% of the world's

cropland is under irrigation and irrigated agriculture contributes about one-third of the world's food supply. Secondary salinization, which is anthropogenic and occurs on irrigated lands, is present in approximately 20% of the irrigated land area and is a major concern for global food production. Soil salinization is one of the major land-degrading threats that influence soil productivity, soil stability, and soil biodiversity. Figure 10.1 and Table 10.1 show the global distribution of salt-affected soils.

Saline soils have issues related to the excess accumulation of soluble salts in the root zone, whereas high levels of sodium relative to other exchangeable cations are the main attribute of sodic soils (Tanji and Wallender, 2011). High salt concentrations have direct effects on plants, whereas sodicity causes the deterioration of soil physical properties (e.g., crusting, reduced infiltration, increased soil strength, and reduced aeration), which can indirectly affect plants.

Salt-affected soils can be classified as saline, sodic, and saline–sodic soils. Briefly, saline soils are plagued by high levels of soluble salts, sodic soils have high levels of exchangeable sodium, and saline–sodic soils have high contents of both soluble salts and exchangeable sodium. These soils will be described more completely later.

Salt-affected soils occur most often in arid and semiarid climates, but they can also be found in areas where the climate and mobility of salts

Environmental Soil Chemistry, Third Edition
https://doi.org/10.1016/B978-0-12-815880-7.00010-9

FIGURE 10.1   Global distribution of salt-affected soils. *From Szabolcs (1989), with permission.*

TABLE 10.1   Global distribution of salt-affected soils.

Continent	Saline	Sodic	Total
	Area in millions of ha		
North America	6.2	9.6	15.8
Central America	2.0	–	2.0
South America	69.4	59.6	129.0
Africa	53.5	27.0	80.5
South Asia	83.3	1.8	85.1
North and Central Asia	91.6	120.1	211.7
Southeast Asia	20.0	–	20.0
Australasia	17.4	340.0	357.4
Europe	7.8	22.9	30.7
Total	351.2	581.0	932.2

From Szabolcs (1989), with permission.

cause saline waters and soils for short periods of time (Tanji and Wallender, 2011). However, for the most part, in humid regions salt-affected soils are not a problem because rainfall is sufficient to leach excess salts out of the soil, into groundwater, and eventually into the ocean. Some salt-affected soils may occur along seacoasts or river delta regions where seawater has inundated the soil (Richards, 1954).

10.2 Salinity and Its Causes

Key points: Inadequate rainfall and excessive evapotranspiration in arid and semiarid regions, poor drainage, application of poor-quality water for irrigation, and anthropogenic activities are the main causes of the accumulation of soluble salts in the soil.

10.2.1 Salinity Constituents

The major cations in saline soils and waters are Na^+, Ca^{2+}, Mg^{2+}, and K^+, and the primary anions of concern are Cl^-, SO_4^{2-}, HCO_3^-, CO_3^{2-}, and NO_3^-. In hypersaline waters or brines ions of B, Sr, Li, SiO_2, Rb, F, Mo, Mn, Ba, and Al (since the pH is high, Al would be in the $Al(OH)_4^-$ form) may also be present (Tanji and Wallender, 2011). Bicarbonate ions result from the reaction of carbon dioxide in water. The source of carbon dioxide is either the atmosphere or respiration from plant roots or soil microorganisms. Carbonate ions are normally found only at pH ≥ 9.5. Boron results from the weathering of boron-containing minerals, such as tourmaline (Richards, 1954).

When soluble salts accumulate, Na^+ often becomes the dominant counterion on the soil exchanger phase, causing the soil to become dispersed. This results in a number of physical problems, such as poor drainage. The predominance of Na^+ on the exchanger phase may occur due to Ca^{2+} and Mg^{2+} precipitating as $CaSO_4$, $CaCO_3$, and $CaMg(CO_3)_2$. Sodium then replaces exchangeable Ca^{2+} and Mg^{2+} on the exchanger phase.

10.2.2 Causes of Soil Salinization

(a) Lack of Water

In arid and semiarid climates there is not enough water to leach soluble salts from the soil. The lack of moisture limits the weathering intensity as well as the movement of weathering products, that is, salts. Consequently, the soluble salts accumulate, resulting in salt-affected soils. In subhumid and humid regions there is adequate rainfall to leach excess salts out of the soil, into groundwater, or surface runoff. Salinity is the concentration of dissolved mineral salts present in waters and soils on a unit volume or weight basis (Tanji and Wallender, 2011).

(b) Excessive Evapotranspiration

An additional factor in causing salt-affected soils is the high potential evapotranspiration in these areas, which increases the concentration of salts in both soils and surface waters. It has been estimated that evaporation losses can range from 50% to 90% in arid regions, resulting in 2- to 20-fold increases in soluble salts (Cope, 1958; Yaalon, 1963).

(c) Poor Drainage

Poor drainage can also cause salinity and may be due to a high water table or low soil permeability caused by the sodicity (high sodium content) of water. Soil permeability is "the ease with which gases, liquids or plant roots penetrate or pass through a bulk mass of soil or a layer of soil" (Glossary of Soil Science Terms, 2008). As a result of the poor drainage, salt lakes can form, like those in the western United States. Irrigation of nonsaline soils with saline water can also cause salinity problems. These soils may be level, well drained, and located near a stream. However, after they are irrigated with saline water, drainage may become poor and the water table may rise.

(d) Irrigation Water Quality

An important factor affecting soil salinity is the quality of irrigation water. If the irrigation water contains high levels of soluble salts, Na, B, and trace elements, serious effects on plants and animals can result (Ayers and Westcot, 1985).

Salinity problems are common in irrigated lands, with approximately one-third of the irrigated land in the United States seriously salt affected (Rhoades, 1993). In some countries it may be as high as 50% (Postel, 1989). Areas affected include humid climate areas, such as the Netherlands, Sweden, Hungary, and Russia, and arid and semiarid regions such as the southwestern United States, Australia, India, Pakistan, and the Middle East. About 40,000 ha of irrigated land each year are no longer productive because of salinity (Yaron, 1981).

One of the major problems in these irrigated areas is that the irrigation waters contain dissolved salts, and when the soils are irrigated, the salts accumulate unless they are leached out. Saline irrigation water, low soil permeability, inadequate drainage, low rainfall, and poor irrigation management, all cause salts to accumulate in soils, which deleteriously affects crop growth and yields. The salts must be leached out for crop production. However, it is the leaching out of these salts, resulting in saline drainage waters, that causes pollution of waters, a major concern in saline environments.

The presence of selenium and other toxic elements (Cr, Hg) in subsurface drainage waters is also a problem in irrigated areas. Selenium (resulting from shale parent material) in drainage waters has caused massive death and deformity to fish and waterfowl in the Kesterson Reservoir of California.

(e) Human Activities

Human activities, such as irrigation with salt-rich waters, road deicing, use of wastewaters for irrigation and disposal on soils, use of fertilizers and other inputs, particularly in soils with low permeability and limited leaching, poor irrigation practices, and improper drainage, can cause the salinization of productive soils.

In coastal areas salinization may be associated with the overexploitation of groundwater caused by the demands of growing urbanization, industry, and agriculture. Overextraction of groundwater aquifers can lower the normal water table, causing seawater intrusion.

Land use change by clearing deep-rooted perennial native vegetation for agriculture has caused widespread salinization in Australia. The shallow-rooted annual crops and pastures are not able to utilize the excess runoff, which seeps past the root zone to enter the groundwater system. Groundwater recharge raises the groundwater table, and when the water table moves closer to the soil surface (e.g., less than 2 m below the surface), groundwater can reach the soil surface via capillary action, and evaporation concentrates salts at the surface.

10.3 Sources of Soluble Salts

Key points: Soluble salts in soils are chiefly derived from the weathering of primary minerals, mobilization of residual fossil salts, atmospheric deposition, saline irrigation, and drainage waters.

The major sources of soluble salts in soils are the weathering of primary minerals and native rocks, residual fossil salts, atmospheric deposition, saline irrigation and drainage waters, saline groundwater, seawater intrusion, additions of inorganic and organic fertilizers, sludges and sewage effluents, brines from natural salt deposits, and brines from oil and gas fields and mining (Suarez and Jurinak, 2011; Tanji and Wallender, 2011).

Because most primary minerals in rocks, formed under high temperature and pressure, are thermodynamically unstable under present atmospheric conditions, minerals weather via hydrolysis, hydration, oxidation, and carbonation reactions, and soluble salts are released. The primary source of soluble salts is fossil salts derived from prior salt deposits or from entrapped solutions found in earlier marine sediments (sedimentary rocks). Salts from these deposits get mobilized by irrigation and rainfall and are the major sources of salt loading in the western US surface waters.

Salts from the atmospheric deposition, both as dry and wet deposition, can range from 100 to 200 kg year^{-1} ha^{-1} along seacoasts and from 10 to 20 kg year^{-1} ha^{-1} in interior areas of low rainfall. The composition of the salt varies with distance from the source. At the coast, it is primarily NaCl. The salts become higher in Ca^{2+} and Mg^{2+} farther inland (Bresler et al., 1982).

10.4 Types of Soil Salinity

Key points: Primary salinity occurs by natural processes, such as the weathering of primary minerals, mobilization of salts from sedimentary deposits, and atmospheric deposition of salts. Secondary salinization is caused by human activities, primarily by increased groundwater recharge after the removal of deep-rooted vegetation in dryland areas and increasing groundwater recharge via the irrigation of agricultural crops under poor drainage conditions in conjunction with marginal quality irrigation waters.

Based on the mechanism of salt accumulation in the soil, soil salinization can be classified as "primary" and "secondary" salinity. Primary salinity is the accumulation of salts by natural processes and is a result of pedogenesis, for example, salt lakes in central Australia. Primary salinity mostly occurs in low rainfall areas.

Secondary salinity is caused by human activities, such as changes to the groundwater balance induced by human activities, particularly the removal of deep-rooted vegetation for agriculture and thereby increasing groundwater recharge. Secondary salinity can also result from restricted leaching associated with poor drainage and input of poor-quality irrigation water and by the addition of chemicals to soils via fertilizers and wastewaters applications.

Secondary salinity can occur in both dryland and irrigated areas. Dryland salinity results from salt build-up in the root zone of plants in nonirrigated areas to the extent that it affects plant growth. In Australia approximately 5.7 million ha of agricultural and pastoral lands are affected by dryland salinity. This has been caused by replacing native deep-rooted perennial vegetation with shallower-rooted crops and pastures, which consume less water, causing increased recharge and rise in the water table, bringing salts to the root zone and the soil surface. In deeply weathered wheat-growing regions of Western Australia a typical 1 ha of land contains between 170 and 950 t of stored salt down to 40 m.

Secondary salinity in irrigated areas is caused by a rise in groundwater tables from excessive irrigation under poor drainage conditions and/or the use of marginal to poor-quality water for irrigation. Additionally, secondary salinity is also caused by seawater intrusion, that is, coastal aquifer systems where seawater replaces groundwater that has been overexploited, and the addition of large loads of salts in effluents derived from intensive agriculture and industrial wastewater.

10.5 Important Salinity and Sodicity Parameters

Key points: Electrical conductivity (EC) is the main parameter used for determining soil salinity. Most threshold values are based on the EC of saturation extracts of soils expressed in dS m^{-1}. Exchangeable sodium percentage (ESP) and sodium adsorption ratio (SAR) are the most frequently used parameters for soil sodicity. Irrigation water quality is also generally assessed using the parameters used for salt-affected soils.

The parameters used to characterize salt-affected soils depend primarily on the concentrations of salts in the soil solution and the amount of exchangeable Na^+ on the soil. Exchangeable Na^+ is determined by exchanging the Na^+ from the soil with another cation such as Ca^{2+} and then measuring the Na^+ in solution by using flame photometry or spectrometry (e.g., atomic absorption or emission spectrometry or inductively coupled plasma emission spectrometry). The concentration of salts in the solution phase can be characterized by several indices (Table 10.2) and can be measured by evaporation or using electroconductometric or spectrometric techniques.

TABLE 10.2   Salinity parameters.

Salinity index	Units of measurement
Total dissolved solids (TDS) or total soluble salt (TSS) concentration	mg L^{-1}
Total concentration of soluble cations (TSC)	$mmol_c$ L^{-1}
Total concentration of soluble anions (TSA)	$mmol_c$ L^{-1}
Electrical conductivity (EC)	dS m^{-1} (high-salinity soils) dS m^{-1} $\times$ 10^3 = μS cm^{-1} (low-salinity soils)

10.5.1 Parameters for Measuring Salinity

(a) Total Dissolved Solids (TDS)

Total dissolved solids (TDS) can be measured by evaporating a known volume of water from the solid material to dryness and weighing the residue. However, this measurement is variable since, in a particular sample, various salts exist in varying hydration states, depending on the amount of drying. Thus if different conditions are employed, different values for TDS will result (Bresler et al., 1982).

The TDS (in mg L^{-1}) can also be estimated by measuring an extremely important salinity index, electrical conductivity (EC_e, measured at 25°C), which is discussed below, to determine the effects of salts on plant growth. The TDS in soil solution extracts of lesser-saline soils (with EC_e between 0.1 and 5.0 dS m^{-1}) may be estimated by using the following relationship (Corwin and Yemoto, 2020):

$$\text{TDS (mg } L^{-1}) = EC_e \text{ (dS } m^{-1}) \times 640$$

(10.1)

For hypersaline soil samples (EC > 5.0 dS m^{-1}), a different conversion factor is used, as given below:

$$\text{TDS (mg } L^{-1}) = EC_e \text{ (dS } m^{-1}) \times 800$$

(10.2)

These empirical relationships for relating EC to TDS in soil solution extracts have been derived from large data sets. The total soluble cations (TSC) concentration or the total soluble anions (TSA) concentration in soil solution can also be estimated from EC_e (dS m^{-1}) across a range of mixed salt concentrations commonly found in soils using the following relationship:

$$\text{TSC or TSA } (mmol_c \text{ } L^{-1}) = EC_e \text{ (dS } m^{-1}) \times 10$$

(10.3)

(b) Electrical Conductivity (EC)

The preferred index to assess soil salinity is electrical conductivity (EC). The EC is based on the concept that the electrical current carried by a salt solution under standard conditions increases as the salt concentration of the solution increases. Electrical conductivity measurements (described in Box 10.1) are reliable, inexpensive to do, and quick. Thus EC is routinely measured in many soil-testing laboratories.

The SI unit of electrical conductivity is Siemens per meter (S m^{-1}); however, the most commonly used unit is deci-Siemens per meter (dS m^{-1}), which is equivalent to an old and discontinued unit, namely, millimho per centimeter (mmho cm^{-1}). Some other EC units in equivalent terms to dS m^{-1} are as follows:

$$1 \text{ dS } m^{-1} = 1 \text{ mS } cm^{-1} = 100 \text{ mS } m^{-1}$$
$$= 1000 \text{ } \mu S \text{ } cm^{-1}$$

BOX 10.1

Electrical conductivity (EC) — Principle and measurement.

Electrical conductivity or specific conductance is a measure of the ability of a substance (e.g., a salt solution or a soil solution) to conduct electric current or to transport an electric charge.

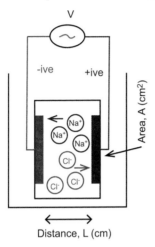

FIGURE B10.1   Principle of the electrical conductivity measurement.

The EC of a solution is measured by applying an alternating electrical current (I) to two electrodes that are immersed in the solution. In response to the applied current, ions are drawn toward the opposing charge end, which generates a very small amount of electric potential difference (V) that is measured. This is called a potentiometric method.

According to Ohm's law, the conductance G is the ratio of current I to voltage V:
G = I/V; The basic unit of conductance is Siemens (S), formerly called mho. Conductance is the reciprocal of resistance (1/R) with the basic unit Ohm.

Specific conductivity (C) is simply the product of measured conductivity (G) and the electrode cell constant (L/A):

$$C = G \times (L/A)$$

where L is the length of the column of liquid (cm) between the electrodes and A is the area (cm^2) of the electrodes Figure B10.1. Since cell geometry affects conductivity values, standardized EC values are expressed in the SI unit (S cm^{-1}) to compensate for variations in electrode dimensions. If the cell constant (K) is 1 cm^{-1}, the specific conductivity is the same as the measured conductivity of the solution.

Temperature also affects conductivity because, at higher temperatures, ions move faster, increasing the conductivity. Since the conductivity of a solution varies with the temperature, the EC values are commonly expressed at 25°C.

Modern EC meters have built-in functions for the cell constant and temperature calibrations; the instrument simply requires a calibration using a standard solution (e.g., 0.01 M KCl).

Electrical conductivity values are expressed according to the method employed: EC$_p$, the EC of the soil paste; EC$_e$, the EC of the extract of a saturated paste of a soil sample; EC$_w$, the EC of a soil solution or water sample; and EC$_a$, the EC of the bulk field soil (Corwin and Yemoto, 2020).

Soil salinity is typically defined based on the EC values of the extract of a saturated soil paste (EC$_e$) because it is not practical to extract soil water under field moisture conditions (Corwin and Yemoto, 2020). The saturation extract method (Sidebar 10.1) accounts for soil texture and soil leaching behavior, and the EC$_e$ has been used as the standard measure for determining the salt tolerance threshold of crops (Bresler et al., 1982). Therefore EC_e has become the standard measure of soil salinity. However, EC_e measurements are tedious and require a certain skill level; soil solution extracts are obtained using higher soil/water ratios, such as 1:1, 1:2, and 1:5 soil/water extracts. Empirical relationships have been derived between EC_e and EC measured using different soil/solution ratios, particularly between EC_e and $EC_{1:5}$. A simple conversion factor based on the soil texture of 344 soil samples in Australia was proposed by Slavich and Petterson (1993). The authors derived a factor of 5.8 for heavy clay, 7.5 for medium clay, 8.6 for textures from clay loam to light-medium clay, 9.5 for loam to sandy clay loam, 13.8 for sandy loam up to sandy clay loam, and 22.7 for sand to clayey sand soils. Soil texture, solubility and composition of salts, and some other factors, including ion exchange, affect the conversion factor to obtain EC_e values from soil/water extracts (Shaw 1994). The software, ExtractChem, allows accurate conversion from one extract ratio to another (Suarez and Taber, 2007). However, adequate chemical data, such as the composition of the major cations and anions and the presence/absence of gypsum, are required for this.

Sidebar 10.1 — How to Obtain Saturation Water Extract?

The electrical conductivity of the extract of a saturated paste of a soil sample (EC_e) is a very common way to measure soil salinity. In this method a saturated soil paste is prepared by adding distilled water to a 200 to 400 g sample of air-dry soil and stirring. The mixture should then stand for several hours so that the water and soil react and the readily soluble salts dissolve. This is necessary so that a uniformly saturated and equilibrated soil paste results. The soil paste should shine as it reflects light, flow some when

the beaker is tipped, slide easily off a spatula, and easily consolidate when the container is tapped after a trench is formed in the paste with the spatula. The extract of the saturation paste can be obtained by suction using a Büchner funnel and a filter paper. The EC of the extract can be measured using a conductivity meter.

It is often desirable to estimate EC based on the soil solution data. Marion and Babcock (1976) developed a relationship between EC_w (dS m^{-1}) and total soluble salt concentration (TSS in mmol$_c$ L^{-1}) and ionic concentration (C in mmol$_c$ L^{-1}), where C is corrected for ion pairs. If there is no ion complexation, TSS = C (Suarez and Jurinak, 2011). The equations of Marion and Babcock (1976) are as follows:

$$\log C = 0.955 + 1.039 \log EC_w \qquad (10.4)$$

$$\log TSS = 0.990 + 1.055 \log EC_w \qquad (10.5)$$

These relationships work well up to 15 dS m^{-1}, which covers the range of EC_e and EC_w for slightly to moderately saline soils (Bresler et al., 1982).

Griffin and Jurinak (1973) also developed an empirical relationship between EC_w and ionic strength (I) at 298 K that corrects for ion pairs and complexes:

$$I \text{ (mM)} = 0.0127 EC_w \qquad (10.6)$$

where EC_w is in dS m^{-1} at 298 K. Figure 10.2 shows the straight-line relationship between I and EC_w predicted by Equation (10.6), as compared to actual values for river waters and soil extracts.

In addition to measuring EC and other salinity indices in the laboratory, it is often important in the management of salt-affected soils, particularly those that are irrigated, to measure, monitor, and map soil salinity of large soil areas (Rhoades, 1999). This would assist in ascertaining the degree of salinity, in determining areas of under- and overirrigation, and in

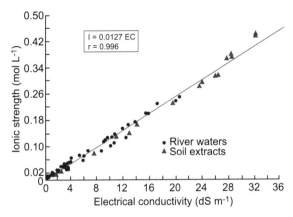

$I = 0.0127\ EC$
$r = 0.996$

• River waters
▲ Soil extracts

FIGURE 10.2   Relationship between the ionic strength and electrical conductivity of natural aqueous solutions. •, River waters; ▲, saturation paste extracts of soils. *From Griffin and Jurinak (1973), with permission.*

predicting trends in salinity. There are a number of rapid instrumental techniques for determining EC and computer-based mapping techniques that allow one to measure soil salinity over large areas. The use of geographic information systems (GISs) and remote sensing techniques also augment these techniques. Box 10.1 describes the principles of EC measurement.

There are three types of soil conductivity sensors that can measure bulk soil electrical conductivity, referred to as apparent soil electrical conductivity (EC_a). Three techniques that measure EC_a are electrical resistivity (ER), electromagnetic induction (EMI), and time domain reflectometry (TDR). Each of these techniques has certain drawbacks and requires some data manipulations and corrections to relate the EC_a to EC_e. These are comprehensively discussed in Rhoades (1999) and Corwin and Yemoto (2020).

10.5.2 Parameters for Measuring the Sodic Hazard

There are several important parameters commonly used to assess the status of Na^+ in the solution and on the exchanger phases. These are the sodium adsorption ratio (SAR), the exchangeable sodium ratio (ESR), and the exchangeable sodium percentage (ESP).

The SAR is measured using the equation:

$$SAR = \frac{[Na^+]}{\sqrt{[Ca^{2+} + Mg^{2+}]}} \qquad (10.7)$$

where the brackets indicate the total concentration of the ions expressed in mmol L^{-1} in the solution phase.

Total concentrations, not activities, are used in Equation (10.7), and thus the SAR expression does not consider decreases in free ion concentrations and activities due to ion pair or complex formation (Sposito and Mattigod, 1977), which can be significant with Ca^{2+} and Mg^{2+}.

One also notes that in Equation (10.7) Ca^{2+} and Mg^{2+} are treated as if they were the same species. There is not a theoretical basis for this other than the observation that ion valence is more important in predicting ion exchange phenomena than ion size. The concentration of Ca^{2+} is much higher than that of Mg^{2+} in many waters (Bresler et al., 1982).

Gapon's exchange reaction has been widely used in salinity and sodicity studies (Suarez and Jurinak, 2011). The original expression for the Gapon equation for Ca to Na exchange is:

$$Ca_{1/2}\text{-X} + Na^- \leftrightarrow Na\text{-X} + 1/2\,Ca^{2+} \qquad (10.8)$$

where cation concentration on exchange (-X) sites are in $mmol_c\ kg^{-1}$ and cation concentrations in solution are in mmol L^{-1}.

$$K_G = \frac{[Na\text{-X}]\sqrt{(Ca^{2+})}}{[Ca_{1/2}\text{-X}](Na^+)} \qquad (10.9)$$

The above equation can be rearranged as

$$\frac{[Na\text{-X}]}{[Ca_{1/2}\text{-X}]} = K_G \cdot \frac{(Na^+)}{\sqrt{(Ca^{2+})}} \qquad (10.10)$$

The U.S. Salinity Laboratory (Richards, 1954) treated a similar behavior for Ca^{2+} and Mg^{2+} in the adsorbed phase and modified the Gapon equation as follows:

$$\frac{[Na\text{-}X]}{\left[Ca_{1/2}\text{-}X + Mg_{1/2}-X\right]} = K'_G \cdot \frac{(Na^+)}{\sqrt{(Ca^{2+} + Mg^{2+})}}$$

$$= K'_G \cdot SAR$$

$$(10.11)$$

where K'_G is the modified Gapon selectivity coefficient (mmol $L^{-1})^{-1/2}$.

Equation (10.11) can be simplified since Na^+, Ca^{2+}, and Mg^{2+} are the most common exchangeable cations in arid soils (Suarez and Jurinak, 2011) and are expressed as:

$$\frac{[Na\text{-}X]}{CEC - [Na\text{-}X]} = K'_G \cdot SAR \qquad (10.12)$$

$$ESR = K'_G \cdot SAR \qquad (10.13)$$

where ESR is the exchangeable Na ratio (Richards, 1954). The U.S. Salinity Lab reported a linear regression equation between ESR and SAR as ESR $= -0.0126 + 0.01475$ SAR with a correlation coefficient of 0.923 for 59 soils from the western United States (Figure 10.3).

Based on the data presented in Figure 10.3, the relationship between the ESP and the SAR is given by the following equation:

$$ESP = \frac{100(-0.0126 + 0.01475 \times SAR)}{1 + (-0.0126 + 0.01475 \times SAR)}$$

$$(10.14)$$

Bower and Hatcher (1964) improved the relationship by adding ranges in the saturation extract salt concentration. The value of K'_G can be determined from the slope of the ESR–SAR linear relationship (Richards, 1954). The K'_G describes Na–Ca exchange well over the range of 0% –40% ESP, where ESP = [Na-X] × 100/CEC and has an average value of 0.015 (mmol $L^{-1})^{-1/2}$ for many irrigated soils from the western United States (Richards, 1954). More recent

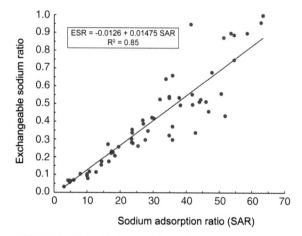

FIGURE 10.3 Exchangeable sodium ratio (ESR) as related to the sodium adsorption ratio (SAR) of the saturation extract. ESR = ES/(CEC − EES), where ES = exchangeable sodium and CEC = cation exchange capacity. *From Richards (1954), with permission.*

studies, however, indicate that various factors, including pH, salt concentration, clay mineralogy, and Na loading, affect the ESR–SAR relationship and the value of K'_G ranges from 0.008 to 0.016 (mmol $L^{-1})^{-1/2}$. It is suggested that K'_G be determined using site-specific data.

Soils with an ESP >30 are very impermeable, which seriously affects plant growth. For many soils, the numerical values of the ESP of the soil and the SAR of the soil solution are approximately equal up to ESP levels of 25–30.

While the ESP is used as a criterion for the classification of sodic soils, the accuracy of the number is often a problem due to errors that may arise in the measurement of CEC and exchangeable Na^+. Therefore the more easily obtained SAR of the saturation extract should be used to diagnose the sodic hazard of soils. However, the quantity and type of clay minerals present in the soil should be considered when assessing how SAR and ESP values affect soil sodicity. For example, a higher SAR value may be of less concern if the soil has a low clay content or contains low quantities of smectite.

Water quality for irrigation purposes is assessed based on the total content of soluble salts (salinity hazard), the relative proportion of Na^+ to Ca^{2+} and Mg^{2+} ions — SAR (sodium hazard), and the concentrations of elements that may cause toxicity or ionic imbalance in plants. Residual sodium carbonates, RSCs (Sidebar 10.2), have also been suggested for the assessment of irrigation water quality.

10.5.3 Parameters for Irrigation Water

Sidebar 10.2 — Residual Sodium Carbonates (RSCs)

The RSC is calculated using the following equation:

$$RSC = (HCO_3^- + CO_3^{2-}) - (Ca^{2+} + Mg^{2+})$$

where the concentrations of all ions in solution are in $mmol_c\ L^{-1}$.

In the water samples containing an excess of $HCO_3^- + CO_3^{2-}$ there is a tendency to precipitate carbonates of Ca and Mg as the soil solution becomes more concentrated; subsequently, Na^+ concentration is increased in the soil solution. This may result in the deterioration of soil physical conditions, such as soil dispersion and reduced infiltration, and soil pH >8.5.

Irrigation water has been classified as (Lloyd and Heathcote, 1985):

Class	RSC
Suitable	<1.25
Marginal	1.25–2.50
Unsuitable	>2.50

The EC_w values for many waters used in irrigation in the western United States are in the range of 0.15–1.50 dS m^{-1}. Soil solutions and drainage waters normally have higher EC_w values (Richards, 1954). General guidelines for salinity hazards based on the EC of irrigation water are given in Table 10.3 (Richards, 1954; Bauder et al., 2011).

Although the salinity of the irrigation water primarily affects plant growth, excess amounts of sodium relative to Ca and Mg in the irrigation water can have serious consequences on crop yield through their effects on soil structure. High Na concentration in the irrigation water causes swelling and dispersion of clay particles, which plugs pore space, causing surface crusting, reduced infiltration rate (IR), and increased runoff. Sodicity in the irrigation water is usually measured by using SAR and irrigation water recommendations in relation to SAR are given in the specific ion section below.

Although SAR is the best index to predict ESP in the context of irrigation water, total cation concentration or EC of the irrigation water is often considered along with the SAR for the classification of Na hazard of irrigation water (Richards, 1954). This is because at the same SAR, the swelling potential of low-salinity (EC_w) irrigation water is greater (i.e., lower IR) than high-EC_w water (Table 10.4) due to the increase in the thickness of the diffuse double layer at low EC_w. Therefore a better evaluation of the water infiltration hazard requires using the EC_w together with the SAR, and the general guidelines are given in Table 10.4.

TABLE 10.3 Salinity hazard based on the electrical conductivity (EC_w) of irrigation water.

EC_w (dS m^{-1})	Salinity hazard and management
<0.75	*Very low hazard*; no detrimental effects on plants, and no salt accumulation in soil is expected.
0.75–1.5	*Some*; may be detrimental to sensitive crops, and good drainage and leaching needed.
1.5–3.0	*Moderate*; may have adverse effects on many crops and leaching required at higher range.
>3.0	*Severe*; generally not acceptable for irrigation, and sensitive plants may have germination problem. Water can be used for salt-tolerant plants on well-drained soils; frequent leaching, and careful management practices needed.

TABLE 10.4   Irrigation water guidelines for the assessment of sodium hazard based on SAR and EC_w.

SAR (mmol L^{-1})$^{-1/2}$	EC_w (dS m^{-1})		
	Potential for water infiltration problem		
	Severe	Slight to moderate	None
0–3	<0.2	0.2–0.7	>0.7
3–6	<0.3	0.3–1.2	>1.2
6–12	<0.5	0.5–1.9	>1.9
12–20	<1.3	1.3–2.9	>2.9
20–40	<2.9	2.9–5.0	>5.0

Modified from Ayers and Westcot (1985), with permission.

TABLE 10.5   Classification of irrigation water in relation to Na, Cl, and B toxicity to plants.

Specific ion or parameter	Degree of restriction on use		
	None	Slight to moderate	Severe
Na – SAR	<3	3–9	>9
Cl (mmol L^{-1})	<4	4–10	>10
B (mg L^{-1})	<0.7	0.7–3.0	>3.0

Adapted from Ayers and Westcot (1985), with permission.

In addition to salinity and sodium hazards, the presence of specific ions (e.g., Na^+, Cl^-, and B) above a threshold concentration in the irrigation water or soil solution may cause direct toxicity to some crop plants. Although the specific role of Na has been established in limiting plant growth, it provides a significant stimulation in the growth of many crop plants at concentrations below the salt-tolerance threshold (Grieve et al., 2012). At concentrations greater than the threshold, Na^+ has both direct and indirect detrimental effects on plants. Indirectly, high Na^+ concentration causes nutritional imbalance and affects soil physical conditions, as discussed before. The direct toxicity of Na^+ generally occurs in woody plant species and there is a wide variation in the tolerance of excessive amounts of Na^+ among species and rootstocks. The toxicity of Na^+ at concentrations in soil water at 5 mmol L^{-1} has been found in avocado, citrus, and stone fruits (Grieve et al., 2012).

Chlorine is an essential micronutrient; however, it can cause toxicity to sensitive crops at high concentrations (Table 10.5). Similar to Na^+, Cl^- is mostly not toxic to nonwoody plants, but woody plants are susceptible to Cl^- toxicity, which varies among varieties and rootstocks within species (Grieve et al., 2012). Boron is also an essential micronutrient that has a narrow optimal concentration range for most crops. Unlike Cl, plant toxicity from B occurs at very low concentrations (Table 10.5). Boron toxicity is more prevalent in arid environments and often occurs in areas with saline soils and waters (Grieve et al., 2012). Similar to salt tolerance, B tolerance fluctuates with climate, soil conditions, and plant variety.

10.6 Classification and Reclamation of Salt-Affected Soils

Key points: Saline soils have an $EC_e > 4$ dS m^{-1} and ESP < 15%, sodic soils have an ESP > 15 (not universal) and $EC_e < 4$ dS m^{-1}, and saline–sodic soils have an $EC_e > 4$ dS m^{-1} and ESP > 15%. The amelioration of saline soils is achieved by leaching soluble salts out of the root zone using fresh water; the leaching requirement depends on the salinity threshold of the crop. Sodic soils are ameliorated by adding a soluble source of Ca, and Ca^{2+} then replaces excess Na^+ from exchange sites. Gypsum is most commonly used to ameliorate sodic soils.

10.6.1 Saline Soils

Saline soils have traditionally been classified as those in which the EC_e of the saturation extract is >4 dS m^{-1} and ESP $< 15\%$. Some scientists have recommended that the EC_e limit for saline soils be lowered to 2 dS m^{-1} as many crops, particularly fruits and ornamentals, can be harmed by salinity in the range of 2–4 dS m^{-1}. Saline soils can be recognized in the field by the presence of white crusts of salts on the soil surface. Saline soils, also called "white alkali" soils correspond to the "Solonchaks" described by the Russian soil scientists. Solonchak is a Reference Soil Group of the World Reference Base for Soil Resources.

Sidebar 10.3 — Solubility (mmol$_c$ L^{-1}) of Salts Commonly Found in Soils and Natural Waters

Low solubility	
Calcium carbonate, $CaCO_3$	0.5
Magnesium carbonate, $MgCO_3$	2.5
Calcium bicarbonate, $Ca(HCO_3)_2$	3–12
Magnesium bicarbonate, $Mg(HCO_3)_2$	15–20
Calcium sulfate, $CaSO_4 \cdot 2H_2O$	30
High solubility	
Sodium sulfate, $NaSO_4 \cdot 10H_2O$	683
Sodium bicarbonate, $NaHCO_3$	1642
Magnesium sulfate, $MgSO_4 \cdot 7H_2O$	5760
Sodium chloride, $NaCl$	6108
Magnesium chloride, $MgCl_2 \cdot 6H_2O$	14,955
Calcium chloride, $CaCl_2 \cdot 6H_2O$	25,470

From Doneen (1975) with permission.

The major problem with saline soils is the presence of soluble salts that control the osmotic potential of the soil solution. Sodium, Ca and Mg are major cations, and the proportion of Na is mostly less than half of the total solute cations in the soil solution. The primary anions in saline soils are Cl^-, SO_4^{2-}, and sometimes NO_3^-. Small amounts of HCO_3^- ions may be present, but soluble carbonates are almost invariably absent. Salts of low solubility (Sidebar 10.3), such as $CaSO_4 \cdot 2H_2O$ (gypsum) and Ca and Mg carbonates, may also be present. Because exchangeable Na^+ is not a problem, saline soils are usually flocculated, and water permeability is good (Richards, 1954).

Saline soils can be reclaimed by leaching them with good-quality (low-electrolyte-concentration) water. The water causes the dissolution of the salts and their removal from the root zone. For successful reclamation, salinity should be reduced in the top 45–60 cm of the soil to below the threshold values for a particular crop being grown (Keren and Miyamoto, 1990). Reclamation can be hampered by several factors (Bresler et al., 1982) such as restricted drainage caused by a high water table, low soil hydraulic conductivity (HC) due to restrictive soil layers, lack of good-quality water, and the high cost of good-quality water.

(a) Water and Salt Balance in the Soil

The accumulation of salts in the root zone depends on the balance of salt input and output from different sources in the root zone. Because salts are carried by water, the extent of salts accumulation in the soil directly depends upon the irrigation water quality, irrigation management, and the adequacy of drainage. The balance between the input and output of water in the root zone can be described as (Ayars et al., 2011):

$$D_s = D_r + D_g + D_i - D_e - D_t - D_d$$

$$(10.15)$$

where D_s = depth of stored soil water; D_r, D_g, and D_i represent the depth of flow of water into the crop's root zone from rainfall, groundwater, and irrigation, respectively; and D_e, D_t, and D_d represent the depth of flow of water out of the crop's root zone due to evaporation, transpiration, and drainage, respectively.

Since salts are moved by water, the change in salt storage (S_s) in the root zone can be calculated as follows:

$$S_s = D_rC_r + D_gC_g + D_iC_i + S_m + S_f$$
$$- D_dC_d - S_p + S_c$$
$$(10.16)$$

where C_r, C_g, C_i, and C_d represent salt concentration in rainwater, upward flowing groundwater, irrigation water, and drainage water, respectively; S_m is the salt dissolved from minerals in soil; S_f is the salt added to soil as a fertilizer or amendment; S_p is the salt precipitated; and S_c is the salt removed in the harvested crop.

Assuming steady-state conditions ($S_s = 0$), the absence of rain and of upward movement of salt from groundwater, and the S_m, S_f, S_p, and S_c are negligible and balance each other and can be ignored, the salt balance in the root zone can be estimated as follows:

$$S_s = D_iC_i - D_dC_d \qquad (10.17)$$

The importance of this concept is greater than its accuracy. It demonstrates that drainage (or leaching) is extremely important in irrigated agriculture; otherwise, soil salinity will ultimately develop in the system.

For maintaining zero salt balance in the root zone, $D_iC_i = D_dC_d$ and it can be written as:

$$\frac{D_d}{D_i} = \frac{C_i}{C_d} \qquad (10.18)$$

which is the leaching fraction (LF) that is discussed in the following section.

Sidebar 10.4 — Leaching Requirement (LR)

The leaching requirement (LR) is defined as the fraction of the irrigation water that must be leached through the root zone to control soil salinity at any specified level (Richards, 1954). In the LR definition and calculations, several assumptions were made that included uniform aerial application of irrigation water, no rainfall; no removal of salt in the harvested crop; and no precipitation of soluble constituents in the soil. Also, the steady-state water-flow rates were assumed, i.e, the total equivalent depths of irrigation and drainage waters over a period of time.

Quantitatively, the LR is simply the ratio of the equivalent depth of the drainage water to the depth of irrigation water and it may be expressed as a fraction or as percent. Further, the ratio is equal to the inverse ratio of the corresponding electrical conductivities, as given below:

$$LR = \frac{D_d}{D_i} = \frac{EC_i}{EC_d}$$

where D_d and D_i are the depth (mm) of drainage and irrigation water (strictly speaking it includes both irrigation and precipitation, i.e., applied water), respectively; and EC_d and EC_i are the electrical conductivity (dS m^{-1}) of drainage and irrigation water, respectively.

Both leaching fraction (LF) and LR are calculated using the same formula, the LF value is expressed as a fraction, whereas LR can be expressed either as a fraction or percentage of the irrigation water, and a threshold EC value is targeted for the drainage water.

Plant roots extract nearly pure water from soils; thus salts present in the irrigation water accumulate over time in the root zone and salt concentration typically increases with depth (Ayars et al. 2011). Researchers at the U.S. Soil Salinity Laboratory formulated the concept of leaching requirement (LR) to address the effects of salinity on plant growth (Sidebar 10.4). Leaching is accomplished by applying an adequate amount of water so that a portion of it infiltrates through and below the entire root zone carrying with it a portion of the accumulated salts. The fraction of the amount of applied water (irrigation and precipitation) that drains beyond the root zone is called the LF.

LR is defined as the fraction of water infiltrating the soil that must move beyond the root zone to prevent soil salinity from exceeding a specified value. The LF is a measure of the level of leaching of salts; with increasing LF, the concentration of salts or EC in the root zone decreases (Ayars et al., 2011), which is defined as:

$$LF = \frac{D_d}{D_i} = \frac{C_i}{C_d} = \frac{EC_i}{EC_d} \tag{10.19}$$

where D_d and D_i are the depth (mm) of drainage water and irrigation water, respectively; C_i and C_d are the salt concentration (mg L^{-1}) of irrigation water and drainage water, respectively; and EC_i and EC_d are the electrical conductivity (dS m^{-1}) of irrigation water and drainage water, respectively.

Since the LR refers to the minimum LF required to leach out salts from the root zone to bring it down to a level that does not measurably affect crop yield, the salinity in the root zone is the maximum permissible salinity level of EC_d (EC_d^*) that results in optimum crop yield. Equation 10.20 can be written as:

$$LR = \frac{C_i}{C_d^*} = \frac{EC_i}{EC_d^*} \tag{10.20}$$

where C_d^* is the maximum permissible salt concentration (mg L^{-1}) and EC_d^* is the maximum permissible electrical conductivity, respectively, of the drainage water.

Different approaches have been used to determine the root zone salinity. Ayers and Westcot (1985) considered the crop water uptake in the root zone in four sections, with the greatest water uptake from the top section of the root zone and decreasing down into the root zone. The average root zone salinity is calculated using the average of five points in the rooting depth. Additionally, the saturation extract EC (EC_e) is assumed to be one-half of the soil-water salinity. Rhoades (1974) introduced a simple estimate of EC_d^*, given as follows:

$$EC_d^* = 5EC_e^* - EC_i \tag{10.21}$$

where EC_e^* is the average EC of the saturaion extract for a given crop appropriate to the tolerable degree of yield depression (usually 10% or less).

Sidebar 10.5 — Volume of Water in Irrigation Water or Precipitation

The volume of water in 1 mm of irrigation water or precipitation is:

$$\frac{1 \text{ mm}^3 \text{ of water}}{1 \text{ mm}^2 \text{ of land area}} = \frac{1 \text{ m}^3 \text{ of water}}{1000 \text{ m}^2 \text{ of land area}}$$

$$= \frac{1 \text{ L}}{\text{m}^2}$$

Thus a rainfall of 1 mm supplies 1 L of water to each square meter of the land area or field.

Following the above, 100 mm irrigation or precipitation add 1 ML (1×10^6) of water to 1 ha (100 m × 100 m) field.

---

where EC_e^* (dS m^{-1}) is the linearly averaged saturation extract EC of the root zone for a given crop with a tolerable degree of yield decline ($\leq 10\%$), which is equivalent to the threshold EC_e values outlined by Mass and Hoffman (1977), and EC_i is the electrical conductivity (dS m^{-1}) of irrigation water. Combining (Equations 10.21 and 10.22), LR is estimated as:

$$LR = \frac{EC_i}{5EC_e^* - EC_i} \tag{10.22}$$

TABLE 10.6   Salt tolerance of agronomic crops.[a]

Crop	Threshold EC_e[b] (dS m^{-1})	Slope[b] (% per dS m^{-1})	Tolerance to salinity[c]
Fiber, grain, and special crops			
Barley (*Hordeum vulgare* L.)	8.0	5.0	T
Canola or rapeseed (*Brassica campestris* L.)	9.7	14.0	T
Canola or rapeseed (*Brassica napus* L.)	11.0	13.0	T
Chickpea (*Cicer arietinum* L.)	—	—	MS
Corn (*Zea mays* L.)	1.7	12.0	MS
Cotton (*Gossypium hirsutum* L.)	7.7	5.2	T
Peanut (*Arachis hypogaea* L.)	3.2	29.0	MS
Rice, paddy (*Oryza sativa* L.)	3.0	12.0	S
Rye (*Secale cereale* L.)	11.4	10.8	T
Sorghum (*Sorghum bicolor* L.) Moench	6.8	16.0	MT
Soybean (*Glycine max* L.) Merrill	5.0	20.0	MT
Sugarcane (*Saccharum officinarum* L.)	1.7	5.9	MS
Wheat (*Triticum aestivum* L.)	6.0	7.1	MT
Grasses and forage crops			
Alfalfa (*Medicago sativa* L.)	2.0	7.3	MS
Clover, alsike (*Trifolium hybridum* L.)	1.5	12.0	MS
Clover, berseem (*Trifolium alexandrinum* L.)	1.5	5.7	MS
Fescue, tall (*Festuca elatior* L.)	3.9	5.3	MT
Kikuyu grass (*Pennisetum clandestinum* L.)	8.0	—	T
Orchard grass (*Dactylis glomerata* L.)	1.5	6.2	MS
Vetch (*Vicia angustifolia* L.)	3.0	11.0	MS

[a] Adopted from Maas (1999) with permission.
[b] Threshold EC_e is the parameter a and slope is the parameter b in Equation 10.31.
[c] These data serve only as a guideline to relative tolerances among crops. Absolute tolerances vary, depending on climate, soil conditions, and cultural practices; S, sensitive; MS, moderately sensitive; MT, moderately tolerant; and T, tolerant.

The above equation, in conjunction with the salinity threshold values (described in Table 10.6), can be used to estimate the LR for crops.

The total annual depth of water (Sidebar 10.5) that needs to be applied to meet both the crop demand and LR can be estimated (Ayers and Westcot 1985) using the following relationship:

$$WR = \frac{ET}{1 - LR} \qquad (10.23)$$

where WR = depth of applied water (mm year^{-1}), ET = total annual crop water demand (mm year^{-1}), and LR = leaching requirement expressed as a fraction.

LRs based on steady-state analyses, such as the models proposed by Rhoades (1974) and Ayers and Westcot (1985), have been considered overly conservative (Corwin and Grattan, 2018). The application of excessive amounts of irrigation water based on overestimated LR values increases salt loads in drainage systems or underlying aquifers, which can have a detrimental impact on the environment and reduce water supplies. However, the newer transient models that provide better estimates of LR are not well developed. The existing models require parameters that are readily available.

10.6.2 Sodic Soils

Sodic soils have an ESP > 15 and an EC$_e$ < 4 dS m^{-1}; the lower limit of the saturation extract SAR is 13. The threshold of ESP for sodic soils is not universal. In Australia sodic soils are defined when the ESP is greater than 6%. The unique clay mineralogy (mixture of phyllosilicates including illite and kaolinite) and very

low ionic strength of the soil solution are the main reasons for the low ESP threshold values used for sodic soils in Australia, South Africa, and other countries. However, universally, Na$^+$ is the major problem in sodic soils. These soils correspond to "Solonetz," a term used by the Russian soil scientists and Solonetz is a Reference Soil Group of the World Reference Base for Soil Resources.

The high amount of Na$^+$ in these soils, along with the low EC$_e$, results in their dispersion (Box 10.2). Clay dispersion occurs when the electrolyte concentration decreases below the flocculation value of the clay (Keren and Miyamoto, 1990). Sodium-affected soils, which contain low levels of salt, have weak structural stability and low HCs and IRs. These poor physical properties result in decreased crop productivity caused by poor aeration and reduced water supply. Low IRs can also cause severe soil erosion (Sumner et al., 1998).

Rengasamy et al. (1984) proposed a scheme for predicting and classifying the dispersive behavior of surface layers of several representative Alfisols from southeastern Australia (Figure 10.4). It was reported that the surface

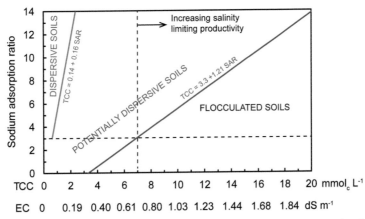

FIGURE 10.4 The classification of dispersive (spontaneous dispersion), potentially dispersive (mechanical dispersion), and flocculated (structurally stable) soils on the basis of total cation concentration (TCC) or electrical conductivity (EC) and sodium adsorption ratio (SAR) of Alfisols in Australia. TCC, EC, and SAR were measured in 1:5 soil-water extract. *From: Rengasamy et al. (1984) with permission.*

soils with SAR > 3 would disperse spontaneously whereas those with SAR < 3 disperse only after mechanical shaking. Because the soil dispersion was dependent on both the SAR and EC (or total cation concentration, TCC) of the solution, both parameters were included in their classification scheme. Rengasamy et al. (1984) divided the soils into three classes: (i) dispersive, (ii) potentially dispersive, and (iii) flocculated soils. Dispersive soils, with SAR > 3 and TCC < (0.14 + 0.16 × SAR), would spontaneously disperse under field conditions. These soils will have severe physical problems, such as crusting and reduced porosity. Potentially dispersive soils are unstable and disperse after mechanical disturbances such as intensive cultivation or raindrop impact.

These soils are separated from the flocculated soils with a line based on the relationship: TCC = 3.3 + 1.21 × SAR. The flocculated soils have more than the minimum required electrolyte levels, which decreases the thickness of the diffuse double layer, and they remain flocculated when subjected to rainfall, irrigation, or other mechanical disturbances. However, excessive soluble salts in soil water may reduce the availability of water to plants, thus limiting productivity, and under such conditions, the selection of a tolerant crop or leaching of salts from appropriate soil layers may become necessary. Such relationships are useful in predicting the dispersive behavior of a soil instead of a single threshold value of SAR or ESP.

Sodic soils, with a pH between 8.5 and 10, have been historically referred to as black alkali soils. The high pH is due to the hydrolysis of Na_2CO_3. Because of the high pH of these soils, a portion of OM dissolves. The dissolved and dispersed OM in the soil solution makes a black layer on the soil surface after evaporation. The major anions in the soil solution of sodic soils are Cl^-, SO_4^{2-}, and HCO_3^-, with lesser amounts of CO_3^{2-}. Since the pH is high and CO_3^{2-} is present, Ca^{2+} and Mg^{2+} carbonates are precipitated, and therefore the soil solution Ca^{2+} and Mg^{2+}

are low. Besides Na^+, another exchangeable and soluble cation that may occur in these soils is K^+ (Richards, 1954).

Sodic soils may be coarser textured on the surface and have higher clay contents on the subsurface horizon due to the leaching of clay material that is Na^+-saturated. Consequently, the subsoil is dispersed, permeability is low, and a prismatic soil structure may result.

For the amelioration of sodic soils, two common approaches using the addition of divalent cations (mainly Ca^{2+}) are: (i) using Ca supplying chemical amendments and (ii) application of successive dilutions of saltwater containing high amounts of Ca (and Mg). Among the chemical amendments, the reclamation of sodic soils is achieved by applying gypsum ($CaSO_4 \cdot 2H_2O$) or a highly soluble and expensive $CaCl_2$ to remove the exchangeable Na^+. The addition of $CaSO_4$ and $CaCl_2$ also increases permeability by increasing electrolyte concentration.

The Ca^{2+} from chemical amendments exchanges with the Na^+, which is then leached out as a soluble salt, Na_2SO_4 or NaCl, as shown below:

$$2Na^+\text{-}X + CaSO_4 \rightarrow Ca^{2+}\text{-}X + Na_2SO_4 \ (\downarrow)$$
$$(10.24)$$

$$2Na^+\text{-}X + CaCl_2 \rightarrow Ca^{2+}\text{-}X + 2NaCl \ (\downarrow)$$
$$(10.25)$$

Sulfur, pyrites, and sulfuric acid can also be applied to ameliorate the Na problem in calcareous soils (where $CaCO_3$ is present). Sulfur and pyrites require an initial phase of microbiological oxidation that produces H_2SO_4, which interacts with $CaCO_3$ to form $CaSO_4$, as given below:

$$CaCO_3 + H_2SO_4 \rightarrow CaSO_4 + CO_2 + H_2O$$
$$(10.26)$$

$$2Na^+ - X + CaSO_4 \leftrightarrow Ca^{2+} - X + Na_2SO_4$$
$$(10.27)$$

However, gypsum is by far the most commonly used amendment for the reclamation of sodic soil because of its low cost, ease of availability, and handling. Gypsum is obtained from mineral deposits as well as a by-product from phosphate fertilizer and power generation industries. The amount of gypsum required to ameliorate a sodic soil, called the "gypsum requirement" (Richards, 1954), is dependent on several factors, including the depth of soil to be ameliorated and the initial and final ESP of the soil. There are different procedures to determine the gypsum requirement to ameliorate sodic soils. The gypsum requirement can be well estimated using the following simple formula (Oster and Jayawardane, 1998):

$$GR = 0.086 \times \rho_b \times d \times F \times CEC \times \left(ESP_i - ESP_f\right) \tag{10.28}$$

where GR is the amount of pure gypsum (Mg ha^{-1}) required to achieve a targeted or final ESP (ESP$_f$) of the soil; 0.086 is the mass of gypsum (g) required to replace a mmol$_c$ of exchangeable Na$^+$, ρ_b is the bulk density (Mg m^{-3}) of the soil; d is the soil depth (m) to be amended; F (unitless) is the Ca–Na exchange efficiency factor with values ranging from 1.1 for an initial ESP (ESP$_i$) of 15 to 1.3 for an ESP$_i$ of 5; and CEC is the cation exchange capacity (mmol$_c$ kg^{-1}).

The estimate of the gypsum requirement based on laboratory procedures is a often large and may be uneconomical. In Australia gypsum application rates in the range of 2.5–5 t ha^{-1} are commonly recommended. The results from field experiments show significant crop yield responses and residual effects for several years from such application rates of gypsum.

The efficiency of applied gypsum in a sodic soil depends on several factors, including the purity of gypsum, particle size, mixing, and soil moisture. The efficiency of the Ca–Na exchange varies with the ESP, being much greater at high ESP and the rate of exchange decreases substantially at ESP levels below 10, when part of the applied Ca displaces exchangeable Mg.

10.6.3 Saline–Sodic Soils

Saline–sodic soils have an EC$_e$ > 4 dS m^{-1} and an ESP > 15. Thus both soluble salts and exchangeable Na$^+$ are high in these soils. Since electrolyte concentration is high, the soil pH is usually <8.5 and the soil is flocculated. However, if the soluble salts are leached out, usually Na$^+$ becomes an even greater problem and the soil pH rises to >8.5 and the soil can become dispersed (Richards, 1954).

In saline–sodic soils reclamation involves the addition of good-quality water to remove excess soluble salts and the use of a Ca^{2+} source (CaSO$_4$·2H$_2$O or CaCl$_2$) to exchange Na$^+$ from the soil as a soluble salt, Na$_2$SO$_4$. In saline–sodic soils a saltwater dilution method is usually effective in reclamation. In this method the soil is rapidly leached with water that has a high electrolyte concentration with large quantities of Ca^{2+} and Mg^{2+}. After leaching, and the removal of Na$^+$ from the exchanger phase of the soil, the soil is leached with water of lower electrolyte concentration to remove the excess salts.

In both saline–sodic and sodic soils the cost and availability of a Ca^{2+} source are major factors in reclamation. It is also important that the Ca^{2+} source fully reacts with the soil. Thus it is better to incorporate the Ca^{2+} source into the soil rather than just putting it on the surface so that Na$^+$ exchange from the soil exchanger phase is enhanced. Gypsum can also be added to irrigation water to increase the Ca/Na ratio of the water and improve reclamation (Keren and Miyamoto, 1990).

10.7 Effects of Soil Salinity and Sodicity on Soil Structural Properties

Soil salinity and sodicity can have a major effect on the structure of soils. Soil structure, or the arrangement of soil particles, is critical in affecting permeability and infiltration. Infiltration refers to the "downward entry of water into the soil through the soil surface" (Glossary of Soil Science Terms, 2008). If a soil has high quantities of Na$^+$ and the EC is low, soil permeability, HC, and the IR are decreased due to the swelling and dispersion of clays and slaking of aggregates (Shainberg, 1990). IR can be defined as "the volume flux of water flowing into the soil profile per unit of surface area" (Shainberg, 1990). Typically, soil IRs are initially high; if the soil is dry, and then they decrease until a steady state is reached. Swelling causes the soil pores to become narrower (McNeal and Coleman, 1966), and slaking decreases the number of macropores through which water and solutes can flow, resulting in the plugging of pores by the dispersed clay. The swelling of clay has a pronounced effect on permeability and is affected by clay mineralogy, the kind of ions adsorbed on the clays, and the electrolyte concentration in solution (Shainberg et al., 1971; Oster et al., 1980; Goldberg and Glaubig, 1987). Swelling is greatest for smectite clays that are Na$^+$-saturated. As the electrolyte concentration decreases, clay swelling increases.

As ESP increases, particularly above 15, swelling clays like montmorillonite (a smectite mineral) retain a greater volume of water (Figure 10.5). HC and permeability decrease as ESP increases and salt concentration decreases (Quirk and Schofield, 1955; McNeal and Coleman, 1966). Permeability can be maintained if the EC of the percolating water is above a threshold level, which is the concentration of salt in the percolating solution, which causes a 10%–15% decrease in soil permeability at a particular ESP (Shainberg, 1990).

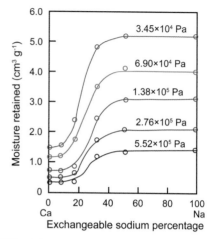

FIGURE 10.5   Water retention as a function of ESP and pressure applied on montmorillonite. *From Shainberg et al. (1971), with permission.*

10.8 Effects of Soil Salinity on Plant Growth

Key points: Salts in saline soils can harm plants by decreasing the osmotic potential; as well, ions, such as Na$^+$, Cl$^-$, or B, can cause direct toxicity to plants. The salinity threshold value, that is, the salinity at which plant yields start to decline, can be estimated from a relationship between yield and EC$_e$ and is a useful parameter in the management of saline soils.

Salinity and sodicity have pronounced effects on the growth of plants (Figure 10.6). The presence of a high concentration of salts in the soil or the irrigation water affects crop yield or quality. On the other hand, sodicity adversely affects soil physical properties, such as crusting, reduced infiltration, and reduced aeration, which may adversely affect plants. In sodic soils high Na$^+$ concentrations may cause deficiencies of other elements, such as K$^+$ or Ca^{2+}, and direct toxicity from Na$^+$ or Cl$^-$ ions may occur in plants.

In saline soils soluble ions such as Cl$^-$, SO$_4^{2-}$, HCO$_3^-$, Na$^+$, Ca^{2+}, Mg^{2+}, and sometimes NO$_3^-$

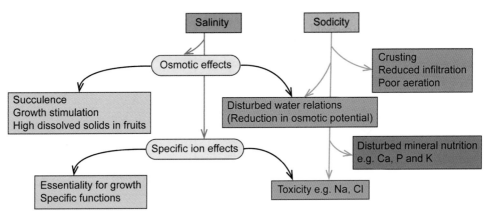

FIGURE 10.6 Effects of salinity and sodicity on plants. The effects of salinity in green boxes are not deleterious to plants. *From Läuchli and Grattan (2011), with permission.*

and K^+ can harm plants by decreasing the osmotic potential. Additionally, some particular ions (e.g., Cl or Na) in the soil solution may have chemical or specific-ion effects on plants (Läuchli and Grattan, 2011). Trace elements such as B, Se, or Mo may be present in high enough concentrations in the soil solution of saline soils, which could cause specific toxicities in some plants. Some effects of salinity on plants are not deleterious (as shown in green boxes in Figure 10.6); however, saline conditions generally adversely affect crops. Plant species, and even different varieties within a particular species, differ in their tolerance to a particular ion (Bresler et al., 1982).

Soil water availability can be expressed as the sum of the matric and osmotic potentials. As the water content decreases, through evaporation and transpiration, both the matric and osmotic potentials decrease and become more negative. The presence of soluble ions in a soil affects the soil's osmotic potential; the osmotic potential (Ψ_p) of the soil solution can be estimated from the saturation extract EC as follows (Corwin and Yemoto, 2020):

$$\psi_p(MP_a) = -0.036\, EC_e\left(dS\ m^{-1} \text{ at } 25°C\right)$$
$$(10.29)$$

However, because of the differences in the molecular compositions of different salts, the above relationship is only an approximation. More accurate estimates of Ψ_p can be made by consideration of the ion composition of the water, such as that presented in the ExtractChem model (Suarez and Taber 2007).

The salinity threshold value, that is, the salinity at which plant yields start to decline, can be estimated from the following relationship between yield and EC_e (Maas and Hoffman, 1977):

$$Y_r = 100 - b(EC_e - a) \qquad (10.30)$$

where Y_r is the relative crop yield (%), that is, the percentage of the yield of the crop grown under saline conditions compared to that obtained under nonsaline but otherwise comparable conditions; EC_e is the mean electrical conductivity (dS m^{-1}) of a saturated soil extract taken from the root zone; a is the threshold level of soil salinity, that is, EC (dS m^{-1}) at which the yield decreases begin; and b is the percentage yield loss per unit increase of salinity in excess of a. A hypothetical example for the fitted model is presented in Sidebar 10.6.

Sidebar 10.6 — Salinity Threshold

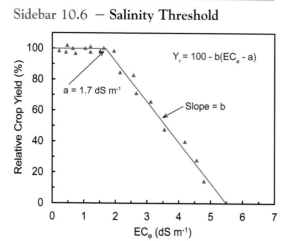

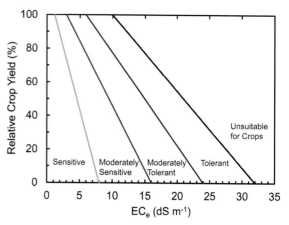

FIGURE 10.7   Divisions for classifying crop tolerance to salinity. *From Maas (1999), with permission.*

A hypothetical fit of the two-piece model of Maas and Hoffman (1977) to determine the salinity threshold for a crop. Δ represents observed yield data and the red line is for fitted model.

The effect of salinity on plant growth is affected by climate, soil conditions, agronomic practices, irrigation management, crop type and variety, stage of growth, and salt composition (Maas, 1999). Salinity does not usually affect the yield of a crop until the EC_e exceeds a certain value for each crop. This is known as the threshold salinity level or the threshold EC_e value, which differs for various crops (Table 10.6). The yields of many crops, for example, most food and fiber crops, will linearly decrease as EC_e increases. Maas, 1999 divided plants into five different tolerance categories based on EC_e (Figure 10.7).

10.9 Effects of Sodicity and Salinity on Environmental Quality

Degradation of soils by salinity and sodicity profoundly affects environmental quality. In particular, the dispersive behavior of sodic soils (Box 10.2), coupled with human activities such as agriculture, forestry, urbanization, and soil contamination, can have dire effects on the environment and humankind. The enhanced dispersion promotes surface crusts or seals, which lead to waterlogging, surface runoff, and erosion. Consequently, high levels of inorganic and organic colloids can be mobilized, which can transport organic and inorganic contaminants such as pesticides, metals, and radionuclides in soils and waters (Sumner et al., 1998).

The enhanced erosion potential of sodic soils also results in increased sediments that can contaminate waters. Suspended sediments in water increase turbidity. This causes less light to pass through, which negatively affects aquatic life. Additionally, increased levels of dissolved organic carbon (DOC) generated in sodic soils can discolor water (Sumner et al., 1998).

Salinization of soils results in soluble salts that can be mobilized in the soil profile, causing land and water degradation. The salts can also effect the release and solubilization of heavy metals (e.g., Cd) into solution, with potential adverse effects on water quality and plant growth (Gambrell et al., 1991; McLaughlin et al., 1994).

BOX 10.2

Dispersion and flocculation of soil clay particles

The charge on soil clay particles is neutralized by hydrated ions of opposite charge that reside in the liquid immediately adjacent to the particles. According to the diffuse double layer model (a more complete description of the diffuse double layer theory is given in Chapter 5), the concentration of the counterions decreases exponentially from away from the clay surface to the bulk solution. Divalent cations are attracted more strongly to the clay surfaces than to monovalent cations. Thus in the presence of a divalent cation like Ca^{2+} the diffuse double layer is more compressed than with the monovalent Na^+ cation, as depicted in the figure (a) below.

According to the Derjaguin–Landau–Verwey–Overbeek (DLVO) theory, two internal forces dominate when colloidal particles interact in an aqueous solution: the electrostatic repulsive force and the van der Waals attractive force. When two clay particles possessing negative charge approach each other, their diffuse counterion atmospheres overlap (b) and the electrical repulsive force, also called swelling pressure, comes into play that keeps the particles apart. On the other hand, the attractive force between clay particles resulting from van der Waals forces decays with the second power of the distance between the particles. Therefore the van der Waals force becomes important when clay colloids are in close proximity to each other. In suspension clay particles are subject to Brownian motion from the action of water molecules, and due to constant motion, they colloid with each other.

The balance between the repulsive and attractive force determines whether the clay particles remain dispersed or flocculated. The minimum electrolyte concentration required to cause the flocculation of a given dispersed system is called the critical flocculation concentration (van Olphen, 1977). When the attractive force is greater than the repulsive force in the diffuse double layer, the particles flocculate (c). This occurs in Ca^{2+}-saturated clays, where the diffuse double layer (or cationic atmosphere) is compressed toward the clay surface and there is a smaller overlap of the swarm of ions for a given distance between the clay particles. Conversely, in Na^+-saturated clays the repulsive force in the double layer is greater than the attractive force that keeps the clay particles in a dispersed state, as depicted in (c).

Rengasamy and Sumner (1998) evaluated the critical flocculation concentrations for several homoionic clays by using chloride solutions of the same cation and illustrated the major role that valence plays. However, even with a given valence, flocculating power differs (K > Na; Ca > Mg). They concluded that compared to Na = 1, the flocculation power of the other cations would be K = 1.8, Mg = 27, and Ca = 45.

Continued

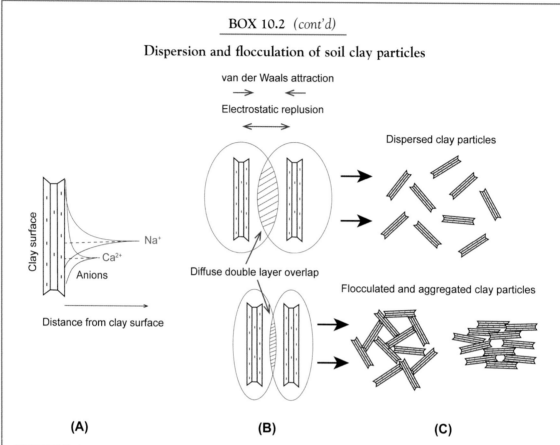

BOX 10.2 (cont'd)

Dispersion and flocculation of soil clay particles

FIGURE B10.2   (A) A schematic representation of the double electric layer on the surface of a negatively charged clay particle in the presence of Na^+ and Ca^{2+} ions. (B) The balance between attractive (van der Waals) and repulsive (electrostatic) forces acting on clay particles determines the (C) dispersion (Na^+-saturated clay) or flocculation (Ca^{2+}-saturated) of clay particles.

Suggested Reading

Ayers, R.S., Westcot, D.W., 1976. Water Quality for Agricultural, Irrigation and Drainage Paper 29. Food and Agriculture Organization of the United Nations, Rome.

Bresler, E., McNeal, B.L., Carter, D.L., 1982. Saline and Sodic Soils. Principles-Dynamics-Modeling. In: Advanced Series in Agricultural Sciences 10. Springer-Verlag, Berlin.

Ghassemi, F., Jakeman, A.J., Nix, H.A., 1995. Salinization of Land and Water Resources: Human Causes, Extent, Management and Case Studies. University of New South Wales Press Ltd, Sydney.

Hopmans, J.W., Qureshi, A.S., Kisekka, I., Munns, R., Grattan, S.R., Rengasamy, P., et al., 2021. Critical knowledge gaps and research priorities in global soil salinity. In: Sparks, D.L. (Ed.), Advances in Agronomy. Elsevier, San Diego, CA.

Lloyd, J.W., Heathcote, J.A., 1985. Natural Inorganic Hydrochemistry in Relation to Groundwater an Introduction. Oxford University Press, New York, NY.

Maas, E.V., Grattan, S.R., 1999. Crop yields as affected by salinity. In: Skaggs, R.W., Van Schilfgaarde, J. (Eds.), Agricultural Drainage. American Society of Agronomy, Crop Science Society of America. Soil Science Society of America, Madison, WI, pp. 55–108.

Richards, L.A. (Ed.), 1954. Diagnosis and Improvement of Saline and Sodic Soils. USDA Agric. Handb. 60. USDA, Washington, DC.

Sumner, M.E., Naidu, R., 1998. Sodic Soils: Distribution, Properties, Management and Environmental Consequences. Oxford University Press, New York, NY.

Wallendar, W.W., Tanji, K.K. (Eds.), 2011. Agricultural Salinity Assessment and Management. 2nd ed., ASCE Manuals and Reports on Engineering Practice No. 71. American Society of Civil Engineers, Reston, United States.

Yaron, D. (Ed.), 1981. Salinity in Irrigation and Water Resources. Dekker, New York, NY.

Problem Set

Q1. The electrical conductivity (EC_e) of the saturation extract of a soil is 2.45 dS m^{-1}. Estimate the total dissolved solids (TDS) and total concentration of soluble cations (TSC) in the soil extract. Does the salinity level in this soil present any problem for field crops?

Q2. An irrigation water has an EC of 3.60 dS m^{-1} and Ca and Mg concentrations of 143 mg L^{-1} and 40 mg L^{-1}, respectively. Calculate the SAR and ESP of the soil in

equilibrium with this water based on the Gapon exchange constant of 0.015 (L mmol^{-1})$^{1/2}$.

Q3. Review the following chemical data for some river and well waters and identify any potential problems to crop plants in the context of salinity, sodicity, and toxicity of specific ions.

Q4. Calculate the residual sodium carbonates (RSCs) for the following well waters. Based on the calculated RSC values, comment on their suitability for their use in irrigation. Speculate on the precipitation of carbonates from the application of these waters for irrigation in a soil.

Q5. A crop is irrigated with a well water of an electrical conductivity (EC_i) of 1.27 dS m^{-1}. The crop is irrigated to achieve a leaching fraction of 0.10. What will be the salinity of the soil-water that is draining from the bottom of the root zone (EC_d).

Q6. A maize crop is irrigated using a river water with an EC of 0.9 dS m^{-1}. The water requirement of the crop (evapotranspiration) for the whole season is 700 mm. The irrigation efficiency is 70%, as 30% of the applied water is lost in lateral runoff, deep drainage, and leaks in the delivery system. How much irrigation

Sample site	EC	Ca^{2+}	Mg^{2+}	Na^+	Cl^-	SO_4^{2-}	HCO_3^-
	(dS m^{-1})			mmol$_c$ L^{-1}			
1. Pecos River, NM, USA	3.37	20.4	6.2	13.3	13.8	23.8	2.3
2. Gila River, AZ, USA	7.42	17.0	12.0	53.1	49.7	28.1	5.5
3. Well in Victoria, Australia	3.40	0.4	2.3	26.2	21.7	1.2	7.7
4. Amazon River, Brazil	0.04	0.2	0.1	0.1	0.1	0.6	0.3
5. Zambezi River, Zimbabwe	0.10	0.5	0.4	0.1	0.1	0.1	0.8
6. Well Tamar, Egypt	2.80	7.6	5.0	16.8	17.5	8.1	4.5

Sample site	Ca^{2+}	Mg^{2+}	HCO_3^-	CO_3^{2-}
		mmol$_c$ L^{-1}		
1. Sample A	2.50	1.23	4.10	0.10
2. Sample B	1.75	1.89	5.20	0
3. Sample C	1.40	3.2	2.0	0

water should be applied to meet the crop requirement as well as the LR to avoid any adverse effect on the yield in the long term?

Q7. Explain the common occurrence of calcite in many sodic soils. What are the main features of such soils?

Q8. Based on the data in Table 10.6, peanut yields decrease approximately 29% per dS m^{-1} when soil salinity exceeds 3.2 dS m^{-1}. Calculate the relative yield in a soil that has an EC$_e$ of 4.15 dS m^{-1}.

Q9. Estimate the gypsum requirement to ameliorate the top 15 cm of a 10 ha uniform field with a sodic soil. The soil has a bulk density of 1.56 Mg m^{-3} and the cation exchange capacity is 245 mmol$_c$ kg^{-1}. The ESP of the soil is 18% and the target is to achieve an ESP of 8% and the gypsum available in the local area is 85% pure.

Q10. A good-quality irrigation water with very low salinity (EC$_i$ = 0.20 dS m^{-1}) is being used for irrigation by a citrus grower. Drainage and water-logging problems have been noticed in the field, which have caused reduced conditions and stress to citrus trees. The local agronomist has recommended an increase of 2 mmol$_c$ L^{-1} Ca in the water using gypsum. The 5 ha orchard needs an irrigation depth of 100 mm. The gypsum available in the local area is 75% pure. Calculate the amount of gypsum required for the recommendation.

References

Ayars, J.E., Hoffman, G.J., Corwin, D.L., 2011. Leaching and rootzone salinity control. In: Wallendar, W.W., Tanji, K.K. (Eds.), Agricultural Salinity Assessment and Management, Second edition. American Society of Civil Engineers, Reston, VA, pp. 371—403.

Ayers, R.S., Westcot, D.W., 1985. Water Quality for Agriculture. FAO Irrigation and Drainage Paper 29 rev 1. Food and Agriculture Organization of the United Nations, Rome, p. 174.

Bauder, T.A., Waskom, R.M., Sutherland, P.L., Davis, J.G., 2011. Irrigation water quality criteria. Colorado State University Extension Publication, Crop Series/Irrigation. Fact sheet no. 0.506, 4 pp.

Bower, C.A., Hatcher, J.T., 1964. Estimation of the exchangeable-sodium percentage of arid zone soils from solution cation composition. In: Western Soc. Soil Sci. Abstr. Vancouver, British Columbia, Canada.

Bresler, E., McNeal, B.L., Carter, D.L., 1982. Saline and Sodic Soils. Springer, New York, NY.

Cope, F., 1958. Catchment salting in Victoria. Soil Conserv. Auth. Victoria Bull. 1, 1—88.

Corwin, D.L., Grattan, S.R., 2018. Are existing irrigation salinity leaching requirement guidelines overly conservative or obsolete? J. Irrig. Drain. Eng. 144 (8), 02518001.

Doneen, L.D., 1975. Water quality for irrigated agriculture. In: Poljakoff-Mayber, A., Gale, J. (Eds.), Plants in Saline Environments. Springer-Verlag, Berlin, pp. 56—66.

Glossary of Soil Science Terms, 2008. Soil Science Glossary Terms Committee and Soil Science Society of America. Soil Sci. Soc. Am, Madison, WI.

FAO, 2021. Food and Agriculture Organization of the United Nations, FAOSTAT, Landuse. www.fao.org/faostat/en/#data/RL/. Accessed 24 June 2021.

Corwin, D.L., Yemoto, S.R., 2020. Salinity: electrical conductivity and total dissolved solids. Soil Sci. Soc. Am. J. 84, 1442—1461.

Gambrell, R.P., Wiesepape, J.B., Patrick Jr., W.H., Duff, M.C., 1991. The effects of pH, redox, and salinity on metal release from a contaminated sediment. Water Air Soil Pollut. 57—58, 359—367.

Goldberg, S., Glaubig, R.A., 1987. Effect of saturating cation, pH and aluminum and iron oxides on the flocculation of kaolinite and montmorillonite. Clays Clay Miner. 35, 220—227.

Grieve, C.M., Grattan, S.R., Maas, E.V., 2012. Plant salt tolerance. In: Wallender, W.W., Tanji, K.K. (Eds.), Agricultural Salinity Assessment and Management. ASCE Manual and Reports on Engineering Prac. No. 71. American Society of Civil Engineers, New York, NY, pp. 405—459.

Griffin, R.A., Jurinak, J.J., 1973. Estimation of acidity coefficients from the electrical conductivity of natural aquatic systems and soil extracts. Soil Sci. 116, 26–30.

Keren, R., Miyamoto, S., 1990. Reclamation of saline, sodic, and boron-affected soils. In: Tanji, K.K. (Ed.), Agricultural Salinity Assessment and Management. ASCE Manuals Prac. No. 71. American Society of Civil Engineers, New York, NY, pp. 410–431.

Läuchli, A., Grattan, S.R., 2011. Plant responses to saline and sodic conditions. In: Wallender, W.W., Tanji, K.K. (Eds.), Agricultural Salinity Assessment and Management. ASCE Manuals and Reports on Engineering Prac. No. 71. American Society of Civil Engineers, New York, NY, pp. 169–205.

Lloyd, J.W., Heathcote, J.A., 1985. Natural inorganic hydrochemistry in relation to groundwater an introduction. Oxford University Press, New York, NY.

Maas, E.V., 1999. Crop yields as affected by salinity. In: Skaggs, R.W., van Schilfgaarde, J. (Eds.), Agricultural Drainage. Agronomy Monograph, Vol. 38. ASA, CSSA, SSSA, Madison, WI, pp. 55–108.

Maas, E.V., Hoffman, G.J., 1977. Crop salt tolerance-current assessment. J. Irrig. Drain. Div 103 (IR 2), 115–134.

Marion, G.M., Babcock, K.L., 1976. Predicting specific conductance and salt concentration of dilute aqueous solution. Soil Sci. 122, 181–187.

McLaughlin, M.A., Palmer, L.T., Tiller, K.G., Beech, T.A., Smart, M.K., 1994. Increased soil salinity causes elevated cadmium concentrations in field-grown potato tubers. J. Environ. Qual 23, 1013–1018.

McNeal, B.L., Coleman, N.T., 1966. Effect of solution composition on soil hydraulic conductivity. Soil Sci. Soc. Am. Proc. 30, 308–312.

Oster, J.D., Jayawardane, N.S., 1998. Agricultural management of sodic soils. In: Summer, M.E., Naidu, R. (Eds.), Sodic Soils: Distribution, Management and Environmental Consequences. Oxford University Press, New York, NY, pp. 125–147.

Oster, J.D., Shainberg, I., Wood, J.D., 1980. Flocculation value and gel structure of Na/Ca montmorillonite and illite suspensions. Soil Sci. Soc. Am. J. 44, 955–959.

Postel, S., 1989. Water for Agriculture: Facing the Limits. Worldwatch Paper 93. Worldwatch Institute, Washington, DC.

Quirk, J.P., Schofield, R.K., 1955. The effect of electrolyte concentration on soil permeability. J. Soil Sci. 6, 163–178.

Rengasamy, A., Greene, R.S.B., Ford, G.W., Mehanni, A.H., 1984. Identification of dispersive behaviour and the management of red-brown earths. Aust. J. Soil Res. 22, 413–431.

Rengasamy, P., Sumner, M.E., 1998. Processes involved in sodic behaviour. In: Sumner, M.E., Naidu, R. (Eds.), Sodic Soils. Distribution, Properties, Management, and Environmental Consequences. New York Press, New York, NY, pp. 35–50.

Richards, L.A. (Ed.), 1954. Diagnosis and Improvement of Saline and Sodic Soils. USDA Agric. Handb. No. 60, Washington, DC.

Rhoades, J.D., 1993. Electrical conductivity methods for measuring and mapping soil salinity. Adv. Agron. 49, 201–251.

Rhoades, J.D., 1999. Use of saline drainage water for irrigation. In: Skaggs, R.W., van Schlfgaarde, J. (Eds.), Agricultural Drainage. Agronomy Monograph, vol. 38. ASA-CSSA-SSSA, Madison, WI, pp. 615–657.

Rhoades, J.D., 1974. Drainage for salinity control. In: van Schilfgaarde, J. (Ed.), Drainage for Agriculture. Monograph, vol. 17. ASA, CSSA, SSSA, Madison, WI, pp. 433–462.

Shainberg, I., 1990. Soil response to saline and sodic conditions. In: Tanji, K.K. (Ed.), Agricultural Salinity Assessment and Management. ASCE Manuals Prac. No. 71. American Society of Civil Engineers, New York, NY, pp. 91–112.

Shainberg, I., Bresler, E., Klausner, Y., 1971. Studies on Na/Ca montmorillonite systems. I: the swelling pressure. Soil Sci. 111, 214–219.

Shaw, R.J., 1994. Estimation of the Electrical Conductivity of Saturation Extracts From the Electrical Conductivity of 1:5 Soil:Water Suspensions and Various Soil Properties, Project Report QO94025. Department of Primary Industries, Brisbane, Queensland.

Shiklomanov, I.A., 2000. Appraisal and assessment of world water resources. Water Int. 25 (1), 11–32.

Slavich, P.G., Petterson, G.H., 1993. Estimating the electrical conductivity of saturated paste extracts from 1:5 soil:water suspensions and texture. Aust. J. Soil Res. 31, 73–81.

Sposito, G., Mattigod, S.V., 1977. On the chemical foundation of the sodium adsorption ratio. Soil Sci. Soc. Am. J 41, 323–329.

Suarez, D.L., Jurinak, J.J., 2011. The chemistry of salt-affected soils and waters. In: Wallender, W.W., Tanji, K.K. (Eds.), Agricultural Salinity Assessment and Management. ASCE Manuals and Reports on Engineering Practice No. 71. American Society of Civil Engineers, New York, NY, pp. 57–88.

Suarez, D.L., Taber, P., 2007. ExtractChem software. Version 1.0.18. US Salinity Lab, Riverside, CA.

Sumner, M.E., Miller, W.P., Kookana, R.S., Hazelton, P., 1998. Sodicity, dispersion, and environmental quality. In: Sumner, M.E., Naidu, R. (Eds.), Sodic Soils: Distribution, Properties, Management, and Environmental Consequences. Oxford University Press, New York, NY, pp. 149–172.

Szabolcs, I., 1989. Salt-Affected Soils. CRC Press, Boca Raton, FL.

Tanji, K.K., Wallendar, W.W., 2011. Nature and extent of agricultural salinity and sodicity. In: Wallendar, W.W.,

Tanji, K.K. (Eds.), Agricultural Salinity Assessment and Management, Second edition. American Society of Civil Engineers, Reston, VA, pp. 1–25.

van Olphen, H., 1977. An Introduction to Clay Colloid Chemistry, Second edition. John Wiley & Sons, New York, NY.

Yaalon, D.H., 1963. The origin and accumulation of salts in groundwater and in soils of Israel. Bull. Res. Coun. Isr., Sect. 11G, 105–131.

Yaron, D. (Ed.), 1981. Salinity in Irrigation and Water Resources. Dekker, New York, NY.

Index

Note: Page numbers followed by "f" indicate figures, "t" indicate tables, and "b" indicate boxes.